Anu Bajaj · Ajith Abraham · Ryotaro Kamimura
Editors

Bio-Inspired Computing

Proceedings of the 14th International Conference on Innovations in Bio-Inspired Computing and Applications (IBICA 2023) held during December 14–15, 2023, Volume 2: Deep Learning and Large Language Models

Springer

Editors
Anu Bajaj
Department of Computer Science and Engineering
Thapar Institute of Engineering and Technology
Patiala, Punjab, India

Ajith Abraham
School of Artificial Intelligence
Sai University
Chennai, India

Ryotaro Kamimura
Tokai University
Tokyo, Japan

ISSN 2367-3370 ISSN 2367-3389 (electronic)
Lecture Notes in Networks and Systems
ISBN 978-3-031-78939-7 ISBN 978-3-031-78940-3 (eBook)
https://doi.org/10.1007/978-3-031-78940-3

This Springer imprint is published by the registered company Springer Nature Switzerland AG
The registered company address is: Gewerbestrasse 11, 6330 Cham, Switzerland

Lecture Notes in Networks and Systems 1229

The series "Lecture Notes in Networks and Systems" publishes the latest developments in Networks and Systems—quickly, informally and with high quality. Original research reported in proceedings and post-proceedings represents the core of LNNS.

Volumes published in LNNS embrace all aspects and subfields of, as well as new challenges in, Networks and Systems.

The series contains proceedings and edited volumes in systems and networks, spanning the areas of Cyber-Physical Systems, Autonomous Systems, Sensor Networks, Control Systems, Energy Systems, Automotive Systems, Biological Systems, Vehicular Networking and Connected Vehicles, Aerospace Systems, Automation, Manufacturing, Smart Grids, Nonlinear Systems, Power Systems, Robotics, Social Systems, Economic Systems and other. Of particular value to both the contributors and the readership are the short publication timeframe and the world-wide distribution and exposure which enable both a wide and rapid dissemination of research output.

The series covers the theory, applications, and perspectives on the state of the art and future developments relevant to systems and networks, decision making, control, complex processes and related areas, as embedded in the fields of interdisciplinary and applied sciences, engineering, computer science, physics, economics, social, and life sciences, as well as the paradigms and methodologies behind them.

Indexed by SCOPUS, EI Compendex, INSPEC, WTI Frankfurt eG, zbMATH, SCImago.

All books published in the series are submitted for consideration in Web of Science.

For proposals from Asia please contact Aninda Bose (aninda.bose@springer.com).

IBICA 2023 Preface – Volume 2

Welcome to the 14th International Conference on Innovations in Bio-Inspired Computing and Applications (IBICA 2023) and 13th World Congress on Information and Communication Technologies (WICT 2023). IBICA-WICT 2023 was held in 5 different cities namely Olten, Switzerland; Porto, Portugal; Kaunas, Lithuania; Greater Noida, India; Kochi, India and in online mode. IBICA-WICT 2023 is hosted and sponsored by the Machine Intelligence Research Labs (MIR Labs), USA and supported by Bennett University, Greater Noida, India.

14th International Conference on Innovations in Bio-Inspired Computing and Applications was organized to provide a platform for world research leaders and practitioners, to discuss the "full spectrum" of current theoretical developments, emerging technologies, and innovative applications of Bio-inspired Computing. Bio-inspired Computing is currently one of the most exciting research areas, and it is continuously demonstrating exceptional strength in solving complex real-life problems. The main driving force of the conference is to further explore the intriguing potential of Bio-inspired Computing.

13th World Congress on Information and Communication Technologies (WICT 2023) was organized to provide an opportunity for the researchers from academia and industry to meet and discuss the latest solutions, scientific results and methods in the usage and applications of ICT in the real world. Last few decades have experienced technologies that are evolving so rapidly, altering the constraints of space and time, and reshaping the way we communicate, learn and think. Rapid advances in information technologies and other digital systems are reshaping our ecosystem. Innovations in ICT allow us to transmit information quickly and widely, propelling the growth of new urban communities, linking distant places and diverse areas of endeavor in productive new ways, which a decade ago was unimaginable. Thus, the theme of this World Congress is "Innovating ICT For Social Revolutions".

IBICA-WICT 2023 received 656 papers from 36 countries and based on the review process we have accepted 263 papers. Each paper was reviewed by at least five or more reviewers, and based on the outcome of the review papers were accepted for inclusion in the conference proceedings. Proceedings were divided into Five Volumes. This Volume is focused on *Deep Learning and Large Language Models.* This Volume comprises of 53 articles focused on advanced machine learning including deep learning methods, large language models and its real-world applications.

First, we would like to thank all the authors for submitting their papers to the conference, for their presentations and discussions during the conference. Our thanks go to program committee members and reviewers, who carried out the most difficult work by carefully evaluating the submitted papers. Our special thanks to the following plenary speakers, for their exciting plenary talks:

- Christine Zarges, Aberystwyth University, UK
- Theresa Schmiedel, University of Applied Sciences and Arts Northwestern Switzerland, Basel, Switzerland

- Ke Feng, Singapore-ETH Centre, The National University of Singapore, Singapore
- Stefka Fidanova, Bulgarian Academy of Sciences, Bulgaria
- Diego Oliva, Universidad de Guadalajara, Mexico
- Sebastian Ventura, University of Cordoba, Spain
- João Pedrosa, INESCTEC, Portugal
- Nuno Bettencourt, Instituto Superior de Engenharia do Porto, Portugal
- Milan Tuba, Singidunum University, Serbia
- Dalia Kriksciuniene, Vilnius University, Lithuania
- Kusum Deep, Indian Institute of Technology Roorkee, India

We express our sincere thanks to the organizing committee chairs for helping us to formulate a rich technical program. Enjoy reading the articles!

Virgilijus Sakalauskas
Thinakaran Perumal
Ajith Abraham
Preethi Thekkath
Dalia Kriksciuniene
Yunchuan Sun
Ryotaro Kamimura
Sreela Sreedhar
Anu Bajaj

IBICA–WICT 2023 Organizing Committees

General Chairs

Virgilijus Sakalauskas	Vilnius University, Lithuania
Thinakaran Perumal	Universiti Putra Malaysia, Malaysia
Ajith Abraham	Sai University, India
Preethi Thekkath	Toc H Institute of Science & Technology, India

Program Chairs

Dalia Kriksciuniene	Vilnius University, Lithuania
Yunchuan Sun	Beijing Normal University, China
Ryotaro Kamimura	Tokai University, Japan
Sreela Sreedhar	Toc H Institute of Science & Technology, India

Web Services Chair

Kun Ma	University of Jinan, China

Publication Chair

Anu Bajaj	Thapar Institute of Engineering and Technology, India

Publicity Chair

Pooja Manghirmalani Mishra	Machine Intelligence Research Labs, India

Local Organizing Committee

Jesna Anver	Toc H Institute of Science & Technology, India
Leda Kamal	Toc H Institute of Science & Technology, India
Elsaba Jacob	Toc H Institute of Science & Technology, India

Mithu Mary George	Toc H Institute of Science & Technology, India
Mima Manual	Toc H Institute of Science & Technology, India
Ashly Jospeh	Toc H Institute of Science & Technology, India
Sajith P. J.	Toc H Institute of Science & Technology, India
Leema Mathew	Toc H Institute of Science & Technology, India
Baazil P. Thampy	Toc H Institute of Science & Technology, India
Sharath P. Raju	Toc H Institute of Science & Technology, India
Philo Sumi	Toc H Institute of Science & Technology, India
Jacob Jose Chemmankuzha	Toc H Institute of Science & Technology, India

International Program Committee

Abdelkrim Haqiq	Hassan 1st University, Morocco
Alberto Bugarín-Diz	Universidade de Santiago de Compostela, Spain
Alberto Guillén	University of Granada, Spain
Alberto Fernandez Hilario	University of Granada, Spain
Alexey Kornaev	Innopolis University, Russia
Alfonso Guarino	University of Foggia, Italy
Alfredo Cuzzocrea	University of Calabria, France
Alicia Troncoso	Pablo de Olavide University, Spain
Ambili P. S.	REVA University, India
Amelia Zafra	University of Cordoba, Spain
Amit Kumar Tyagi	Vellore Institute of Technology, India
Amutha S.	Vellore Institute of Technology, India
Ana de Almeida	Instituto Universitário de Lisboa, Portugal
Ana Madureira	Polytechnique of Porto, Portugal
Anazida Zainal	Universiti Teknologi Malaysia, Malaysia
Andre C. P. L. F. de Carvalho	University of Sao Paulo, Brazil
André S. Santos	ISEP, Portugal
Andrea Soltoggio	Loughborough University, UK
Andries Engelbrecht	University of Stellenbosch, South Africa
Anitha Kumari K.	PSG College of Technology, India
Antonello Florio	Politecnico di Bari, Italy
Antonio Arauzo-Azofra	University of Cordoba, Spain
Antonio Salmeron	Universidad de Almeria, Spain
Anurag Rana	Shoolini University, India
Arijit Ghosal	St. Thomas' College of Engineering and Technology, India
Arthi B.	SRM Institute of Science and Technology, India
Aryan Chaudhary	Bio Tech Sphere Research, India
Aureli Soria-Frisch	Starlab Barcelona, Spain

Azah Kamilah binti Muda	UTeM, Malaysia
Bac Le	University of Science, Vietnam
Bay Vo	Ho Chi Minh City University of Technology, Vietnam
Bernard De Baets	Ghent University, Belgium
Bernardete Ribeiro	CISUC, University of Coimbra, Portugal
Bhawna Goyal	Chandigarh University, India
Carlos Cotta	Universidad de Málaga, Spain
Carlos Pereira	Universidade de Santiago de Compostela, Portugal
Carlos Serro	ISCTE - Instituto Universitário de Lisboa, Portugal
Catarina I. Reis	Polytechnic of Leiria, Portugal
Cesar Hervas-Martinez	University of Cordoba, Spain
Chang-Yong Lee	Kongju National University, Korea
Chao-Chun Chen	National Cheng Kung University, Taiwan
Chun-Hao Chen	National Taipei University of Technology, Taiwan
Corrado Mencar	University of Bari Aldo Moro, Italy
David A. Pelta	Universidad de Granada, Spain
Debdatta Kandar	North-Eastern Hill University, India
Deepika Koundal	UPES, India
Deepti Chaudhary	Kurukshetra University, India
Devi Priya R.	KPR Institute of Engineering and Technology, India
Rajashree Shettar	RV College of Engineering, India
Amit Kumar Mishra	Jain University, India
Iqbal Sarker	Edith Cowan University, Australia
Saira Varghese	Toc H Institute of Science & Technology, India
Durai Raj Vincent P. M.	VIT Vellore, India
Durgesh Srivastava	Chitkara University, India
Eiji Uchino	Yamaguchi University, Japan
Elakkiya R.	BITS Pilani Dubai Campus, UAE
Elizabeth F. G. Goldbarg	Universidade Federal do Rio Grande do Norte, Brazil
Enrique De la Cal	University of Oviedo, Spain
Enrique Herrera-Viedma	University of Granada, Spain
Fabio Zambetta	RMIT University, Australia
Fernando Batista	INESC-ID and ISCTE-IUL, Portugal
Fernando Gomide	University of Campinas, Brazil
Francisco Herrera	Universidad de Granada, Spain
Gabriella Casalino	University of Bari, Italy
Gahangir Hossain	University of North Texas, USA

Gaurav Gupta	Shoolini University, India
George Papakostas	International Hellenic University, Greece
Gianluca Zaza	University of Bari, Italy
Giovanna Castellano	University of Bari, Italy
Gregorio Sainz-Palmero	Universidad de Valladolid, Spain
Guenther Raidl	Vienna University of Technology, Austria
Hajime Nobuhara	University of Tsukuba, Japan
Heba Afify	Biomedical Engineering, Egypt
Hsueh-Chan Lu	National Cheng Kung University, Taiwan
Hudson G. Medeiros	Federal University of Rio Grande do Norte, Brazil
Irwin King	The Chinese University of Hong Kong, Hongkong
Isaac Chairez	Tecnológico de Monterrey, Mexico
Isabel Jesus	Institute of Engineering of Porto, Portugal
Ivo Pereira	University Fernando Pessoa, Portugal
Ja-Hwung Su	National University of Kaohsiung, Taiwan
Janos Botzheim	Eötvös Loránd University Faculty of Informatics, Hungary
Janusz Kacprzyk	Polish Academy of Sciences, Poland
Jayaraman Valadi	FLAME University, India
Jerry Chun-Wei Lin	Western Norway University of Applied Sciences, Norway
Jesus S. Aguilar-Ruiz	Universidad Pablo de Olavide, Spain
Jesus Alcala-Fdez	University of Granada, Spain
Jimmy Ming-Tai Wu	Shandong University of Science and Technology, China
Jolanta Mizera-Pietraszko	Military University of Land Forces, Poland
Jose A. Gamez	Universidad de Castilla-La Mancha, Spain
Jose Jerez	Universidad de Malaga, Spain
Jose A. Lozano	UPV/EHU, Spain
Jose Maria-Luna	University of Cordoba, Spain
Jose Puerta	Universidad de Castilla-La Mancha, Spain
José Everardo B. Maia	UECE, Brazil
Juan Vidal	University of Santiago de Compostela, Spain
Ju-Chin Chen	National Kaohsiung University of Science and Technology, Taiwan
Julia Flores	University of Castilla - La Mancha (UCLM), Spain
K. Reddy Madhavi	Sree Vidyanikethan Engineering College, India
Kawuu W. Lin	National Kaohsiung University of Science and Technology, Taiwan
Kingsley Okoye	Tecnologico de Monterrey, Mexico
Kolla Bhanu Prakash	Koneru Lakshmaiah Education Foundation, India

Lalitha K.	Kongu Engineering College, India
Leandro M. Almeida	Universidade Federal de Pernambuco, Brazil
Leandro Dos Santos Coelho	Pontifical Catholic University of Parana, Brazil
Leonilde Varela	University of Minho, Portugal
Loan Nguyen	International University, Vietnam
Luca Di Gaspero	DPIA, University of Udine, Italy
Luciano Sánchez	Oviedo University, Spain
Luis Magdalena	Universidad Politecnica de Madrid, Spain
Luís B. Elvas	Instituto Universitário de Lisboa, Portugal
Luiz Zárate	Pontifícia Universidade Católica de Minas Gerais, Brazil
Manikandan V. M.	SRM University AP, India
Mansi Sharma	Amrita Vishwa Vidyapeetham, India
Manuel Fernandez Delgado	University of Santiago de Compostela, Spain
Manuel Lama	University of Santiago de Compostela, Spain
Manuel Mucientes	University of Santiago de Compostela, Spain
Maria Jose Del Jesus	Universidad de Jaén, Spain
Mário Antunes	Instituto Politécnico de Leiria, Portugal
Marley M. B. R. Vellasco	PUC Rio de Janeiro, Brazil
Matteo Gaeta	Università degli Studi di Salerno, Italy
Mauricio Ayala-Rincon	Universidade de Brasilia, Brazil
Mauro Birattari	Université Libre de Bruxelles, Belgium
Meera Ramadas	MIR Labs, USA
Michael Blumenstein	University of Technology Sydney, Australia
Michael N. Vrahatis	University of Patras, Greece
Michele Mastroianni	Università degli Studi di Salerno, Italy
Mika Sato-Ilic	University of Tsukuba, Japan
Nikola Kasabov	Auckland University of Technology, New Zealand
Olgierd Unold	Wroclaw University of Science and Technology, Poland
Oscar Castillo	Tijuana Institute of Technology, Mexico
Oscar Corcho	Universidad Politécnica de Madrid, Spain
Pascal Bouvry	University of Luxembourg, Luxembourg
Pascal Van Hentenryck	University of Michigan, USA
Patricia Melin	Tijuana Institute of Technology, Mexico
Paulo Carrasco	Alagrve University, Portugal
Paulo Moura Oliveira	UTAD University, Portugal
Paulo Novais	Universidade do Minho, Portugal
Pedro Larranaga	Technical University of Madrid, Spain
Quang Minh Tran	Ho Chi Minh City University of Technology, Vietnam
Rabel Guharoy	Bharati Vidyapeeth Deemed University, India

Radhika S.	Sathyabama Institute of Science and Technology, India
Radu-Emil Precup	Politehnica University of Timisoara, Romania
RajaniKanth Aluvalu	Vardhaman College of Engineering, India
Richard Weber	University of Chile, Chile
Roberto Iglesias Rodriguez	Universidade de Santiago de Compostela, Spain
Rohit Anand	GB Pant Govt. Engineering College, India
Ronald Yager	Machine Learning Institute, Iona College, USA
Rudolf Kruse	University of Magdeburg, Germany
Rutuparna Panda	VSS University of Technology, India
Saman Halgamuge	University of Melbourne, Australia
Sancho Salcedo-Sanz	Universidad de Alcalá, Spain
Sandeep Sreekumar	IndsutryApps, Singapore
Santosh Kumar Srivastava	Galgotias University, India
Sarina Sulaiman	Universiti Teknologi Malaysia, Malaysia
Sebastian Basterrech	VSB-Technical University of Ostrava, Czech Republic
Sebastián Ventura Soto	University of Cordoba, Spain
Shabeer Puthukkudi	British Telecom, United Kingdom
Shali Rani	Chitkara University, India
Shankru Guggari	KLE Technological University, India
Shilpa Gupta	Maharaja Agrasen University, India
Shing Chiang Tan	Multimedia University, Malaysia
Silviu Ionita	University of Pitesti, Romania
Sirine Marrakchi	University of Sfax, Tunisia
Stéphane Grieu	Université de Perpignan, France
Steven D. Prestwich	University College Cork, Ireland
Sujata Dash	Nagaland University, India
Sung-Bae Cho	Yonsei University, Korea
Teresa Ludermir	Federal University of Pernambuco, Brazil
Thillaiarasu Nadesan	Reva University, India
Thinagaran Perumal	Universiti Putra Malaysia, Malaysia
Tho Quan	Ho Chi Minh City University of Technology, Vietnam
Thomas Hanne	University of Applied Sciences Northwestern, Switzerland
Tom Gedeon	The Australian National University, Australia
Tzung-Pei Hong	National University of Kaohsiung, Taiwan
Van Vo	Ho Chi Minh University of Industry, Vietnam
Victor M. Rivas	University of Jaen, Spain
Vijay P.	Vellore Institute of Technology, India
Vishwanath N.	Toc H Institute of Science & Technology

Wen-Yang Lin	National University of Kaohsiung, Taiwan
Wilfried Elmenreich	Alpen-Adria-Universitat, Austria
Witold Pedrycz	University of Alberta, Canada
Xiao-Zhi Gao	University of Eastern Finland, Finland
Yaochu Jin	University of Bielefeld, Germany
Yilun Shang	Northumbria University, UK
Yukio Ohsawa	University of Tokyo, Japan

Contents

A Comparative Analysis of Ensemble Learning Methods for Multi-Label Classification on Bioinformatics

Sonia Guehria[1(✉)], Habiba Belleili[2], and Nabiha Azizi[2]

[1] LRI Laboratory- Badji Mokhtar University, 12-23000 Annaba, BP, Algeria
sn.guehria@gmail.com

[2] LabGED -Badji Mokhtar Annaba University, 12-23000 Annaba, BP, Algeria

Abstract. Multi-label classification is a fundamental task in bioinformatics, where each instance can be associated with multiple labels simultaneously. Recently, ensemble learning has gained popularity to address the complex nature of bioinformatics data. This paper presents a comprehensive performance comparison of popular state-of-the-art ensemble methods in the realm of bioinformatics for Multi-Label Classification. The study assesses the effectiveness of ensemble multi-label classifiers on various datasets, considering their characteristics, such as label dependencies, imbalanced data, and the high dimensionality of the output space. Through extensive experimentation, we analyze the performance of these ensemble methods and their applicability to diverse bioinformatics data scenarios. The findings revealed that none of the methods exhibited unequivocal superiority over the others in effectively addressing the majority of challenges arising from multi-label classification tasks, emphasizing the need for choosing methods based on the specific characteristics of the data.

Keywords: Bioinformatics · Ensemble learning · Multi-label Classification

1 Introduction

In bioinformatics, diverse and extensive datasets often require the assignment of multiple labels for accurate referencing of an instance. One prominent example lies within functional genomics, where a gene can be associated with a multitude of functions encompassing metabolism, protein synthesis, and transcription. Similarly, in clinical text data, such as Electronic Health Records (EHRs), a wide range of labels can be attributed, including diagnostic codes, medical history, problem lists, and details about procedures performed on patients. These complexities highlight the need for Multi-Label Classification (MLC) in bioinformatics to effectively analyze and process the valuable information present in various datasets.

Multi-Label Classification (MLC) problem has emerged as a powerful way in bioinformatics, enabling researchers to analyze complex biological data and gain valuable insights into the relationships between different biological phenomena. This classification paradigm has been requested in various real-world applications within bioinformatics, including clinical text data [1], gene data [2], and protein data [2, 3]. By modeling

A. Bajaj et al. (Eds.): IBICA 2023, LNNS 1229, pp. 1–11, 2025.
https://doi.org/10.1007/978-3-031-78940-3_1

complex relationships between various biological factors, MLC allows a richer representation of the data and captures multiple aspects simultaneously. On the other hand, MLC enables the efficient processing of diverse datasets with valuable information by assigning relevant labels to instances across different dimensions. In addition, it demonstrates its predictive impact in multi-label learning systems, as the inclusion of multiple labels enhances accurate predictions and a better understanding of the relationships between different biological objects.

Although MLC had proven significant progress in the field of bioinformatics [3–8], it has also introduced new challenges that need to be addressed for effective analysis of complex biological data. One of these issues is the high dimensionality of the output space [9], which can increase the computational complexity of models, lead to overfitting, and require careful feature selection with robust dimensionality reduction techniques. Another problem often encountered in bioinformatics is unbalanced data [10, 11], where certain labels are more frequently represented in some datasets than others. This problem can affect the performance of MLC models, which may be biased in favour of the majority of labels while neglecting the minority ones. Modeling label dependencies is an additional complex problem that needs to be tackled [12], as many biological phenomena are interconnected. Thus, neglecting label dependencies or addressing them independently can result in suboptimal results and data loss [13].

To address the aforementioned issues, an ensemble approach for MLC has been developed to improve the generalization ability of learning systems and achieve higher accuracy compared to single multi-label classifiers [14]. This approach has gained significant popularity and achieved notable success, particularly in the field of bioinformatics [7, 15, 16]. While numerous research papers in MLC domain implicitly reference the ensemble approach [17–19], there is a relatively limited number of research articles that offer broad views [20, 21] comparative studies [22, 23], and surveys [24] on Ensemble Multi-Label Classifiers (EMLCs). The main contribution of this paper is to offer a comprehensive performance comparison of popular state-of-the-art EMLCs on bioinformatics data. The objective is to identify the EMLC that demonstrates superior performance in various scenarios and provide guidelines for selecting the most suitable algorithm based on the characteristics of the datasets.

The rest of the paper is organized as follows. Section 2, introduces the problem of MLC and discusses its approaches. Section 3, presents an overview of the state-of-the-art EMLCs used in the experimental study. In Sect. 4, the experimental design is described, including details about the datasets used and the evaluation metrics employed. The results of the comparative analysis of EMLCs are presented and discussed in Sect. 5. Finally, in Sect. 6, the conclusion of this work is drawn, along with future perspectives.

2 Background

In this section, we will introduce the task of MLC and provide a brief overview of popular algorithms used in the field, with a specific focus on EMLCs.

2.1 Multi-Label Classification Problem

MLC is a machine learning task where each instance can be associated with a set of class labels simultaneously. It aims to construct a predictive model that can accurately classify a new instance by partitioning a sets of relevant and irrelevant labels. The formal description of the MLC task, as presented by Tsoumakas et al. [1] uses $L = \{y_j : j = 1 \dots q\}$ denotes the finite set of labels, and $D = \{(x_i, Y_i), i = 1 \dots m\}$ represents a set of multi-label training examples, where x_i is the feature vector associated with the i-th example, and $Y_i \subseteq L$ presents the set of labels associated with that instance.

Further, EMLC is a collection of m multi-label classifiers, denoted as $MLC_1, MLC_2 \dots MLC_m$. Each individual multi-label classifier MLC_i produces a multi-label prediction $P_{i1}, P_{i2} \dots P_{in}$. To predict the label subset for the new instance x_i, the multi-label predictions generated by all individual classifiers are combined by taking the average value of the predictions for each label.

2.2 Multi-Label Classification Approaches

Madjarov et al. [22], who conducted an extensive comparison of multi-label learning methods on various benchmark datasets, categorized these methods into three approaches: *Problem Transformation*, *Algorithm Adaptation*, and *Ensemble*.

Problem Transformation approach involves utilizing independent algorithms to transform the multi-label task into one or more single-label tasks, where each new problem is solved by traditional algorithms. Some methods of this approach convert the multi-label problem into several independent binary classification problems, such as Binary Relevance (BR) [25], and Chain Classifier (CC)) [26]. Others transform the multi-label problem into a multi-class problem, such as Label Power Set (LP) [27] and Pruned set (PS) [28].

Algorithm Adaptation approach extends single-label learning algorithms such as DT, SVM, ANN, and K-NN to handle Multi-Label Data (MLD) directly. The methods belonging to this category of approach include: Multi-Label C4.5 (ML-C4.5) [29], Predictive Clustering Trees (PCT) [30], Rank-SVM [31], Back Propagation for Multi label learning (BP-MLL) [32], and Multi-Label k-nearest neighbor (ML-kNN) [33].

Ensemble approach, developed on top of the *Problem Transformation* or *Algorithm Adaptation* methods, aims to address the limitations of single multi-label classifiers by combining multiple base classifiers. The goal is to create a diverse and accurate ensemble of classifiers that can collectively overcome the drawbacks associated with individual classifiers.

3 Popular Ensemble Multi-Label Classifiers

Ensemble methods based on problem transformation involve constructing an ensemble of single-label learners, where the base classifiers are obtained from the multi-label classifiers. The best-known algorithms in this category include EBR, ECC, EPS, RAKEL, HOMER, and MLS.

Ensemble of Binary Relevance classifiers (EBR) [34] constructs multiple BR classifiers using a random sub-sample of the training dataset. By leveraging the diversity among the base classifiers, this ensemble method effectively enhances the performance of BR. Ensemble of Classifier Chains (ECC) [26] employs multiple CCs as base classifier, trained on a random subset of instances. It considers label dependencies by expanding the feature space of each classifier with the label predictions from previous classifiers. Ensemble of Pruned Sets (EPS) [28] constructs multiple PS classifiers using a random subset of instances. It addresses the issue of imbalanced data by pruning infrequently occurring label sets. RAndom k-labELsets (RAKEL) [27] creates multiple LP classifiers trained on a small random subset of labelsets with a size k. This method aims to reduce the sensitivity of models to class-imbalance by including a more balanced distribution of classes in the base classifiers. Hierarchy Of Multi-label classifiER (HOMER) [35] transforms a MLC problem into a hierarchy, where each binary classifier focuses on a smaller number of labels. This method efficiently handles MLC problems with a large number of labels by constructing a tree-shaped hierarchy of simpler MLC problems. Multi-label Stacking (MLS) [36] utilizes BR twice, incorporating predictions from the previous level as inputs to the meta-level to generate final predictions for each corresponding label. This method considers label correlations at the meta-level, enhancing the overall prediction process.

Ensemble methods based on algorithm adaptation involve utilizing ensembles that are composed of base classifiers adapted from existing classification algorithms. Typical algorithms in this category include RF-PCT and TREMLC. Random Forest of Predictive Clustering Trees (RF-PCT [37] uses PCT [30] as the base classifier, which are constructed using a double random selection process. In the first selection, a subset of instances is chosen per bagging for each model. The second selection occurs at each tree node, where a subset of original features is randomly selected, with the selection of optimal features. Triple Random Ensemble for Multi-label Classification (TREMLC) [38] is constructed by incorporating three random selection subsets of labels, features, and instances drawn without replacement. Each base classifier LP in TREMLC is trained using a specific portion of the data, randomly. LP classifiers predict k-sets of labels using only a subset of features.

4 Experimental Design

In this section, we present the experimental design used to compare popular EMLCs, which were trained using their default parameter settings. The experiment section is divided into three parts. Firstly, we describe the experimental multi-label datasets. We then present evaluation measures for MLC. Finally, we discuss experimental results.

4.1 Experimental Multi-Label Datasets

Seven datasets considered in our comparative study are selected from the Mulan Repository[1] to assess the effectiveness of state-of-the-art ensemble methods for MLC in various bioinformatics application areas such as biology and medicine. The datasets from the biology domain such as Genbase [2], HumanPse [39], PlantPse [39], and Yeast [31]. The

[1] https://www.uco.es/kdis/mllresources/.

datasets from the medical domain are CHD_49 [5] and Medical dataset [4]. Each multi-label dataset is described with some basic meta-features, such as the number of instances (m), number of labels (q), number of attributes (d), label cardinality per instance (card), average imbalance ratio per label (avgIR) and pairs label dependency (rDep). These datasets were obtained in their original format and were pre-divided into a learning set, comprising approximately 2/3 of the data, and a test set containing the remaining 1/3 of the dataset. An overview of the different datasets used in our experimentation, along with their respective properties is provided in Table 1.

Table 1. Properties of different multi-label datasets used in our experimentation.

Datasets	Domain	m	q	d	card	Dens	avgIR	rDep
CHD_49	Medicine	555	6	49	2.580	0.430	5.766	0.267
Genbase	Biology	662	27	1 186	1.252	0.046	37.315	0.157
HumanPse	Biology	3106	14	440	1.185	0.085	15.289	4.418
Medical	Text	978	45	1 449	1.245	0.028	89.501	0.039
PlantPse	Biology	978	12	440	1.079	0.090	6.690	0.318
VirusGo	Biology	207	6	749	1.217	0.203	4.041	0.400
Yeast	Biology	2 417	14	103	4.237	0.303	7.197	0.670

4.2 Evaluation Measures

To assess the performance of various EMLCs in our comparative experimentation, we utilized five common evaluation metrics [40], which are categorized into two groups: Example-based metrics and Label-based metrics. These metrics encompassed three example-based measures (Hloss, Accuracy, and F1_score), defined by Eq. (1) to (3), along with two label-based measures (Micro-F1, and Macro-F1), and defined by Eq. (4) and (5), as outlined in Table 2. For each evaluation metric, the testing dataset consists of pairs (x_i, y_i) where $1 \leq i \leq n$. Here, $y_i \in \{0, 1\}$ represents the actual labels of the ith test example, and $\widetilde{y_i} = f(x_i)$ represents its predicted labels.

5 Results and Discussion

Based on the results presented in Tables 3 and 4, it is evident that the performance of the algorithms varies across different datasets. Notably, EPS and RAKEL consistently achieve low Hloss values on most datasets. On the other hand, RF-PCT algorithm stands out as one of the best performers in terms of Accuracy.

Table 5 presents the predictive performance of EMLCS based on F1_score, which assesses the multi-label predictions for each instance as a unified entity. This metric provides valuable insights into how well the models handle MLC task, considering the relationship between labels. Based on the obtained results, we can see that on datasets

Table 2. Multi-label evaluation metrics used in our experiments.

Metric	Formula
$Hloss \downarrow$	$= \frac{1}{N} \sum_{i=1}^{N} \frac{1}{Q} \lvert h(x_i) \Delta Y_i \rvert$ (1)
$Accuracy \uparrow$	$= \frac{1}{N} \sum_{i=1}^{N} \frac{\lvert h(x_i) \cap Y_i \rvert}{\lvert h(x_i) \cup Y_i \rvert}$ (2)
$F1\_score \uparrow$	$= \frac{1}{N} \sum_{i=1}^{N} \frac{2 \times \lvert h(x_i) \cap Y_i \rvert}{\lvert h(x_i) \rvert + \lvert Y_I \rvert}$ (3)
$Micro - F1 \uparrow$	$= \frac{2 \times Micro\_precision \times Micro\_recall}{Micro\_precision + Micro\_recall}$ (4)
$Macro - F1 \uparrow$	$= \frac{1}{M} \sum_{j=1}^{M} \frac{2 \times p_j \times r_j}{p_j + r_j}$ (5)

with a low degree of correlation between labels, such as Medical, Genbase, and CHD_49 datasets, ECC, and MLS algorithms demonstrate superior performance compared to the others. For moderately correlated datasets like VirusGo, HumanPse, and PlantPse datasets, RAKEL and RF-PCT algorithms outperform the others. On the other hand, on highly correlated datasets like Yeast, HOMER algorithm proves to be the most effective.

Moreover, for datasets with high dimensionality, which is the product of the number of instances (m), features (d), and labels (q), such as HumanPse and Medical datasets, we observe that the EPS and RAKEL algorithms achieve closer and higher scores in terms of Accuracy and F1_score compared to the majority of the other algorithms. This is attributed to the fact that EPS and RAKEL can effectively reduce the complex nature of high dimensionality of the datasets differently. EPS employs a pruning process to eliminate samples with infrequent labelsets, while RAKEL splits the output space into random small subsets.

It is evident that both Micro-F1 and Macro-F1 metrics are used in cases with both balanced and unbalanced data. Macro-F1 offers equal importance to each label and is useful for obtaining balanced performance for all labels. On the other hand, Micro-F1 focuses on overall performance based on the total number of correctly classified instances. The results presented in Tables 6 and 7 demonstrate that different algorithms perform better depending on the level of imbalance datasets. For datasets with a low imbalance level, such as VirusGo and CHD_49 datasets, RF-PCT is the best algorithm. For datasets with a medium imbalance level, like PlantPse and Yeast datasets, MLS and HOMER are the top-performing algorithms. Finally, for datasets with a high imbalance level, such as HumanPse, Genbase, and Medical datasets. ECC and RAKEL algorithms outperform others.

Based on the above results, it becomes evident that none of the methods exhibited unequivocal superiority over the others in effectively addressing the majority of challenges arising from multi-label classification tasks. Furthermore, the choice of the most suitable ensemble method for MLC in bioinformatics field highly depends on the

Table 3. Predictive performance of EMLCs in terms of Hloss measure.

Datasets	EBR	ECC	EPS	RAKEL	HOMER	MLS	RF-PCT	TREMLC
CHD_49	**0.294**	0.329	0.302	0.320	0.321	0.321	0.307	0.329
Genbase	0.002	**0.001**	0.002	**0.001**	**0.001**	**0.001**	0.049	0.010
HumanPse	**0.081**	0.084	0.082	0.092	0.121	0.114	0.085	0.089
Medical	0.013	**0.010**	0.012	0.011	0.126	**0.010**	0.025	0.014
PlantPse	0.088	0.093	0.089	0.104	0.138	0.132	0.089	0.102
VirusGo	0.041	0.041	0.042	0.038	0.126	0.038	**0.035**	0.143
Yeast	**0.201**	0.204	0.208	0.243	0.254	0.244	0.214	0.225

Table 4. Predictive performance of EMLCs in terms of Accuracy measure.

Datasets	EBR	ECC	EPS	RAKEL	HOMER	MLS	RF-PCT	TREMLC
CHD_49	0.508	**0.535**	0.525	0,522	0.490	0.452	0.530	0.518
Genbase	0.172	0.218	0.167	0.240	0.198	**0.243**	0.171	0.194
HumanPse	0.108	0.151	0.160	0.162	0.138	**0.189**	0.113	0.145
Medical	0.756	**0.768**	0.767	**0.768**	0.740	0.749	0.593	0.625
PlantPse	0.720	**0.734**	0.718	0.732	0.681	0.689	0.572	0.631
VirusGo	0.856	0.844	0.867	0.856	0.761	0.856	**0.891**	0.734
Yeast	0.488	0.546	0.491	0.534	**0.564**	0.434	0.529	0.438

Table 5. Predictive performance of EMLCs in terms of F1 measure.

Datasets	EBR	ECC	EPS	RAKEL	HOMER	MLS	RF-PCT	TREMLC
CHD_49	0.625	0.640	0.640	0.584	0.611	0.584	0.647	0.635
Genbase	0.177	0.228	0.173	0.259	0.231	**0.273**	0.180	0.207
HumanPse	0.107	0.151	0.204	**0.209**	0.159	0.168	0.109	0.149
Medical	0.780	**0.793**	0.779	0.781	0.759	0.781	0.788	0.659
PlantPse	0.526	0.551	0.518	0.543	0.510	0.517	**0.658**	0.459
VirusGo	0.880	0.876	0.890	0.877	0.934	0.877	**0.935**	0.920
Yeast	0.584	0.667	0.584	0.658	**0.684**	0.577	0.611	0.609

characteristics of tested dataset, including label dependencies, imbalance data, and high dimensionality of output space. These findings serve as valuable indicators to select the

Table 6. Predictive performance of EMLCs in terms of Micro-F1 measure.

Datasets	EBR	ECC	EPS	RAKEL	HOMER	MLS	RF-PCT	TREMLC
CHD_49	0.645	0.658	0.654	0.610	0.635	0.610	**0.667**	0.656
Genbase	0.979	0.979	0.968	**0.980**	0.978	0.976	0.010	0.893
HumanPse	0.237	0.283	0.231	**0.307**	0.261	0.291	0.239	0.257
Medical	0.805	**0.810**	0.775	0.808	0.768	0.808	0.185	0.719
PlantPse	0.166	0.210	0.153	0.223	0.208	**0.246**	0.166	0.205
VirusGo	0.890	0.890	0.881	0.897	**0.912**	0.897	**0.912**	0.901
Yeast	0.617	0.628	0.616	0.572	**0.664**	0.572	0.608	0.601

Table 7. Predictive performance of EMLCs in terms of Macro-F1 measure.

Datasets	EBR	ECC	EPS	RAKEL	HOMER	MLS	RF-PCT	TREMLC
CHD_49	0.488	0.504	0.499	0.460	0.482	0.460	**0.511**	0.494
Genbase	0.730	0.735	0.668	**0.739**	0.736	0.736	0.001	0.611
HumanPse	0.096	0.113	0.085	**0.155**	0.134	0.138	0.078	0.117
Medical	0.643	**0.659**	0.606	**0.659**	0.333	0.659	0.026	0.274
PlantPse	0.079	0.095	0.063	0.115	0.141	**0.158**	0.057	0.105
VirusGo	0.785	0.822	0.833	0.784	0.766	0.784	**0.834**	0.633
Yeast	0.385	0.398	0.374	0.383	**0.401**	0.384	0.393	0.385

appropriate EMLCs, tailored to the specific bioinformatics data, aiming to improve classification accuracy and effectively manage the complexities associated with multi-label tasks.

6 Conclusion and Future Work

In this paper, we have conducted a comprehensive performance comparison of popular state-of-the-art ensemble methods for MLC in the field of bioinformatics using five different evaluation metrics for MLC. The study focused on addressing the challenges posed by MLC task by considering specific characteristics of each dataset, such as label dependencies, unbalanced data, and the high dimensionality of the output space. Thus, the predictive performance of different EMLCs depends on the specific characteristics of each dataset, and the strengths and weaknesses of each algorithm. Furthermore, the obtained results on the bioinformatics datasets proved that no method was able to verify its superiority over the others in solving the majority of problems arising from MLC.

As future research, we propose to investigate the scalability and efficiency of these algorithms on large-scale data, as bioinformatics datasets continue to grow in both quantity and complexity. In addition, the development of specialized ensemble techniques can efficiently manage the imbalances commonly found in biological data, thereby ensuring accurate predictions for all labels.

References

1. Tsoumakas, G., Katakis, I., Vlahavas, et I.: Mining Multi-label Data. In : Data Mining and Knowledge Discovery Handbook, pp. 667–685. Springer US (2009)
2. Diplaris, S., Tsoumakas, G., Mitkas, P.A., Vlahavas, et I.: Protein Classification with Multiple Algorithms. In: Advances in Informatics, vol. 3746, pp. 448–456. Berlin (2005)
3. Wang, X., Li, G.-Z., Liu, J.-M., Zhao, et R.-W.: Multi-label Learning for Protein Subcellular Location Prediction. In: 2011 IEEE International Conference on Bioinformatics and Bio medicine, pp. 282–285. Atlanta (2011). https://doi.org/10.1109/BIBM.2011.36
4. Pestian, J.P. et al.: A shared task involving multi-label classification of clinical free text. In: Proceedings of the Workshop on BioNLP 2007 Biological, p. 97. Prague (2007)
5. Shao, H., Li, G., Liu, G., Wang, et Y. : Symptom selection for multi-label data of inquiry diagnosis in traditional Chinese medicine. Sci. China Inf. Sci. **56**(5), 1–13 (2013)
6. Tanaka, E.A., Nozawa, S.R., Macedo, A.A., Baranauskas, et J.A.: A multi-label approach using binary relevance and decision trees applied to functional genomics. Journal of Bio medical Informatics **54**, 85–95 (2015). https://doi.org/10.1016/j.jbi.2014.12.011
7. Du, J., Chen, Q., Peng, Y., Xiang, Y., Tao, et Z. C.: ML-Net: multi-label classification of biomedical texts with deep neural networks. J. American Medical Informatics Asso. https://doi.org/10.1093/jamia/ocz085
8. Kaur, R., Ginige, et J.A.: Analysing Effectiveness of Multi-Label Classification in Clinical Coding. In: Proceedings of the Australasian Computer Science Week Multiconference, p. 1–9. ACM, Sydney NSW Australia (2019). https://doi.org/10.1145/3290688.3290728
9. Zhang, J., et al.: Multi-label Feature Selection via Global Relevance and Redundancy Optimization. In : Proceedings of the 29th International Joint Conference on AI, pp. 2512–2518. Japan (2020). https://doi.org/10.24963/ijcai.2020/348
10. Rana, P., Sowmya, A., Meijering, E., Song, et Y.: Imbalanced classification for protein subcellular localization with multilabel oversampling. Bioinformatics **39**(1), btac841 (2023). https://doi.org/10.1093/bioinformatics/btac841
11. Pang, Y., Yao, L., Xu, J., Wang, Z., Lee, et T.-Y.: Integrating transformer and imbalanced multi-label learning to identify antimicrobial peptides and their functional activities. Bioinformatics **38**(24), 5368–5374 (2022). https://doi.org/10.1093/bioinformatics/btac711
12. Zhang, Q., et al.: Accurate prediction of multi-label protein subcellular localization through multi-view feature learning with RBRL classifier. Briefings in Bioinformatics, bbab012 (2021). https://doi.org/10.1093/bib/bbab012
13. Xia, Y., Chen, K., Yang, et Y.: Multi-label classification with weighted classifier selection and stacked ensemble. Information Sciences **557**, 421–442 (2021)
14. Mahdavi-Shahri, A., Houshmand, M., Yaghoobi, M., Jalali, et M. : Applying an ensemble learning method for improving multi-label classification performance. In : 2nd International Conference of Signal Processing and Intelligent Systems (ICSPIS), pp. 1–6. Iran (2016). https://doi.org/10.1109/ICSPIS.2016.7869900
15. Papanikolaou, Y., Tsoumakas, G., Laliotis, M., Markantonatos, N., Vlahavas, et I.: Large-Scale Online Semantic Indexing of Biomedical Articles via an Ensemble of Multi-Label Classification Models. Journal of Biomedical Semantics **8** (2017)
16. Li, R., Liu, W., Lin, Y., Zhao, H., Zhang, et C.: An Ensemble Multilabel Classification for Disease Risk Prediction. Journal of Healthcare Engineering, 1–10 (2017)

17. Bogatinovski, J., Todorovski, L., Džeroski, S., Kocev, et D.: Comprehensive comparative study of multi-label classification methods. Expert Systems with Applications **203**, 117215 (2022). https://doi.org/10.1016/j.eswa.2022.117215
18. Sorower, M.S.: A Literature Survey on Algorithms for Multi-label Learning. In : Computer Science, p. 26 (2010)
19. Aljedani, N., Alotaibi, R., Taileb, et M.: Multi-Label Arabic Text Classification: An Over view. IJACSA **11**(10) (2020). https://doi.org/10.14569/IJACSA.2020.0111086
20. Miller, V.: Multi-label classification. J. Biomed. Inform. **59**, 15–29 (2015)
21. Moyano, J.M., Gibaja, E.L., Cios, K.J., Ventura, et S.: Review of ensembles of multi-label classifiers: Models, experimental study and prospects. Information Fusion **44**, 33–45 (2018). https://doi.org/10.1016/j.inffus.2017.12.001
22. Madjarov, G., Kocev, D., Gjorgjevikj, D., Džeroski, et S.: An extensive experimental comparison of methods for multi-label learning. Pattern Recognition **45**(9), 3084–3104 (2012)
23. Gharroudi, O., Elghazel, H., Aussem, et A.: Ensemble Multi-label Classification: A Comparative Study on Threshold Selection and Voting Methods. In: 27th IEEE International Conference on Tools with Artificial Intelligence (ICTAI), pp. 377–384 (2015)
24. Guehria, S., Belleili, H., Azizi, et N.: A Survey on Ensemble Multi-label Classifiers. 2022. 14th International Conference on Soft Computing and Pattern Recognition (SoCPaR 2022), pp. 100–109. Lecture Notes in Networks and Systems (2022)
25. Tsoumakas, G., Katakis, et I.: Multi-Label Classification: An Overview. Int. J. Data Warehousing and Mining **3**(3), pp. 1–13 (2007)
26. Read, J., Pfahringer, B., Holmes, G., Frank, et E.: Classifier Chains for Multi-label Classification. In : Machine Learning and Knowledge Discovery in Databases, pp. 254–269. Berlin (2009). https://doi.org/10.1007/978-3-642-04174-7_17
27. Tsoumakas, G., Vlahavas, et I.: Random k-Labelsets: An Ensemble Method for Multilabel Classification. In: Machine Learning: ECML 2007, vol. 4701, pp. 406–417. Berlin (2007)
28. Read, J., Pfahringer, B., Holmes, et G.: Multi-label Classification Using Ensembles of Pruned Sets. In : 8th IEEE International Conference on Data Mining, p. 995–1000. Pisa (2008)
29. Clare, A., King, et R.D.: Knowledge Discovery in Multi-label Phenotype Data. In: Principles of Data Mining and Knowledge Discovery, vol. 2168, pp. 42–53. Berlin (2001)
30. Blockeel, H., De Raedt, L., Ramon, et J.: Top-down induction of clustering trees. In: Proceedings of the 5th International Conference on Machine Learning (ICML), p. 55–63 (2000)
31. Elisseeff, A., Weston, et J.: A kernel method for multi-labelled classification. In : Proceedings of the 14th International Conference on Neural Information: (NIPS'01), pp. 681–687 (2001)
32. Zhang, M.-L., Zhou, et Z.-H.: Multilabel Neural Networks with Applications to Functional Genomics and Text Categorization. IEEE Trans. Knowl. Data Eng. **18**(10), 1338–1351 (2006). https://doi.org/10.1109/TKDE.2006.162
33. Zhang, M.-L., Zhou, et Z.-H.: ML-KNN: A lazy learning approach to multi-label learning. Pattern Recognition **40**(7), 2038–2048 (2007)
34. Read, J., Pfahringer, B., Holmes, G., Frank, et E.: Classifier chains for multi-label classification. Mach Learn **85**(3), 333–359 (2011)
35. Tsoumakas, G., Katakis, I., Vlahavas, et I.: Effective and Efficient Multilabel Classification in Domains with Large Number of Labels. In : Proceedings of the ECML/PKDD 2008 Workshop on Mining Multidimensional Data (MMD'08), pp. 53–59 (2008)
36. Tsoumakas, G., Dimou, A., Spyromitros, E., Kompatsiaris, I., Vlahavas, et I.: Correlation-Based Pruning of Stacked Binary Relevance Models for Multi-Label Learning. In : Proceedings of the 1st International Workshop on Learning from Multi-Label Data, p. 17 (2009)
37. Kocev, D., Vens, C., Struyf, J., Džeroski, et S.: Ensembles of Multi-Objective Decision Trees. In : Machine Learning: ECML 2007, vol. 4701, pp. 624–631. Berlin (2007)

38. Nasierding, G., Kouzani, A.Z., Tsoumakas, et G.: A triple-random ensemble classification method for mining multi-label data. In: 2010 IEEE International Conference on Data Mining Workshops, pp. 49–56. Sydney (2010)
39. Xu, J., Liu, J., Yin, J., Sun, et C.: A multi-label feature extraction algorithm via maximizing feature variance and feature-label dependence simultaneously. Knowledge-Based Systems **98**, 172–184 (2016). https://doi.org/10.1016/j.knosys.2016.01.032
40. Gibaja, E., Ventura, et S.: Multi-label learning: a review of the state of the art and ongoing research: a review on multi-label learning. WIREs Data Mining Knowl Discov **4**(6), 411–444 (2014). https://doi.org/10.1002/widm.1139

Image Classification Based on Federated Learning Using PETA Dataset

José Pereira, Joana Rosa, Nuno Antunes, Pedro Santos, and João C. Ferreira(✉)

Faculty of Logistics, Molde University College, NO-6410 Molde, Norway
joam@himolde.no

Abstract. Federated Learning (FL) is an emerging machine learning technique enabling collaborative model training across decentralised devices without central data aggregation. This study applies FL to image classification using the diverse Pedestrian Attribute Recognition at a Distance (PETA) dataset, aiming to create an effective FL framework for training image classification models while focusing on data privacy and communication efficiency.

The motivations behind the adoption of FL in image classification, emphasising the importance of privacy preservation. The proposed FL pipeline involves decentralised edge devices, a central server orchestrating training, and a collaborative learning protocol. Each edge device processes a local subset of the PETA dataset and communicates only model updates to the central server, reducing raw data exchange.

The FL approach is evaluated on the PETA dataset, comparing its performance with various pre-trained machine learning models. Different FL strategies, that are approaches used to combine the weights of different models, were explored, analysing their impact on model convergence and performance.

The study confirms FedProx as the most effective Federated Learning (FL) strategy for accuracy and precision. It also highlights that FL-trained models achieve better performance in terms of accuracy and precision than centrally trained models.

Moreover, potential limitations of FL in image classification is analysed, including the demand for substantial computational resources, particularly GPU RAM.

Finally, this paper showcases FL's applicability and presents the advantages in image classification tasks, emphasising collaborative model training on pedestrian image data while preserving data privacy and without requiring the exchange of confidential information between entities.

Keywords: Federated Learning · Computer Vision · Image Classification · Artificial Intelligence

1 Introduction

Federated learning (FL) is a machine learning decentralised technique, where the algorithm is trained across multiple devices without the necessity of direct data

A. Bajaj et al. (Eds.): IBICA 2023, LNNS 1229, pp. 12–36, 2025.
https://doi.org/10.1007/978-3-031-78940-3_2

exchange [1]. This stands in contrast to conventional machine learning methods that rely on centralising datasets for training purposes. The primary advantage of our FL-based approach is its ability to empower diverse stakeholders to collaboratively contribute to the development of robust machine learning models [2]. This collaborative effort occurs while concurrently addressing critical concerns such as data privacy, security, and access rights.

In the pursuit of advancing machine learning capabilities for image classification, we have developed a FL-based framework designed to provide unique training models able to perform for real-time person attribute detection.

This study addresses the research question: "Can a Federated Learning framework improve the performance of image classification models in a public space context?".

With this research question in mind, the defined objective is to retrain pre-trained machine learning models, both with centralised data and within a Federated Learning framework, employing various FL strategies.

This research aligns with the Cross-Industry Standard Process for Data Mining (CRISP-DM) methodology [3], systematically progressing through following stages: Business Understanding, Data Understanding, Data Preparation, Modelling, Evaluation, and Deployment. All of these stages are covered in Sect. 3. By following to this structured approach, we ensure a comprehensive exploration of the data, ensuring the development of resilient machine learning models within the FL framework. This methodology not only enhance the reliability of our models but also augments contributions to a diverse spectrum of potential technological advancements in the field of image classification.

2 Related Work

The paper by McMahan et al. [4] was the first to introduce the concept of FL, described as an alternative that leaves the training data distributed on edge devices, learning the shared model by aggregating locally-computed upgrades. This approach comes to solve the limitation in which rich data, generated by devices, is very large in quantity. Transmitting such data to a central server would consume high bandwidth. Additionally, the information may be privacy-sensitive, so it is not ideal to transmit it from the client devices into a central server for training. Existing distributed optimisation approaches are only satisfied by a data centre-grade network fabric, which often isn't available or is costly.

The authors introduce the "Federated Averaging" algorithm, combining client-level SGD training with communication rounds, where the server performs model averaging. According to the tests, the algorithm is robust to unbalanced and non-IID data distributions. It can be used for unsupervised or supervised tasks, on which the supervised task labels can be inferred by the user interaction with the device.

The FL privacy has three main aspects:

- Regarding sharing the global model with clients, one cannot rely on security to mitigate attacks of this style. However, since the model is comprised of the aggregate updates from multiple clients, such attacks are more difficult;
- Regarding the update messages shared by clients, the first step for securing these messages is having capable encryption and other standard security protocols on the central server. The client updates are noised, in order to prevent the central server from making inferences about certain clients;
- Considering the information transmitted is the minimal update necessary to improve a particular model, the updates should last a very short time. Since the source of updates isn't needed for aggregation, one can ensure total client anonymity with an appropriate network architecture (such as The Onion Routing (TOR)).

The authors propose the "Federated Averaging" strategy, which includes three main parameters to control the computational resources in each round:

- C, the fraction of clients that perform computation on each round;
- E, the number of training passes each client performs over the local dataset on each round;
- B, the minibatch size used for client updates.

The algorithm was tested with image classification and language model tasks, training over 2000 individual models for the experiments, over the MNIST digit recognition dataset, and the Shakespeare Dataset.

According to [5], FL is a distributed method that involves training models using a vast amount of data that is dispersed across various locations. It is important to decide whether the training algorithms should work synchronously or asynchronously, which changes according to the aggregation methods applied. The authors mention the recent trend towards synchronous large batch training, which is employed, e.g. by the Federated Averaging algorithm [6].

The importance of privacy in data averaging is also stated, with contributions made by [7], with differential privacy, and by Secure Aggregation, [8].

The authors have developed a scalable and production-ready solution for training models with FL for mobile devices, in a synchronous manner. The solution has into consideration the limitations of computation and power draw associated with mobile devices, so as to enable the mobile devices to compute the weights, without the training data ever leaving the device.

The proposed solution is able to run large-batch SGD-style algorithms. The primary algorithm used for the aggregation of weights is the Federated Averaging, which is computed in the cloud, constructing a global model. This model is then pushed into all client devices for inference.

Considering the potential tens of thousands of devices connected to this FL architecture, the server only selects a subsect of a few hundred for an FL round, which corresponds to working collaboratively on a FL task. These devices perform the FL plan requested by the server, which consists of applying their local dataset, in the global model and sending the updated checkpoint to the central server, which is then aggregated into the global model by the server.

The authors identify 6 main steps, span across three main stages of the FL protocol: selection, configuration and reporting. The selection stage searches for devices that are charging and connected to an unmetered network. From all the devices connected, a subset is picked based on several criteria. The configuration stage corresponds to the distribution of the FL plans and FL checkpoint (central model) to each device, according to the selected aggregation mechanism. In reporting stage, devices report updates and as they are received, the server aggregates them and instructs the devices when to reconnect. The round is abandoned if there are not enough device reports in time. The device is required to have locally collected data for training and evaluation. The server is designed around the Actor Programming model, by [9], where different devices act as top-level synchronization coordinators, device connection selectors, master round-aggregators and aggregators, to which work is delegated. This is important for large federated-learning architectures. Analytic is used to control the pipeline, gathering information about what is happening in the field and monitoring device health.

The secure aggregation technique mentioned by the authors ensures data in individual devices remains encrypted, even when loaded in memory. It includes a four-round interactive protocol, where the first two rounds correspond to sharing the established secrets on the device level to ensure security. In the third round, devices commit model updates to the central server, which are aggregated. The last round is the finalisation phase, where the central server uses device-wise secrets to unmask the aggregated model update.

The FL technique can be used for on-device item ranking, content suggestion for on-device keyboards and next-word prediction. This solution answers questions regarding device availability, as smartphones have distinct operation times, correlated to the different time zones. It also addresses unreliable device connectivity and interrupted execution, orchestration of lock-step execution across devices with varying availability and limited device storage and computational resources. On the other hand, this solution is at an early stage and no discussion is presented on the required components associated with mobile devices.

In [10], the FL technique is integrated into a computer vision scenario, allowing data parties to build and train Machine Learning models collaboratively, whilst keeping their data secure and private. The premise lies in the fact that data is mostly generated by edge devices. As such, moving training tasks of computer vision tasks to these devices, only sharing the model parameters offers multiple advantages in terms of data privacy, security and efficiency, as fewer data is transmitted and we take advantage of distributed computing for model training and validation.

Current works in the computer-vision field still depend on pre-existing public datasets, due to the lack of public annotated data for several applications. This has been causing a lag in developments in this area of research. The authors present a dataset of real-world images from 26 street cameras and 7 annotated object categories. Results on model efficiency, performance and communication are provided from training a YOLOv3 and Faster R-CNN.

The presented dataset has twenty-six 704×576 video feeds, captured at different times of the day. These video feeds were used to collect 956 legible images, meaningfully distinguishable from one another. The labels represented in the data are basket, carton, chair, electro-mobile, gastank, sunshade and table, across 2544 selected items in these images.

Two data divisions are presented in the paper: the first corresponds to interpreting each camera as a unique client of the network. The second approach consisted in grouping nearby cameras into a single client, creating 5 clients of camera clusters.

Two state-of-the-art object detection algorithms were implemented and integrated with the FL technique: Faster R-CNN (with pre-trained VGG1) SGD with a fixed learning rate of 1e-4 and a momentum of 0.9 and YOLO via Adam with a learning rate of 1e−3.

The authors used a modified "FederatedAveraging" strategy so it better suits the problem of evaluating the effect of different data divisions. The modified version saves and restores checkpoints on hard devices, instead of communicating with the central server.

The FL hyperparameters are the number of clients that participate in each round, the number of times each client performs over its local dataset per round and the minibatch size. An Intersection Over Union (IOU) chart was created with the results of both architectures on different parameters. The minibatch size was set to 1. With a threshold of 0.5, authors have observed that 5 clients participating in each round have a comparable performance to centralised training, whereas using 20 clients showed a lower performance. A larger number of steps over the local dataset per client also showed that the models would converge faster. Training with a pre-trained model showed faster convergence.

In [6], the authors propose a solution for detecting anomalies of sensors in Industrial IoT applications, by applying a FL framework. This framework works to collaboratively train a deep anomaly detection model. The authors also propose a Convolutional Neural Network-Long Short Term Memory (CNN-LSTM) that can accurately and timely detect anomalies. The solution is supported by the premise that edge devices, such as robots, collect time-series sensing data that captures the behaviour and condition of the IoT nodes. With this data, edge devices can train a global anomaly detection model, keeping the training datasets locally and without compromising privacy. The trained models and framework were evaluated on two real-world datasets: power demand and space shuttle.

The proposed framework included a cloud aggregator and several edge devices. The cloud aggregator was responsible for performing two main actions: first, it initialised the global model and sends it to all edge devices; then, the server aggregated the gradients uploaded by edge devices so that the model converged. Edge devices are agents with local models and functional mechanisms. These devices use their local dataset to train the global model, before uploading these updated gradients to the cloud aggregator. These devices contain a deep anomaly detection mechanism, which reduced economic losses by detecting anomalies. Additionally, each edge device was equipped with a gradient compression mechanism, which compressed local gradients, in order to reduce the number of gradients exchanged with the cloud aggregator. For model aggregation, it used "FederatedAveraging".

The CNN-LSTM model used normalised data as input for the CNN to capture input features. These features were used as input for the LSTM models, in order to predict future time series data. The output was an anomaly detection score, that can help to detect anomalies.

The gradient compression mechanism was built on the premise that only gradients with the largest absolute value were useful for gradient updates. The smaller parameter nudges usually do not matter. According to the authors, 99.9% of model training can generally be reached with the 0.1% gradients with the largest absolute value. For this reason, small gradients were not sent directly to the central server, only gradients larger than a defined threshold were transmitted. Instead, small gradients were accumulated in the edge devices until they reached the threshold value. The threshold value was adjusted by testing a simple CNN on MNIST and CIFAR-10 datasets, and the value of 0.3 was selected as the best trade-off between gradient exchanges and achieved accuracy.

The framework performance was compared to LSTM, GRU, Stacked Auto Encoder, and SVM methods with identical configurations. In these experiments, the proposed approach achieved the highest accuracy on both datasets. According to the authors, this model also improved communication efficiency by compressing gradients by 300 times without losing accuracy. In [11], the authors propose a new aggregation method for FL, the Personalised Conditional FedAvg (PC-FedAvg), which allows for weight communication and aggregation control, thus personalising the resulting models to each client and mitigating the problem of non-IID data distribution. A one-class SVM is trained for anomaly detection, using SGD, instead of the more commonly used Sequential Minimal Optimisation for SVMs. This is due to the fact that, according to the authors, this strategy is fitter for single-class problems. In the proposed PC-FedAvg method, only clients whose local loss value is below the overall median loss are included in the aggregation at the central model. The remaining clients are dropped until the next round so that no bad weights are aggregated. The algorithm is validated with two datasets collected from structural bridges, in order to detect potential damage, achieving accuracies of 96% and 91% for each of the damage-detection tasks.

In [12], the authors explore the usage of FL for IoT cybersecurity, by monitoring devices for anomalies and network intrusion attacks. The authors propose a FedIoT platform that includes FedDetect for anomaly detection and a system design for the evaluation of the efficacy of FL recognising attack types on IoT devices. The IoT devices communicate directly with the FL server and the standard FL communication and rounds are applied.

The training data used was synthesised from the N-BaIoT and LANDER datasets from which 5000 benign and 5000 malign data samples were randomly selected from a wide range of attack types. The anomaly detection module uses a deep autoencoder, but other models, such as the variational autoencoder or attention-based CNN-LSTM can also be applied. This algorithm is trained on the benign dataset, in order for the algorithm to be able to extract and reconstruct features on benign samples but fails to do so on abnormal samples with unseen features. The MSE was calculated, which represented the error of reconstruction and a threshold was inferred, in order to prevent abnormal samples from being classified as benign. The FedDetect algorithm used the Adam optimiser, which was proven to outperform SGD and SGD with momentum. It also used a cross-round learning rate scheduler, the cosine scheduler, rather than "FederatedAveraging" locally.

In [13], the authors propose a new approach that leverages a pre-trained anomaly detection model to detect abnormal client behaviours and eliminate their adverse impacts. An anomaly detection model runs at the server over the local model weight updates received from the clients. The technique of "Federated Averaging" is applied from the server side to aggregate the model weight updates. Some dimensionality reduction techniques are used to generate surrogates of the local model weight updates at the server for anomaly detection.

In terms of problem definition, a horizontal FL system consisting of one server and K clients, that collaboratively train a model using the "Federated Averaging" algorithm. Due to some space limitations, the study focuses on the case where the clients send to the server the local model weight updates without encryption. On the server side, an anomaly detection model is employed to detect anomalous clients and eliminate their impact on the federated model training.

A mechanism was implemented to assign a credit score to each client of the FL system, based on the anomaly score produced by the anomaly detection model. The anomaly score is derived from a pre-trained autoencoder anomaly detection model. This autoencoder is deployed at the server to detect abnormal model weight updates from the clients (anomalous clients).

As for deep learning models, the dimension of the model weight can be extremely large which may lead to very high computational complexity for training the autoencoder model for anomaly detection. For this reason and to bind the computational complexity, some dimensionality reduction techniques were applied, namely to reduce the weight vector coming from the FL clients. For this work for example some weights were selected, namely the weights of the last convolutional layer of the neural network.

After obtaining the anomaly scores from the clients, the credit scores are computed and then the aggregation operation is performed. This paper defined a threshold for the credit scores: if the anomaly score is strictly higher than a specific threshold then the credit score is set to zero. Otherwise, the credit score is set to the ratio between the number of data points of that specific FL client and the total number of data points.

The authors have evaluated the solution proposed with a 62-class image classification task over Federated Extended MNIST (FEMNIST) dataset. There were in total 801,074 samples distributed unevenly among 3,500 writers in FEMNIST. Each writer has an average of 229 samples with a standard deviation of 89.6. Each writer in the FEMNIST dataset represents a client, resulting in a heterogeneous federated setting. During each global round, K = 20 writers were selected and have run mini-batch gradient descent locally with a batch size of 16 and a learning rate of 0.06. Each client performed local model weight update for 20 epochs. The percentage of abnormal clients was set to 30% in every global round. The image classification model trained in FL was a CNN model with two convolutional layers. The autoencoder model used at the server for anomaly detection had four hidden layers. The input of the autoencoder was a vector with a dimensionality of 3000 randomly taken from the last convolutional. ReLu activation was used in the hidden layers. The autoencoder model was trained with accumulated local model weight updates at the server, with the loss function being the MSE, a batch size of 32, and a dropout rate of 0.2. Three adversarial attack models were considered, namely sign-flipping, additive noise and gradient ascent. Sign-flipping attack flipped the sign of the model weight, the additive noise attack added Gaussian noise to the model weight, and in the gradient ascent attack, the anomalous clients run gradient ascent locally, instead of gradient descent. The detection methods proposed by the authors demonstrated a superior performance (over 10%) when compared with defense-based methods (GeoMed, Krum, and Trimmed Mean).

The [14] evaluates the practical feasibility of auditing a federated deep learning process by using blockchains, demonstrated in a realistic use case on intrusion detection and a real-world dataset (i.e. the Intrusion Detection Evaluation Dataset, CICIDS2017). The authors used the MultiChain open-source blockchain platform, as it had a limited relative performance impact and network overhead while providing full transparency over the distributed deep learning process. The motivation behind this study was a network intrusion detection scenario with multiple interconnected devices, possibly spanning multiple collaborating enterprises.

The CICIDS2017 dataset contained network traffic information from 12 machines including benign traffic and common attacks collected throughout a period of five days. The malicious attacks included brute force FTP, brute force SSH, DoS, heartbleed, web attacks, infiltration, botnet, and DDoS. The authors have used an autoencoder which has a fairly simple architecture with three hidden layers to perform the intrusion detection. The "Federated Averaging" technique was then used to combine the weights sent by the local FL clients. The federated deep learning framework was then integrated with permissioned blockchains to help audit the distributed learning process. The results revealed that the blockchain technology used, MultiChain, conducted a limited performance overhead of about 5–15% for this particular federated intrusion detection case, while providing full transparency over the distributed learning process.

The results also revealed that the performance overhead caused by the blockchain is closely tied to the complexity of the (deep) neural network at the core of the FL algorithm. The total amount of weights of the neural network will define the performance impact of the proposed approach. For this work, when considering an autoencoder with three hidden layers as the anomaly detection method, a total amount of 3000 weights was considered.

In the [15] an implementation of a FL framework in a smart healthcare ecosystem is proposed. Nowadays, there is a proliferation of smart medical devices and applications used in smart hospitals, smart-home-based care, and nursing homes. These devices are used to sense patients' vital body parameters, monitor health conditions and generate multivariate data to support just-in-time health services. This data is commonly analysed in centralised servers and with this fact come inherent privacy issues associated with sending patients' personal health data to the server. In this study, an Anomaly Detection (AD) model integrated into a FL (FL) framework is proposed. This framework utilises edge cloudlets to run AD models locally without sharing patients' data. The proposed distributed learning system consists of a hierarchical FL that allows aggregation at different levels enabling multi-party collaboration: it is a disease-based grouping mechanism where different AD models are grouped based on specific types of diseases.

It is proposed a Federated Time Distributed (FEDTIMEDIS) Long Short-Term Memory (LSTM) framework to build a local AD model for identifying anomalies using hierarchical federation employing Digital Twin (DT) and Edge Cloudlet Computing (ECC).

This approach allows passing multiple input samples to the model simultaneously for many participants. The architecture utilised four LSTM cells and two stacked layers before the Fully Connected (FC) or dense layer, to ensure fast processing, less computational resources requirement, and optimal solution with better performance.

The proposed framework provides data security and privacy through edge cloudlet deployment of FL-based AD model, model integrity, and statistical features through distributed model training and only sharing gradients, reduced computation and communication costs via local training and disease-based grouping and hierarchical FL enabling high accuracy.

This approach utilises parameter sharing and aggregation as opposed to patient data sharing for detecting anomalous traits and for this reason it is robust, scalable, and capable to learn long-term dependencies inherent to Internet of Medical Things (IoMT) sensor data. This study presents a Remote Patient Monitoring (RPM) use case and implementation using DT and ECC. The disease-based grouping of AD models improves model accuracy, and hierarchical FL enables easier collaboration among multiple healthcare organisations but the authors refer that future work needs to be carried on to evaluate detailed metrics including detection accuracy, false positive rate, recall, F1 score, training time and communication cost of the proposed model.

In the [16], the authors addressed the need to leverage real-world health data for machine learning tasks, considering this data is sensitive and the resources may be scarce. The proposed framework doesn't move or share data across sites or with the central server during model training and uses a differential privacy mechanism. The FL process works like the initially proposed architecture, with a centralised server, clients, and training rounds. The models trained to validate the architecture are the perceptron, SVM, and logistic regression. Differential privacy works with the addition of noise as an exponential mechanism, an input, output, or objective perturbation. The authors have decided, supported by literature, which was proven to outperform other methods, to add noise to the objective function.

In [17], the authors designed and implemented an FL approach to address challenges in person re-identification using attribute recognition. To enhance the learning capabilities of the edge nodes, the authors employed federated learning and transfer learning methods. Federated learning allowed edge nodes to independently learn from local data, while transfer learning was used to retrain low-quality local models and improve overall model accuracy. The models were trained using PETA and RAP datasets. The authors proposed a novel federated aggregation strategy named "FedTransferLoss". This strategy aimed to achieve optimal global accuracy by using transfer learning to re-train low-quality local models. FedTransferLoss was designed to outperform traditional federated aggregation algorithms, such as FedAvg. The results demonstrated that FedTransferLoss achieved higher accuracy, recall, and precision values compared to the traditional FedAvg algorithm, validating the efficacy of their methodology.

In [18], the authors addressed a problem in Federated Learning (FL) where achieving personalised models for individual clients faced challenges due to statistical differences in data. They proposed a solution called RepPer, a two-stage framework that separates the learning of data representations from the classification process. This approach overcomes issues associated with non-uniformly distributed data, a common problem in FL. In the first stage, client-specific models are developed, resilient to these challenges, and aggregated into a shared representation. In the second stage, personalised classifiers are trained based on this common representation. The authors demonstrated the effectiveness of RepPer through experiments, showing superior flexibility and personalisation, especially when dealing with non-uniform data. Notably, they highlighted its potential for lightweight edge computing, making it applicable to devices with limited computing resources.

In [19], the authors of this work addressed challenges related to model training in medical image diagnosis applications, specifically focusing on the difficulties posed by the quantities and diversities of datasets in real medical scenes. The problems identified were twofold: the unavailability of required data in a single institution due to factors such as the number of patients or the type of pathology, and the impracticality of sharing patient data due to medical data privacy regulations. These challenges hindered the fusion of data from multiple entities to train a medical model while ensuring the privacy of sensitive patient information.

To address these issues, the authors developed a federated learning architecture. The authors conducted experiments using a breast cancer histopathological dataset (BreakHis) to validate the effectiveness of their proposed federated learning framework. The results of their experiments demonstrated that the federated learning system performed comparably well to centralised learning approaches. This suggests that the proposed framework could effectively address challenges related to limited data accessibility and privacy constraints in medical image diagnosis applications.

In [20], the authors reviewed and analysed the application of federated learning (FL) in deep neural network (DNN) medical image processing. They referred how the potential of DNNs for clinical diagnostics is limited by security and privacy issues while exchanging local medical data across several facilities. The study addresses this by introducing FL, which maintains data confidentiality and privacy while enabling collaborative model training across decentralised medical facilities. Reviewing FL applications in medical image processing, talking about security issues, attempting to enhance FL model performance, and describing obstacles and potential research approaches are the main topics of discussion. The goal of the study is to research the application of FL to protect the sensitive nature of medical data while improving the precision and effectiveness of DNNs in medical picture analysis.

In [21], the FedProx framework was presented by the authors as a solution to the problems associated with heterogeneity in federated learning, a situation in which devices within a network reveal variability in data distribution and system characteristics. With minor modifications, FedProx is a generalisation of FedAvg, the state-of-the-art technique at the moment. In the presence of system-level variations and non-identically distributed data, the framework ensures convergence. Averaging a 22% improvement in absolute test accuracy, FedProx outperformed FedAvg in real-world testing on federated datasets, exhibiting more robust convergence in highly heterogeneous cases. In federated learning applications, FedProx proves to be a potential method for obtaining accuracy and stability.

In [22], the authors examined the impact of non-identical data distributions in Federated Learning, a privacy-preserving method for training visual models with real-world data from mobile devices. They suggested a technique to generate datasets with distinct levels of similarity and evaluated the Federated Averaging algorithm. The results showed that as data distributions became more asymmetric, performance decreased. To address this, the authors introduced a mitigation strategy using server momentum, Federated Averaging with Server Momentum (FedAvgM). In a variety of non-identical data scenarios, experiments on CIFAR-10 showed improved classification performance; in the most imbalanced settings, accuracy increased from 30.1% to 76.9%.

3 Federated Learning Framework

The main goal is to implement a FL framework in order to evaluate if this mechanism can improve the Machine Learning algorithms that may be running in a decentralised manner to perform classification tasks. Regarding the FL strategy itself, a centralised one was employed, due to its simplicity. This way, a centralised server is configured to be responsible to send the general model to the clients and to aggregate all the model updates from the local nodes according to a specific strategy of FL.

As one of the main goals is to get and analyse the camera feeds in order to identify persons of interest when monitoring the security of a specific event, extensive research for public datasets was made to identify sets of annotated images of pedestrians both in indoor and outdoor scenarios. Each image display a pedestrian and is annotated with his/her physical characteristics (for example the colour of the upper body fabric he/she is wearing, the colour of the lower body fabric he/she is wearing the accessory that he/she is carrying (if any), etc.). Figure 2 presents an example of an image in PETA dataset folder [23] (Fig. 1).

Fig. 1. Example of image in PETA Dataset [23].

These image datasets are then used to calibrate an algorithm baseline, that will be spread across the FL clients. The clients will use this algorithm baseline, retrain it with their local datasets and send the result of the training to the central server to be combined according to a specific FL strategy. After the framework implementation, several rounds of FL have run and some statistics are collected to assess if this framework can revise the algorithm weights in order to leverage the performance of the machine learning algorithms running locally in the clients.

3.1 Data Understanding

The PETA dataset is a dataset for the recognition of pedestrian attributes, and it consists of 19000 images from 8705 distinct people. Each image can be segmented into a set of categories: Gender; Age; Upperbody (refers to the part of the body from the waist up, including the chest, shoulders, arms, and back); Lowerbody (refers to the part of the body from the waist down, particularly the legs); Footwear; Hair.

The categories Upperbody and Lowerbody can be split into three subcategories: clothes (one of the classes could be for example a T-shirt or Jacket), style (could be formal or casual) and color; Footwear can be with two subcategories clothes and color and finally Hair with two subcategories style and colour.

The categories Upperbody, Lowerbody, Footwear, and Hair are multi-label categories and each one of these categories can have several category values. For example, a person may have for Upperbody "Blue, Black, Red" which means that these three colours are present in his/her Upperbody fabric, as it is shown on Figs. 2.

It is also relevant to highlight that this dataset captures images of pedestrians in real-world scenarios, providing a distinct advantage over controlled environments. This diversity includes both indoor and outdoor images, ensuring that the trained models are exposed to a wide range of contextual variations.

The PETA dataset is organised in four folders: CUHK (Indoor images; PNG format), DataForModel (Diverse Images; JPG, BMP and PNG formats), TownCentre (Outdoor Images; JPG format), and GRID (Indoor Images; JPEG format) as shown in Fig. 3.

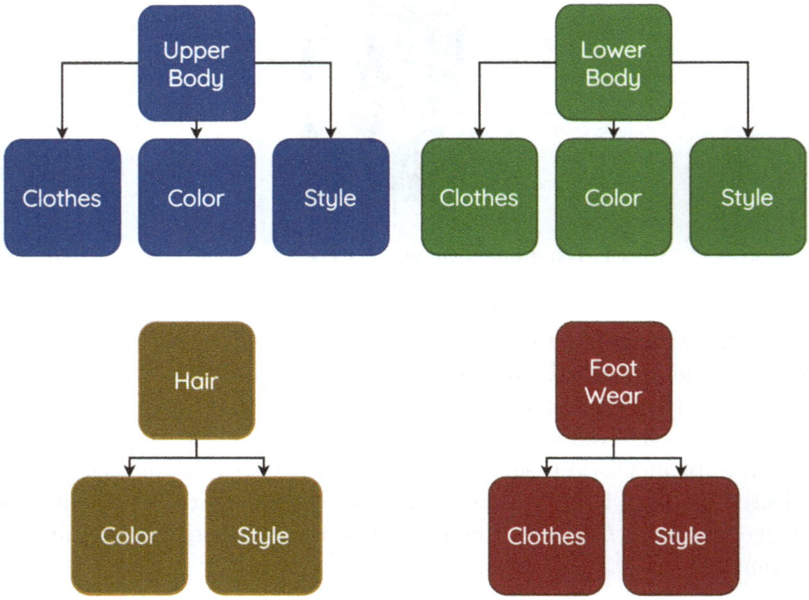

Fig. 2. Hair and Footwear characteristics included in the the PETA dataset.

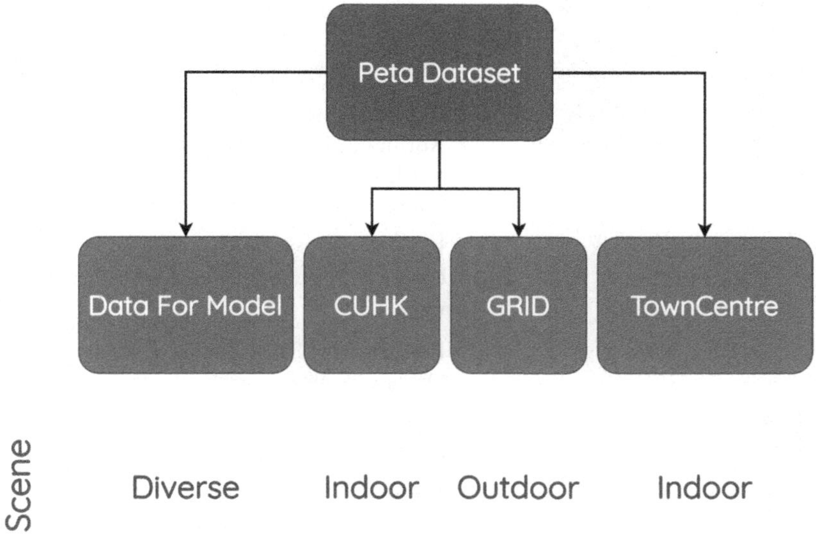

Fig. 3. PETA dataset folders used in the FL framework.

3.2 Data Preparation

The first stage of data preparation consisted of carrying out an analysis of the representativeness of the categories. The dataset was unbalanced, some category values had a very low number of associated images. Since an unbalanced dataset could lead to bad results regarding precision and recall levels, feature engineering was performed. More specifically, some category values with few images were grouped into new category values with a more relevant number of associated images compared to the other category values within the same category as shown in Table 1. This process is called binning (grouping category values into new ones to make them more relevant). This transformation addresses some concerns with the original data. By agglomerating category values with very low representativity in the data, it is possible to achieve two main goals: (1) significantly reduce the number of category values in the dataset; and (2) mitigate the problem with highly imbalanced data, as the newly created categories values include examples from several different old category values. The problem with highly imbalanced data was also resolved with data binning technique mentioned once clustering the categories, creating new categories, gives them the same level of representativeness.

Table 1. New Category Values created after Feature Engineering

Category	New Category Value	Old Category Value
Carrying	Backpack	Backpack
	No Backpack	Nothing, MessengeBag, LuggageCase, SuitCase, PlasticBags, Umbrella, ShoppingTro, BabyBuggy, Folder
UpperBody Color	Black	Black
	Grey	Grey
	White	White
	Other	Red, Brown, Blue, Green, Purple, Pink, Orange, Yellow
LowerBody Color	Black	Black
	Grey	Grey
	White	White
	Other	Red, Brown, Blue, Green, Purple, Pink, Orange, Yellow

There were categories whose category values present in the images were difficult to identify because they had a small size in the images or are difficult to distinguish from other categories. These categories were eventually discarded. Once the previously mentioned categories were removed, only category values belonging to the following categories remained: upperbodyColor, lowerbodyColor, and carrying. To ensure that the model learned each category equally, the number of images per category value with better representativeness was reduced in a way that all category values would have the same number of images as shown in Tables 2, 3 and 4.

Table 2. UpperBodyColor - Number of images in the dataset before (Unbalanced Dataset) and after Data Preparation (Balanced Dataset)

UpperBodyColor – Number of Images		
Colors	Unbalanced dataset	Balanced Dataset
Black	2908	891
Grey	1323	891
Blue	1218	891
Other	1998	891
Total	7447	3564

Table 3. LowerBodyColor - Number of images in the dataset before (Unbalanced Dataset) and after Data Preparation (Balanced Dataset)

LowerBodyColor – Number of Images		
Colors	Unbalanced dataset	Balanced Dataset
Black	4168	744
Grey	2162	744
Blue	1368	744
Other	798	744
Total	8496	2976

Table 4. Carrying - Number of images in the dataset before (Unbalanced Dataset) and after Data Preparation (Balanced Dataset)

Carrying - Number of Images		
Object	Unbalanced dataset	Balanced Dataset
Backpack	5605	2148
No Backpack	2148	2148
Total	7753	4296

Table 5 presents the number of images after data balancing for each PETA dataset folder.

3.3 Modelling

In the computer-vision field, several types of pre-trained models can be integrated into custom models to help achieve better results. In the scope of the implementation of the FL framework, two types of pre-trained models were selected: InceptionV3 and EfficientNetV2M. It is important to stress that the scope of this work is to implement and evaluate a decentralised training framework and not to develop a machine learning model for image classification from the ground up, which motivated the selection of a pre-trained model.

Table 6 presents the Keras models that were analysed along with their respective characteristics.

The selection of the pre-trained models was made based on the Top-1 accuracy (only considers the most likely predicted category value for each test sample), and Top-5 accuracy (considers the top 5 predicted category values for each test sample) achieved by each model and its low memory size needed. Despite the models InceptionResNetV2 and NASNetLarge being referred to have good accuracy values, they were discarded for this prototype due to computational resource limitations (both models consume a high amount of memory). In the table below are presented the accuracies for each model and the details of the

Table 5. Number of images after data balancing for LowerBodyColor, UpperBodyColor and Carrying

LowerBody Color	PETA Dataset	Flower					
		Baseline		FL Model -Train		FL Model -Test	
		DT	CUHK	GRID	TC	GRID	TC
Black	744	361	277	56	18	24	8
Grey	744	361	277	56	18	24	8
Blue	744	361	277	56	18	24	8
Other	744	361	277	56	18	24	8
Total	2976	1444	1108	224	72	96	32

UpperBody Color	PETA Dataset	Flower					
		Baseline		FL Model -Train		FL Model -Test	
		DT	CUHK	GRID	TC	GRID	TC
Black	891	316	450	75	13	32	5
Grey	891	316	450	75	13	32	5
Blue	891	316	450	75	13	32	5
Other	891	316	450	75	13	32	5
Total	3564	1264	1800	300	52	128	20

Carrying	PETA Dataset	Flower					
		Baseline		FL Model -Train		FL Model -Test	
		DT	CUHK	GRID	TC	GRID	TC
Black	891	316	450	75	13	32	5
Grey	891	316	450	75	13	32	5
Blue	891	316	450	75	13	32	5
Other	891	316	450	75	13	32	5
Total	3564	1264	1800	300	52	128	20

models in terms of number of parameters, depth (topological depth of the network, including activation layers, batch normalisation layers, etc.).

In machine learning, a baseline model serves as a starting point for comparison with more complex models, allowing researchers to assess how well more sophisticated models perform and determine whether they add any useful information. For each category, two baseline models were developed, one based on models InceptionV3 and EfficientNetV2M pre-trained with the Imagenet dataset. All baseline models were then re-trained in a centralised approach with local data and tested with the same training set and testing set. On the local data, each model was retrained within 20 epochs, using an SGD optimiser with learning rate 0.01 and momentum 0.9. The batch size was 8. In this step,

Table 6. Accuracy reference values for a set of pre-trained models available from the Python library, Keras.

Model	Size (MB)	Top-1 Accuracy	Top-5 Accuracy	Parameters	Depth	Time (ms) per inference step (CPU)	Time (ms) per inference step (GPU)
ResNet 50	98	74.9%	92.1%	25.6M	107	58.2	4.6
InceptionV3	92	77.9%	93.7%	23.9M	189	42.2	6.9
InceptionResNetV2	215	80.3%	95.3%	55.9M	449	130.2	10.0
MobileNetV2	14	71.3%	90.1%	3.5M	105	25.9	3.8
DenseNet121	33	75.0%	92.3%	8.1M	242	77.1	5.4
DenseNet169	57	76.2%	93.2%	14.3M	338	96.4	6.3
DenseNet201	80	77.3%	93.6%	20.2M	402	127.2	6.7
NASNetMobile	23	74.4%	91.9%	5.3M	389	27.0	6.7
NASNetLarge	343	82.5%	96.0%	88.9M	533	344.5	20.0
EfficientNetB0	29	77.1%	93.3%	5.3M	132	46.0	4.9
EfficientNetV2M	220	85.3%	97.4%	54.4M	–	–	–

only two folders from the PETA dataset were used, the DataForModel folder and the CUHK folder. Since federate learning is a technique that requires a lot of memory, it was decided to use the two folders with more images (CUHK folder and DataForModel Folder) for the baseline model and the two folders with fewer images (GRID folder and TownCentre Folder) for FL clients.

Both folders were split into a training set (70% of images) and a testing set (30% of images). Afterward, the two training sets were merged into one training set and the two test sets were merged into one test set, as illustrated in Fig. 4.

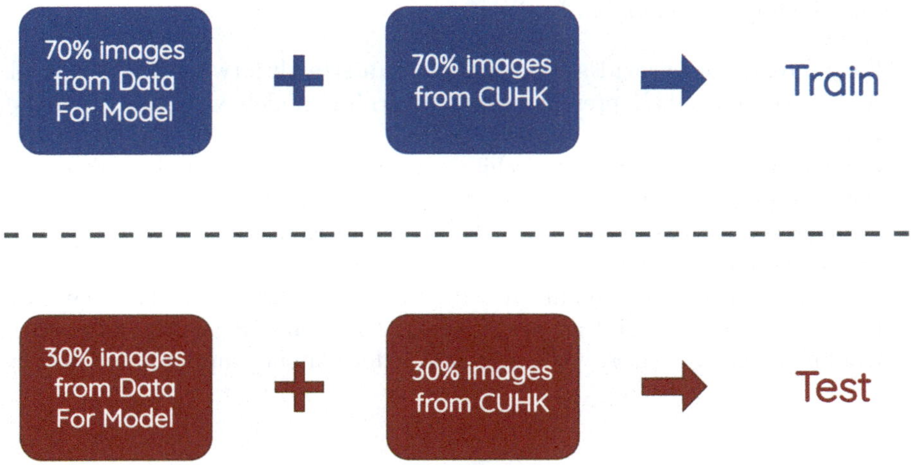

Fig. 4. PETA dataset used in Baseline models training and testing procedures.

In order to implement the FL framework, three main strategies were used: the FedAvg [6], the most common strategy that simply combines equally the weights of the models from all the FL clients, the FedProx [21] which is a method more suited to cope with the heterogeneity of the data that may exist among the FL clients and the FedAvgM [22] which adds a momentum term that helps to smooth the model updates and avoid oscillations.

Figure 5 provides a comparative analysis of the technical features of FL implementation tools as outlined in [24].

	TFF	Syft	FedScale	LEAF	Flower
Single-node simulation	√	√	√	√	√
Multi-node execution	*	√	(√)***		√
Scalability	*		**		√
Heterogeneous clients		(√)***	**		√
ML framework-agnostic		****	****		√
Communication-agnostic					√
Language-agnostic					√
Baselines			√	√	*

Labels: * Planned / ** Only simulated
*** Only Python-based / **** Only PyTorch and/or TF/Keras

Fig. 5. Comparison of technical characteristics of the tools to implement FL [24].

We opted for the Flower Framework, whose architecture is showcased in Fig. 6, following thorough research to identify the most suitable solution for testing the application of FL. It allows:

- Easy integration: provides developers with a straightforward and adaptable API to combine their present machine learning models with the FL framework;
- Flexibility: provides a flexible architecture that may be adapted to any organisation's particular requirements. Moreover, a variety of machine learning frameworks, such as TensorFlow, PyTorch, and Keras are supported;
- Decentralised Learning: this allows for decentralised machine learning by enabling devices to communicate with each other directly, rather than having to communicate with a central server. This makes it possible to perform training on edge devices, which can help reduce latency and improve privacy.

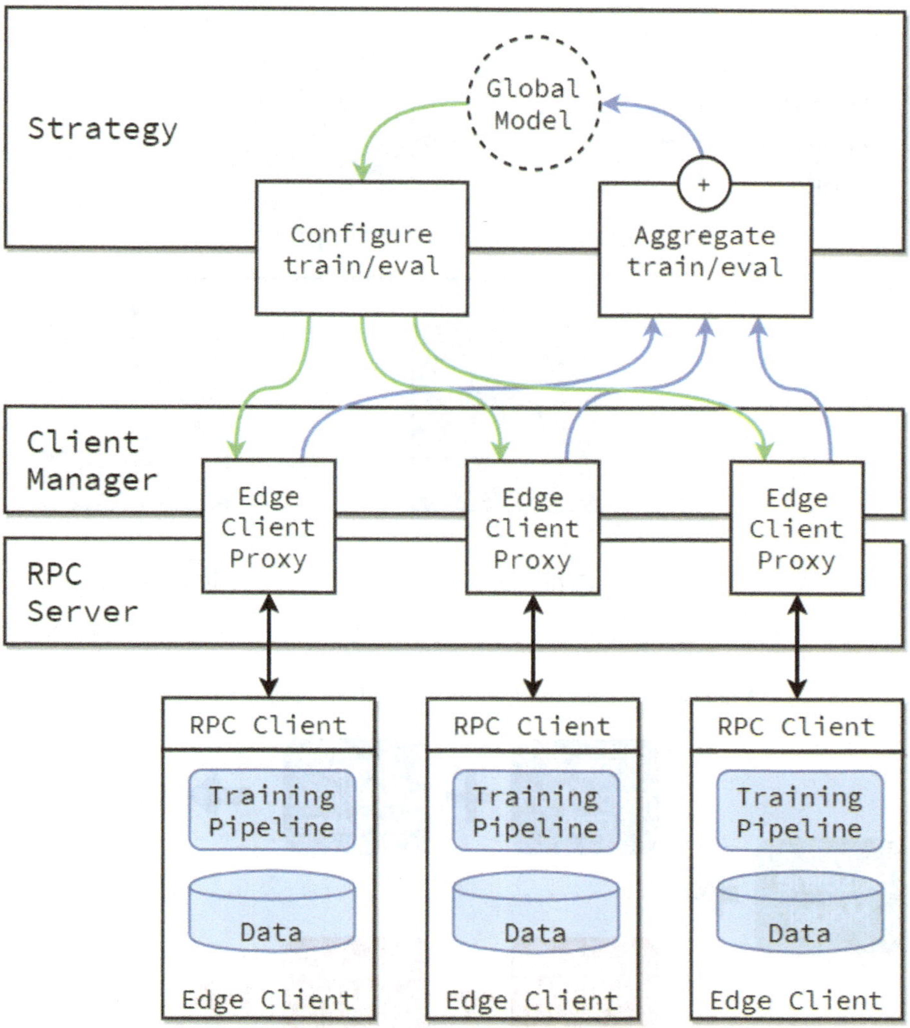

Fig. 6. Flower Core Framework Architecture with Virtual Client Engine [24].

Due to the fact that the FL framework is in general terms a technique that consumes a high amount of memory, for the proof of concept in the context of the project, it was decided to use just two FL clients as shown in Fig. 7. These two clients will be parametrised with two different models: InceptionV3 and EfficientNetV2M so it is possible to compare their performance metrics and check which one is more suited in this context.

Each client's model was retrained with a different dataset: client 1's model was retrained with 70% of images from the GRID folder and tested with 30% of images from the same folder; Client 2's model was retrained with 70% of images from TownCentre folder and tested with 30% of images from the same folder, as shown in Fig. 8.

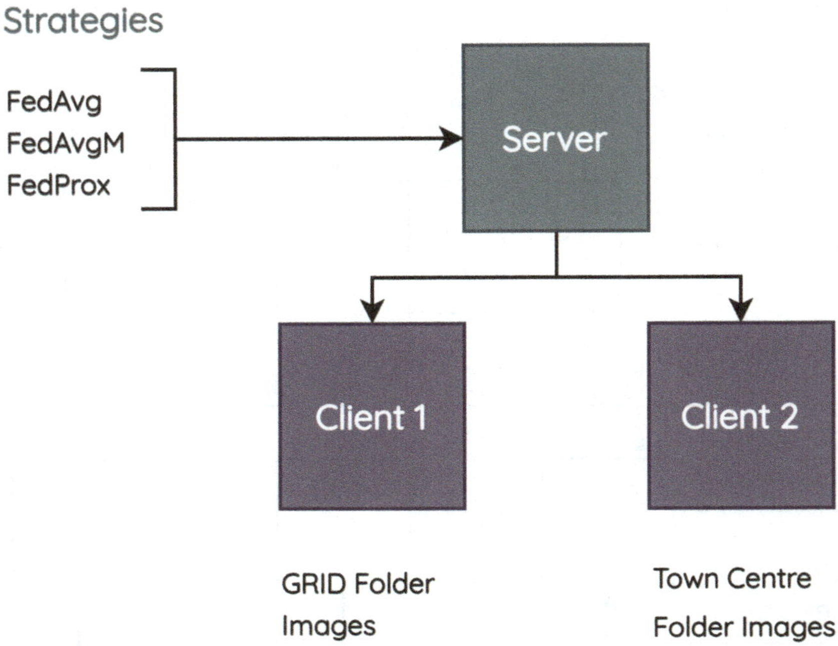

Fig. 7. FL Client used and federated Strategies applied.

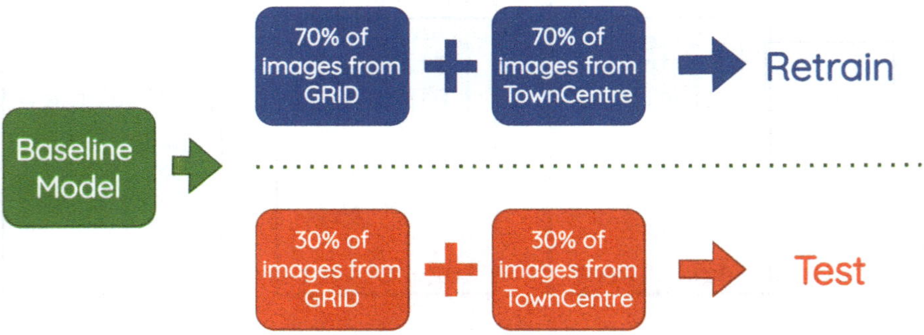

Fig. 8. PETA dataset folders used in the FL framework.

4 Results

In general, the FL strategy which provided the best results is the strategy FedProx, while FedAvg was the strategy with the second-best results. Comparing the accuracy and precision rates of baseline models revealed that EfficientNetV2M outperforms InceptionV3 when it comes to accuracy and precision. However, after applying FL, InceptionV3 performed better when compared to EfficientNetV2M as shown in Tables 7, 8 and 9. LowerbodyColor was the category with the worst outcomes and carrying was the category with the best results. This may be due to the fact that carrying has only two classes and upperbodyColor and lowerbodyColor have both four classes, since in the process of identifying a large number of classes, there's a higher likelihood of classes that are visually similar or share common characteristics. This can lead to confusion for the models, making it harder to differentiate between classes.

Client 2, which was trained with images from the TownCentre folder showed worse results than client 1, trained with the GRID folder. Given that the GRID folder contains more images than the TownCentre folder, client 1 may have outperformed client 2 in terms of accuracy and precision rates because of the larger number of images in the GRID folder which may have improved the knowledge collected by client 1's machine learning algorithm. The models retrain with FL revealed better performance compared to the baseline models associated. For instance, the accuracy and precision of InceptionV3 in the prediction of upperbodyColor and lowerbodyColor were higher in federated learned models using the FedProx strategy than in the pre-trained InceptionV3 baseline model of each category.

Note: I corresponds to InceptionV3, and E corresponds to EfficientNetV2M

Table 7. LowerBodyColor - Performance values for baseline and FL models

Baseline		Flower													
		Rounds	Metric	FedAvg				FedAvgM				FedProx			
				GRID		TC		GRID		TC		GRID		TC	
E	I			E	I	E	I	E	I	E	I	E	I	E	I
0.62	0.59	2	Acc	0.47	0.74	0.28	0.38	0.4	0.46	0.28	0.41	0.6	0.69	0.31	0.38
			Prec	0.71	0.76	0.38	0.62	0.34	0.39	0.2	0.4	0.64	0.72	0.4	0.44
		5	Acc	0.52	0.74	0.28	0.47	0.4	0.45	0.28	0.41	0.64	0.68	0.3	0.38
0.69	0.67		Prec	0.57	0.76	0.39	0.69	0.35	0.38	0.19	0.39	0.67	0.71	0.45	0.44
		10	Acc	0.52	0.74	0.31	0.47	0.41	0.48	0.31	0.38	0.63	0.68	0.34	0.38
			Prec	0.57	0.76	0.41	0.69	0.36	0.39	0.26	0.34	0.66	0.71	0.46	0.44

Table 8. UpperBodyColor - Performance values for baseline and FL models

Baseline		Flower													
		Rounds	Metric	FedAvg				FedAvgM				FedProx			
				GRID		TC		GRID		TC		GRID		TC	
E	I			E	I	E	I	E	I	E	I	E	I	E	I
0.7	0.66	2	Acc	0.51	0.70	0.4	0.55	0.42	0.42	0.65	0.65	0.58	0.72	0.7	0.6
			Prec	0.50	0.72	0.40	0.6	0.51	0.51	0.56	0.56	0.59	0.72	0.66	0.75
		5	Acc	0.51	0.70	0.4	0.5	0.47	0.46	0.65	0.65	0.58	0.72	0.55	0.6
0.7	0.68		Prec	0.52	0.72	0.43	0.6	0.48	0.46	0.54	0.54	0.59	0.72	0.59	0.75
		10	Acc	0.5	0.70	0.4	0.55	0.54	0.52	0.7	0.65	0.59	0.72	0.6	0.6
			Prec	0.49	0.72	0.40	0.6	0.5	0.49	0.55	0.54	0.59	0.72	0.64	0.75

Table 9. Carrying - Performance values for baseline and FL models

Baseline		Flower													
		Rounds	Metric	FedAvg				FedAvgM				FedProx			
				GRID		TC		GRID		TC		GRID		TC	
E	I			E	I	E	I	E	I	E	I	E	I	E	I
0.62	0.59	2	Acc	0.69	0.61	0.8	0.5	0.57	0.61	0.7	0.8	0.74	0.69	0.8	0.4
			Prec	0.69	0.63	0.8	0.25	0.72	0.74	0.81	0.86	0.76	0.69	0.86	0.38
		5	Acc	0.69	0.51	0.8	0.5	0.57	0.64	0.7	0.8	0.76	0.7	0.8	0.4
0.69	0.67		Prec	0.69	0.63	0.8	0.25	0.68	0.76	0.81	0.86	0.78	0.7	0.86	0.38
		10	Acc	0.71	0.61	0.8	0.5	0.61	0.66	0.7	0.7	0.74	0.7	0.8	0.4
			Prec	0.71	0.63	0.8	0.25	0.68	0.75	0.81	0.81	0.77	0.7	0.86	0.38

Note: I corresponds to InceptionV3, and E corresponds to EfficientNetV2M.

5 Conclusions

In summary, our research into Federated Learning (FL) reveals its consistent capability to generate highly accurate models when leveraging decentralised datasets. The comprehensive analysis conducted demonstrates a notable improvement in the accuracy of machine learning models within the FL framework compared to baseline models (centralised trained models). While FL proves to be a powerful approach, it does come with a computational resource challenge, particularly in terms of demanding GPU RAM usage.

It's worth highlighting that the FL strategy FedProx stands out, showcasing superior accuracy and precision rates when compared to alternative approaches such as FedAvg and FedAvgM. In the terms of image classification tasks, our findings indicate that EfficientNetV2M outperforms InceptionV3, exhibiting not only enhanced performance but also greater efficiency in model training.

One of the aspects to consider is the variation of model performance that relies on factors such as the specific dataset characteristics and the choice of evaluation metrics. Our observations suggest that augmenting the training dataset with an increased number of images is a promising solution for optimising model performance. Furthermore, the inference that training models with images featuring box-bounding annotations may lead to superior results compared to those with only multiple categories and labels adds an additional layer of insight.

In general terms and in response to the research question, FL can improve the performance of the machine learning algorithms while preserving data privacy as the training and testing datasets themselves do not need to be exchanged.

Federated Learning's unique feature, where only model weights are shared, not the actual datasets, promotes collaboration without exposing sensitive information. This smart approach not only improves how models are shared with clients but also boosts cooperation among entities working towards common goals, such as securing specific events and improving healthcare between hospitals.

References

1. Aledhari, M., Razzak, R., Parizi, R.M., Saeed, F.: Federated learning: a survey on enabling technologies, protocols, and applications. IEEE Access **8**, 140699–140725 (2020)
2. Rieke, N., et al: The future of digital health with federated learning. NPJ Digit. Med. **3**(1) (2020). https://doi.org/10.1038/s41746-020-00323-1
3. Chapman, P.: CRISP-DM 1.0: Step-by-step data mining guide (2000). https://api.semanticscholar.org/CorpusID:59777418
4. McMahan, H.B., Moore, E., Ramage, D., Agüera y Arcas, B.: Federated learning of deep networks using model averaging. ArXiv, abs/1602.05629 (2016). https://api.semanticscholar.org/CorpusID:16861557
5. Bonawitz, K., et al.: Towards federated learning at scale: system design. arXiv:1902.01046 [cs.LG] (2019)
6. McMahan, H.B., Moore, E., Ramage, D., Hampson, S., Agüera y Arcas, B.: Communication-efficient learning of deep networks from decentralized data. arXiv:1602.05629 [cs.LG] (2023)
7. McMahan, H., Ramage, D., Talwar, K., Zhang, L.: Learning differentially private language models without losing accuracy. ArXiv (2017)
8. Bonawitz, K., et al.: Practical secure aggregation for privacy-preserving machine learning. In: Proceedings of the 2017 ACM SIGSAC Conference on Computer and Communications Security (2017). https://doi.org/10.1145/3133956.3133982
9. Hewitt, C.: Actor model for discretionary, adaptive concurrency. CoRR, abs/1008.1459 (2010). http://arxiv.org/abs/1008.1459
10. Luo, J., et al.: Real-world image datasets for federated learning. arXiv:1910.11089 [cs.CV] (2021)
11. Anaissi, A., Suleiman, B.: A personalized federated learning algorithm: an application in anomaly detection. arXiv:2111.02627 [cs.LG] (2021)
12. Zhang, T., He, C., Ma, T., Gao, L., Ma, M., Avestimehr, S.: Federated learning for internet of things. In: Proceedings of the 19th ACM Conference on Embedded Networked Sensor Systems (2021). https://doi.org/10.1145/3485730.3493444
13. Li, S., Cheng, Y., Liu, Y., Wang, W., Chen, T.: Abnormal client behavior detection in federated learning. arXiv:1910.09933 [cs.LG] (2019)
14. Preuveneers, D., Rimmer, V., Tsingenopoulos, I., Spooren, J., Joosen, W., Ilie-Zudor, E.: Chained anomaly detection models for federated learning: an intrusion detection case study. Appl. Sci., **8**(12), 2663 (2018). https://doi.org/10.3390/app8122663
15. Gupta, D., Kayode, O., Bhatt, S., Gupta, M., Tosun, A.S.: Hierarchical federated learning based anomaly detection using digital twins for smart healthcare. arXiv:2111.12241 [cs.CR] (2021)
16. Choudhury, O., et al.: Differential privacy-enabled federated learning for sensitive health data. arXiv:1910.02578 [cs.LG] (2020)
17. Girija, S., et al.: Attribute recognition for person re-identification using federated learning at all-in-edge. Internet Things **22**, 100793 (2023). https://doi.org/10.1016/j.iot.2023.100793
18. Jing, C., et al.: A closer look at personalization in federated image classification. ArXiv (2022). https://api.semanticscholar.org/CorpusID:248392299
19. Li, L., Xie, N., Yuan, S.: A federated learning framework for breast cancer histopathological image classification. Electronics (2022)

20. Nazir, S., Kaleem, M.: Federated learning for medical image analysis with deep neural networks. Diagnostics (2023)
21. Li, T., Sahu, A.K., Zaheer, M., Sanjabi, M., Talwalkar, A., Smith, V.: Federated optimization in heterogeneous networks. ArXiv (2018). https://arxiv.org/abs/1812.06127
22. Hsu, T.-M.H., Qi, H., Brown, M.: Measuring the effects of non-identical data distribution for federated visual classification. ArXiv (2019). https://arxiv.org/abs/1909.06335
23. Deng, Y., Luo, P., Loy, C.C., Tang, X.: Pedestrian attribute recognition at far distance. In: Proceedings of ACM Multimedia (ACM MM) (2014)
24. Beutel, D.J., et al.: Flower: a friendly federated learning research framework. arXiv:2007.14390 [cs.LG] (2022). https://arxiv.org/abs/2007.14390

Doctors' Handwriting Recognition Using CNN and BLSTM Models

Santosh Khanal[1], Rabindra Bista[1](✉), and Joao C. Ferreira[2,3,4]

[1] Department of Computer Science and Engineering, Kathmandu University, Dhulikhel-4, Kavre, Nepal
rbista@ku.edu.np

[2] Inov Inesc Inovação—Instituto de Novas Tecnologias, 1000-029 Lisbon, Portugal

[3] Instituto Universitário de Lisboa (ISCTE-IUL), ISTAR, 1649-026 Lisbon, Portugal

[4] Faculty of Logistics, Molde University College, NO-6410 Molde, Norway

Abstract. Handwritten Text image Recognition (HTR) in medical profession has become the major challenge due to the complexity of doctors' handwriting styles and the critical need for precise and efficient text recognition. This study proposes an innovative approach that combines the power of Bidirectional Long Short-Term Memory networks (BLSTM) and Convolutional Neural Networks (CNN) to address these challenges effectively. The proposed model harnesses the bidirectional capabilities of BLSTMs to capture contextual dependencies within doctors' handwritten notes, enabling it to understand and interpret handwriting more accurately. Additionally, CNNs are employed for feature extraction, enabling the model to recognize salient patterns and representations within handwritten text images. This paper presents comprehensive experiments conducted on a diverse dataset of doctors' handwritten notes, demonstrating the model's superior performance compared to conventional approaches and outperform the existing work in terms of accuracy. Two different crucial metric word error rate and character error rate are accessed through the combination of BLSTM and CNN that yields state-of-the-art accuracy, robustness to variations in handwriting styles, and remarkable adaptability across various medical document in context of Nepal.

Keywords: Handwritten text image · bidirectional long short-term memory network · convolution neural network · word error rate · character error rate · accuracy

1 Introduction

Handwriting recognition systems have witnessed significant advancements in recent years, offering promising solutions for various applications, from digitizing historical manuscripts to automating data entry. Among these applications, one area of critical importance is the recognition of handwritten notes and prescriptions authored by medical professionals. The deciphering of doctor's handwriting, often characterized by its complexity and idiosyncrasies, poses a unique challenge to the field of optical character recognition (OCR) and machine learning. Medical practitioners frequently rely on

A. Bajaj et al. (Eds.): IBICA 2023, LNNS 1229, pp. 37–47, 2025.
https://doi.org/10.1007/978-3-031-78940-3_3

handwritten records for patient care, including prescriptions, treatment plans, and clinical notes. However, the inherent variability in doctors' handwriting can introduce errors in data interpretation, leading to potential patient safety risks. Even in the most developed country like USA has records of Death [1] of the patient because of the misread of the doctor's handwriting by the pharmacist. Since medical practitioners have very fast writing with broken incomplete words which makes very difficult to understand, it needs to detect through the OCR [2] technology. When dealing with cursive and mixed cursive words, recognition can be tricky and challenging when dealing with a single character. Unique writing techniques, alphabetic forms, and font sizes provide the difficulty of identifying the characters [3].

The integration of deep learning techniques, particularly Convolutional Neural Networks (CNNs) and Recurrent Neural Networks (RNNs), [4] has done a significant work in handwriting text recognition system. CNNs are employed for character and word image recognition, while RNNs, such as Long Short-Term Memory (LSTM) networks and Transformer models [5], are adept at handling sequential data and capturing contextual information. Recent advancements in neural network architectures have led to substantial improvements in HTR accuracy, yet the recognition doctor's handwriting have not been achieved as what have been claimed in normal HTR accuracy.

In this research, bidirectional long short-term memory networks has been proposed as an advancement in sequence modeling, addressing the limitations of traditional RNNs by capturing information in both forward and backward directions. BLSTMs have proven to be especially effective in deciphering complex and cursive handwriting styles, capturing long-range dependencies, and providing robust recognition [6].

2 Literature Review

Handwriting recognition in the medical field is a critical task for various applications, including digitizing patient records, prescription validation, and data extraction. Detecting the handwriting of doctors poses unique challenges due to the diverse styles of handwriting and the need for high accuracy to ensure patient safety. In recent years, machine learning and deep learning techniques have been applied to this problem with promising results.

One of the earliest uses of CNNs for the problem of generic handwriting recognition was proposed in combination with the hidden model. In this work, a CNN-HMM [7] model was applied using the tandem approach. Convolutional architecture was employed for feature extraction, and it was influenced by the LeNet-5 model. Better than those produced with handmade features and on par with those from LSTM networks, the results obtained using CNN were also more accurate. The CNN model was then employed by other studies to extract the features and apply them to determine the handwriting text recognition's closest neighbors. A Deep Convolutional Recurrent Neural Network was developed to recognize text in photographs of prescriptions written by doctors and translate cursive handwriting into understandable text [8]. By calculating the accuracy, F1-score, and Aggregated of two models—CRNN and CRNN with model-based normalization scheme—the researchers were able to compare how well they performed. A approach for altering any fully convolutional text recognition model [9] (at the line level) (at the paragraph level) is developed and this adaptation has an advantage over earlier page-level adaptations in that it can be completed without dividing the transcription into lines. DeepDoc [10], a deep learning framework for medical document analysis, including handwriting recognition has been proposed that utilizes convolutional neural networks (CNNs) for text extraction and recognition, achieving state-of-the-art results in detecting doctors' handwriting from medical documents. CNN-LSTM-CTC architectures have seen to be effective in a variety of text recognition tasks in the recent years. The implementation the CNN based MDLSTM network architecture [11] which starts working with the images pixel directly with creating numbers of convolution layers. A new model CNN-LSTM-CTC [12] has been proposed with the architecture that implementation run time trained data augmentation and the batch normalization techniques to address the text image classification problem. This architecture is a trainable layer that applies geometrical alterations to the input in the same way as random hand movements do in handwriting. The normalized line is then transcribed using a CNN-LSTM-CTC network. A new neural design [13] called Convolutional Recurrent Neural Network that incorporates convolution, pooling, and recurrence in a single framework (CRNN). The CRNN can be thought of as an upgraded version of the RNN, with a local connection and shared weights similar to a convolutional kernel replacing the direct connection between the input and hidden layer.

3 Methodology

The work involves developing an OCR for recognizing and extracting the text from a text image. The text image is an image of doctor's prescription card. It contains text related to medical parameters. The process in developing an OCR involves four different stages, namely; preprocessing, segmentation, feature extraction and recognition. After developing the system, it is trained and tested with different image dataset. The image dataset used in training and testing process were collected by capturing image of patient's treatment reports via visiting hospitals of Nepal (Fig. 1).

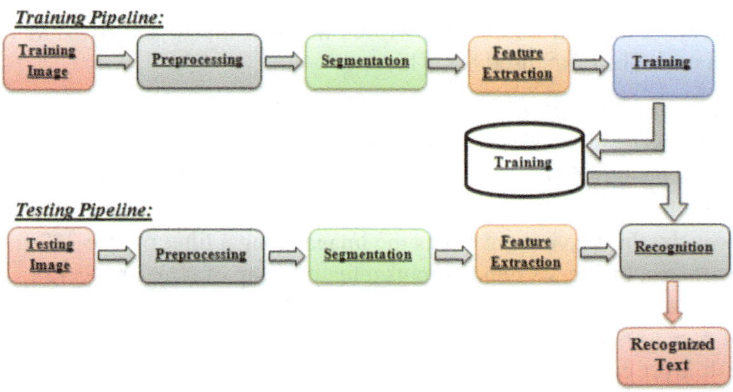

Fig. 1. Overview of developed OCR system.

3.1 Preprocessing

The inputted image is preprocessed for improving the quality of the image so that the efficiency of the system in recognizing the text can be increased. The various preprocessing activities involved with the system were image resizing, gray-scale conversion, normalization, contrast stretching and binarization. The input image is resized to a size 800*64 so that the recognition model can easily recognize the text from the extracted features. Inter cube interpolation method has been used resizing the input image. To create uniform distribution of light intensities values in the image, contrast stretching has been done using min-max method. A thresholding method has been used for binarization of the image.

3.2 Segmentation

Segmentation process has always been a crucial step in optical character recognition because correct segmentation always leads to increase in recognition accuracy of an OCR system. This work involves carrying line segmentation to segment the text image into lines. A morphology-based text line extraction algorithm has been used for segmenting the text image into lines for further processing. The binarized image is subjected to a dilation process for line segmentation. The dilation process causes the text areas grow in size. This effect eventually leads to an increase in the boundaries of foreground pixels. The dilated image is then subjected for finding contours. Using the contours, bounding boxes are identified and based on these bounding boxes, the text image is segmented into lines.

3.3 Feature Extraction and Recognition

This step involves using a deep learning based model for extraction of features and recognition of text from a text image. The model has been developed by combining Convolutional Neural Network (CNN) with Bidirectional Long Short-Term Memory (BLSTM) based Recurrent Neural Networks (RNNs). CNN has been used for extracting multi-scale features from the handwritten line text image and BLSTM works as a recognition unit. The CNN used has 7 layers.

- First layer convolutes the inputted image using a filter of size 5*5 followed by pooling matrix of size 2*2. The first layer outputs a feature map of size 400*32*64.
- The second layer of CNN convolutes the image using a filter of size 5*5 followed by pooling matrix of size 1*2 and outputs a feature map of size 400*16*128.
- Similarly, the third layer convolutes the image using a filter of size 3*3 followed by pooling matrix of size 2*2 and outputs a feature map of size 200*8*128.
- The fourth layer convolutes the image using a filter of size 3*3 and outputs a feature map of size 200*8*256.
- The fifth layer convolutes the image using a filter of size 3*3 followed by pooling matrix of size 2*2 and outputs a feature map of size 100*4*256.
- The sixth layer convolutes the image using a filter of size 3*3 followed by pooling matrix of size 1*2 and outputs a feature map of size 100*2*512.

- The seventh layer convolutes the image using a filter of size 3*3 followed by pooling matrix of size 1*2 and outputs a feature map of size 100*1*512.

The output of the CNN layer is then subjected to a collapse layer for dimension reduction (i.e. 100*1*512- > 100*512). The output before CNN FC layer is passed to the BLSTM for sequence dependency and time-sequence operations. The final output of the bidirectional LSTM layers is fed into Connectionist Temporal Classification (CTC) layer to convert the sequence into feature vector ready for word classification. Figure 2. Illustrates the working mechanism of the model used in the work.

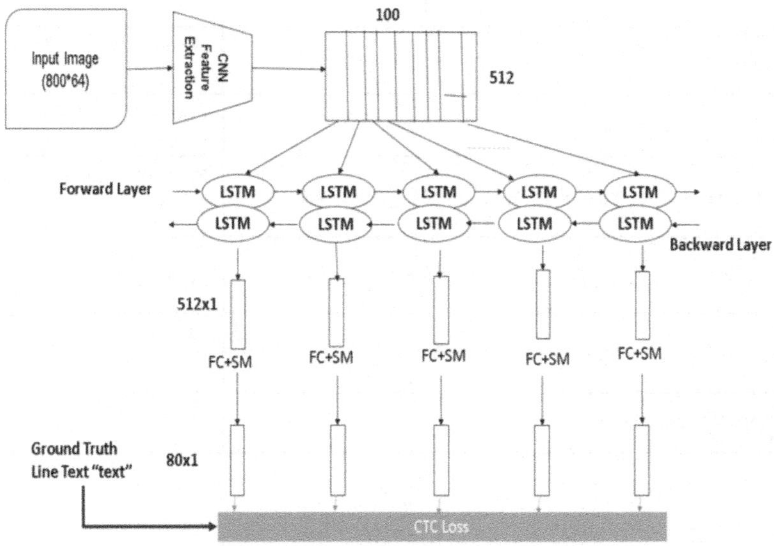

Fig. 2. Convolutional Neural Network Model

CTC loss function, given by conditional probability [14] is used to train the RNN which eliminate the Alignment problem in Handwritten, since handwritten have different alignment of every writers. The CTC loss function takes the BLSTM output and the Ground Truth (GT) and calculates the loss simply as '-log (GT)' to minimize negative maximum likelihood path. It finds the possible paths from the given labels and finally CTC decode is used to decode the output during Prediction.

3.4 Training

The model has been trained from scratch for line, word and character recognition. The model needs line images (x) along with their transcriptions (y) for training. Thus, the data required for training the model were collected from various sources. One way that is chosen is to capture the image of hand written text image and manually annotate them. Thus, this work involves visiting hospitals of Nepal for collecting the image dataset of Doctor's prescription and manually annotating them for training the model. In the actual situation, medical practitioners from various departments were asked to record medical notes on outpatient department (OPD) cards. This was done with the consent of both the hospital administration and the patients themselves. Subsequently, these notes were captured by taking photos using mobile cameras. Table 1 shows the labeling of the datasets used in the training process.

Table 1. Examples of Labeled datasets

Dataset	Label
	Ibuprofen – 150 mg
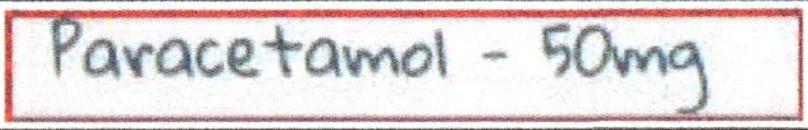	Paracetamol – 50mg
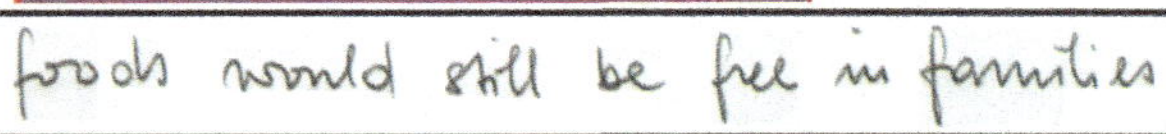	Foods would still be free in families
	Pain in epigastric region x 1 week.
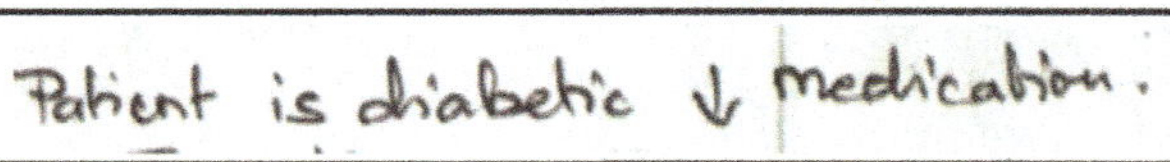	Patient is diabetic L medication
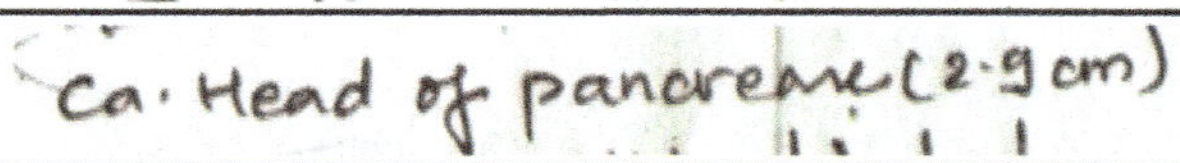	ca . Head of pancreases (2.9 cm)
	Hypertension

4 Experiment

4.1 Data

The data used in the experiment were by capturing the image of the different doctor's writing via visiting hospitals of Nepal. About 3000 images with 15290 lines were captured through the mobile phones camera. Among them, 10700 line images were used for training the model and 4590 line images were used for testing the model. Around 70% of the collected data were used for training the developed model and 30% of the collected data were used for testing the model.

4.2 Experimental Environment

The test experiment of the work has been conducted on a system which have following configuration.

Table 2. Experimental Environment

Title	Description
Computer System	Hp Envy 4710HQ, Intel ® Octa-Core ™ i7 Processor @ 2.50 GHz, 8 GB RAM, 2 GB Graphics, running windows 8.1
Platform	Python 3.7, tensorflow 1.8.0, OpenCV 3, Numpy, Flask

4.3 Performance Evaluation

To fully evaluate the effectiveness of a model, this work involves use of Precision, Recall, F1 Score and Accuracy values as an evaluating criterion. For getting those values, a confusion matrix was constructed. Table 4. Below shows an illustration of Confusion Matrix (Table 3).

Table 3. Experimental Evaluation matrix

	Positive Prediction	Negative Prediction
True Class	True Positives (TP)	False Negatives (FN)
False Class	False Positives (FP)	True Negatives (TN)

5 Result and Discussion

Table 2 shows the result of training and testing the model. It presents the line-level Word Error Rate (WER) and Character Error Rate (CER) of the BLSTM-based system.

Table 4. Experimental Result

	WER	*CER*
Training	15.74%	8.65%
Testing	18.18%	12.37%

Above table shows that the word error rate is greater than character error in both training and testing process. It is due to the size of characters in the text in an image. Different writers write the characters, words and sentences with different font size and font style. Because of this, a word in the recognition does not overlap with any words in the reference. As a result, an error occurs in character and word recognition process and hence word level error rate is greater than character level error rate.

Graph 1 shows the accuracy of training and testing the BLSTM-based system for each epoch (Fig. 3).

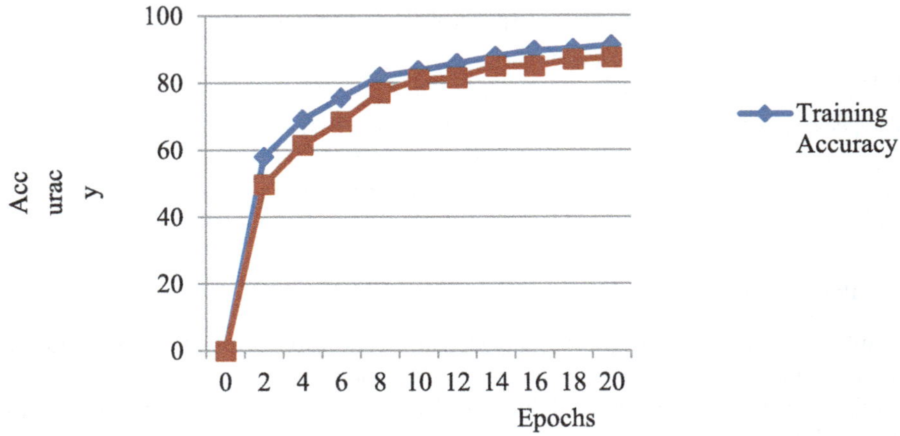

Fig. 3. Accuracy of training and testing the BLSTM-based system for each epoch

The HTR system achieved an accuracy of 91.35% during the training phase and 87.63% during testing in character level and 84.26% and 81.82% in word level. However, the quality of text images, including issues like uneven lighting conditions and low scanning resolution, significantly incomplete, illegible words impacted the HTR system's recognition efficiency. This results outperforms the result obtained from [15] which is 76%in training phase in terms of accuracy. Additionally, the presence of connected characters within the text images also challenged in accuracy to identify individual characters. While the system did not attain 100% accuracy, its performance, which approaches near-perfection, suggests its suitability for further processing and utilization.

The BLSTM-based system's training phase was notably time-consuming. In an effort to reduce the training duration, it passed through training with a limited number of epochs. Initially, the system passed training for only 2 epochs, followed by a subsequent training session of 4 epochs. A comparative analysis was performed between these training sessions, and based on the results, the number of epochs was incrementally increased to achieve optimal results within a reasonable timeframe. The training process was eventually capped at a maximum of 20 epochs. Graph 2 illustrates the progressive reduction in loss as the number of epochs increased, demonstrating the system's efficiency during testing." (Fig. 4).

The initial epochs, the CTC loss is typically high. This is because the model starts with random weights and does not accurately predict the sequence of characters in the handwritten text. As training progresses, the model gradually learns to align its predictions with the ground truth. The CTC loss steadily decreases, indicating that the model is improving its ability to recognize and transcribe characters. With more epochs, the CTC loss continues to decrease until it reaches a point of convergence. This indicates that the model has learned to align the predicted and ground truth sequences effectively, resulting in accurate text recognition. The loss value stays at about 12.37 without significant drop later. It may be due to the under fitting problem of the model (Table 5).

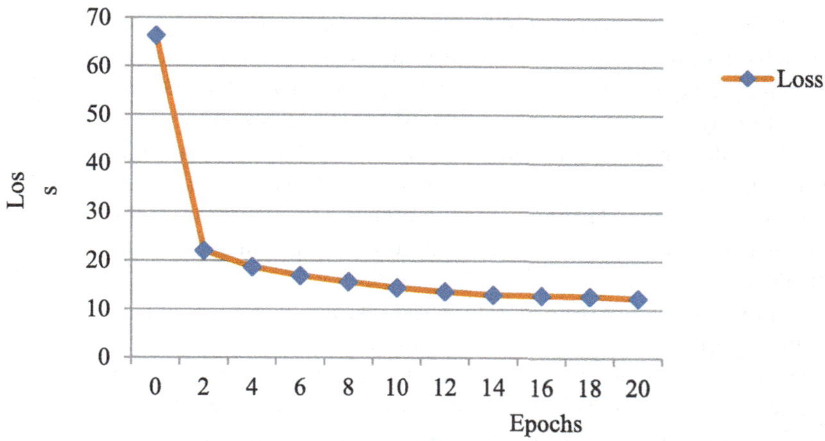

Fig. 4. Illustration of the progressive reduction in loss versus number of epochs

Table 5. Line level precision, recall and f1 score of recognition system.

	Precision	Recall	F1 Score
Training	0.817	0.781	0.778
Testing	0.732	0.725	0.722

The HTR system developed in this work has considerable Precision, Recall and F1 score values. These values can be increased by increasing the volume of datasets and by balancing the ratio of training and testing data.

In a developing nation like Nepal, where the doctor-to-patient ratio falls short of WHO standards, physicians often face challenges in dedicating sufficient time to each patient. This situation often leads to hurried note-taking and incomplete words, which can introduce a higher risk of errors when transcribed manually. The HTR system developed in this research holds the potential to bring about substantial improvements in deciphering doctors' handwritten notes, thus enhancing the clarity and accuracy of their writing.

6 Conclusion

In conclusion, the Doctors Handwriting Recognition System, which was created utilizing a fusion of CNN (Convolutional Neural Network) and BLSTM (Bidirectional Long Short-Term Memory) models, has produced encouraging results regardless of the very curvy and incomplete, illegible words. The doctor's Handwriting Text Recognition (HTR) system's accuracy throughout training and testing phases was 91.35% and 87.63%, respectively, demonstrating its great capacity to reliably transcribe handwritten medical papers. For the recognition of doctors' handwriting, combining the BLSTM and CNN models has been effective. BLSTM's bidirectional nature enables it to record both past and future context, while CNNs are excellent at extracting features from pictures, making them suitable for this purpose.

These findings are encouraging for the healthcare sector since a reliable and effective HTR system may improve data retrieval, administration of medical records, and patient care by guaranteeing the accuracy and accessibility of doctors' handwritten notes. The system may be improved and optimized further to achieve even greater accuracy rates and a wider range of applications in healthcare document management. The suggested method offers a noteworthy and intriguing contribution to the field of Handwritten Text Image Recognition. Given the vital importance of accurately deciphering doctors' handwritten notes in the medical domain, any progress in this domain carries great significance.

In contrast to other existing literature, this research paper has added value by showcasing the superior performance of the proposed model, as demonstrated through extensive experiments conducted on a diverse dataset of doctors' handwritten notes. The specific focus on the context of Nepal also underscores the adaptability of the proposed approach in a particular regional setting, which could be crucial for the effective implementation of Handwritten Text Recognition systems in various regions around the world. Furthermore, the emphasis on evaluating the model's performance using metrics such as word error rate and character error rate suggests a comprehensive assessment, contributing to a more nuanced understanding of the model's effectiveness in recognizing complex handwritten medical text."

Acknowledgements. This work was supported by the project BLOCKCHAIN.PT (RE-C05-AQ2 i01.01 – Mobilizing Agendas/Alliances for Reindustrialization, within Portugal's Recovery and Resilience Plan, component 5 – Business Capitalization and Innovation, and under the Incentive System Regulation "Agendas for Business Innovation", approved by Ordinance No. 43-A/2022 of January 19, 2022). We would like to express our sincere gratitude to the Nepalese Education in E-health Masters (NEEM) project under GA101083048 at the Health Informatics Lab of Kathmandu University for supporting this research.

References

1. Brits, H., et al.: Illegible handwriting and other prescription errors on prescriptions at national district hospital, Bloemfontein. South African Family Practice **59**, 1–4 (2017)
2. Memon, J., Sami, M., Ahmed Khan, R., Uddin, M.: Handwritten Optical Character Recognition (OCR): A Comprehensive Systematic Literature Review (SLR) (2020)
3. Perwej, Y., Chaturvedi, A.: Neural Networks for Handwritten English Alphabet Recognition. Int. J. Comput. Appl. **20**(7), 1–5 (2011). https://doi.org/10.5120/2449-2824
4. Bluche, L., Jégou, H., Collobert, R.: A Deep Neural Network for Handwriting Recognition. In: ICDAR (2014)
5. Liu, Y., Jin, L.: A Flexible Framework for End-to-End Handwritten Text Recognition with Transformers (2020). arXiv preprint arXiv:2012.06379
6. Gupta, A., Shivakumara, P., Pal, U.: Handwritten Text Recognition Using BLSTM and Feature Engineering (2019)
7. Bluche, T., Ney, H., Kermorvant, C.: Tandem HMM with convolutional neural network for handwritten word recognition. In: ICASSP, IEEE International Conference on Acoustics, Speech and Signal Processing - Proceedings, pp. 2390–2394. IEEE. (2013). isbn: 9781479903566. https://doi.org/10.1109/ICASSP.2013.6638083

8. Joy Fajardo, L., et al.: Doctor's Cursive Handwriting Recognition System Using Deep Learning. In: IEEE (2019). https://doi.org/10.1109/HNICEM48295.2019.9073521
9. Yousef, M., Hussain, K.F., Mohammed, U.S.: Accurate, data efficient, unconstrained text recognition with convolutional neural networks. In: arXiv 108, p. 107482 (2018). issn: 23318422. arXiv: 1812.11894
10. Zhang, L., Zhang, Z., Liu, Z., Liu, L.: DeepDoc: A Deep Learning-Based Framework for Medical Document Analysis (2021)
11. Graves, A., Schmidhuber, J.: Oine handwriting recognition with multidimensional recurrent neural networks. In: Advances in Neural Information Processing Systems 21 - Proceedings of the 2008 Conference, pp. 545–552 (2009). isbn: 9781605609492
12. Puigcerver, J.: A Multidimensional Recurrent Layers Really Necessary for Handwritten Text Recognition. In: Proceedings of the International Conference on Document Analysis and Recognition, ICDAR. Volume 1. IEE., pp. 67–72 (2017). isbn: 9781538635865. https://doi.org/10.1109/ICDAR.2017.20
13. Huang, X., et al.: End-to-end sequence labeling via convolutional recurrent neural network with a connectionist temporal classifcation layer. In: International Journal of Computational Intelligence Systems **13**(1), 341–351 (2020). issn: 18756883. https://doi.org/10.2991/ijcis.d.200316.001
14. Graves, A., Fernández, S., Gomez, F., Schmidhuber, J.: Connectionist Temporal Classification: Labelling Unsegmented Sequence Data with Recurrent Neural Networks. ICML 2006, Pittsburgh, USA, pp. 369–376
15. Fajardo, L.J., et al.: Doctor's Cursive Handwriting Recognition System Using Deep Learning. 2019 IEEE 11th International Conference on Humanoid, Nanotechnology, Information Technology, Communication and Control, Environment, and Management (HNICEM), pp. 1–6. Laoag, Philippines (2019). https://doi.org/10.1109/HNICEM48295.2019.9073521

Unveiling the Viability of Integrating ChatGPT into the Realm of Intelligent Libraries

Khushboo Jain[1(✉)], Arun Agarwal[2], Sheetal Singh[2], Sandhya Avasthi[3], and Ajith Abraham[4,5(✉)]

[1] School of Computer Science, University of Petroleum and Energy Studies, Dehradun, UK, India
khushboojain2806@gmail.com
[2] Department of Computer Science, Ramanujan College, University of Delhi, Delhi, India
arun.agarwal@ramanujan.du.ac.in
[3] Department of CSE, ABES Engineering College, Ghaziabad, UP, India
[4] School of Computer Science, Engineering and Technology, Bennett University, Greater Noida, Uttar Pradesh 201310, India
ajith.abraham@ieee.org
[5] Machine Intelligence Research Labs (MIR Labs), Auburn, WA 98071, USA

Abstract. As a standout among the leading artificial intelligence (AI) models, ChatGPT has showcased exceptional learning prowess and impressive adaptability across diverse domains. Among these realms of exploration, the library domain holds a significant place. This piece offers a comprehensive introduction to the nuances, benefits, and research implications of ChatGPT. Moreover, it delves into the practicality and prudent suggestions for deploying ChatGPT within the sphere of intelligent libraries. The article aspires to furnish valuable perspectives for the real-world integration of ChatGPT into library settings, consequently expediting the evolution of sophisticated smart library systems.

Keywords: Artificial Intelligence · ChatGPT · Intelligent libraries

1 Introduction

In recent times, the realm of artificial intelligence (AI) has experienced an exponential surge, captivating significant interest in the form of ChatGPT [1]. This AI-powered language model stands at the forefront, leveraging extensive textual datasets to mimic the art of generating conversational language akin to human interactions [2]. The prowess of ChatGPT has attracted substantial attention not only from the corporate sector but also within academic circles, and its influence has transcended borders, gaining recognition at national and international levels alike. Within this landscape, the library domain emerges as a promising arena for the application of ChatGPT's capabilities, poised to effectively grapple with the challenges and intricacies encountered by libraries in their evolving roles.

A. Bajaj et al. (Eds.): IBICA 2023, LNNS 1229, pp. 48–61, 2025.
https://doi.org/10.1007/978-3-031-78940-3_4

At its core, this research endeavors to unfurl the potential horizons where ChatGPT can be harnessed within the library context. By illuminating various conceivable application scenarios, the aim is to not just decode the present challenges but to illuminate a pathway that charts the future trajectory of libraries' technological evolution [3]. This journey extends beyond the conventional bounds, offering innovative service paradigms and novel concepts that can be seamlessly integrated within the library ecosystem [4]. Through a judicious blend of imagination and practicality, this exploration seeks to foster a fresh realm of ideas and service models that could redefine the essence of modern libraries [5].

The remaining of this book chapter is organized as follows: The overview of ChatGPT is explained and presented in Sect. 2. The Advantages and Drawbacks of ChatGPT are presented in Sect. 3. The Potential Applications of ChatGPT are illustrated in Sect. 4. The Glimpse into the Future of Empowering Smart Libraries through ChatGPT is highlighted in Sect. 5. Finally, the work is concluded and imminent work is exposed to researchers in Sect. 6.

2 Outline of ChatGPT

2.1 Unveiling the Essence of ChatGPT

Referred to as Chat Generative Pre-Trained Transformer, ChatGPT stands as a technological marvel at the intersection of artificial intelligence and human language [6]. Created under the auspices of OpenAI, this chatbot was unveiled on the notable date of November 30, 2022, heralding a new era in conversational AI [7]. At the present juncture, it commands a distinguished position as one of the foremost artificial intelligence-powered tools within the expansive realm of natural language processing, showcasing its unrivaled prowess in this arena.

ChatGPT's remarkable capability to not just mimic but deeply comprehend human language serves as a testament to its sophistication. Beyond mere mimicry, it delves into the intricacies of language, enabling it to partake in seamless and coherent conversations. The extent of its capabilities transcends these exchanges, stretching into domains as diverse as code generation, copywriting, and language translation. This multifaceted nature empowers ChatGPT to be more than just a conversational entity; it transforms into a versatile tool capable of contributing across a spectrum of language-related tasks [8].

Emanating from its pre-trained foundation as a "Transformer," ChatGPT harnesses its understanding of context and semantics, facilitating its aptitude to process, interpret, and generate human-like text [9]. In effect, it emerges as a virtual collaborator, deftly navigating through linguistic nuances to produce meaningful and contextually relevant responses. This transformative potential significantly extends its reach beyond the boundaries of conventional chatbots, positioning it as a pioneering force within the landscape of AI-driven language interaction.

The journey of ChatGPT encapsulates not only its technological brilliance but also the glimpse it offers into the future of human-AI synergy. As it continues to evolve and expand its capabilities, this extraordinary creation ushers in an era where language is no longer a barrier, but a bridge connecting human ingenuity with the marvels of artificial intelligence [10].

2.2 Global and Local Reverberations

In the wake of OpenAI's unveiling of ChatGPT, the resonance has been both widespread and profound, captivating not only local stakeholders but also global players in the corporate and academic domains. This phenomenon has orchestrated strategic maneuvers that underscore the seismic potential of ChatGPT across borders.

Embarking on this journey, notable industry titans and tech trailblazers have surged forward with innovative ventures. Microsoft, for instance, unveiled the New Bing, an ingenious search engine fortified by the remarkable capabilities of ChatGPT technology [11]. This seminal move is but a harbinger of a broader vision, with plans afoot to seamlessly weave ChatGPT into the very fabric of Microsoft's Office software suite. Echoing this innovative spirit, Google ventured into the realm of AI-driven chat-bots with Bard, signaling a novel foray into harnessing the nuances of human interaction through artificial intelligence.

The allure of ChatGPT's potential has extended its tendrils across corporate giants spanning diverse sectors. The intentions of industry powerhouses like Apple, Amazon, Meta, and formidable entities like Baidu, Alibaba, Tencent, and Huawei are clear indications of the scope and scale of exploration they envisage in integrating ChatGPT's capabilities into their offerings [12]. A convergence of technological finesse and ambition, these endeavors collectively paint a picture of how the future might be shaped by the marriage of human ingenuity with AI capabilities.

Stepping beyond the corporate realm, education and academia have embraced ChatGPT's transformative potential [13]. Ne-tease's strategic implementation of ChatGPT within the education sector echoes the broader sentiment of molding AI-driven tools to catalyze learning experiences. Likewise, JD.com's imminent introduction of an industrial iteration of ChatGPT across the finance and retail sectors showcases a practical alignment with industry needs, injecting AI-infused innovation into daily operations.

At the crossroads of innovation and national recognition, ChatGPT's prowess has not gone unnoticed. A series of pivotal press conferences orchestrated by the State Council Information Office on February 24th, 2023, reverberated with insights from influential figures. Wang Zhigang, the Minister of Science and Technology, lauded ChatGPT's strides in natural language understanding and processing, underscored by the potent fusion of algorithms, data, and computational might. Chen Jiachang, the Director-General of the Department of High and New Technology at the Ministry of Science and Technology, sheds light on the phenomenon-level impact of ChatGPT's application, highlighting its potential to emulate facets of general artificial intelligence, unfurling boundless applications across industries [14].

In essence, ChatGPT's meteoric rise has galvanized an international symphony of innovation, punctuated by the acknowledgment of its potency at the highest levels of national governance. This confluence paints a vivid canvas of the transformative journey that AI, led by ChatGPT, is steering across industries and borders.

3 Advantages and Drawbacks of ChatGPT

3.1 Advantages

Within the realm of language generation, ChatGPT emerges as a trailblazing model with an array of distinct advantages that redefine the landscape of conversational AI. Noteworthy among these attributes is its innate ability to seamlessly craft coherent and contextually relevant textual responses, a cornerstone of engaging and natural conversations. What sets ChatGPT apart is its remarkable versatility, enabling it to effortlessly traverse an expansive spectrum of tasks. From answering queries and offering explanations to crafting imaginative content and aiding in a multitude of language-centric endeavors, its adaptive prowess shines brightly [15, 16].

A hallmark of ChatGPT's capabilities lies in its dynamic learning mechanism. Through interaction with voluminous datasets, it evolves its understanding and adapts to diverse domains and topics, thus augmenting its knowledge reservoir organically. This learning extends hand in hand with scalability, allowing for the efficient handling of intricate language patterns and larger datasets. The accessibility afforded by APIs grants developers the power to seamlessly integrate ChatGPT into their applications and services, enriching user experiences with its profound language generation abilities [17, 18].

In comparison to conventional chatbots, ChatGPT's prowess becomes apparent in its ability to furnish responses that not only encompass accuracy but also flow with fluency. The underpinning of this acumen lies in its adeptness at comprehending multi-turn dialogues, enabling the model to encapsulate conversation history and thus engage in continuous, meaningful interactions. Moreover, ChatGPT's utilization of reinforcement learning, guided by human feedback, fine-tunes its output preferences, culminating in responses that resonate harmoniously with human expectations [19].

3.2 Drawbacks

Notwithstanding its substantial contributions, ChatGPT is not immune to limitations that cast shadows upon its transformative potential. One such facet revolves around its real-time context handling, which occasionally translates into responses lacking the persistence of memory across interactions, yielding outcomes that appear inconsistent and forgetful. Furthermore, its reliance on training data places it on the precipice of reproducing biases and inaccuracies embedded within the very datasets it learns from [15, 17].

A notable challenge ChatGPT grapples with lies in deciphering ambiguity inherent in queries, potentially leading to responses that might not entirely encapsulate users' intended meanings. While it excels in generating contextually relevant text, its grasp of the nuanced intricacies of a conversation might fall short, impinging on the depth of its comprehension. Ethical concerns emerge due to the potential for misuse, spotlighting the pressing need to confront and mitigate the risks associated with generating deceptive or harmful content [16, 18].

The facet of credibility introduces another layer of complexity, as ChatGPT's responses may appear reasonable while harboring factual inaccuracies. Its scope of knowledge is circumscribed by its last training update, creating limitations in addressing current and emerging questions. The expense of training and deploying ChatGPT, coupled with possible latency issues, demands prudent consideration when contemplating its implementation. It's imperative to note that the bedrock of ChatGPT, its pre-trained language model, is forged from real-world language data, hence exposing it to biases that warrant careful navigation. The intricacies of these biases are also influenced by labeling strategies, further necessitating a vigilant approach to ensure fairness and inclusivity in language generation [20].

Also, ChatGPT cannot do this for researchers:

- Provide original research ideas
- Interpret complex data analysis
- Engage in academic peer review
- Provide real time feedback
- Generate comprehensive literature review
- Develop experimental mythologies
- Make ethical decisions
- Write research proposals of grant applications
- Contribute to scientific breakthrough

4 Potential Applications of ChatGPT

In recent years, the ascendancy of ChatGPT has heralded an era of transformative application and intensive research, positioning it as a pioneering Chatbot program armed with natural language processing prowess. This technological marvel has rippled across a myriad of fields and scenarios, redefining the way we interact, communicate, and access information [21]. Herein, we delve into a comprehensive exploration of ChatGPT's multifaceted roles and the vast expanse of its impact.

Elevating Conversations Through Chat-Bots: At the heart of its capabilities lies the capacity to engender conversational chat-bots that not only decipher user inputs but respond with a human-like cadence [22]. These bots, while cost-effective and efficient, harbor the potential to address common reference questions and guide users to pertinent services within libraries. Nonetheless, the complexities inherent in human interaction, both cognitive and emotional, may still pose a challenge. These Chatbot, notably, foster meaningful connections with the younger generation, simultaneously catering to multiple user inquiries in real-time.

Empowering Multilingual Communication through Language Translation: ChatGPT emerges as a beacon in the realm of language translation, capable of being fine-tuned to facilitate seamless multilingual communication [1]. Within the domain of libraries, this facet holds tremendous significance. By bridging linguistic barriers, ChatGPT empowers document utilization, proving instrumental in translating articles from unfamiliar languages into the user's preferred linguistic canvas.

Revolutionizing Library Services: Libraries find newfound allies in ChatGPT, which revolutionizes service paradigms. From automating user services and content creation to generating insightful reports, ChatGPT's natural language processing acumen paves the way for question-answering systems, virtual assistants, and interactive chat interfaces that redefine the library experience [4].

Amplifying Intelligence in Finance: The financial sector echoes the resonance of ChatGPT's potential. Its infusion breathes life into risk control, intelligent customer service, investment advisories, and real-time news dissemination [7]. By elevating the intelligence quotient of financial services, ChatGPT reshapes the contours of personalized financial experiences.

Engineering Excellence in Production and Manufacturing: Industries steeped in production and manufacturing reap the benefits of ChatGPT's acumen [23]. It lends its capabilities to quality control, equipment maintenance, optimized production planning, and process refinement. This infusion of intelligence propels the efficiency of enterprise production lines and supply chains, amplifying business outcomes.

Navigating Healthcare with Precision: The healthcare realm finds a stalwart ally in ChatGPT. It assists medical practitioners in consultations, diagnoses, and medical image recognition. Additionally, it affords prowess in health data management, enriching medical practices with operational efficiency and refining patient-centric treatment paradigms [24].

Learning Transformed in Education: The educational landscape is enriched by ChatGPT's role as a multifaceted teaching assistant [25]. From grading assignments and offering elucidations to recommending resources and supporting paper composition, ChatGPT redefines education through personalized and intelligent engagement.

Enhancing Customer Support: ChatGPT's versatility extends to customer support, where it metamorphoses into an unceasing online presence [7]. Through this transformation, it curates immersive and satisfying customer interactions, transcending temporal constraints and enhancing satisfaction levels.

Reshaping Search Engines and social media: The fusion of ChatGPT into search engines has unleashed new dimensions of user experiences [26]. Microsoft's integration into Bing and Baidu's "Wenxin Yiyan" exemplifies how the amalgamation of search and question-answering cultivates superior search outcomes. Furthermore, within the realms of social media, ChatGPT emerges as a creative powerhouse, catalysing content creation, especially in niches like travel, food, and fashion.

Infusing Entertainment with Intelligence: The sphere of entertainment reaps rewards through ChatGPT's creative capacities [1, 7]. Within interactive live streaming, ChatGPT introduces refined virtual personas, intensifying engagement, and interaction with participants.

In summation, the breadth of ChatGPT's potential is staggering, transcending industries and domains. Its profound influence reverberates through each application, catalyzing heightened intelligence and operational efficiency, ultimately affording users unparalleled convenience and efficacious service experiences (Fig. 1).

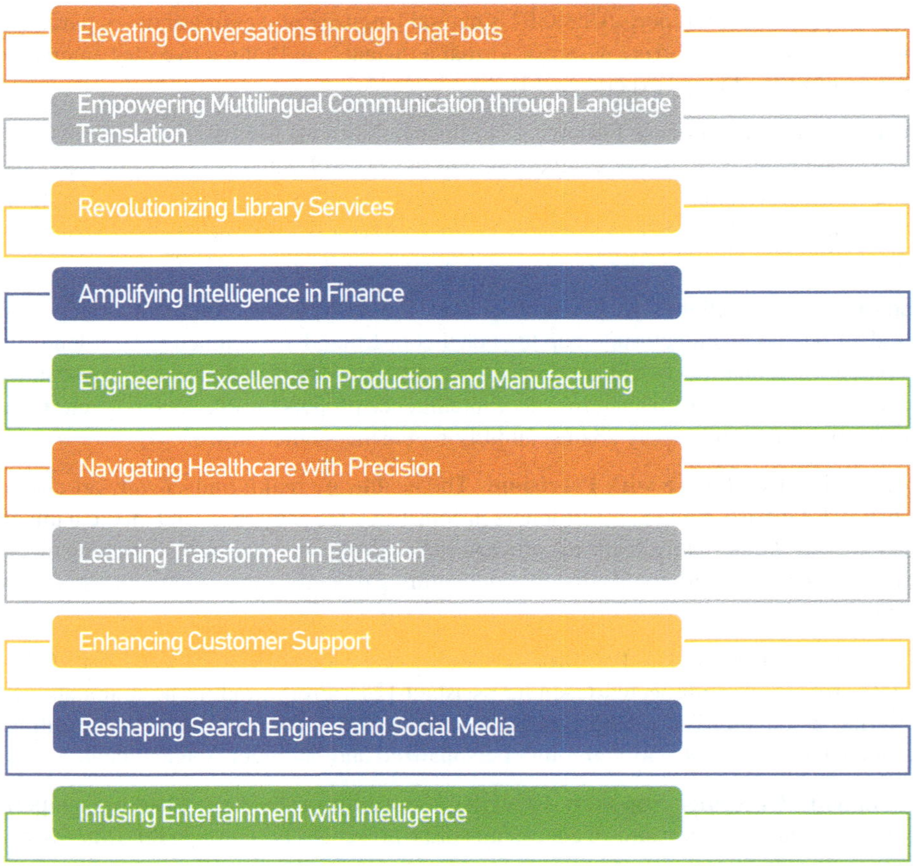

Fig. 1. Potential Applications of ChatGPT

5 Empowering Smart Libraries Through ChatGPT: A Glimpse into the Future

In the evolving landscape of intelligent libraries, ChatGPT emerges as a catalyst for transformative change, unveiling a tapestry of innovative applications that redefine user experiences and elevate library functionalities. Scholars have fervently explored avenues where ChatGPT's capabilities can be harnessed, illuminating a path toward enriched recommendations, astute searching, automated summarization and classification, seamless translation, and even comprehensive writing and editing support [27]. The profound implications of this technology reverberate through the diverse dimensions of library operations, fostering innovation across academia and information management.

The realm of academic libraries, for instance, witnesses a metamorphosis across a spectrum of domains. Research endeavors are infused with ChatGPT's prowess, amplifying the scope of reference consultations, teaching interactions, and information literacy initiatives. By embarking on an innovative journey, academic libraries stand poised to revolutionize resource development, enrich specialized research services, and provide tailored consulting experiences, thereby democratizing access to educational opportunities [1, 4, 7].

Moreover, the transformative footprint of ChatGPT extends to literature and information work, where its influence is akin to a seismic shift. This transformation unfurls on various fronts, beginning with the reimagining of data organization. ChatGPT transcends surface-level information classification, ushering in the era of semantic content organization, thereby augmenting the effectiveness of literature and information systems [28]. The paradigm shift continues as knowledge service models pivot from traditional information retrieval mechanisms to the dynamic realm of knowledge-based question-answering, enhancing the depth of user engagement.

Intricacies in the analysis methods of literature and information work undergo a renaissance as well. Manual processes cede ground to large-scale intelligent analyses, transforming the landscape of insights extraction and interpretation. This metamorphosis, however, doesn't come without challenges. As ChatGPT becomes the bedrock of literature and information services, concerns about data security and integrity surface, prompting the imperative establishment of robust risk management mechanisms [29].

As user experiences undergo a revolution, the very fabric of reading habits is rewoven. Collaborative reading, fostered by the seamless interaction of humans and machines, carves out a new trajectory. Conventional processes in library information work are reshaped, prompting a comprehensive re-evaluation of team capabilities and the design of job systems that are both responsive and adaptive.

Peering into the future, the trajectory of ChatGPT's influence in intelligent libraries appears destined to reshape the landscape [1, 7, 9]. Its capabilities promise to emancipate libraries from routine tasks, catapulting efficiency and quality to new heights. Equipped with ChatGPT, libraries are poised to sculpt personalized experiences, orchestrating service journeys that are more precise, user-centered, and ultimately transformative. In an era where information is power, ChatGPT stands as a beacon, illuminating the path toward a smarter, more empowered library ecosystem [1, 21].

Please note that the first paragraph of a section or subsection is not indented. The first paragraphs that follows a table, figure, equation etc. does not have an indent, either (Fig. 2).

Elevating Efficiency through Intelligent Retrieval Services: In the landscape of modern libraries, the infusion of ChatGPT technology heralds a new era of operational efficiency. This transformative potential finds its first expression in the realm of retrieval services, where automation becomes the cornerstone of expeditious user experiences. By harnessing the capabilities of ChatGPT, libraries stand to revolutionize routine tasks, such as book retrieval and reservation management, thereby heightening the quality of user services [30].

Innovative retrieval strategies, driven by ChatGPT, pave the way for nimble searches and swift resource allocation. Through automated mechanisms, the process of book retrieval is streamlined, accelerating users' access to desired resources. Notably, the landscape of literature retrieval undergoes a paradigm shift, as ChatGPT's natural language prowess enables users to conduct searches through fluid, conversational queries.

Fig. 2. Empowering Smart Libraries through ChatGPT

No longer bound by specific keywords or search syntax, readers engage with the library catalog more intuitively, enhancing the accessibility of resources.

Moreover, ChatGPT's transformative impact extends to user reservations, where diverse scenarios are adeptly supported. The technology seamlessly allocates reservation resources based on user specifications, while automated reminders ensure users are informed ahead of their reservation time. Through these interventions, libraries not only ensure optimal utilization of resources but also deliver a superlative user experience, punctuated by prompt service and seamless interactions.

Intelligence Unleashed through Question-Answering Services: Libraries, once the bastions of knowledge, now adopt a futuristic veneer, propelled by ChatGPT's intelligence. The advent of intelligent question-answering services heralds a paradigm shift in user engagement, rendering library staff availability a secondary concern. Readers are empowered to interact with a chatbot that furnishes immediate responses to their inquiries, irrespective of staff availability [31].

The spectrum of questions handled by ChatGPT is diverse, spanning from basic queries about library services, policies, and collections to nuanced research inquiries. The technology's natural language understanding capabilities enable it to comprehend user questions and offer informative responses. This engagement transcends mere information dissemination, as ChatGPT assumes the role of a virtual reference consultant, directing users to relevant resources and refining their research strategies.

Intriguingly, intelligent consultation extends beyond textual exchanges. ChatGPT's adeptness in understanding conversational nuances creates an interactive experience, fostering deeper user engagement. Moreover, by tapping into user interests and learning trends, ChatGPT is poised to recommend resources aligned with users' preferences, amplifying the library's efficacy in meeting evolving needs.

Navigating Literary Terrain with Intelligent Recommendations: ChatGPT's transformative impact reverberates through the corridors of book recommendations. Embracing machine learning techniques and natural language processing, libraries redefine their role as literary curators. By analyzing readers' borrowing histories, preferences, and responses, ChatGPT extrapolates insights into individual reading habits, propelling it to offer personalized recommendations [32].

The reader's journey becomes an intricate tapestry woven by ChatGPT's analytical prowess. Insights gleaned from past interactions shape recommendations, leading readers to discover books and authors aligned with their preferences. This personalized touch transcends mere reading experiences, catalyzing higher engagement and satisfaction rates.

With every interaction, ChatGPT evolves, honing its ability to provide accurate recommendations. The technology's ever-learning nature augments the accuracy of its suggestions, enabling it to predict users' evolving preferences and enhancing the relevance of its counsel.

Guiding Collection Planning with Insightful Analytics: In the evolving landscape of collection planning, ChatGPT's analytical acumen emerges as a beacon guiding librarians' decisions. By parsing reader queries, ChatGPT discerns trends and interests, unveiling insights that shape collection expansion. Analyzing circulation data and trending themes, the technology illuminates the types of books that captivate readers' imaginations [33].

The interplay of ChatGPT's data-driven insights and librarians' expertise ushers in an era of informed collection management. Libraries are better equipped to procure resources that resonate with readers' needs, ensuring their collections remain contemporary and appealing. This symbiotic relationship between artificial intelligence and human curation translates into enriched user experiences and a library ecosystem attuned to evolving preferences.

Personalization: Forging Connections through User Profiles: The dawn of personalization in smart libraries is catalyzed by ChatGPT's seamless integration. Libraries can craft user profiles by amalgamating personal information, borrowing history, and user data. By delving into these profiles, libraries glean insights into users' predilections, enabling them to offer customized recommendations and services [34].

User engagement reaches new heights as intelligent recommendations are tailored to individual interests and preferences. ChatGPT deciphers borrowing histories, enhancing the library experience with resource suggestions that resonate with each user. The dynamic interplay between user profiles and ChatGPT's insights ensures a user-centric approach that empowers readers to explore their literary passions.

Empowering Librarians: Knowledge Amplification: ChatGPT extends its impact beyond user interactions, assuming the role of a knowledge repository for librarians [12]. Through data analysis of borrowing habits, user behavior, and participation in library activities, ChatGPT generates insights and reports that assist librarians in informed decision-making. This data-driven approach informs resource procurement, service planning, and activity design, underpinning a library ecosystem that thrives on evidence-based strategies.

Unleashing Equality: A Vision for Inclusive Services: In the pursuit of equitable library services, ChatGPT emerges as a force that transcends biases, upholds impartiality, and champion's accessibility. The technology dismantles geographic barriers, ensuring remote and marginalized communities access library services. The egalitarian ethos is preserved, as every individual benefits from unbiased service, untainted by factors like identity, gender, or location [35].

With ChatGPT at the helm, libraries pioneer equal service, defying disparities that once plagued the library landscape. Collaborative efforts culminate in an ecosystem where artificial intelligence products like ChatGPT break barriers, disseminate knowledge, and foster a library experience that transcends boundaries and emboldens inclusivity.

6 Conclusion

This comprehensive article delves into the multifaceted world of ChatGPT, unraveling its essence, global and domestic recognition, and its profound impact across diverse industries. As the narrative unfolds, it navigates the uncharted territory of ChatGPT's potential within the realm of smart libraries, propelling us into a future teeming with possibilities.

The journey commences by dissecting the very concept of ChatGPT, unveiling its intricate layers and revolutionary capabilities. The spotlight then shifts to the global stage, where the technology's compelling allure has garnered attention both near and far. Its application landscape spans industries, painting a picture of transformation that transcends borders and industries alike.

Embarking on the quest to illuminate its potential in the realm of smart libraries, the article meticulously charts its course. As it treads the path of foresight, the spotlight focuses on the synergy between ChatGPT and libraries, illuminating a landscape where the technology can be harnessed to propel services into the future.

However, this journey is not without its challenges. The tapestry of ChatGPT's evolution necessitates addressing critical issues that touch on accuracy, reliability, and the vital matter of data security and privacy. Furthermore, the article underscores the need to integrate diverse technological realms, crafting a mosaic of services that cater to an array of user needs. The library landscape, traditionally a bastion of human interaction, must metamorphose to accommodate the integration of artificial intelligence.

In the face of these challenges, libraries are beckoned to embrace innovation. It's a clarion call to recalibrate approaches, to adapt and respond with agility, weaving AI into the very fabric of their operations. The roadmap for the future lies in a harmonious coexistence, where libraries symbiotically dance with ChatGPT to create intelligent and personalized reading experiences.

In essence, this article serves as a navigational guide through the uncharted seas of ChatGPT's potential, drawing us closer to the shores of a smarter, more connected library landscape. It's a vision where innovation and tradition converge, where intelligence enhances human endeavors, and where libraries stand as beacons of progress in an ever-evolving world.

Funding. This research received no specific grant from any funding agency.

Conflict of Interest. The authors have no conflict of interest.

Compliance with Ethical Standards. This work does not contain any studies with human participants or animals performed by any of the authors.

Data Availability Statement. Data sharing is not applicable to this article as no new data were created or analyzed in this study.

References

1. Bozkurt, A., et al.: Speculative futures on ChatGPT and generative artificial intelligence (AI): A collective reflection from the educational landscape. Asian Journal of Distance Education **18**(1) (2023)
2. Gimpel, H., et al.: Unlocking the power of generative AI models and systems such as GPT-4 and ChatGPT for higher education: A guide for students and lecturers (No. 02–2023). Hohenheim Discussion Papers in Business. Economics and Social Sciences (2023)
3. Sheth, K., Patel, K., Shah, H., Tanwar, S., Gupta, R., Kumar, N.: A taxonomy of AI techniques for 6G communication networks. Comput. Commun. **161**, 279–303 (2020)
4. Jain, K., Agarwal, A.: Novel Library Innovations and Advancements After the COVID-19 Pandemic Era. In: Handbook of Research on Advancements of Contactless Technology and Service Innovation in Library and Information Science, pp. 191–210. IGI Global (2023)
5. Pastor, A.B., Canniffe, E., Jiménez, C.J.R.: Learning from Letchworth and Welwyn Garden City: Garden cities' policies for the development of existing settlements in the contemporary world. Land Use Policy **132**, 106759 (2023)
6. de Siles, E.L.: Artificial Intelligence Bias and Discrimination: Will We Pull the Arc of Moral Universe towards Justice?. Revista Forumul Judecatorilor 40 (2022)
7. Ray, P.P.: ChatGPT: A comprehensive review on background, applications, key challenges, bias, ethics, limitations and future scope. Internet of Things and Cyber-Physical Systems (2023)
8. Waddington, J.: Developing primary school students' foreign language learner self-concept. System **82**, 39–49 (2019)
9. Peters, M.A., et al.: AI and the future of humanity: ChatGPT-4, philosophy and education–Critical responses. Educational Philosophy and Theory, 1–35 (2023)
10. Sarangi, S., Sharma, P.: Artificial intelligence: evolution, ethics and public policy. Taylor & Francis (2018)
11. Garon, J.: A Practical Introduction to Generative AI, Synthetic Media, and the Messages Found in the Latest Medium. Synthetic Media, and the Messages Found in the Latest Medium (2023)

12. Dwivedi, Y.K., et al.: So what if ChatGPT wrote it? Multidisciplinary perspectives on opportunities, challenges and implications of generative conversational AI for research, practice and policy. Int. J. Inf. Manage. **71**, 102642 (2023)
13. Adiguzel, T., Kaya, M.H., Cansu, F.K.: Revolutionizing education with AI: Exploring the transformative potential of ChatGPT. Contemporary Educational Technology **15**(3), ep429 (2023)
14. Xu, X.: Exploration of the Feasibility of Applying ChatGPT in the Field of Smart Libraries
15. Sallam, M., Salim, N., Barakat, M., Al-Tammemi, A.: ChatGPT applications in medical, dental, pharmacy, and public health education: a descriptive study highlighting the advantages and limitations. Narra J. **3**(1), e103–e103 (2023)
16. Biswas, S.S.: Role of chat gpt in public health. Ann. Biomed. Eng. **51**(5), 868–869 (2023)
17. Cao, J., Li, M., Wen, M., Cheung, S.C.: A study on prompt design, advantages and limitations of chatgpt for deep learning program repair (2023). arXiv preprint arXiv:2304.08191
18. Kalla, D., Smith, N.: Study and Analysis of Chat GPT and its Impact on Different Fields of Study. Int. J. Innov. Sci. Res. Technol. **8**(3) (2023)
19. Azaria, A.: ChatGPT usage and limitations (2022)
20. Cribben, I., Zeinali, Y.: The benefits and limitations of ChatGPT in business education and research: a focus on management science, operations management and data analytics. Operations Management and Data Analytics (2023)
21. Dave, T., Athaluri, S.A., Singh, S.: ChatGPT in medicine: an overview of its applications, advantages, limitations, future prospects, and ethical considerations. Frontiers in Artificial Intelligence **6**, 1169595 (2023)
22. Bilquise, G., Ibrahim, S., Shaalan, K.: Emotionally Intelligent Chatbots: A Systematic Literature Review. Human Behavior and Emerging Technologies (2022)
23. Recanati, A.M.: Future. In: AI Battle Royale: How to Protect Your Job from Disruption in the 4th Industrial Revolution, pp. 11–44. Springer International Publishing, Cham (2023)
24. Agarwal, A., Jain, K., Yadav, R.K.: A mathematical model based on modified ID3 algorithm for healthcare diagnostics model. Int. J. Sys. Assur. Eng. Manage. 1–11 (2023)
25. Grassini, S.: Shaping the future of education: exploring the potential and consequences of AI and ChatGPT in educational settings. Education Sciences **13**(7), 692 (2023)
26. Tlili, A., et al.: What if the devil is my guardian angel: ChatGPT as a case study of using chatbots in education? Smart Learning Environments **10**(1), 15 (2023)
27. Jain, K., Agarwal, A., Jain, A., Abraham, A.: A Multi-layer Deep Learning Model for ECG-Based Arrhythmia Classification. In: International Conference on Intelligent Systems Design and Applications, pp. 44–52. Springer Nature Switzerland, Cham (2022)
28. Jain, K., Gupta, M., Patel, S., Abraham, A.: Object Classification Using ECOC Multi-class SVM and HOG Characteristics. In: International Conference on Intelligent Systems Design and Applications, pp. 23–33. Springer Nature Switzerland, Cham (2022)
29. Jain, K., Gupta, M., Abraham, A.: A review on privacy and security assessment of cloud computing. J. Inf. Assur. Secur **16**, 161–168 (2021)
30. Ruixue, Z.H.A.O., Yongwen, H.U.A.N.G., Weilu, M.A., Wenjia, D.O.N.G., Guojian, X.I.A.N., Tan, S.U.N.: Insights and reflections of the impact of ChatGPT on intelligent knowledge services in libraries. Journal of Library and Information Sciences in Agriculture **35**(1), 29 (2023)
31. Lund, B.D., Wang, T.: Chatting about ChatGPT: how may AI and GPT impact academia and libraries? Library Hi Tech News **40**(3), 26–29 (2023)
32. AlAfnan, M.A., Dishari, S., Jovic, M., Lomidze, K.: Chatgpt as an educational tool: Opportunities, challenges, and recommendations for communication, business writing, and composition courses. J. Artif. Intel. Technol. **3**(2), 60–68 (2023)
33. Biswas, S.S.: Potential use of chat gpt in global warming. Ann. Biomed. Eng. **51**(6), 1126–1127 (2023)

34. Megahed, F.M., et al.: How generative ai models such as chatgpt can be (mis) used in spc practice, education, and research? an exploratory study. Quality Engineering, 1–29 (2023)
35. Gill, S.S., Kaur, R.: ChatGPT: Vision and challenges. Internet of Things and Cyber-Physical Systems **3**, 262–271 (2023)

Pseudo-learning to Identify Prototype Networks for Interpreting Multi-layered Neural Networks

Ryotaro Kamimura(✉)

Tokai University, Kitakaname, Hiratsuka, Kanagawa 259-1292, Japan
ryotarokami@gmail.com, ryo@keyaki.cc.u-tokai.ac.jp

Abstract. The present paper aims to propose a method for evaluating and interpreting prototype networks hidden within multi-layered neural networks. The prototype network can be obtained by compressing as many different types of multi-layered networks as possible. However, the final prototype networks are naturally realized by losing information in the original multi-layered neural networks. To preserve the information in the original multi-layered neural networks, we introduce pseudo-learning, where learning essentially involves not learning, thereby retaining the original internal information of multi-layered neural networks as much as possible. Pseudo-learning is implemented by adopting the concept of the conventional learning approach, such as entropy minimization, or in our terms, potentiality minimization. By keeping the weights fixed in prototype networks, we aim to change only the potentiality to simulate a learning process by modifying a parameter for potentiality. The method was preliminarily applied to a simple qualitative bankruptcy dataset. The results confirmed that the compressed networks or prototype networks could be altered with fixed connection weights to evaluate generalization performance. This enabled the interpretation of the internal information transferred from the original multi-layered neural networks.

Keywords: potentiality · prototype network · pseudo-learning · compression

1 Introduction

1.1 Problems of Computation and Interpretation

Neural networks have been applied to many practical problems, particularly due to their improved generalization performance. Nowadays, the network size of neural networks becomes larger and larger for dealing with a large number of data sets. When the data sets and network size become larger, we must face many problems for learning, for example, catastrophic forgetting [1], adversarial attacks [2], and so on. In particular, the heavy computation of those larger networks has forced us to develop simpler and faster learning. Though there have been many attempts to develop simpler and faster learning methods such as extreme learning [3], one-shot learning [4], other conventional acceleration methods [5], the majority of methods have been applied to improve generalization performance, without considering how better generalization is realized inside neural networks. Even if learning is faster with better generalization, those methods cannot necessarily be applied to actual and practical problems due to the corresponding safety or ethical issues, inherent to any applications [6].

A. Bajaj et al. (Eds.): IBICA 2023, LNNS 1229, pp. 62–73, 2025.
https://doi.org/10.1007/978-3-031-78940-3_5

1.2 Prototype Network for Interpretation

The present paper aims not to propose a method for achieving better generalization but to interpret how neural networks produce the final outputs, even though due attention is given to improving generalization. To address the interpretation of neural networks, we assume the existence of prototype networks underlying various types of surface multi-layered neural networks. These prototype networks can generate a multitude of complex surface networks. Consequently, the interpretation process is reframed as the discovery of these prototype networks within the intricate networks. The challenging problem of interpretation is effectively transformed into a more tangible task of identifying simple and transparent prototype networks.

1.3 Pseudo-learning

The prototype networks are obtained, as mentioned above, by condensing as much information as possible from the original and many multi-layered neural networks. However, the compression of the multi-layered neural network is necessarily accompanied by the loss of original information. In particular, one of the main problems is that the final prototype networks do not necessarily lead to improved generalization. To address this issue, it is possible to re-train or fine-tune the prototype networks to achieve better generalization. However, for faithful interpretation, we need to interpret the main mechanism of multi-layered neural networks as they are. During fine-tuning, it may occur that the information from the original internal representations is lost in this re-training. This emphasizes the need to develop a method for improving generalization while preserving the information of the original multi-layered networks as much as possible.

For faithful interpretation with enhanced generalization performance, we introduce pseudo-learning, in which learning essentially entails not learning. This implies that a learning process is compelled to be realized without actually engaging in a learning process, albeit ideally. This paper aims to propose a method to emulate a learning process through a much simpler approach without compromising the original information from multi-layered neural networks. Although the method adopted in the experiments of this paper involves specific aspects of learning or partial learning, in reality, partial learning can be achieved through a straightforward parameter choice in the end.

1.4 Main Contributions

Then, we should outline the primary contributions of this paper to provide a clearer understanding of its objectives. The main contributions are summarized as follows:

- The present paper proposes a method for identifying prototype networks within complex networks. The interpretation process is conceptualized as the task of discovering these prototype networks.
- The method is implemented by drawing inspiration from the conventional concept that learning involves a process of mutual information maximization or entropy minimization, although in this case, entropy is replaced by a more straightforward potentiality for efficient computation.

- The prototype networks, acquired by compressing as many diverse networks as possible, serve as coarse approximations to the actual prototype networks.
- To interpret and accurately assess their performance, these networks are not fine-tuned through full learning but are instead adjusted by modifying a parameter related to the potentiality of the networks, referred to as pseudo-learning.
- The paper presents preliminary experimental results in which multi-layered neural networks were compressed into simpler prototype networks. The simplicity of internal representation was controlled using pseudo-learning to evaluate their generalization and interpretation performance.

2 Related Works

The proposed method can be related to many methods in neural networks as well as machine learning. In particular, we should mention the relations to faster learning, stable learning, compression, and multi-view learning among many others.

First, our method can be considered one of the new methods to realize a process of learning, ideally, without learning. Due to the complexity of neural networks and datasets, the problem of heavy computation to reach the final solutions has been one of the main problems of neural networks. To make learning faster, there have been many different types of learning methods [3,4,7]. However, they have mainly focused on the improvement of prediction performance, with little attention paid to the properties of internal representation, as mentioned in the introductory section.

Our approach is based on information-theoretic methods, where learning can be considered as a process of mutual information maximization or as a process of conditional entropy minimization in the simplest form [8,9]. One of the main characteristics is to replace a process of error minimization with a process of entropy minimization or potentiality minimization in our terminology. This makes it possible to achieve, though ideally, learning without learning.

Second, our method involves compressing as many multi-layered neural networks as possible. The prototype network can be identified through a process of compression from multi-layered neural networks to less complicated ones. As the network size of neural networks becomes larger, many compression methods have been proposed [10,11]. However, they mainly focus on prediction performance without paying due attention to internal representations. Our method aims to preserve the original information in multi-layered neural networks while giving proper consideration to generalization.

Third, our method aims to consider the instability of neural networks as one of the major properties to achieve the stability of interpretation. Interpretation methods have advanced significantly in recent times [12–15]. However, the majority of methods have attempted to explain only an instance of many created by a neural network, even if they try to interpret the inner mechanism globally and logically [16,17]. This is because they have not addressed the instability as one of the main characteristics of neural learning. Neural networks can produce a variety of different final solutions even for simple problems, and interpretation should be inherently unstable. So, how can we globally understand those unstable final results? Our method seeks to make full use of the instability of neural networks, which allows us to view an input from multiple perspectives.

This instability can be related to the recent approach to multi-viewpoint learning. Multi-view learning has received much attention due to the diversity among features in datasets [18–21]. In the simplest case, data are taken from different viewpoints and for different purposes. If it is possible to consider these different viewpoints collectively, generalization performance may be improved. This improvement can be achieved by reducing differences among different viewpoints, naturally handling all different viewpoints in unified and combined ways, and seeking features common to all viewpoints. However, the inference mechanism to handle these multi-viewpoints has not received the attention it deserves because multi-viewpoints are given from external sources. In many datasets commonly used in the human, social sciences, and business data domains, it is often impossible to identify which viewpoint is adopted, as many viewpoints are implicitly included in the data creation process. The present paper aims to demonstrate that multi-viewpoints should not be imposed from external sources but generated from inside.

3 Theory and Computational Methods

This paper aims to propose an interpretation method in which the process of interpretation is replaced by finding a prototype network that inherits the main information contained in the original multi-layered neural networks. The method is composed of compression to create a prototype network, followed by pseudo-learning to evaluate and interpret the performance and internal representations. As illustrated in Fig. 1, initially, a multi-layered neural network is systematically compressed into the simplest prototype network (b). Subsequently, the performance of this prototype network can be assessed using pseudo-learning (c). In pseudo-learning, all connection weights, except those associated with bias, are held fixed, and only the bias is updated, a practice termed "partial learning" in pseudo-learning. It should be noted that this partial learning can be readily substituted by a simple parameter change. We utilize partial learning for the sake of convenience in implementing learning procedures for our experiments.

3.1 Compression

Any multi-layered neural network can be compressed into its simplest form, devoid of hidden layers, to create a prototype network for interpretation. For stable compression for interpretation, it is essential to compress as many different networks as possible, generated from diverse initial conditions and input patterns. This can be viewed as learning from multiple perspectives, as discussed in the related works section.

Now, let us assume that a five-hidden-layered neural network can be compressed into a network without hidden layers, as depicted in Fig. 1. First, we compress the connection weights from the first to the second layer, denoted as (1, 2), and from the second to the third layer (2, 3) using an initial condition and a subset of a data set. Note that, for simplicity, the layer numbering proceeds from the first layer (input layer) to the final and sixth layer (output layer). Subsequently, we obtain the compressed weights between the first and third layer, denoted as (1, 3).

$$w_{ik}^{(1,3)} = \sum_{j=1}^{n_2} w_{ij}^{(1,2)} w_{jk}^{(2,3)}. \tag{1}$$

These compressed weights are further combined with weights from higher layers. By repeating these processes, we arrive at the compressed weights between the first and fifth layer, denoted as $w_{iq}^{(1,5)}$. Using these connection weights, we derive the final and fully compressed weights (1, 6).

$$w_{ir}^{(1,6)} = \sum_{q=1}^{n_5} w_{iq}^{(1,5)} w_{qr}^{(5,6)}. \tag{2}$$

These compressed weights are used in the corresponding prototype network.

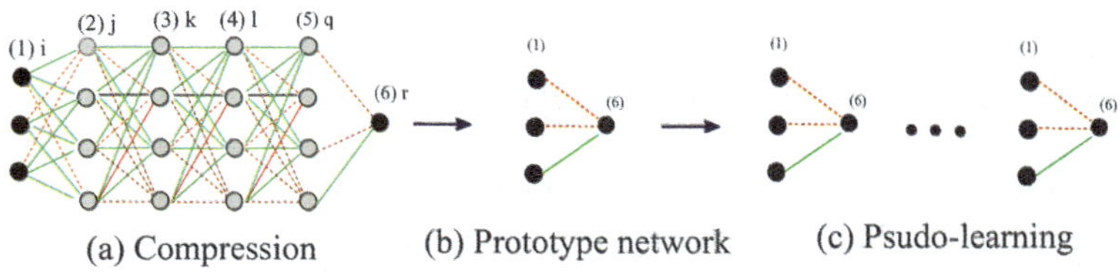

Fig. 1. Network compression (a) from the initial state to the simplest and compressed network as a prototype network (b), and pseudo-learning (c) to evaluate generalization and interpretation performance.

3.2 Pseudo-learning

Then, the prototype network may not lead to improved generalization due to compression, which corresponds to the information loss during the compression process. In this scenario, we need to restore the information from the original multi-layered neural networks. One possibility is to retrain the prototype network. However, this may result in the loss of connection weights that contain information from the original multi-layered neural networks. In this paper, we aim to interpret the information in the original multi-layered neural networks, and to address this information loss problem, we propose a new and straightforward method to train the prototype network with minimal learning. A key aspect is that we leverage the property of learning as entropy minimization or information maximization [8], and instead of training the network to minimize errors between outputs and targets, we seek to reduce entropy, which is replaced by a simpler form of potentiality.

Now, let us describe how to define and compute the potentiality for connection weights. For simplicity, we consider weights from the tth hidden layer to the $t+1$th hidden layer, denoted as $(t, t+1)$, and the absolute weight from the jth neuron to the kth neuron is defined as:

$$u_{jk}^{(t,t+1)} = \mid w_{jk}^{(t,t+1)} \mid. \tag{3}$$

For each layer, we can compute the individual potentiality as follows:

$$
{}^{\alpha}g_{jk}^{(t,t+1)} = \left[\frac{u_{jk}^{(t,t+1)}}{\max_{j'k'} u_{j'k'}^{(t',t'+1)}} \right]^{\alpha}, \tag{4}
$$

where the max operation is over all connection weights in the layer, and the parameter α should be greater than zero for the stability of learning.

For each layer, we have the corresponding potentiality:

$$
{}^{\alpha}G^{(t,t+1)} = \frac{1}{n_t \, n_{t+1}} \sum_{j}^{n_t} \sum_{k}^{n_{t+1}} \left[\frac{u_{jk}^{(t,t+1)}}{\max_{j'k'} \, u_{j'k'}^{(t,t+1)}} \right]^{\alpha}. \tag{5}
$$

For simplicity, the number of input neurons is greater than one, and at least one connection weight should be greater than zero. This average potentiality becomes maximum when all connection weights have the same values. When only one connection weight is greater than zero, while all the others are zero, the corresponding coefficient becomes minimum.

As mentioned, learning can be viewed as a process of potentiality reduction. One of the simplest methods to achieve this is to increase the parameter α. When the parameter α is increased, the potentiality naturally tends to decrease. Thus, the primary idea behind pseudo-learning is to replace learning as changes in connection weights with learning as a simple adjustment of the parameter α. In practical experiments, this can be accomplished by increasing the parameter α while keeping connection weights fixed.

4 Results and Discussion

4.1 Experimental Outline

To illustrate the efficacy of our pseudo-learning method, we provide a straightforward example using a qualitative bankruptcy dataset [22, 23]. In this dataset, we aim to compress the knowledge from multi-layered neural networks and transfer it to the simplest network referred to as "prototype networks". The dataset consisted of 250 input patterns with only six input variables, and our goal was to predict bankruptcy based on these six variables. To better show the performance of networks with many hidden layers, we limited the number of learning epochs to 100 for all methods. This limitation was imposed because networks without many hidden layers may struggle to learn the relationships between inputs and targets even for the training dataset. In our experiments, the number of hidden neurons in each hidden layer was set to ten, and the number of hidden layers ranged from 0 to 20. Beyond this point, the generalization performance deteriorated significantly.

4.2 Multi-layered Neural Networks

The results confirmed that potentiality could be reduced, except for networks without hidden layers, and consequently, generalization performance could be improved by

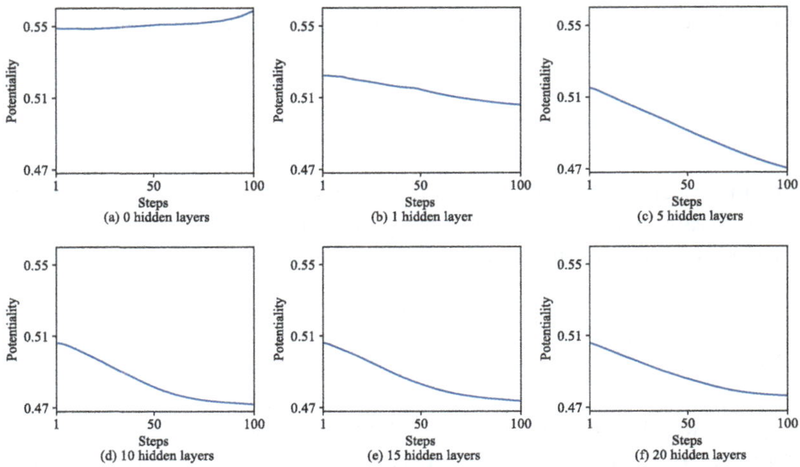

Fig. 2. Potentiality as a function of the number of learning steps, when the number of hidden layers increased from 0 (a) to 20 (f) for the qualitative bankruptcy data set.

increasing the number of hidden layers under the condition that the number of training epochs was constant (100 steps). This suggests that learning can be viewed as a process of potentiality reduction.

Figure 2 illustrates the potentiality as the number of hidden layers increases from 0 (a) to 20 (f). When there were no hidden layers, as shown in Fig. 2(a), the potentiality was close to the midpoint of 0.5. However, as the number of hidden layers increased from 1 (a) to 20 (f), the potentiality tended to decrease with an increase in the number of learning epochs, indicating that an increase in the number of hidden layers resulted in reduced potentiality.

Then, the improvement in generalization followed the reduction in potentiality. Figure 3 depicts the generalization performance of neural networks as the number of hidden layers increased from 0 (a) to 20 (f). As evident in the figure, generalization performance gradually improved as the number of hidden layers increased from 0 (a) to 1 (b). When the number of hidden layers increased from 5 (c) to 15 (e), generalization performance approached its maximum values. However, when the number of hidden layers further increased to 20 (f), generalization performance showed a slight decrease. This indicates that generalization performance could be enhanced with an increase in the number of hidden layers, while correspondingly, potentiality decreased.

4.3 Prototype Networks

Now, let us turn our attention to the results for the corresponding prototype networks. First, we observed a decrease in potentiality as we increased the parameter α. Figure 4 displays the potentiality as the parameter α is increased from 0.1 to 2.0, with the number of hidden layers ranging from 0 (a) to 20 (f). As evident in the figures, the potentiality gradually decreased as the parameter α increased from 0.1 to 2.0. Notably, when the number of hidden layers was 10 (d) and 15 (e), the potentiality exhibited a significant decrease.

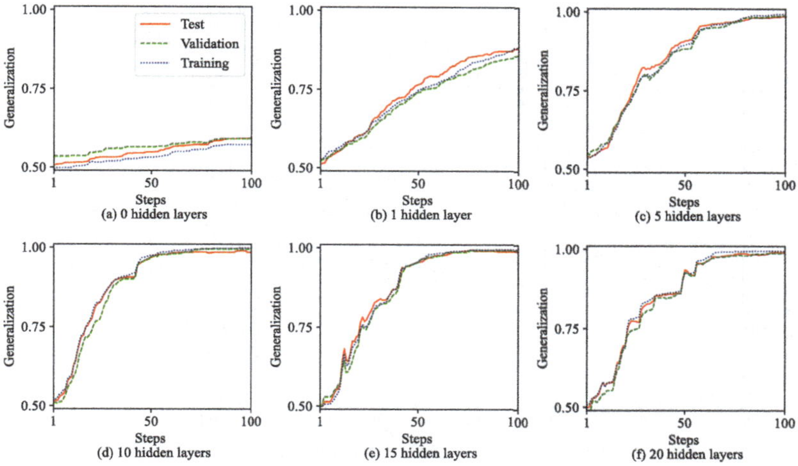

Fig. 3. Generalization accuracy as a function of the number of learning steps, when the number of hidden layers increased from 0 (a) to 20 (f) for the qualitative bankruptcy data set.

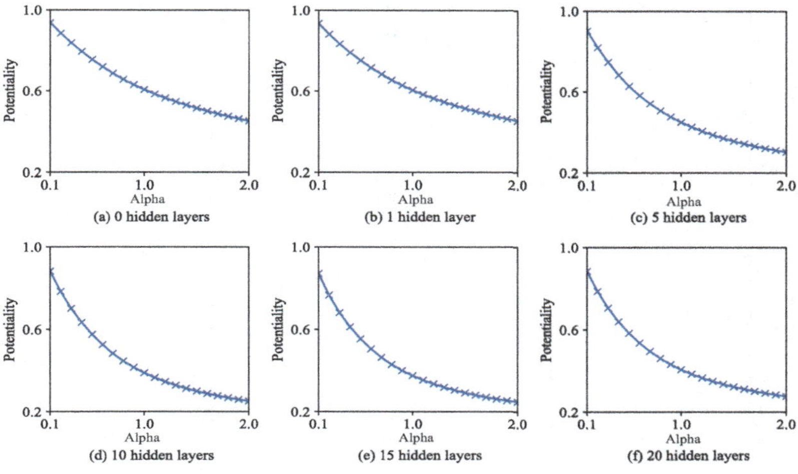

Fig. 4. Potentiality decreased, when the parameter α increased from 0.1 to 2.0 and when the number of hidden layers increased from 0 (a) to 20 (f) for the qualitative bankruptcy data set.

Correspondingly, generalization could be increased by changing the parameter α. Figure 5 shows generalization performance when the parameter α increased from 0.1 to 2.0 and when the number of hidden layers increased from 0 (a) to 20 (f). When the number of hidden layers was 0 in Fig. 5(a), generalization performance decreased when the parameter α increased. When the number of hidden layer increased from 1 (b), generalization performance increased gradually, when the parameter α increased. In particular, when the number of hidden layers increased to 15 in Fig. 5(e), generalization performance increased and became close to 0.96.

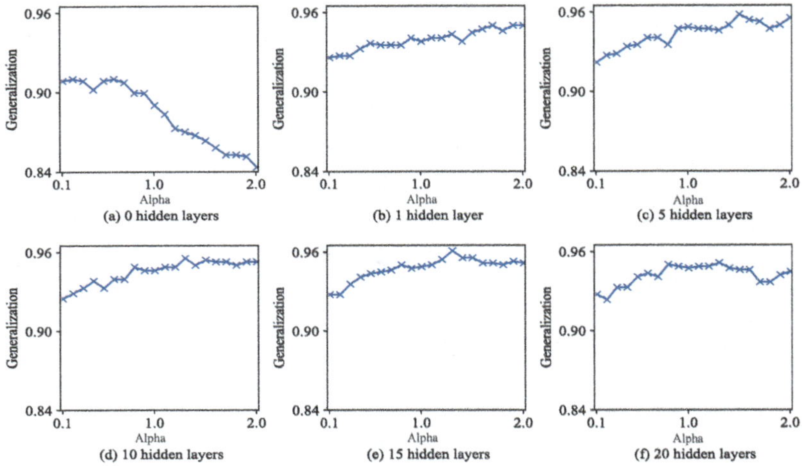

Fig. 5. Generalization accuracy, when the parameter α increased from 0.1 to 2.0, and when the number of hidden layers increased from 0 (a) to 20 (f) for the qualitative data set.

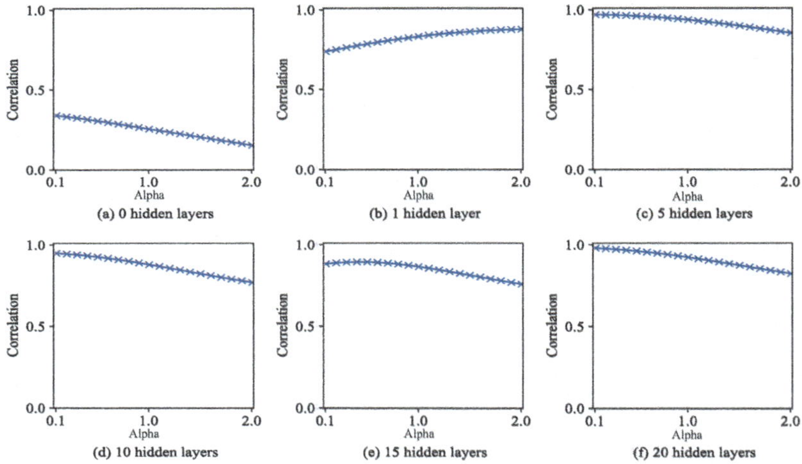

Fig. 6. Correlation coefficients between weights and the original correlation coefficients, when the parameter α increased from 0.1 to 2.0, and when the number of hidden layers increased from 0 (a) to 20 (f) for the qualitative bankruptcy data set.

Furthermore, we observed that the correlation coefficients between weights and the corresponding original correlation coefficients of the training data sets were higher as the number of hidden layers increased. This suggests that the prototype networks aim to represent relationships among inputs and outputs in a more linear and separate manner. Figure 6 illustrates the correlation coefficients between weights and the actual original correlation coefficients. As seen in the figure, when there were no hidden layers 0(a), the correlation coefficients were lower and decreased gradually. However, for other cases,

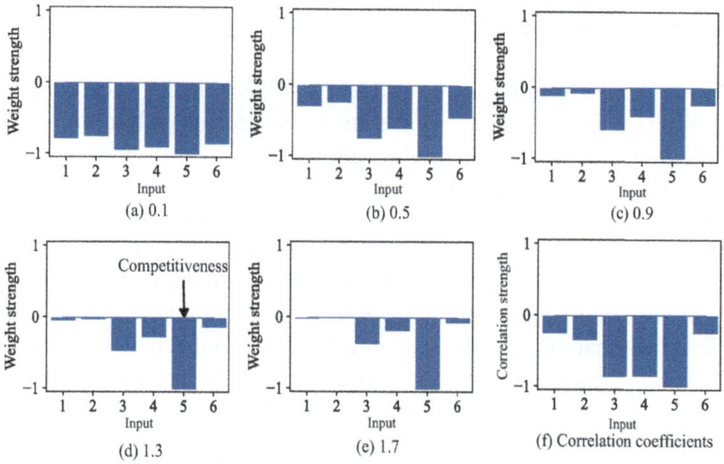

Fig. 7. Connection weights when the parameter θ increased from 1.0 (b) to 1.9 (s) and the original correlation coefficients (t) for the qualitative bankruptcy data set.

the correlation coefficients significantly increased and approached the maximum value. When there were 15 hidden layers 15 (e), the correlation coefficient was slightly lower, which can be attributed to achieving better generalization performance at the expense of a slight degradation in the correlation coefficients.

The results regarding connection weights indicated that as the parameter α increased, the number of strong weights naturally decreased, making the significance of specific inputs more apparent. Figure 7 illustrates the changes in connection weights of prototype networks as the parameter α increased from 0.1 (a) to 1.7 (e), alongside the original correlation coefficients of the training data sets (f). As the parameter α increased, the number of stronger weights gradually diminished, and ultimately, only the fifth input became significantly stronger. This indicates that the prototype network attempted to predict the outputs by placing greater emphasis on the fifth input variable. The fifth input variable represents "competitiveness", signifying that bankruptcy was naturally linked to the degree of competitiveness of companies.

4.4 Summary of Correlation and Generalization

Table 1 presents a summary of the experimental results. The results demonstrated that the optimal generalization performance was achieved when the number of hidden layers was set to 15, both for multi-layered neural networks and their corresponding prototype networks. Interestingly, the highest generalization accuracy of 0.995 was attained when there were 15 hidden layers. Furthermore, when the number of hidden layers was 15, the prototype network achieved the highest generalization accuracy of 0.961, significantly surpassing the results obtained with fewer hidden layers. It's worth noting that the correlation coefficients were higher with a greater number of hidden layers. Nevertheless, by slightly reducing the correlation coefficients, the best generalization performance was achieved.

5 Conclusion and Future Works

The aim of the present paper was to introduce a novel learning method for prototype networks. Prototype networks are believed to be embedded within multi-layered neural networks, and their simplicity enables us to estimate and interpret the inference mechanisms of the original multi-layered neural networks. These prototype networks are derived by compressing the original multi-layered neural networks, but the compressed versions do not necessarily yield better performance, particularly in terms of generalization, due to information loss. When fine-tuning the connection weights of a prototype network, the weights often undergo substantial changes, leading to the loss of the original information. In extreme cases, the final weights may be entirely different from the originally compressed ones. To address this issue, we introduced the concept of pseudo-learning, which represents an ideal form of learning without learning.

The method was applied to a simple bankruptcy dataset. The experimental results demonstrate that knowledge from multi-layered neural networks can be effectively transferred to the corresponding prototype networks. Moreover, the application of pseudo-learning showed that the generalization performance of the prototype networks did not change significantly when compared to the performance of the original multi-layered neural networks.

Table 1. Summary of experimental results on averaged correlation coefficients and generalization performance by our methods for the qualitative bankruptcy data set. The notation of prototype networks, for example, "0(15)" means that the prototype network (0) is obtained by compressing 15 hidden layered networks. Bold type letters indicate the maximum values.

Methods	Hidden	Alpha	Generalization	Correlation
Multi-layered	0		0.587	
	1		0.881	
	5		0.985	
	10		0.987	
	15		**0.995**	
	20		0.989	
Prototype	0(00)	0.2	0.911	0.340
	0(01)	1.7	0.951	0.878
	0(05)	1.5	0.959	**0.906**
	0(10)	1.3	0.956	0.849
	0(15)	1.3	**0.961**	0.837
	0(20)	1.3	0.952	0.898

The experiments in this paper were conducted on small datasets under the conditions to show the capabilities of prototype network evaluation. However, it is imperative to apply this method to larger datasets to confirm that knowledge from multi-layered neural networks can be transferred to much simpler networks for interpretation. Furthermore, the need arises to develop more advanced pseudo-learning techniques where learning can truly occur without traditional learning processes.

References

1. French, R.M.: Catastrophic forgetting in connectionist networks. Trends Cogn. Sci. **3**(4), 128–135 (1999)
2. Ilyas, A., Santurkar, S., Tsipras, D., Engstrom, L., Tran, B., Madry, A.: Adversarial examples are not bugs, they are features. In: Advances in Neural Information Processing Systems, vol. 32 (2019)
3. Huang, G.-B., Zhu, Q.-Y., Siew, C.-K.: Extreme learning machine: theory and applications. Neurocomputing **70**(1–3), 489–501 (2006)

4. Vinyals, O., Blundell, C., Lillicrap, T., Wierstra, D., et al.: Matching networks for one shot learning. In: Advances in Neural Information Processing Systems, vol. 29 (2016)
5. Liang, T., Glossner, J., Wang, L., Shi, S., Zhang, X.: Pruning and quantization for deep neural network acceleration: A survey. Neurocomputing **461**, 370–403 (2021)
6. Arrieta, A.B., et al.: Explainable artificial intelligence (XAI): concepts, taxonomies, opportunities and challenges toward responsible AI. Inf. Fusion **58**, 82–115 (2020)
7. Kim, J., Oh, T.-H., Lee, S., Pan, F., Kweon, I.S.: Variational prototyping-encoder: one-shot learning with prototypical images. In: Proceedings of the IEEE/CVF Conference on Computer Vision and Pattern Recognition, pp. 9462–9470 (2019)
8. Watanabe, S.: Knowing and Guessing: A Quantitative Study of Inference and Information. Wiley, New York (1969)
9. Linsker, R.: Self-organization in a perceptual network. Computer **21**(3), 105–117 (1988)
10. Buciluǎ, C., Caruana, R., Niculescu-Mizil, A.: Model compression. In: Proceedings of the 12th ACM SIGKDD International Conference on Knowledge Discovery and Data Mining, pp. 535–541. ACM (2006)
11. Hinton, G., Vinyals, O., Dean, J.: Distilling the knowledge in a neural network. Stat **1050**, 9 (2015)
12. Montavon, G., Samek, W., Müller, K.-R.: Methods for interpreting and understanding deep neural networks. Digit. Signal Process. **73**, 1–15 (2018)
13. Zhou, B., Bau, D., Oliva, A., Torralba, A.: Interpreting deep visual representations via network dissection. IEEE Trans. Pattern Anal. Mach. Intell. **41**(9), 2131–2145 (2018)
14. Fong, R., Vedaldi, A.: Explanations for attributing deep neural network predictions. In: Explainable AI: Interpreting, Explaining and Visualizing Deep Learning, pp. 149–167. Springer (2019)
15. Liang, Yu., Li, S., Yan, C., Li, M., Jiang, C.: Explaining the black-box model: a survey of local interpretation methods for deep neural networks. Neurocomputing **419**, 168–182 (2021)
16. Yang, C., Rangarajan, A., Ranka, S.: Global model interpretation via recursive partitioning. In: 2018 IEEE 20th International Conference on High Performance Computing and Communications; IEEE 16th International Conference on Smart City; IEEE 4th International Conference on Data Science and Systems (HPCC/SmartCity/DSS), pp. 1563–1570. IEEE (2018)
17. Lundberg, S.M., et al.: From local explanations to global understanding with explainable AI for trees. Nat. Mach. Intell. **2**(1), 2522–5839 (2020)
18. Sun, S.: A survey of multi-view machine learning. Neural Comput. Appl. **23**, 2031–2038 (2013)
19. Li, Y., Yang, M., Zhang, Z.: A survey of multi-view representation learning. IEEE Trans. Knowl. Data Eng. **31**(10), 1863–1883 (2018)
20. Nguyen, N.D., Wang, D.: Multiview learning for understanding functional multiomics. PLoS Comput. Biol. **16**(4), e1007677 (2020)
21. Zhang, C., Cui, Y., Han, Z., Zhou, J.T., Fu, H., Hu, Q.: Deep partial multi-view learning. IEEE Trans. Pattern Anal. Mach. Intell. **44**(5), 2402–2415 (2020)
22. Qualitative Bankruptcy. UCI Machine Learning Repository (2014). https://doi.org/10.24432/C52889
23. Kim, M.-J., Han, I.: The discovery of experts' decision rules from qualitative bankruptcy data using genetic algorithms. Expert Syst. Appl. **25**(4), 637–646 (2003)

Reshaping Security: Adversarial Defense in Machine Learning Application

Yajnaseni Dash[1(✉)], Ajith Abraham[2], and Manish Raj[3]

[1] School of Artificial Intelligence, Bennett University, Greater Noida, Uttar Pradesh, India
yajnasenidash@gmail.com
[2] School of Artificial Intelligence, Sai University, Chennai, India
ajith.abraham@ieee.org
[3] School of Computer Science Engineering and Technology, Bennett University, Greater Noida, Uttar Pradesh, India
Manish.Raj@bennett.edu.in

Abstract. The rapid growth of machine learning (ML) has produced remarkable achievements across varied research areas. However, this progress has brought forth a novel challenge known as adversarial attacks. An ML approach called Adversarial Machine Learning (AML) is used to trick models using hostile input, frequently leading to system failures. This technique may be used for different data and model sets as it can be utilized to alter input data, interpret data incorrectly, and cause the system to inaccurately identify data when the application is being implemented. The integrity and performance of ML systems can be seriously jeopardized by adversarial attacks, which include targeted, non-targeted, white-box, black-box, and gray-box attacks, etc. While certain defenses, such as adversarial training, robust optimization frameworks, and adversarial prediction problems, attempt to improve model robustness, real robustness is still difficult to achieve because of large computing costs and possible accuracy trade-offs. Although machine learning's explosive growth has produced many technological advances, it also raises questions about how susceptible it is to hostile attacks. In addition to describing AML concepts, current research, attack methods, and practical applications, this article addresses the security vulnerabilities associated with malevolent attackers influencing and eluding ML systems.

Keywords: Machine Learning · Adversarial machine learning · Adversarial Attacks · Generative Adversarial Networks · Deep Learning

1 Introduction

Recent years have seen a notable increase in the success and expansion of machine learning (ML), which has advanced tasks such as image recognition, machine learning for natural language processing, and self-driving automobile advancements [1, 2]. Attackers can control machine learning models through the use of carefully crafted inputs, a strategy known as adversarial machine learning (AML) [2]. Adversarial attacks can produce alerts that are undetectable by individuals or computers, which makes ML models

A. Bajaj et al. (Eds.): IBICA 2023, LNNS 1229, pp. 74–82, 2025.
https://doi.org/10.1007/978-3-031-78940-3_6

vulnerable to manipulation. A class of AML defenses known as Adversarial Prediction Problems (APPs) seeks to enhance learning systems by altering the nature of training. Assuming an opponent controls the data production process during operations, APPs concentrate on developing adversarially robust prediction models. As a two-agent game, they simulate the interactions between a hypothetical adversary and a predictor during training. The adversary's goal is to maximize their estimated cost under the unknown distribution, while the predictor is to minimize the expected cost under the operations data distribution. The ultimate optimization issues vary depending on the circumstance. Recent studies on the generation of adversarial examples have included novel goals and techniques to improve the process, to minimize disturbances, and accommodate various distance measurements. Table 1 summarises the advantages, and disadvantages of major scientific contributions [3–12] in AML.

Other defenses adapt training techniques to account for the existence of adversaries during operations. Attackers accomplish this by mimicking real, possibly attacked operational data using a variety of heuristics applied to simulated data. Understanding and reducing hostile threats are crucial as ML grows to permeate many industries. To ensure the security and dependability of ML systems, researchers and practitioners have to keep creating strong ML models and defenses. This article [13] highlights existing issues, constraints, and possible future research approaches related to deep learning technology's susceptibility to undetectable adversary perturbations in safety-critical settings.

2 Types of Adversarial Attacks

This work has surveyed attack strategies and practical implications, highlighting the requirement for strong machine-learning models and creative defenses. Adversarial machine learning is a specialized area within machine learning that focuses on analyzing and countering both offensive and defensive attacks on machine learning algorithms. It involves the deliberate creation of deceptive inputs to mislead the target decision models.

This section explores the many kinds of AML attacks [14–17]. It is necessary to understand the attack types for safeguarding the ML models. The attacks can be categorized into targeted, non-targeted, white-box, black-box, gray-box, poisoning, and evasion attacks based on outcome, knowledge, and stage of attacks [17] as shown in Fig. 1.

1) Targeted attacks: An ML model is forced to classify certain instances into an incorrect class.
2) Non-targeted attacks: An ML model is forced to classify certain instances into an incorrect class however, don't specify the misclassified class.
3) White Box Attack: The attacker has complete knowledge of the ML model.
4) Black Box Attacks: The attacker has the least knowledge about the ML model.
5) Grey Box Attacks: The attacker has partial knowledge about the ML model, for instance, the model parameters, architecture, training data, and training strategies.

   This discussion below focuses on stage-based attacks, including poisoning attacks and evasion attacks.
6) *Poisoning Attacks:* Poisoning attacks modify an ML model's training data to degrade its performance. Attackers insert fraudulent data points into a training dataset to

Table 1. Major contributions to the area of adversarial machine learning research.

Authors and references	Research Contribution	Techniques Used	Advantages	Disadvantages
Goodfellow et al. [3]	Introduces GANs (Generative Adversarial Networks)	Deep Learning, Neural Networks	Effective for data generation, image synthesis	Vulnerable to adversarial attacks, training instability
Szegedy et al. [4]	Discovers adversarial examples, small, imperceptible changes to input data causing misclassification	Gradient-based attacks, Deep Neural Networks (DNNs)	Highlights security concerns in ML, better model robustness needed	Requires additional defenses, may affect model performance
Papernot et al. [5]	Proposes adversarial training to improve model robustness against attacks	Adversarial training, Deep Neural Networks, black-box strategy	Enhances model security, mitigates adversarial impact	With increased computational cost, the model may become overly conservative
Madry et al [6]	Develops a robust optimization framework (PGD) for training deep neural networks against adversarial attacks	Projected Gradient Descent (PGD), Deep Neural Networks	Achieves state-of-the-art robustness, formalized adversarial training	Training can be computationally expensive, loss of model accuracy
Carlini & Wagner [7]	Proposes the C&W attack, a versatile and effective white-box adversarial attack	Optimization techniques, targeted attacks	Demonstrates the effectiveness of adversarial attacks	Raises concerns about model vulnerability, necessitates defenses
Athalye et al. [8]	Introduces the "obfuscated gradients" problem, showing that some defenses can be bypassed	Defensive methods analysis, obfuscated gradients	Highlights limitations in adversarial defense techniques	Challenges in achieving true model robustness, need for more robust defenses

(*continued*)

Table 1. *(continued)*

Authors and references	Research Contribution	Techniques Used	Advantages	Disadvantages
Wang et al. [9]	Shown improvement of adversarial robustness requires revisiting misclassified examples	Misclassification Aware Adversarial Training (MART) was proposed to differentiate between misclassified and correctly classified examples	Achieves robustness without sacrificing model accuracy	A computationally intensive, trade-off between robustness and accuracy
Eykholt et al. [10]	Demonstrates real-world adversarial attacks on physical objects on deep learning visual classification	Physical attacks, Transferability of attacks	Raises concerns about physical-world ML safety	Limited real-world impact, challenges in large-scale deployment
Lecuyer et al. [11]	Presents Differential Privacy for Robust Machine Learning to protect against privacy and adversarial attacks	PixelDP is a verified defense that extends to massive networks and datasets, utilizing differential privacy and resilience against adversarial cases	Improved privacy and robustness guarantee	Potential accuracy loss, parameter tuning challenges
Wu et al. [12]	Introduces the concept of "Adversarial Weight Perturbation" for improved adversarial robustness	Adversarial Weight Perturbation (AWP), a double-perturbation mechanism that improves adversarial robustness in training methods	Enhanced adversarial defense, reduced overfitting	Requires fine-tuning, additional computational overhead

add bias to a model's learned decision limits. Poisoning attacks have been shown in spam filtering and recommendation systems [15]. The attack point's progressive movement towards a local maximum is depicted. The fact that the hinge loss is a trustworthy estimate of the classification error is highlighted between the hinge loss and the classification error [15].

7) *Evasion Attacks:* Evasion attacks, often referred to as adversarial perturbations, involve deceiving an ML model by introducing subtle modifications to the input data. The authors [14] introduced a simple yet powerful gradient-based method for assessing the vulnerability of various widely-used classification algorithms to evasion

attempts in a comprehensive manner. The development of reliable communication in the IoT is guided by many important principles. The authors depict multiple scenarios for gradient descent escape strategies [14]. The learned classifier's function g(x) is displayed as a color map, with high values suggesting the malignant class (in red, orange, and yellow) and low values showing the right class (in green, cyan, and blue). The attacker has no assurance that an evasion example generated by the gradient descent algorithm will correctly evade the appropriate classifier. To increase the likelihood of succeeding in evasion, the attacker should choose attack sites in areas with a high density of valid data points, where the estimation of g(x) is more accurate (closer to the true g(x)) and has a negative value.

The attacker's actions stay undetected in these areas, and while these perturbations frequently elude human observation, they can lead to erroneous forecasts. The Fast Gradient Sign Method (FGSM) [2] and Projected Gradient Descent (PGD) [6] are two notable evasion attacks. There is a continual reduction in adversarial case training loss during the training procedure in comparison to the PGD adversary [6]. The loss of cross-entropy on adversarial cases throughout the training phase. It shows how adversarial loss on samples for training evolves when the MNIST and CIFAR10 networks are trained against a PGD adversary. The CIFAR10 curve exhibits sharp declines in the initial learning rate, indicating the possibility of continuously lowering loss to improve a more powerful classifier.

8) *Model Inversion Attacks:* Model inversion attacks are a type of attack that presents a danger in privacy-sensitive sectors by retrieving sensitive data from ML models using carefully selected inputs.

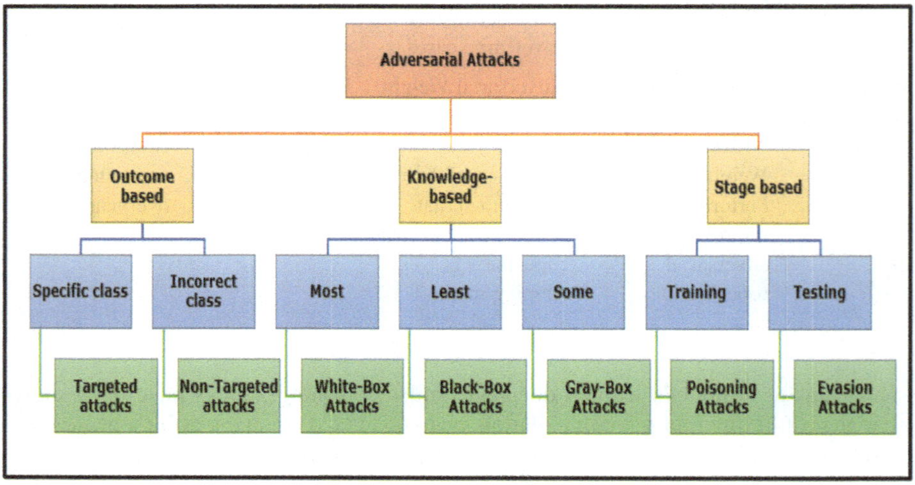

Fig. 1. Adversarial attacks

Adversarial attacks comprise a variety of strategies used to discover imperfections or manipulate ML models, sometimes with malevolent intent. Targeted and non-targeted attacks misclassify data, whereas outcome-based attacks alter a model's output to produce a certain outcome. Using parameters from models and design information, knowledge-based attacks generate adversarial instances. Zhou et al. [16] presented an enhanced model inversion attack methodology, leveraging semantic loss regularization and adversarial example injection to improve reconstruction accuracy in black-box settings. Experimental results demonstrate significant performance gains, underscoring the severity of privacy threats and the need for robust defenses in machine learning models. Attacks using white boxes are more successful than those using black boxes as the former can access internal model data. Attacks using gray-box techniques only comprehend parts of the paradigm. At several points throughout a model's lifespan, attacks take place. Building strong defenses against adversarial manipulation in ML systems requires an understanding of various attack types. The attacker's familiarity with the ML system determines the classification of the attack. White-box, black-box, and gray-box are the three primary varieties of knowledge-based attacks. White-box attacks presuppose that the training set, model architecture, and hyper-parameters of the ML system are all fully understood. Attackers can access the model but not the training process due to black-box attacks, which require little information. While black-box and white-box attacks assume similar access to data distribution and feature representation, gray-box attacks capture adversary knowledge between them. In fields like healthcare, finance, and cybersecurity, these attacks are crucial.

Machine learning algorithms can have either reactive or proactive defenses. Guards are pre-processing measures that identify and minimize adversarial inputs, whereas defenses by design incorporate new models, regularisation approaches, or training procedures. Formal methods are employed to certify the security of models that have been certified, although this is only done for the training data set. These defensive measures can be used at any stage in the processing chain.

3 Real-World Inferences

Adversarial examples can be generated for different applications of Machine Learning and Deep Learning, but each domain has a distinct input format. Initially, strategies were based on the assumption that domains are continuous. However, in practical applications, it is common to encounter discrete input spaces. Limitations on the input data, such as the process of extracting features, can introduce further limitations on the techniques used to carry out adversarial attacks. The process of mapping perturbations from the feature space to the issue space in the opposite direction is complex and requires careful consideration to ensure that the original meaning of the unperturbed sample is maintained. This section addresses the possible implications of actual AML attacks on healthcare, autonomous automobiles, and finance. Figure 2 visually maps how AML threats affect different domains—Healthcare, Autonomous Vehicles, and Finance—by linking each to common attack types like Evasion, Poisoning, and Model Inversion. Each arrow highlights a specific real-world vulnerability, emphasizing the distinct and shared challenges AML poses across critical sectors.

a) Healthcare: Healthcare systems are vulnerable to AML attacks, which can result in inaccurate medical image classification, prognosis, and patient safety. In radiology especially, strong security protocols are essential. It is difficult to strike a compromise

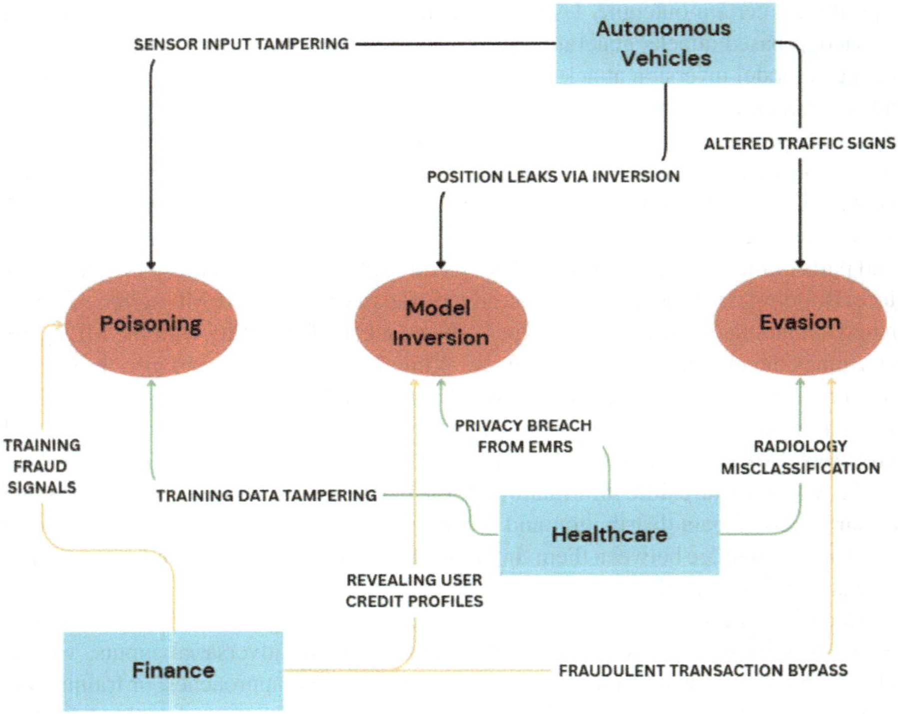

Fig. 2. Mapping AML Vulnerabilities: Healthcare, Autonomous Vehicles, and Finance

between model interpretability and robustness in the AML domain since deep learning in healthcare informatics demands enormous training data sets [18, 19]. The incorporation of machine learning in the health information industry has resulted in the emergence of adversarial attacks, which can manipulate a system's output and consequently generate erroneous findings. The attacks generally target computer scientists, but the conflicting interests in healthcare and the substantial financial stakes in system outputs pose considerable obstacles. Machine-learning algorithms in the US have a direct influence on the approval of insurance claims, which is influenced by conflicting financial interests. The consequences of algorithmic vulnerabilities are significant and have wide applicability, particularly in the fee-for-service economy.

b) Autonomous vehicles: The security and privacy of autonomous cars are seriously threatened by AML attacks, thus thorough knowledge and defense are necessary to guarantee the broad use of this technology [20]. There is a growing use of deep neural networks (DNNs) in the automobile industry, which necessitates further research to develop secure and robust applications. Adversaries can modify initial inputs by introducing small changes that are difficult for humans to detect but can cause a well-trained model to generate inaccurate results. The DNNs are susceptible to adversarial perturbations, which exploit the non-linear nature of DNNs along with the inadequate regularisation strategies commonly employed in most learning tasks. Research on adversarial attacks has devised methods to target image classification models, and various computer vision tasks such as object detection, object tracking, and semantic segmentation. Evasion attacks manipulate test data to intentionally misclassify it, whereas poisoning attacks use vulnerabilities in the perception layer to disrupt its functionality and produce a denial of service. Poisoning attacks in transfer learning are particularly relevant to autonomous vehicles (AVs) since a DNN that has been trained for image classification can be subsequently trained to identify traffic signs. Attackers can simply tamper with adversary attacks for AVs, hence altering the behaviour of the vehicle. The attacker attempts to discern vulnerabilities inside the AV system, including the data obtained from various sensors (such as Cameras, Lidar, Radar, IMU/GPS), 3D printed items designed to exploit weaknesses, and the methods employed for data processing.

c) Finance: Attackers can design fraudulent transactions that elude fraud detection systems, which might cause both parties to suffer substantial financial losses. Adversarial attacks pose a major problem in deep learning, as machine learning models are susceptible to these attacks because of their extensive parameterization. DNN models are employed to identify instances of credit defaults and fraud. However, their susceptibility to adversarial attacks is a consequence of their intricate nature. The deep finance models are susceptible to adversarial attacks in genuine black-box scenarios, which highlights the importance of adversarial learning in enhancing their effectiveness. The attacks manipulate or introduce tokens with minimal monetary values for each token to deceive the model. The defences make the majority of these attacks ineffectual, with the exception of one attack that relies on sampling from a language model. Adversarial attacks and defenses against them are highly significant in financial applications.

Thus, industries including healthcare, autonomous vehicles, and banking require strong machine learning models and creative defenses to fight against AML attacks. To secure the security and dependability of ML systems, this study highlights the necessity of taking preventative measures and underlines how crucial it is to comprehend and mitigate these risks for the technology's ongoing success.

4 Conclusion

The upsurge of machine learning has ushered in a generation of unparalleled technological progress, encompassing its multifaceted aspects. However, this exceptional progress has underscored a crucial concern i.e. the susceptibility of ML methods to adversarial attacks. Researchers and practitioners should continue to be vigilant and prepared to confront the evolving challenges posed by using AML to protect the future of this transformative technology—adversarial attacks are a serious concern. So, in this study, we have explored attack strategies and discussed real-world inferences in critical domains. There is a need for the development of strong ML models and innovative defenses to tackle AML attacks.

References

1. Song, Y., Kim, T., Nowozin, S., Ermon, S., Kushman, N.: Pixeldefend: leveraging generative models to understand and defend against adversarial examples (2017). arXiv preprint arXiv: 1710.10766

2. Goodfellow, I.J., Shlens, J., Szegedy, C.: Explaining and harnessing adversarial examples (2014). arXiv preprint arXiv:1412.6572
3. Goodfellow, I., et al.: Generative adversarial networks. Commun. ACM **63**(11), 139–144 (2020)
4. Szegedy, C., et al.: Intriguing properties of neural networks (2013). arXiv preprint arXiv: 1312.6199
5. Papernot, N., et al.: Practical black-box attacks against machine learning. In: Proceedings of the 2017 ACM on Asia conference on computer and communications security, pp. 506–519 (2017)
6. Madry, A., Makelov, A., Schmidt, L., Tsipras, D., Vladu, A.: Towards deep learning models resistant to adversarial attacks (2017). arXiv preprint arXiv:1706.06083
7. Carlini, N., Wagner, D.: Towards evaluating the robustness of neural networks. In: 2017 IEEE symposium on security and privacy (sp), pp. 39–57. IEEE (2017)
8. Athalye, A., Carlini, N., Wagner, D.: Obfuscated gradients give a false sense of security: Circumventing defenses to adversarial examples. In: International conference on machine learning, pp. 274–283. PMLR (2018)
9. Wang, Y., Zou, D., Yi, J., Bailey, J., Ma, X., Gu, Q.: Improving adversarial robustness requires revisiting misclassified examples. In: International conference on learning representations (2019)
10. Eykholt, K., et al.: Robust physical-world attacks on deep learning visual classification. In: Proceedings of the IEEE conference on computer vision and pattern recognition, pp. 1625–1634 (2018)
11. Lecuyer, M., Atlidakis, V., Geambasu, R., Hsu, D., Jana, S.: On the connection between differential privacy and adversarial robustness in machine learning. Stat **1050**, 9 (2018)
12. Wu, D., Xia, S.T., Wang, Y.: Adversarial weight perturbation helps robust generalization. Adv. Neural. Inf. Process. Syst. **33**, 2958–2969 (2020)
13. Han, S., Lin, C., Shen, C., Wang, Q., Guan, X.: Interpreting Adversarial Examples in Deep Learning: A Review. ACM Computing Surveys (2023)
14. Biggio, B., et al.: Evasion attacks against machine learning at test time. In: Machine Learning and Knowledge Discovery in Databases: European Conference, ECML PKDD 2013, Prague, Czech Republic, September 23–27, 2013, Proceedings, Part III, 13 pp. 387–402. Springer Berlin Heidelberg (2013)
15. Biggio, B., Nelson, B., Laskov, P.: Poisoning attacks against support vector machines (2012). arXiv preprint arXiv:1206.6389
16. Zhou, S., Zhu, T., Ye, D., Yu, X., Zhou, W.: Boosting Model Inversion Attacks with Adversarial Examples. IEEE Transactions on Dependable and Secure Computing (2023)
17. Wu, D., et al.: Adversarial attacks and defenses in physiological computing: a systematic review. National Science Open **2**(1), 20220023 (2023)
18. Dong, J., Chen, J., Xie, X., Lai, J., Chen, H.: Adversarial Attack and Defense for Medical Image Analysis: Methods and Applications (2023). arXiv preprint arXiv:2303.14133
19. Tomsett, R., et al.: Why the failure? how adversarial examples can provide insights for interpretable machine learning. In: 2018 21st International Conference on Information Fusion (FUSION), pp. 838–845. IEEE
20. Amirkhani, A., Karimi, M.P., Banitalebi-Dehkordi, A.: A survey on adversarial attacks and defenses for object detection and their applications in autonomous vehicles. Vis. Comput. **39**(11), 5293–5307 (2023)

Skin-Deep AI: Convolutional Neural Networks for Predicting Dermatological Conditions

Manav Gupta[1], Vaibhav Pushpad[1], Yajnaseni Dash[2(✉)], and Ajith Abraham[3]

[1] School of Computer Science Engineering and Technology, Bennett University, Greater Noida, UP, India
{e23mcag0067,e23mcag0079}@bennett.edu.in

[2] School of Artificial Intelligence, Bennett University, Greater Noida, UP, India
yajnasenidash@gmail.com

[3] School of Artificial Intelligence, Sai University, Chennai, India
abraham.ajith@gmail.com

Abstract. Skin diseases impact over 900 million individuals globally, presenting significant healthcare challenges, especially in underserved regions with limited access to dermatological services. This paper aims to develop Convolutional Neural Networks (CNNs) for the early diagnosis and prediction of common skin diseases, focusing on melanoma, psoriasis, dermatitis, and herpes. The primary goal is to provide an accessible tool to improve public health outcomes in areas with constrained medical services. The methodology involves compiling a comprehensive dataset of various images of diseased skin, covering different conditions. Steps include image preprocessing, augmentation, and feature extraction using CNN models, leading to disease classification. Our approach enhances the proficiency and accuracy of skin disease diagnosis, proving particularly useful where access to medical professionals is restricted. This research demonstrates that fine-tuning CNN architectures with the Adam optimizer significantly improves the accuracy of skin disease detection, offering a valuable tool for real-time diagnosis and severity assessment of skin conditions.

Keywords: Skin disease · Dermatology · Machine learning · Deep learning · Convolutional Neural Networks (CNNs)

1 Introduction

One of the largest and most visible organs in the body, the skin serves as a barrier against heat, injury, and UV rays damage, all of which are vital to maintaining survival. Unfortunately, skin diseases are a common health concern, affecting millions of people globally. The severity of skin disorders varies; some are situational and caused by environmental factors such as pollution, while others may have a genetic basis. As these conditions range in severity from mild to fatal, accurate identification is crucial. Though it's not always curable, treating symptoms is the goal of treatment, The intricacy of skin features, such as texture, tone, and presence of hair, makes it difficult to properly examine

A. Bajaj et al. (Eds.): IBICA 2023, LNNS 1229, pp. 83–90, 2025.
https://doi.org/10.1007/978-3-031-78940-3_7

and analyze skin conditions [1, 2]. Researchers are investigating several methods using machine learning (ML) and deep learning for the diagnosis of skin disorders because of the notable intensification in the occurrence of these conditions among people. A method for diagnosing psoriasis was presented by authors [3]. A technique for intelligently detecting skin illnesses using sensors and video analysis was presented by Połap et al. [4]. At 82.4% training accuracy, their Convolutional Neural Networks (CNNs) architecture demonstrated promising outcomes. The automated eczema detection and severity measuring model developed by [5] addresses the time and expense restrictions related to skin disease diagnosis. This approach may identify eczema-infected areas by using texture and color cues from images. An analogous application of ML classifiers has been made to the detection of chronic renal disease [6], where Naive Bayes has shown the greatest accuracy on a filtered dataset. For the diagnosis of eczema, Abdulbaki et al. [7] suggested a cloud computing-based platform paired with a back-propagation neural network. Janoria et al. [8] presented a transfer learning-based CNN model for the identification of skin cancer. The CNN technique outperforms other ML algorithms in Bhadula et al.'s assessment [9]. For determining skin diseases [10] the diagnosis of renal illness [11], heart condition [12], rice disease of the plants [13], and Alzheimer's disease [14], CNN techniques have been proposed. The objective of this work is to improve the detection of skin diseases by employing various deep-learning techniques [15]. Several studies have looked into the feature and texture selection procedure for identifying the different shapes and colors connected to different skin illnesses. It's interesting to see that most CNN methods now achieve comparable levels of accuracy.

2 Methodology

The specifics of the CNN model and the process for diagnosing skin diseases are covered in this section under the relevant sections. Figure 1 shows the specifics of the operation process.

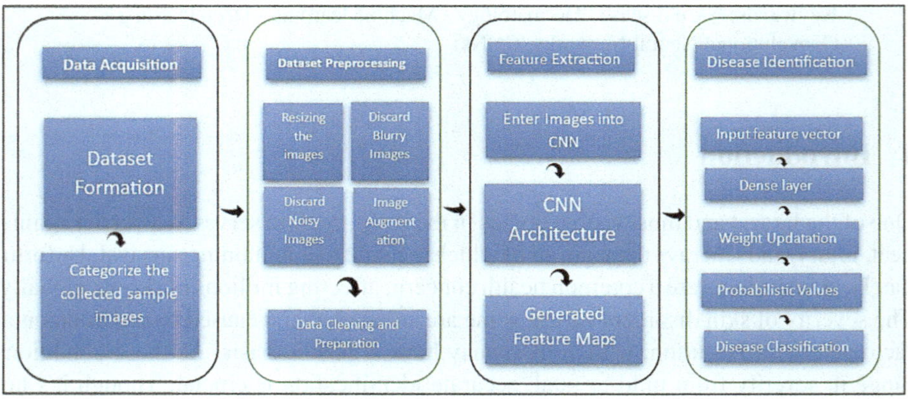

Fig. 1. The process of operation

The process begins with gathering data, and then the gathered images are categorized. After that, these are downsized, and the noisy data is removed. The purpose of image augmentation is to expand the dataset. Following that, CNN models are used to extract the features, and classification is then carried out.

2.1 Dataset Collection

The skin, a vital and prominent organ essential for survival, is unfortunately afflicted by various skin diseases affecting millions of people. These conditions range from minor inconveniences to life-threatening issues. Our research focuses on four prevalent types of skin diseases: Eczema, Psoriasis, Herpes, and Melanoma. To ensure representation across different body parts, we collected sample images for each disease class, resulting in a comprehensive set of images displayed in Fig. 2. A dataset was formed using the Skin Diseases Image Dataset from Kaggle [16], specifically extracting four classes: Eczema, Psoriasis, Atopic Dermatitis, and Melanoma. After data pruning, the final dataset comprised 1805 images.

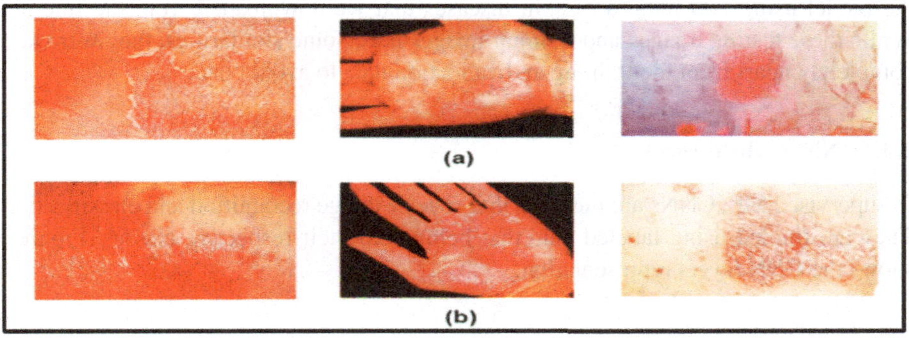

Fig. 2. Sample images of our skin disease dataset Eczema and Psoriasis

Images of diseased skin were sourced from various outlets. Following image collection, we manually curated the dataset by identifying and excluding noisy, low-contrast images. The final dataset comprises 1805 images of diseased skin, including 397 Eczema images, 480 Psoriasis images, 276 Atopic Dermatitis images, and 652 Melanoma images as presented in Table 1.

CNN architectures are used to extract features after a variety of data pre-processing and cleaning procedures. Lastly, CNN's dense layer is used to carry out the categorization task.

2.2 Data Augmentation

To address the issue of overfitting, the incorporation of data augmentation plays a crucial role. This technique allows to significantly enhance the diversity of available data for training models without the need for additional data collection. In our dataset, we applied various data augmentation methods, including a 20% increase in brightness, a 10%

Table 1. Total training and validation images after augmentation

Disease types	Before augmentation	After augmentation	Train data	Validation data
Eczema	397	1985	1588	397
Psoriasis	480	2400	1920	480
Atopic Dermatitis	276	1380	1104	276
Melanoma	652	3260	2608	652
Total	1805	9025	7328	1805

enhancement in contrast, random rotation of 5 degrees Celsius, and horizontal flipping and vertical flipping. This approach aims to improve interaction and account for all potential variations, resulting in the generation of five distinct augmented versions for each sample image. The augmentation process is detailed in Table 1, which illustrates the increased number of dataset images. We used an 80–20 data split for the training and validation sets, yielding 1,805 validation images and 7,328 training images. We tested the model using 200 images of skin diseases that we downloaded from the Internet. While these testing images and training images have some similarities, they were kept completely apart from the training and validation data to avoid bias.

2.3 CNN Architectures

In supervised ML, CNNs are methods that specify image recognition and classification. They are trained using labeled data that belongs to their respective classes. Figure 3 shows the CNN layers representation.

2.3.1 CNN Layers

The input layer displays the input form as (224, 224, 3), signifying a 224*224-pixel image with three RGB color channels.

i) *Layers of convolution:* The first layer of convolution has 16 filters and a kernel size of (3, 3). The padding value is set to 'same,' which adds zero padding to the input to make the output's height and width equal. It makes use of the Rectified Linear Unit (ReLU) activation function.
ii) *Second and third convolutional layers:* 32 filters with a (3,3) kernel size are used in this instance. There is 'same' padding and ReLU activation.

iii) *MaxPooling layers:* MaxPooling is applied with a pool size of (2, 2) after each convolutional layer, reducing the input's spatial dimensions (height and breadth) by a factor of 2.
iv) *Flatten layer:* Before sending the input to the fully connected layers, it is flattened to a one-dimensional array.
v) *Fully linked (Dense) layers:* ReLU activation functions and two fully linked layers with 1024 neurons are introduced. The output layer, a fourth Dense layer with a

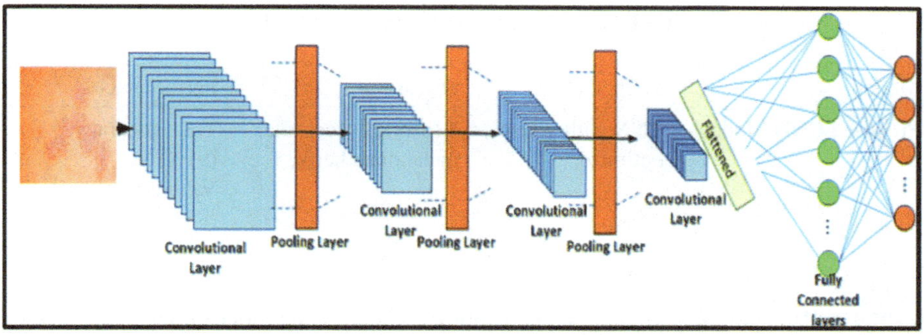

Fig. 3. CNN implementation design

softmax activation function is added. Softmax makes sure that the total of these probabilities equals one. The output layer shows the likelihood that the input image belongs to each of the four classes.

2.4 Experimental Setup

The system used for the experiment has an Intel Core i5 processor, 8GB of RAM, and Nvidia GTX-1060 graphics. The models were trained collaboratively using the Keras framework and TensorFlow backend to do the classification. For training and validation, the dataset has been split 80/20, respectively. To test the models, 200 images have been taken into consideration. Throughout the training and validation phase, the cross-validation method with 20 epochs has been taken into consideration. The weights have been updated using the Adam optimizer technique. We have employed the categorical_crossentropy loss function because our dataset only includes four classes. The hyperparameters used in our experiment are input image size (224 × 224), epoch (20), activation (ReLu, Softmax), dropout function, optimizer (Adam), and metrics (accuracy).

3 Result and Discussion

First, the input images are processed using CNN architectures, which are made up of several layers for pooling and convolution. These layers create feature vectors from the inputs. A fine-tuning strategy was used for the categorization procedure. It is required that every layer in the CNN structures be frozen, except the top dense layer. The weights were used in the final dense layer and modified using the Adam optimizer.

Numerous scientists have proposed several methods for differentiating between different skin disorders. Table 2 presents our proposed results. The following findings were obtained using experimental analysis of a dataset about skin diseases.

1) On a particular dataset, fine-tuning dramatically improves any CNN model's accuracy compared to the transfer learning approach.
2) The Adam optimizer performed faster in terms of execution time than the RMSprop optimizer, which required more time to complete the classification task.

Table 2. Assessment of the proposed model

Model	Method used	Disease	Performance
Proposed CNN Model	CNN with Adam optimizer	4 (psoriasis, eczema, melanoma, herpes)	Train accuracy (91.11%) Validation accuracy (80.93%) Test Accuracy (81.25%)

A CNN model, optimized with the Adam optimizer, was proposed for classifying four skin diseases: psoriasis, eczema, melanoma, and herpes. The model achieved a training accuracy of 91.11%, indicating strong learning capability. Its validation and test accuracies were 80.93% and 80.25%, respectively, demonstrating reasonable generalizability and potential for accurate skin disease classification, with scope for further improvement.

3) The Adam and RMSprop optimizers both performed similarly in terms of accuracy, indicating that either optimizer can be used for tasks involving classification.
4) When updating weights, a learning rate of 0.001 with a momentum of 0.9 worked better. However, the performance of learning rates of 0.01 and 0.0001 was comparable.

Medical science will advance in a larger sense with the aid of computational support and disease recognition. To identify four common skin diseases (psoriasis, melanoma, herpes, and eczema) a deep learning-based method was proposed for skin disease detection in this paper using CNN techniques. The CNN architecture was trained using the fine-tuning technique and higher accuracy was obtained in the proposed method. This useful application can be put to good use in the diagnosis and severity assessment of skin conditions in real-time.

4 Conclusion

In this study, we developed a Convolutional Neural Network (CNN) model aimed at the early diagnosis and prediction of common skin diseases, focusing on melanoma, psoriasis, dermatitis, and herpes. Specifically, the proposed model achieved a training accuracy of 91.11%, a validation accuracy of 80.93%, and a testing accuracy of 80.25%. The motivation behind this research is to provide an accessible diagnostic tool that can significantly enhance public health outcomes, particularly in regions with limited access to dermatological services. Our methodology involved compiling a comprehensive dataset of skin disease images, followed by meticulous image preprocessing, augmentation, and feature extraction using CNN models. The final classification task was performed using a dense layer in the CNN architecture. Data augmentation played a critical role in expanding the dataset and addressing overfitting, while our CNN architecture, optimized with the Adam optimizer, achieved notable performance metrics.

The experimental results demonstrated that fine-tuning the CNN model significantly improved accuracy compared to transfer learning approaches. Specifically, the proposed model achieved a training accuracy of 91.11% and a validation accuracy of 80.93%. The Adam optimizer was found to be more efficient in terms of execution time compared to the RMSprop optimizer, although both optimizers yielded similar accuracy levels. The current findings suggest that deep learning techniques, particularly CNNs, hold great promise for the automated diagnosis of skin diseases. The proposed method provides a robust framework for accurately identifying and classifying skin conditions, which could be highly beneficial in clinical settings, especially in underserved regions. This research contributes to the advancement of medical science by leveraging computational support for disease recognition, ultimately aiming to improve diagnostic accuracy and patient outcomes in dermatology. Future work will focus on expanding the dataset, exploring additional deep-learning architectures, and integrating this diagnostic tool into real-world clinical applications. By continuing to refine and validate our approach, we aim to enhance its practical utility and reliability for widespread adoption in healthcare systems.

References

1. Ruzicka, T., Ring, J., Przybilla, B. (eds.): Handbook of atopic eczema. Springer Science and Business Media (2013)
2. Lowes, M.A., Suárez-Fariñas, M., Krueger, J.G.: Immunology of psoriasis. Annu. Rev. Immunol. **32**, 227–255 (2014). https://doi.org/10.1146/annurev-immunol-032713-120225
3. Shrivastava, V.K., Londhe, N.D., Sonawane, R.S., Suri, J.S.: Computer-aided diagnosis of psoriasis skin images with HOS, texture and color features: a first comparative study of its kind. Comput. Methods Programs Biomed. **126**, 98–109 (2016)
4. Połap, D., Winnicka, A., Serwata, K., Kęsik, K., Woźniak, M.: An intelligent system for monitoring skin diseases. Sensors **18**(8), 2552 (2018)
5. Alam, M.N., Munia, T.T.K., Tavakolian, K., Vasefi, F., MacKinnon, N., Fazel-Rezai, R.: Automatic detection and severity measurement of eczema using image processing. In: 2016 38th Annual International Conference of the IEEE Engineering in Medicine and Biology Society (EMBC) (pp. 1365–1368). (2016) IEEE
6. Khan, M., Prottasha, M.S.I., Nasim, T., Mehedi, A., Pranto, M.A.M., Asad, N.: Performance analysis of various machine learning classifiers on reduced chronic kidney disease dataset. Int. J. Recent Res. Rev. **13**, 16–21 (2020)
7. Shawkat Abdulbaki, A., Najim, S.A.M., Abdulrasoul Khadim, S.: Eczema disease detection and recognition in cloud computing. Res. J. Appl. Sci. **13**, 182–188 (2019)
8. Janoria, H., Minj, J., Patre, P.: Classification of skin disease from skin images using transfer learning technique. In: 2020 4th International Conference on Electronics, Communication and Aerospace Technology (ICECA) IEEE pp. 888–895. (2020)
9. Bhadula, S., Sharma, S., Juyal, P., Kulshrestha, C.: Machine learning algorithms-based skin disease detection. Int. J. Innovative Technol. Exploring Eng. (IJITEE) **9**(2), 4044–4049 (2019)
10. Abbas, M., Imran, M., Majid, A., Ahmad, N.: Skin diseases diagnosis system based on machine learning. J. Comput. Biomed. Informatics **4**(01), 37–53 (2022)
11. Bhaskar, N., Suchetha, M.: A computationally efficient correlational neural network for automated prediction of chronic kidney disease. IRBM **42**(4), 268–276 (2021)
12. Shankar, V., Kumar, V., Devagade, U., Karanth, V., Rohitaksha, K.: Heart disease prediction using CNN algorithm. SN Comput. Sci. **1**(3), 170 (2020)
13. Prottasha, M.S.I., Hossain, A.B.M., Rahman, M.Z., Reza, S.M., Hossain, D.A.: Identification of various rice plant diseases using an optimized convolutional neural network. Int. J. Comput. Digit. Syst. (2021)

14. Farooq, A., Anwar, S., Awais, M., Rehman, S.: A deep CNN-based multi-class classification of Alzheimer's disease using MRI. In: 2017 IEEE International Conference on Imaging Systems and Techniques (IST) IEEE, pp. 1–6 (2017)
15. Dash, Y., et al.: Insights into dermatological disorders: Understanding skin diseases through medical image analysis. Int. J. Comput. Inf. Syst. Ind. Manage. Appl. **16**(3), 10 (2024)
16. https://www.kaggle.com/datasets/ismailpromus/skin-diseases-image-dataset/data

A Deep Dive into Southern India's Rainfall: LSTM Perspectives

Megha Dhaka[1], Jeevika Rajput[1], Yajnaseni Dash[2(✉)], and Ajith Abraham[3]

[1] School of Computer Science Engineering and Technology, Bennett University, Greater Noida, UP, India
simmydhaka06@gmail.com, jeevikarajput0003@gmail.com
[2] School of Artificial Intelligence, Bennett University, Greater Noida, UP, India
yajnasenidash@gmail.com
[3] School of Artificial Intelligence, Sai University, Chennai, India
abraham.ajith@gmail.com

Abstract. This study investigates rainfall patterns in the subdivisions of Southern India from 1901 to 2015, focusing on Kerala, Telangana, Tamil Nadu, coastal Karnataka, and Andhra Pradesh. Using deep Long Short-Term Memory (LSTM) neural networks, we analyzed historical climate data to uncover complex patterns and variations in rainfall. The results provide insights into the intricacies of rainfall distribution in these regions, highlighting both consistent and fluctuating trends. Coastal Karnataka and Kerala exhibit higher and more consistent rainfall, while Tamil Nadu and Telangana show greater variability. The study employs advanced statistical methods and data visualization tools aiding in the development of informed strategies for agriculture, hydrology, and environmental management. These findings are crucial for policymakers and researchers in crafting sustainable water management practices and resilient agricultural strategies amidst climate change. The current study emphasizes the need for adaptive techniques to address specific climate conditions and adaptive strategies to achieve resilient growth in the aspect of a changing climate.

Keywords: Machine Learning · Rainfall · South India · Predictive analytics · Regression · LSTM

1 Introduction

The objective of this study is to make a comprehensive study of the rainfall pattern in Southern India subdivisions from 1901 to 2015. These areas (Kerala, Telangana, Tamil Nadu, coastal Karnataka, and coastal Andhra Pradesh) are very important in the Indian subcontinent and have varied climates. Understanding the historical pattern of rainfall in these areas is important for diverse sectors such as agriculture, hydrology, and environmental management. Using historical climatic data, this study attempts to predict long-term changes in precipitation patterns in several subdivisions of South India. Using advanced statistical modeling, data visualization tools, and large datasets, this

A. Bajaj et al. (Eds.): IBICA 2023, LNNS 1229, pp. 91–97, 2025.
https://doi.org/10.1007/978-3-031-78940-3_8

paper aims to identify complex patterns, anomalies, and rainfall variability. The findings are important for policymakers, environmental researchers, and agricultural planners to develop informed strategies and approaches for adaptation in response to a changing climate, and to support resource management well and sustainable development in South Indian subdivisions. This study focuses on the analysis of past rainfall in southern India, a region heavily dependent on agriculture. Variations in rainfall significantly affect crop production, soil moisture, and irrigation. The objective is to identify trends and anomalies to help farmers and policymakers develop more efficient water management strategies. Sustainable water management and resilience in agricultural practices depend on understanding long-term rainfall patterns, especially in the aspect of climate change Water allocation in residential, commercial, and agricultural areas, rainfall variability directly impacts. The data obtained is useful for many projects and environmental protection measures.

Several studies contribute to the understanding of rainfall patterns in South India. The dynamics of the Indian summer monsoon were presented by Goswami and Chakravorty [1]. Authors [2] explored the teleconnections and variability of Northeast monsoon rainfall in South Peninsular India. Indian summer monsoon variability has been discussed [3]. The article [4] presented India's flood risk and the impact of climate change on heavy rainfall events. Authors [5] have shown the climate variability in the Southern Indian Ocean using self-organizing maps. Climate variability and the circulation of the Indian Ocean are crucial factors that significantly influence global weather patterns [6]. A satellite-based study [7] has enhanced the rainfall analysis using remote sensing. The study utilizes NCEP CFSv2 to analyze the seasonal forecasting of Indian summer monsoon rainfall, focusing on forecast and predictability error [8]. The study [9] investigates the impact of climate change, monsoon irregularities, and the resulting disparities in South India. Authors [10] examined the impact of urbanization on heavy rainfall climatology during the Indian summer monsoon, revealing a rising trend of heavy rainfall over non-urban areas. Dash et al. [11–21] have used artificial intelligence and machine learning to estimate precipitation, providing a basis for predictive models.

2 Data and Method

The present study has selected a dataset from Kaggle [22] to study rainfall patterns in the southern region of India. We have extracted and processed the raw dataset for the specified subdivisions. The processed dataset can be made available upon request. The dataset provides detailed information on rainfall amounts for different seasons and locations, aiding researchers in understanding climate, water availability, agricultural development, and its impact on South India. The method applied here is Long Short-Term Memory (LSTM). The LSTM technique is used for sequential data processing in time series prediction. It has a single unit that is specific to regression tasks. The layers of the model are stacked linearly in a sequential model's construction. We have employed the Adam optimizer, and root mean squared error is used to assess the model's performance. Training entails using a batch (size = 32) process to apply the fit algorithm across 100 epochs to the given training data (train_X and train_Y).

3 Result and Analytical Evaluations

This study examines rainfall patterns from 1901 to 2015, focused on the subdivisions of Southern India in detail. It assesses the historical climate data to find complex patterns and variations in rainfall of Kerala, Telangana, Tamil Nadu, coastal Karnataka, and Andhra Pradesh using deep LSTM neural network technique. The outcomes of the study are expected to provide new insights into the complexity of rainfall patterns in Southern Indian subdivisions, and imperative information can be extracted for rational decisions in various sectors by using advanced statistical methods and data visualization software. Understanding how rainfall patterns vary across the aforesaid regions is important for making intelligent decisions about agriculture, infrastructure development, and water resource management. Identification of excess rainfall whether flood or drought is necessary to develop a flexible framework for the analysis of potential climate-related problems. The region-wise RMSE scores of different regions of India provide important insights for various patterns.

3.1 Estimating the Average Quantity of Rainfall

The study compares annual precipitation in different subdivisions compared to the national average. Andhra Pradesh has 1052.90 mm of rainfall, Telangana has 953.58 mm of rainfall, and Tamil Nadu has 943.71 mm of rainfall in an annual average estimate. However, Karnataka has 3408.40 mm, and Kerala has an above-average rainfall of 2922.65 mm, indicating a wetter climate. Coastal Karnataka consistently exhibits the highest annual rainfall among the selected subdivisions throughout the years, displaying a wider range of fluctuation compared to other regions. It's followed by Kerala, which also demonstrates relatively high and consistent rainfall volumes over the analyzed period. Conversely, Tamil Nadu and Telangana have lower annual rainfall as compared to Coastal Karnataka and Kerala. Over the years, fluctuations in rainfall patterns are noticeable across all subdivisions, emphasizing the variability in rainfall distribution. Coastal Karnataka and Kerala appear to experience more consistent and higher rainfall, while Tamil Nadu, and Telangana, show more variability and occasional lower rainfall periods.

These comparisons highlight the diverse rainfall patterns across different subdivisions in India, with some areas experiencing notably lower rainfall than the national average while others exhibit significantly higher rainfall.

4 Rainfall Metrics Across Subdivisions

Table 1 presents a comparison study and a description of the rainfall (mm) data for the selected subdivisions. The rainfall statistics for various subdivisions in South India provide a comprehensive overview of the regional precipitation patterns. Coastal Karnataka records the highest mean rainfall at 3408.4 mm, with a median of 3360.7 mm, a maximum of 5553.9 mm, and a minimum of 2510.9 mm, indicating significant rainfall levels compared to other regions. In contrast, Tamil Nadu exhibits the lowest mean rainfall at 943.7 mm, with a median of 944.5 mm, a maximum of 1365.3 mm, and a minimum

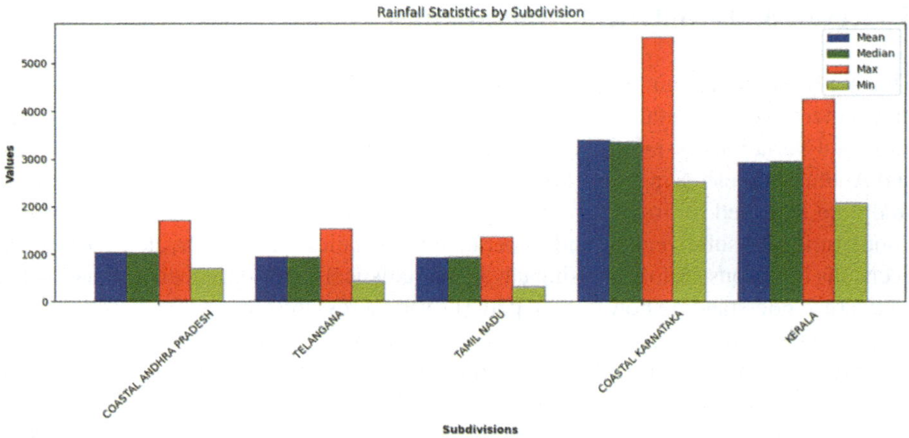

Fig. 1. Visualization of rainfall statistics

of 318.0 mm, reflecting its relatively drier climate. Coastal Andhra Pradesh and Telangana also experience considerable rainfall, with means of 1052.9 mm and 953.3 mm, respectively, and maximum rainfall values reaching 1712.4 mm and 1544.9 mm. Kerala, known for its high rainfall, has a mean of 2922.6 mm, a median of 2934.3 mm, a maximum of 4257.8 mm, and a minimum of 2068.8 mm. These statistics highlight the significant variation in rainfall across the subdivisions, with Coastal Karnataka receiving the highest and Tamil Nadu the lowest rainfall.

Table 1. Rainfall Statistics by South India Subdivisions

Subdivision	Mean	Median	Max	Min
Coastal Andhra Pradesh	1052.9	1026.5	1712.4	703.2
Telangana	953.3	942.1	1544.9	437.0
Tamil Nadu	943.7	944.5	1365.3	318.0
Coastal Karnataka	3408.4	3360.7	5553.9	2510.9
Kerala	2922.6	2934.3	4257.8	2068.8

The mean, median, maximum, and minimum rainfall (mm) as shown in Fig. 1 for each measured region is also calculated and presented in Table 2.

5 RMSE Scores Using LSTM

The actual and forecast rainfall for each of the states in South India is displayed with the Root Mean Squared Error (RMSE) scores for the individual states in Table 3. The RMSE values indicate the model's accuracy in predicting rainfall, with coastal Andhra Pradesh achieving the lowest RMSE.

Table 2. Rainfall statistics

Statistics	Highest	Lowest
Mean	Coastal Karnataka (3408.41)	Tamil Nadu (943.7)
Median	Coastal Karnataka (3360.7)	Telangana (942.1)
Maximum	Coastal Karnataka (5553.9)	Tamil Nadu (1365.3)
Minimum	Coastal Karnataka (2510.9)	Tamil Nadu (318.0)

Table 3. RMSE values of South India Subdivisions using LSTM

Subdivisions	RMSE
Coastal Andhra Pradesh	0.25
Telangana	0.28
Tamil Nadu	0.27
Coastal Karnataka	0.33
Kerala	0.31

The current study evaluates precipitation trends in Southern India from 1901 to 2015 using a sophisticated deep LSTM neural network approach, with a focus on assessing model accuracy through RMSE values. Table 3 presents the RMSE scores for different Southern Indian subdivisions, demonstrating the model's precision in predicting rainfall. Coastal Andhra Pradesh achieved the lowest RMSE (0.25), indicating highly accurate rainfall forecasts closely matching actual measurements. Telangana and Tamil Nadu followed with RMSE values of 0.28 and 0.27, respectively, showcasing minimal discrepancies between forecasted and actual rainfall. Coastal Karnataka and Kerala exhibited slightly higher RMSE values of 0.33 and 0.31, suggesting moderate accuracy in wetter climates. These RMSE values highlight the model's capability to understand and simulate rainfall patterns, providing valuable insights for decision-making across various sectors. This comprehensive analysis underscores the effectiveness of the LSTM model in capturing the complex rainfall dynamics in Southern India, utilizing advanced statistical techniques and data visualization tools to inform policy and planning.

6 Conclusion

The study employed a sophisticated Long Short-Term Memory (LSTM) neural network methodology to examine rainfall patterns in Southern India, uncovering notable disparities in yearly precipitation. The coastal regions of Karnataka and Kerala received the highest and most consistent precipitation, whilst Tamil Nadu and Telangana saw lower and more erratic rainfall patterns. The LSTM approach demonstrated efficacy in rainfall forecasting, with low RMSE scores, so confirming its suitability for climatic research. An in-depth comprehension of past precipitation patterns is essential for the fields of agriculture, hydrology, and environmental management. This knowledge assists policy-makers in formulating sustainable development policies.

References

1. Goswami, B.N., Chakravorty, S.: Dynamics of the Indian summer monsoon climate. In: Oxford Research Encyclopedia of Climate Science (2017)
2. Sreekala, P.P., Rao, S.V.B., Rajeevan, M.: Northeast monsoon rainfall variability over south peninsular India and its teleconnections. Theor. Appl. Climatol. **108**, 73–83 (2012)
3. Sahastrabuddhe, R., Ghausi, S.A., Joseph, J., Ghosh, S.: Indian Summer Monsoon Rainfall in a changing climate: a review. J. Water Clim. Change **14**(4), 1061–1088 (2023)
4. Guhathakurta, P., Sreejith, O.P., Menon, P.A.: Impact of climate change on extreme rainfall events and flood risk in India. J. Earth Syst. Sci. **120**, 359–373 (2011)
5. Morioka, Y., Tozuka, T., Yamagata, T.: Climate variability in the southern Indian Ocean as revealed by self-organizing maps. Clim. Dyn. **35**, 1059–1072 (2010)
6. Schott, F.A., Xie, S.P., McCreary, J.P., Jr.: Indian Ocean circulation and climate variability. Rev. Geophys. (2009). https://doi.org/10.1029/2007RG000245
7. Prakash, S., Sathiyamoorthy, V., Mahesh, C., Gairola, R.M.: An evaluation of high-resolution multi-satellite rainfall products over the Indian monsoon region. Int. J. Remote Sens. **35**(9), 3018–3035 (2014)
8. Pokhrel, S., et al.: Seasonal prediction of Indian summer monsoon rainfall in NCEP CFSv2: forecast and predictability error. Clim. Dyn. **46**, 2305–2326 (2016)
9. Dagdeviren, H., Elangovan, A., Parimalavalli, R.: Climate change, monsoon failures and inequality of impacts in South India. J. Environ. Manag. **299**, 113555 (2021)
10. Kishtawal, C.M., et al.: Urbanization signature in the observed heavy rainfall climatology over India. Int. J. Climatol. **30**(13), 1908–1916 (2010)
11. Dash, Y., Mishra, S.K., Sahany, S., Panigrahi, B.K.: Indian summer monsoon rainfall prediction: a comparison of iterative and non-iterative approaches. Appl. Soft Comput. J. ASOC-4445, **70**, 1122–1134 (2018)
12. Dash, Y., Mishra, S.K., Panigrahi, B.K.: Rainfall prediction for Kerala State of India using artificial intelligence approaches. Comput. Electr. Eng. **70**, 66–73 (2018)
13. Dash, Y., Mishra, S.K., Panigrahi, B.K.: Predictability assessment of northeast Monsoon Rainfall in India using Sea surface temperature anomaly through statistical and machine learning techniques. Environmetrics (2018). https://doi.org/10.1002/env.2533
14. Dash, Y., Mishra, S.K., Panigrahi, B.K.: Neural Network-Based Approaches for Prediction of the Indian Summer Monsoon Rainfall. In: 2nd ICICICT, Kannur, Kerala, India, IEEE, pp. 550–554 (2019)
15. Dash, Y., Mishra, S.K., Panigrahi, B.K.: Prediction of southwest monsoon rainfall for Kerala, India using ε-SVR model. In: 2nd ICICICT, Kannur, Kerala, India, IEEE, pp. 746–748 (2019)
16. Dash, Y., Mishra, S.K., Panigrahi, B.K.: Rainfall prediction of a Maritime State (Kerala), India using SLFN and ELM techniques. In: ICICICT, Kannur, India, IEEE, pp. 1714–1718 (2017)
17. Dash, Y., Mishra, S.K., Panigrahi, B.K.: NEMR predictability assessment over Indian peninsula using ELM. CI 2017, NCAR Technical Note NCAR/TN-536+PROC, pp. 77–80 (2017)
18. Dash, Y., Mishra, S.K., Panigrahi, B.K.: Indian summer monsoon prediction by deep long short-term memory neural network using time series based approach. AGU Fall Meeting Abstracts, A33M-2968 (2019)

19. Dash, Y., Abraham, A.: Integrating empirical analysis and deep learning for accurate monsoon prediction in Kerala, India. Appl. Comput. Geosci. 14 (2024). https://doi.org/10.1016/j.acags.2024.100211
20. Dash, Y.: The Role of Machine Learning for Indian Monsoon Prediction. In: Seventh WMO International Workshop on Monsoons (IWM-7), abstract ID-84, p. 163 (2022)
21. Dash, Y., Kumar, N., Raj, M., Abraham, A.: Deep learning approach for predicting monsoon dynamics of regional climate zones of India. Appl. Comput. Geosci. **23**, 100176 (2024)
22. Kaggle datasets: https://www.kaggle.com/datasets/kkhandekar/statewise-rainfall-1901-to-2015/data

Pre-Owned Automotive Price Prediction Using Machine Learning Technique

K. Logeswaran[1(✉)], S. Savitha[2], P. Suresh[3], K. R. Prasanna Kumar[4], M. Gunasekar[5], M. Vasugi[1], J. Ruthranayaki[1], T. Akilesh[1], and C. Harish[1]

[1] Department of Artificial Intelligence and Data Science, Kongu Engineering College, Erode, Tamilnadu 638060, India
klogesbtech@gmail.com

[2] Department of Computer Science and Engineering, K.S.R College of Engineering College, Tiruchengode, Tamilnadu 637215, India

[3] Department of Database Systems, School of Computer Science and Engineering, Vellore Institute of Technology, Vellore, Tamilnadu 632014, India

[4] Department of Computer Science and Design, Kongu Engineering College, Erode 638060, India

[5] School of Computer Science and Engineering, Vellore Institute of Technology, Vellore 632014, Tamil Nadu, India

Abstract. A car's price estimate has drawn a lot of attention because it calls for a lot of work and the expert's understanding in the topic. For the dependable and accurate prediction, a large number of unique attributes are considered. We employ the Ridge, Lasso, and Random Forest Regressor in Linear Regression, Gradient Boosting Regressor to develop a model for predicting used automobile prices. However, the aforementioned methods were used in concert. Before beginning the process of constructing the model, in order to assist people, comprehend the dataset, this project displayed the data. This data collection was partitioned and altered to suit the regression in order to guarantee the model's performance. Each regression's performance was assessed using R-square. The final model has a higher forecast accuracy and more components related to used cars than the earlier research. Despite the decline in the market for new cars, the used car market has kept growing. Because used automobiles are more affordable and may be sold again after a few years of use for a profit, the majority of people choose to purchase them. The price of used automobiles is influenced by factors including fuel type, colour, model, mileage, engine, gearbox, number of seats, etc. The cost of used cars on the market will continue to fluctuate. As a result, to predict the cost of used cars, an evaluation model is required.

Keywords: Car price prediction · Random Forest Regressor · Lasso Regressor · Linear Regression · Gradient Boosting Regressor

1 Introduction

Predicting car prices is an intriguing and common problem. Due to the fact that the cost of a car often depends on a number of distinctive characteristics and factors, accurate automobile price prediction requires specialist expertise [1]. The most significant ones

A. Bajaj et al. (Eds.): IBICA 2023, LNNS 1229, pp. 98–106, 2025.
https://doi.org/10.1007/978-3-031-78940-3_9

are often brand and model, age, horsepower, and miles [2]. Due to frequent changes in gasoline prices, the price of a car is significantly influenced by the kind of gasoline used in the vehicle and fuel consumption per mile. A car's price will also be determined by a variety of factors, such as the exterior colour., door count, transmission type, dimensions, safety, air conditioning, interior, and whether or not it includes navigation.

In this study, we explored various approaches and methodologies to improve the used car price prediction's accuracy [3]. A number of factors will be taken into account when estimating the car's price. We can predict the exact cost of a car rather than merely its price range thanks to regression algorithms since they provide us a continuous number rather than a segmented value as an output. A user interface that displays a car's cost based on input from any user and takes input from any user has also been developed. Here, there are three different fuel data set types. Here, diesel, gasoline, and LPG are all used. We employed the Regressor for Random Forest, Ridge, Lasso, and Linear Regression, Gradient Boosting Regressor to create a model for estimating the cost of second-hand cars.

2 Literature Survey

[4] predicts the price of second-hand car is for Linear Regression. Based on the values of another variable, linear regression analysis can be used to predict the value of a variable. You want to be able to predict the dependent variable. You generate a prediction about the value of the other variables using the independent variable. With the help of one or more independent variables that can most accurately predict the value of the dependent variable, this type of analysis calculates the coefficients of the linear equation. The differences between expected and actual output values are minimised by linear regression by fitting a straight line or surface. The best-fit line for a set of paired data may be found using straightforward linear regression calculators that employ the "least squares" technique. Next, you determine X's value from Y's (an independent variable).

Lasso Regression algorithm is used in [5] to predict the price of used car. A regularisation method is lasso regression. Regression methods are preferred for a more precise forecast. This model makes advantage of shrinkage. When data values shrink toward the mean, this is referred to as shrinkage. Simple, sparse models are encouraged by the lasso approach (i.e., models with fewer parameters).

Gradient boost regression algorithm is used in [6] to forecast price of second-hand car. An analytical method called gradient boosting regression is intended to investigate the link between two or more variables (X, and Y). The results of the analysis show which significant variables (Xi) have an effect on the dependent variable (y) and what kind of connection exists between each of these variables and the dependent variable.

3 Methodology

3.1 Dataset Description

For our project the data that used was "car_data" which was loaded from Kaggle. It completely contains the information about details of the used cars. The data set contain 9 columns and 301 rows respectively. The column name that the data set contains were "car_Name", "year", "selling_price", "present_price', "Kms_Driven", "Fuel_Type", "Seller_type", "Transmission", 'owner" (Fig. 1).

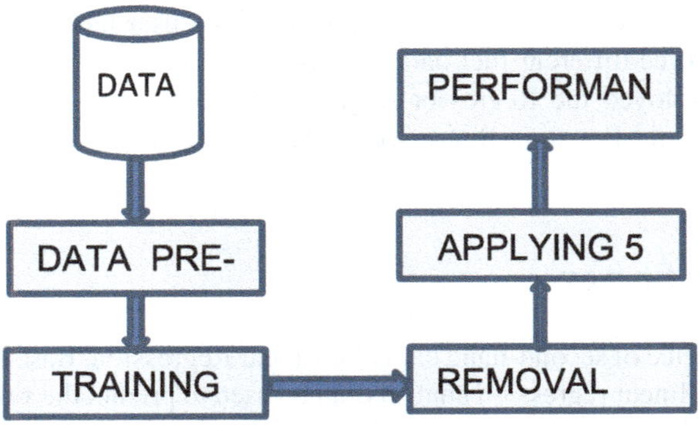

Fig. 1. Methodology Model

- Pre-processing the data

  Encoding for labels. There are predictors in the dataset. They are a numerical variable and also categorical. Machine learning models must be used; hence we require a numerical representation of the features. Consequently, all non-numeric properties were converted to their numerical equivalents. In our dataset during preprocessing with the help of the year column age of the car is calculated with the formula 2022 – year, then with the help of pivot table other non-numerical columns like Fuel_Type, Transmission_Type, seller_type, are changed to numerical values. Then some columns are renamed as Selling_Price to Seling_Price(lacs), Present_Price to 'Present_Price(lacs) and owner to Past_Owners. Finally, the Car_Name contain different car model names we cannot change it to numerical data so drop it. And finally, the table is converted into all numerical valued columns.

  - Feature extraction
  - Machine leaning models

- Linear Regression

A model for estimating the cost of a second-hand car that would be created using artificial neural networks. In contrast to earlier models that used conventional linear regression techniques, the new model was created to deal with nonlinear data interactions. Compared to other linear models, the non-linear model proved more accurate in predicting car prices. Results from their projection model were comparable to those from the basic regression model. Because of the high demand for car dealers to deliver the cars at the conclusion of the lease year, they have created a specialised programme called ODAV (Optimal Distribution Of Auction Vehicles).

- Lasso Regression

When we develop our statistical models, both overfitting and underfitting are taken into consideration. The models might not perform well on the test data set because they are overly biased toward the training set of data. Overfitting is the term for this. We first choose the subset of attributes that results in the best/least sum of squared error when predicting the price using Lasso regression on the training data set. The best subset of attributes is "lassoed" using 10-fold cross-validation. L1 regularisation is employed.

- Random Forest

Built on decision trees, the random forest modelling approach is used for behaviour analysis and modelling predictions. It has several decision trees, each of which represents a different instance of how the categorization of data entered into the random forest is done. The forecast chosen by the random forest approach is the one that receives the most votes after taking into account each case separately.

- Ridge Regression

In instances when linearly independent variables have a high degree of correlation, ridge regression is a technique for predicting the coefficients of multiple-regression models. It has been used to a wide range of disciplines, including engineering, chemistry, and econometrics. Ridge regression is a type of linear regression that includes a little amount of bias to improve long-term predictions. Ridge regression is a method of model normalisation that simplifies the model. Another term for it is L2 normalisation. By inserting a penalty term, this technique improves the cost function. The level of bias induced into the model is evaluated by the Ridge Regression penalty. By splitting the squared weight of each unique label by lambda, we may determine it.

- Gradient Boost Regression

An analytical method called gradient boosting regression is intended to investigate the link between two or more variables (X, and Y). The results of the analysis show which significant factors (X I have an effect on the dependent variable (y) and what kind of connection exists between each of these factors and the dependent variable. In order to discover the optimum tree model, the Gradient Boosting approach is specifically implemented by XG Boost, which makes use of more precise approximations. It uses a variety of clever techniques that greatly enhance its success, especially with structured data. Additional benefits of XG Boost include its quick training speed and ability to disperse training across clusters. As a result, XG Boost was another model used in this research. Table displays XG Boost's performance.

4 Performance Evaluation

The performance of machine algorithm applied on the car dataset was measured using metrics namely R Squared and Cross Validation Score mean.

- LINEAR REGRESSION:

  Train R^{2-}score: 0.88.
  Test R^2-score: 0.86.
  Train CV scores: [0.89912189 0.89838688 0.82980069 0.81769076 0.76113925].
  Train CV mean: 0.84.

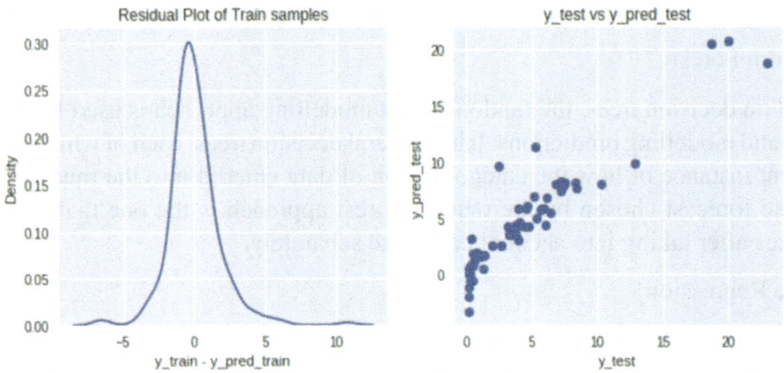

Fig. 2. R^2 Value and CV score for Linear Regression

Figure 2 illustrate the R^2 value and CV score obtained for Linear Regression ML model.

- RIDGE REGRESSION:

  Train R2-score: 0.88.
  Test R2-score: 0.86.
  Train CV scores: [0.89912009 0.8983878 0.82980715 0.81769155 0.73377277].
  Train CV mean: 0.84.
  Figure 3 illustrate the R^2 value and CV score obtained for Ridge Regression ML model.

- LASSO RIGRESSION:

  Train R2-score: 0.88.
  Test R2-score: 0.86.
  Train CV scores: [0.8990973 0.89975361 0.8301628 0.81769086 0.73120322].
  Train CV mean: 0.84.

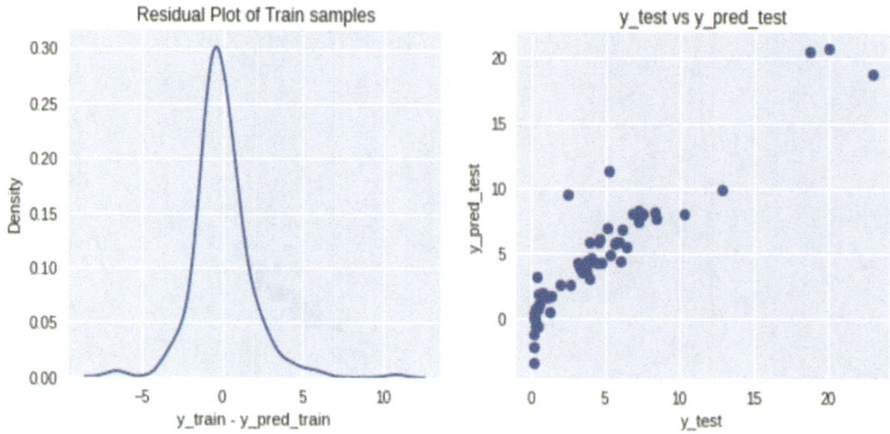

Fig. 3. R^2 Value and CV score for Ridge Regression

Figure 4 illustrate the R2 value and CV score obtained for Lasso Regression ML model.

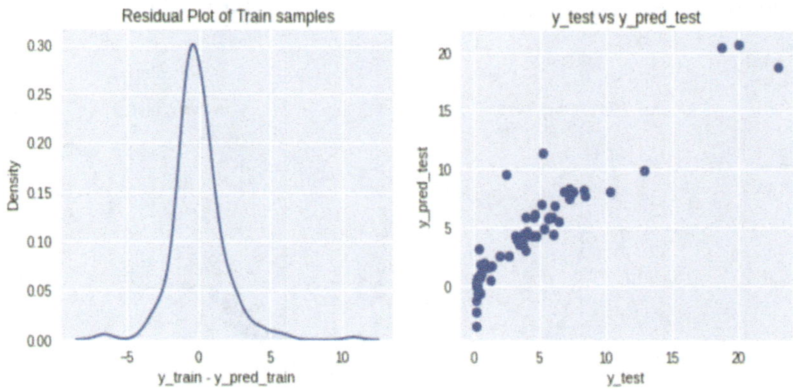

Fig. 4. R^2 Value and CV score for Lasso Regression

- RANDOM FOREST:

  Train R2-score: 0.98.
  Test R2-score: 0.88.
  Train CV scores: [0.84357772 0.96581665 0.84727422 0.88956666 0.73871238].
  Train CV mean: 0.86.
  Figure 5 illustrate the R2 value and CV score obtained for Random Forest ML model.

- GRADIENT BOOSTER REGRESSION:

  Train R2-score: 1.0

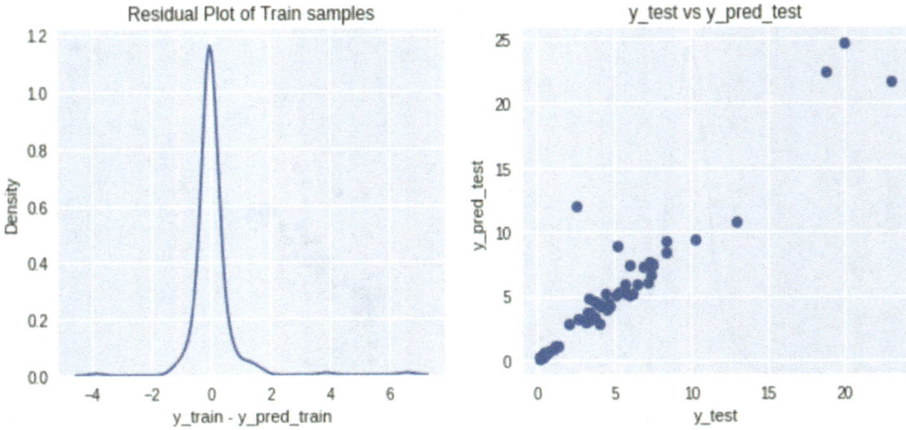

Fig. 5. R^2 Value and CV score for Random Forest

Test R2-score: 0.94.

Train CV scores: [0.95122331 0.90545026 0.89227483 0.94305868 0.68744002].

Train CV mean: 0.88.

Figure 6 illustrate the R2 value and CV score obtained for Gradient Booster ML model.

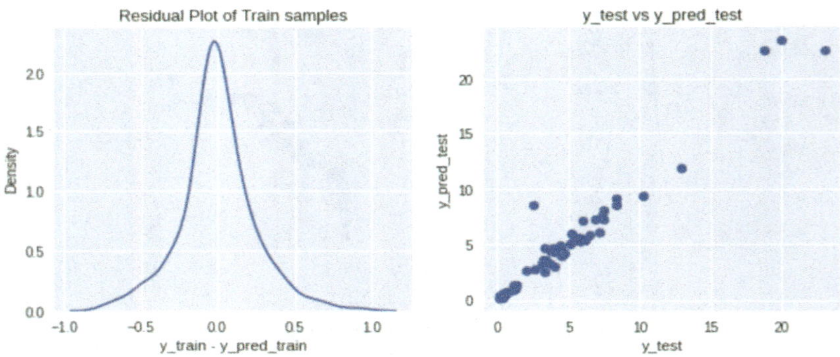

Fig. 6. R^2 Value and CV score for Gradient Booster

Figure 7 illustrates that Gradient Boosting ML model performs better when compared with other ML models used namely Linear Regression, Ridge Regression, Lasso Regression and Random Forest.

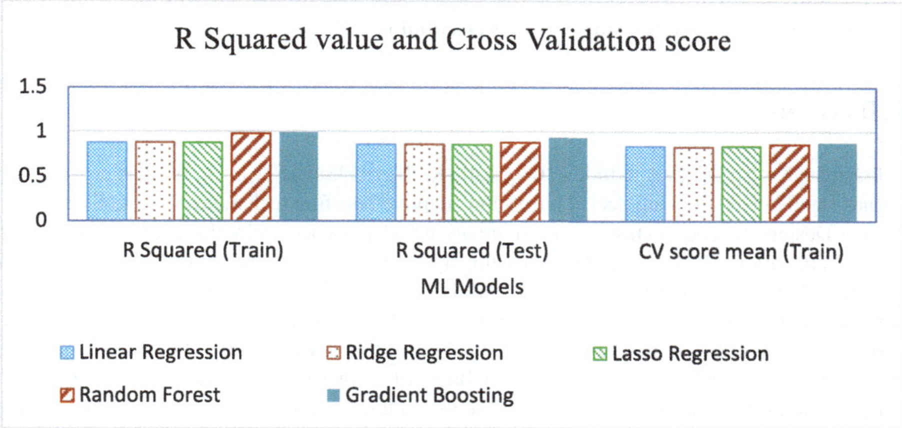

Fig. 7. Performance comparison of ML models used

Table 1 describes the tabulated R2 value and CV scores obtained from different ML models used.

Table 1. R2 value and CV score of ML models used

Model Name	R Squared (Train)	R Squared (Test)	CV score mean (Train)
Linear Regression	0.88	0.86	0.84
Ridge Regression	0.88	0.86	0.83
Lasso Regression	0.88	0.86	0.84
Random Forest	0.98	0.88	0.86
Gradient Boosting	1	0.94	0.88

5 Conclusion

The proposed study aimed to predict used automobile prices by employing a combination of regression techniques, namely Ridge, Lasso, Random Forest Regressor, Linear Regression, and Gradient Boosting Regressor. Prior to constructing the model, we focused on aiding users' understanding of the dataset, thereby facilitating a more comprehensive analysis. To facilitate effective model development, we meticulously partitioned and tailored the dataset to align with the requirements of regression analysis. This careful preparation was pivotal in ensuring optimal model performance and robustness. Through rigorous evaluation, we assessed the performance of each regression technique using the R-squared metric, a widely recognized measure of the model's ability to explain the variance in the target variable. Our comprehensive analysis culminated among all five techniques used, Gradient Boosting Regression ML model is identified as best one because it has R square test value is nearly equal to 94%.

References

1. Prasanna Kumar, K.R., Aravind, S., Gopinath, K., Navienkumar, P., et al.: Enhancing the credit card fraud detection using decision tree and adaptive boosting techniques. In: Intelligent Systems Design and Applications: 22nd International Conference on Intelligent Systems Design and Applications (ISDA 2022) Held December 12-14, 2022 - Volume 3, pp. 358–365. Springer Nature Switzerland, Cham (2023). https://doi.org/10.1007/978-3-031-35501-1_36
2. Logeswaran, K., Suresh, P., Anandamurugan, S.: Particle swarm optimization method combined with off policy reinforcement learning algorithm for the discovery of high utility itemset. Inf. Technol. Control **52**(1), 25–36 (2023). https://doi.org/10.5755/j01.itc.52.1.31949
3. Logeswaran, K., Suresh, P.: High utility itemset mining using genetic algorithm assimilated with off policy reinforcement learning to adaptively calibrate crossover operation. Comput. Intell. (2021). https://doi.org/10.1111/coin.12490
4. Çelik, Ö., Ömer Osmanoğlu, U.: İkinci el araba fiyatlarinin tahmini. Eur. J. Sci. Technol. (2019). https://doi.org/10.31590/ejosat.542884
5. Nasiboglu, R., Akdogan, A.: Estimation of the second hand car prices from data extracted via web scraping techniques. J. Mod. Technol. Eng. **5**(2), 157–166 (2020)
6. Yu, L., Yan, H., Liu, L., Yi, S., et al.: Car trading cycle prediction based on random forest algorithm. In: 2022 7th International Conference on Intelligent Computing and Signal Processing, ICSP 2022, pp. 639–642 (2022). https://doi.org/10.1109/ICSP54964.2022.9778603

A Comparison of Machine Learning Algorithms on Handwritten Digit Recognition

Neha[1(✉)] and Meenakshi[2]

[1] Galgotias University, Greater Noida, Uttar Pradesh 203201, India
nehabishnoi93@gmail.com
[2] Department of Electronics and Communication Engineering, Chandigarh University, Mohali, India

Abstract. Handwritten digit recognition is a crucial issue in the study of computer vision and machine learning, with applications ranging from postal code identification in mail sorting to digitizing old documents. This paper weighs up the accomplishments and execution of numerous classifier calculations on database of manually written digits. A interesting zone of design acknowledgment is the advancement of Handwritten character recognition (HCR). For character recognition, a few techniques have been proposed. It too centers on troubles and obstacles included in extraction and acknowledgment of handwritten digits. This errand is performed utilizing calculations like SVM (Support Neural Systems), Decision Tree, Random Forest, artificial Neural Network, K-Nearest Neighbor (KNN). Indeed, so, there are sufficient inquire about and papers that portray the strategies for changing paper record content into machine-readable frame. Character recognition innovation may be vital within the close future in arrange to prepare and filter existing paper documents in arrange to set up a paperless environment. This survey paper offers a careful investigation of the handwritten character recognition field.

Keywords: Handwritten Digit Recognition · CNN · SVM · Deep Learning · MNIST Dataset

1 Introduction

This fast progress has come about from a combination of a number of advancements counting the multiplication of effective, reasonable computers, the innovation of modern calculations that take advantage of these computers, and the accessibility of huge databases of characters that can be utilized for preparing and testing. In this paper we differentiate the relative merits of each of the calculations [1]. In expansion to exactness, we see at measures influencing execution, such as preparing time, run time, and memory prerequisites. Since numerous classifiers cannot process efficiently the crude pictures or information, include extraction may be a preprocessing step that points at diminishing the measurement of the information whereas extricating important data. The execution of a classifier can depend as much on the quality of the highlights as on the classifier itself there's a require for handwritten recognition frameworks for perusing and documenting ancient archives [2].

A. Bajaj et al. (Eds.): IBICA 2023, LNNS 1229, pp. 107–113, 2025.
https://doi.org/10.1007/978-3-031-78940-3_10

The capacity of a computer to acknowledge and get it coherent written by hand information some time recently dissecting it for utilize in various automated process systems is portrayed as Handwritten Character recognition (HCR) [3]. In computer vision and machine learning, handwritten digit recognition may be a critical issue. It incorporates recognizing handwritten numbers from filtered photos and encompasses a number of employments, counting keeping money, postal mechanization, and signature verification [4]. Off-line and online handwriting recognition strategies are the two categories beneath which Handwriting recognition is categorized. Be that as it may, within the online framework, the arrange of the writer's strokes as well as the two-dimensional facilitates of succeeding focuses are spoken to as a work of time [5]. The HCR framework has a few steps, including pre-processing, feature extraction, classification, and at long last genuine recognition. In this study, tests were performed on the MNIST written by hand digit information set with SVM, Decision Trees, Random Forests, ANN, KNN, K-Means Algorithm. The victory rates of the calculations within the field of handwritten recognition were compared. Within the strategies segment of this article, brief data is given around handwritten recognition and compared machine learning strategies. Within the third area, the values gotten as a result of the consider were compared [6] (Figs. 1 and 2)

2 Handwritten Digit Recognition Process

The handwritten digit recognition process is devided into different steps:

- Words are divided into parts like digits and words.
- Data is processed and defined.
- Words are prepared for recognition.
- Last step is to eliminate errors. Use od dictionaries is also possible.

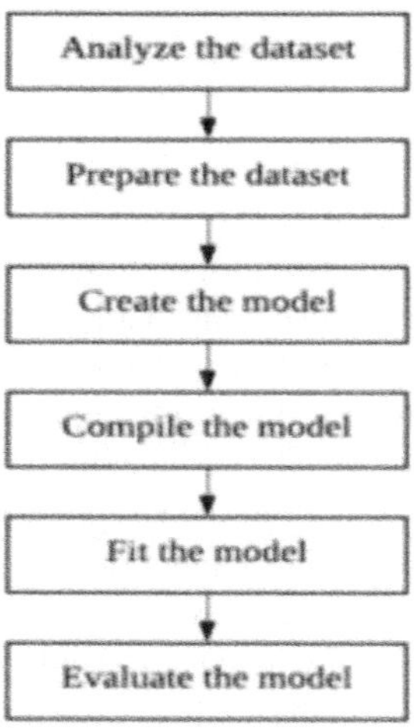

Fig. 1. Steps involved in handwritten digit recognition

Pre-processing: The pre-processing of the picture in corporate changing over the picture to a standard organize and decreasing commotion as the introductory organize in handwritten digit acknowledgment [7]. Pre-processing strategies incorporate smoothing, binarization, and thresholding, among others. Setting an edge esteem causes all pixels underneath it to be set to dark, and all pixels over it to be set to white. This strategy is known as thresholding. Another strategy is known as "binarization," which involves changing the grayscale picture into a binary image in which all pixels underneath a particular limit esteem are allotted to dark and all pixels over that limit esteem are turned to white. Using various methods, smoothing could be a prepare that includes decreasing commotion from the picture [8].

Feature Extraction: It is the procedure utilized to discover the pertinent highlights in a picture that will be used for classification after pre-processing. HOG (Histogram of Oriented Gradients), SIFT (Scale-Invariant Feature Transform), and LBP are a couple of well-liked feature extraction strategies [9]. The histogram of slopes for each person cell is computed utilizing the HOG feature extraction approach, which partitions the picture into little cells. The SIFT procedure entails locating the image's key points and calculating the descriptors for each of them. Using the LBP procedure, the picture is isolated into littler areas, and the neighborhood parallel design is calculated for each locale.

Classification: The method of classifying a handwritten digit picture into one of the 10 conceivable digits employments the extricated highlights as its final step (0 to 9) [10]. The classification techniques k-nearest neighbors (KNN), support vector machines (SVM), and artificial neural networks (ANN) can all be utilized. Finding the test image's k-nearest neighbors and placing it within the course with the lion's share of those neighbors constitute the essential KNN classification calculation. SVM, a more modern classification method, looks for the hyperplane that maximally separates the different classes [11].

Deep learning has been a powerful strategy for handwritten digit recognition in later a long time. Convolutional neural systems (CNNs), a sort of deep learning algorithm, have illustrated momentous execution in precisely recognizing handwritten digits. Convolutional, pooling, and completely associated layers are a few of the layers that make up CNNs, a frame of neural network utilized to translate pictures. The ultimate classification is carried out utilizing completely associated layers after the features have been diminished in measurement utilizing pooling layers and extricated from the picture using convolutional layers [12].

3 Convolutional Neural Network

A convolutional neural network could be a shape of deep neural network with numerous employments in computer vision, face recognition, signal processing, recommendation systems, object recognition, and picture categorization [13]. CNN is a supervised deep learning multi-layered neural arrange that will be thought of as a two-part framework that combines a programmed include extractor with a trainable classifier. Basic neural systems called convolutional networks utilize convolution on at slight stone layer instead of standard framework duplication. Input and output layers as well as a number of covered up layers make up convolutional neural networks [14]. It is particularly made to perceive two-dimensional shapes that are exceptionally safe to twists like translation, scaling, and skewing [15]. They (CNNs) are more viable since of their capacity to naturally distinguish key properties of a thing (here, an object can be a picture, a handwritten character. A CNN requires small pre-processing and include feature extraction work since it combines the feature extraction and classification operations. A CNN can naturally extricate rich and associated components from photographs. Furthermore, even with scant training data, a CNN can provide respectable recognition accuracy, etc.) without human supervision or assistance [16].

Support Vector Machines
SVM implements two class classifications in its simplest form. It has been used as of late as an elective to well-liked procedures including neural network. The advantage of Support Vector Machine is that, based on the thought of basic threat minimization, it considers both experimental data and structural conduct for improved generalization capabilities (SRM) [17, 18]. For datasets that can be isolated directly, the basic SVM formulation is utilized. It can be connected by implication to nonlinear datasets by changing over to the linear feature space from the nonlinear inputs [19, 20].

4 Performance Metrics

Different performance metrics are utilized to assess the execution of handwritten digit recognition models, counting accuracy, confusion matrices, recall, and F1-score. Accuracy is the most commonly utilized metric and is characterized as the rate of accurately classified images. Recent progressions in performance evaluations include the utilization of class-wise accuracy, confusion matrices, and receiver operating characteristic (ROC) curves.

Datasets: A few datasets have been utilized for assessing handwritten digit recognition models. The foremost broadly utilized dataset is the MNIST dataset, which includes 70 thousand pictures of manually written digits. Other datasets incorporate the EMNIST dataset, which includes 280 thousand pictures of transcribed digits and characters, and the SVHN dataset, which contains street see pictures of house numbers See Tables 1 and 2.

Fig. 2. Sample pictures of MNIS information Comes about

5 Results

Table 1. Report values of compared algorithms

Algorithm	Accuracy	Training Time	Test Time
Support Vector Machine	90%	372.294 s	97.769 s
Decision Tree	87%	21.142 s	0.645 s
Random Forest	97%	63.055 s	1.106 s
Artificial Nueral Network	97%	178.390 s	1.014 s
K-Nearest Neighbour	98%	36.396 s	865.93 s
K-Means Algorithm	98%	8.350 s	0.569 s

Table 2. Accuracy of compared algorithms

Algorithm	Precision	Recall	FI-Score
Support Vector Machine	0.97	0.98	0.97
Decision Tree	0.85	0.90	0.86
Random Forest	0.97	0.97	0.97
Artificial Nueral Network	0.98	0.98	0.97
K-Nearest Neighbour	0.96	0.98	0.96

6 Conclusion

Handwritten digit recognition has immense applications in the field of medical, banking, student management, and taxation process etc. Many classifiers like KNN, SVM, CNN are used to identify the digit from the handwritten image. As per the review, CNN is providing better performance than others. All the algorithms of machine learning are discussed in this paper. MNIST dataset consist of handwritten numbers from 0–9 and it is a standard dataset used to find performance of classifiers. Machine learning algorithms consists of three different stages. First is preprocessing where dataset is converted into binary form and image processing has been applied on it. Second stage is segmentation where the image is converted into multiple segments. Third stage is feature extraction where features of image are identified. Last stage is classification where classifiers like KNN, SVM, CNN are used. Results of KNN is improved a lot by using CNN classifier but it can be improved further in terms of complexity, duration of execution and accuracy of results by making combination of classifiers or using some additional algorithm with it.In computer vision and machine learning, handwritten digit recognition could be a pivotal issue that can be settled using a variety of strategies. The three processes in handwritten digit recognition are pre-processing, feature extraction, and classification. Deep learning has led to the development of more potent methods that can recognize handwritten digits with great accuracy. There is still potential of development, therefore further study might concentrate on creating more effective algorithms and enhancing the functionality of already existing ones.

References

1. Mohamad A., wadi Hassan, H., Dewi, N., Habibollah, H.: An audit on highlight extraction and highlight choice for manually written character recognition muhammad.
2. Purohit, A., Chauhan, S.S.: A writing overview on manually written character acknowledgment.
3. LaxmiSahu, V., Kubde, B.: Offline written by hand character acknowledgment procedures utilizing neural organize: a survey
4. Goodfellow, I., Bengio, Y., Courville, A.: Deep learning. MIT Press (2016)

5. Ahlawat, S., Choudhary, A., Nayyar, A., Singh, S., Yoon, B.: Improved written by hand digit acknowledgment utilizing convolutional neural systems (CNN)
6. Handwritten Digit Acknowledgment using Convolutional Neural Systems and Gabor filters Andr´ es Caldero´ n, Sergio Roa and Jorge Victorino Grupo Simbiosis Bogota´, Colombia
7. Gaganashree, J., Padmashali, S., Kumari, D.: Handwritten Digit Acknowledgment Utilizing Profound Learning Student, Division of Computer Science and Building, Srinivas Founded of Innovation, Mangaluru, India
8. Handwritten digit Acknowledgment utilizing Bolster Vector Machine Anshuman Sharma
9. Liu, S., Chen, Y., Peng, L., Zhang, Q.: (2021). Manually written Digit Acknowledgment Utilizing Capsule Systems. IEEE Get to, **9**, 5083–5093.
10. Liu, Y., Zhu, X., Li, C., Li, J.: A progressed written by hand digit recognition system using consideration component based convolutional neural arrange. diary of encompassing insights and humanized computing pp. 1–16 (2022).
11. Farhan, M.S., Jang, J., Kim, H., Lee, S.: Handwritten digit acknowledgment through gated repetitive units with spatial consideration. Knowl. Based Frameworks **236**, 107191 (2021).
12. Xie, Y., Xiong, G., Xiong, S., Du, J.: A novel profound learning system for manually written digit acknowledgment utilizing stochastic slope plunge and ROC bends. Connected Sci. **11**(10), 4572 (2022)
13. Rabia, K., Serap, K.: Manually written digit acknowledgment utilizing machine learning. Sakarya Coll. Diary Sci. 25(1), 65–71 (2021). https://doi.org/10.16984/saufenbilder.801684 Get to interface http://www.saujs.sakarya.edu.tr/en/pub/issue/58068/801684
14. Inquire about [Best of the Net]. In: IEEE Flag Handling Magazine, vol. 29, no. 6, pp. 141–142, Nov. 2012, https://doi.org/10.1109/MSP.2012.2211477. https://iq.opengenus.org/mnist-handwritten-recognition-dataset/
15. LeCun Yann, M.L.A., et al.: Comparison of learning calculations for transcribed digit acknowledgment. Worldwide Conference on Fake Neural Networks (1995)
16. Le Cun, Y., et al.: Comparison of learning calculations for written by hand digit acknowledgment. In International conference on counterfeit neural systems (Vol. 60, No. 1, pp. 53–60) (1995)
17. Yann, L., et al. Comparison of learning calculations for handwritten digit recognition. In: International conference on manufactured neural networks pp. 53–60 (1995)
18. Lauer, F., Suen, C.Y., Bloch, G.: A trainable include extractor for handwritten digit acknowledgment. Des. Acknowledgment **40**(6), 1816–1824 (2007)
19. Lauer, F., Suen, C.Y., Bloch, G.: A rainable highlight extractor for written by hand digit acknowledgment. Des. Acknowledgment **40**(6), 1816–1824 (2007)
20. Fabien, L., Ching, S.Y., Gerard, B.: A trainable highlight extractor for written by hand digit recognition. Des. Acknowledgment **40**(6), 1816–1818 (2000)

Hybrid Deep Neural Network for Detecting Fake Traffic Data in Vehicular Social Network

M. S. Kaniskaa[1], P. Priakanth[2], and K. Sangeetha[3](✉)

[1] PSG Institute of Technology and Applied Research, Coimbatore, India
[2] Department of Computer Science and Design, Kongu Engineering College, Perundurai, India
[3] Department of Computer Science and Engineering, Kongu Engineering College, Perundurai, India
sangeetha_k.cse@kongu.edu

Abstract. Vehicular social networks (VSNs) are a new type of communication where essential ideas from two distinct fields—mobile social networks (MSNs) and Vehicular Ad-hoc Networks (VANETs) are combined. In the virtual community of cars, passengers, and drivers on the roads, VSNs involve social interactions of vehicles with similar goals, interests, or movement patterns. Possessing a good ability for fraudulent content detection is useful in establishing improved network performance in VSN. In this paper, a hybrid Deep Neural Network-based model for fraudulent content detection for long-term VSN is proposed. Recurrent Neural Networks (RNNs) and Generative Adversarial Networks (GANs) are combined to capture two parts of semantics, using Graphical Neural Network (GNN) as the foundational architecture. Several tests are run on social network dataset for evaluation, and the findings showed that they could increase the detection effect by 8% to 15% over the baseline techniques.

Keywords: Vehicular Social Network · Deep neural network · Recurrent Neural Networks · Generative Adversarial Networks traffic data · network performance

1 Introduction

The idea of combining potential societalactivities of thedevices in mobile contexts is known as vehicular social networking. In order to interact within the virtual community of vehicles, Road Side Units (RSUs), drivers', travelers' and pedestrians' smart devices, VSNs develop social networking. This communication is entrenched on societal attributes, analogous safety, societal activities and aims. The research area is considering social networking and its applications since they have demonstrated how social information acquired through node interaction can improve the functionality of mobile communication systems. It is believed that a variety of applications could be developed by utilizing the social behaviors of network elements.

Vehicular social networks are derived from well-known VANETs, where social connections are made as they occur based on the patterns of mobility, routines, and common interests of the vehicles [1]. A node is chosen as a next-hop forwarder in VANETs using

A. Bajaj et al. (Eds.): IBICA 2023, LNNS 1229, pp. 114–123, 2025.
https://doi.org/10.1007/978-3-031-78940-3_11

a greedy approach. Mobility models as well as physical characteristics like channel quality, transmission range, inter-vehicle distances, and others are taken into consideration. The architecture (Fig. 1) shows how different VSN components are mapped out. As the vehicle-to-vehicle (VTV) and vehicle-to-infrastructure (VTI) communication links are based on the store-carry-and-forward scheme, vehicles in the vehicular social network differentiate the importance of a vehicle in the network based on social aspects. Finding a social vehicle increases network performance since it has the greatest chance of spreading a message widely [3–5].

Position, content, and relationship are the key determinants of a VSN's formation [2]. For instance, cars driving toward a school are more likely to have interests in common with other cars in a school environment. Therefore, it is most likely that a VSN will be constructed amid these vehicles. Other VSNs shown in Fig. 1 that are related to public transit, including the bus and the interested nearby automobiles, should be given the same considerations. There is a center (social) node in each VSN. Consider a VSN that is constructed based on information content supplied by drivers who have similar interests, such as a VSN of cars heading towards a stadium to watch a football game. Similarly cars and school buses moving towards the school are forming a VSN and they are sharing the common information among the social network formed. The availability of parking spaces, parking tickets, and other mutual interests can be shared by adjacent automobiles.

There is a high possibility that fake news is disseminated into network by any of the nodes in the VSN or the Road Side Unit itself when it is compromised. To identify the fake news in the Vehicular Social Network, the work is carried out.

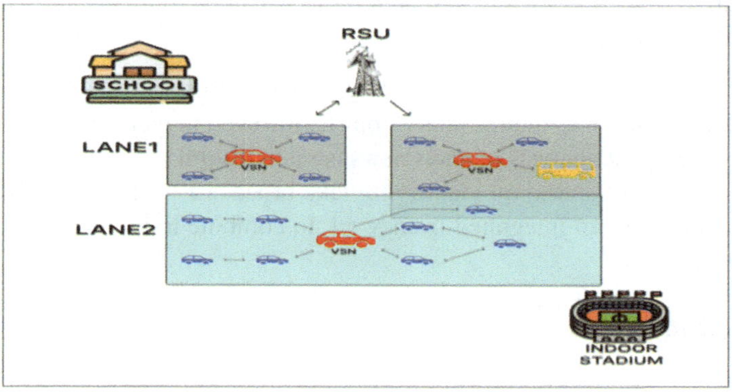

Fig. 1. Architecture of Vehicular Social Network (VSN)

2 Related Works

It is necessary to give the VSN a high ability to detect fraudulent content in order to build a trustworthy and long-lasting VSN [6]. Further dangerously, the lack of data regonization may compromise the VSN's ability to run consistently [7]. Inaccurate information may cause user behavior to be misguided and lead to major road accidents. Because of this, the question of how to build a viable VSN while immediately identifying fraudulent contents is still an important one for research. [8].

It is difficult to achieve direct detection of instantaneous small false information in VSN due to the lack of appropriate information as the discriminative basis [9]. It is challenging to determine the veracity of small information by straightforward semantic analysis and modeling from a semantics-only perspective [10]. As a result, the current researches rely on ancillary data like context and actions. Detecting gossip dealings, which are set of communal actions, comments, and performance, has naturally been the focus of recent research [11–15]. However, manipulating attributes for event-level false information identification is time and money-consuming and results in a clear time lag. [16, 17].

The identification of false information events sometimes isn't adaptable to the time-changeable VSN atmosphere since it takes too long to collect the data and train the models. Short news, on the other hand, can reduce overhead and speed up detection. Quick verification of the veracity of a single brief piece of news inside the VSN is therefore urgently needed.

In fact, there are some research already done in the field of False Information Recognition (FIR), which seeks to spot the purposefully false and biased content. The FIR is an example of a speedy Natural Language Processing (NLP) task where semantic encoding and representation are crucial elements. Convolution Neural Network (CNN) approaches [18] and recurrent neural network (RNN) methods are two categories into which the currently conducted research can be divided.

In summary, there are some study gaps in the VSN scenario that need to be filled by future research. More specifically, research projects can be routinely carried out taking into account the following factors:

- In order to gain more features for modeling, existing efforts must use event-level false broadcastrecognition as their primary object. However, in many instances, quick and accurate intelligent analysis of short news is really useful.
- Previous research focused on identifying fake broadcast from the standpoint of one element. However, an integrated framework to combine traits from several fields is still lacking.

3 Methodology

Based on the extensive literature survey made, Lexical Semantic Analysis can be adopted into the model for classifying the word based on the true meaning of the words and also taking into account for the variability of meaning from context to context.

The dimensionality of the feature space is maintained when a brief text is represented using the combined lexical and semantic features. Short writings' vocabulary is associated with studied subjects. Both the text's context and the subject-word distribution learned from prior knowledge have an impact on matching a word to a corresponding topic. Upon completion of the word-topic assignment, a brief text's semantic representation can be obtained by substituting corresponding topics for the original words. As a result, a vector might be used to represent all of the messages, with each element standing for the allocated topic's time. As a result, the number of subjects determines how dimensional the feature space is maintained.

3.1 Classification Models

In the work, the three kinds of learning representations are used to classify and detect the fake new in the Vehicular Social Network. They are the three different Deep Neural Network systems like MultiLayer Perceptron (MLP), Convolutional Neural Network (CNN)and CNN with Recurrent Neural Network (RNN) and GAN for text classification.

MLP for Text Classification. The Multi Layer Perceptron (MLP) is a kind of feed-forward artificial neural network. An MLP model is the most basic type of deep neural network and it has several completely linked layers. The internal states of the MLP, xi(k) and a weighted sum over i for an n-layer MLP encoder on word counts ci are used. The weights will be learned by the downstream classifier. Specifically, it is proposed to add weighted combinations of the internal states of our variational encoder to the vector representations for each document for any sequence-to-vector encoder, f(sv). On the weighted combination, the model learns a supervised classifier.

$$t^{n} = \Omega\theta + \sum\nolimits_{k=1}^{n} xi(k) \tag{1}$$

$$W(yi|xi) = fn\left[t^{n} f(sv) v(xi)\right] \tag{2}$$

where fn is a neural classifier and $\{\Omega_0, \ldots, \Omega n\}$ are softmax-normalized trainable parameters.

C for Text Classification. CNN is a feed forward neural network made up of pooling dotted between convolutional layers. The region features are to be learned by a convolutional layer.

Every CNN model employs a number of filters of varioumeasurements lengthwise given the filter length l, the convolution takes l-gram features into account. Though, these l-gram features only take into account local viewpoints of the content and not the document's overall perspective, which is required for a number of transitional instances like negation in sentiment analysis.

When CNN models are trained, using different random seeds may result in behavior changes, occasionally by in excess of 1%. This is a result of the randomness in deep learning, including the dataset shuffle, initialization of the weights, and tensor drop-out rate. By using 10 different random seeds to train each model, we were able to do a group analysis and lessen the effect of the random seed on our results while still capturing the basic characteristics of the model.

RNN for Text Classification. An artificial neural network called a Recurrent Neural Network (RNN) has components that organize an connotation between node edges. RNNs have the ability to jointly improve the categorization groups using recurrent techniques. In the suggested model, RNN is used to track several articles and out perform CNN. The existing input sign xt and the preceding unseen state ht-1 are used to calculate the activation of the unseen condition ht at time-step t.A recurrent neural network (RNN) may accommodate any length of sequence by repeatedly concerning a transition task to the interior unseen state vector ht of the input series. The activation of the unseen situation ht at time-step t is determined using the task f of the existing input representation xt and the earlier unseen state ht-1.

$$h = 0 \quad if\ t = 0 \tag{3}$$

$$h = f(ht - 1, xt) \quad Otherwise \tag{4}$$

The state-to-state transition function f is frequently used and is made up of an element-wise nonlinearity with xt and ht-1 transformations. Unfortunately, a drawback of RNNs with this type of transition function is that during training, the gradient vector's components may expand or decay exponentially over lengthy sequences. It is challenging for the RNN model to study exnded distance correlations in a series because of the issue with bursting or vanishing gradients.

CNN with RNN and GAN for Text Classification. This study makes an effort to integrate the confined feature communication capabilities of CNN and the linear feature expression capabilities of RNN in order to close this gap. In order to create a mixed GNN structure, they are integrated into Generative Adversarial Network (GAN) architecture. Since the information propagation process is a part of GNN [29], it is presumed that every word in a sentence is connected to the process. Thus, the building of graph networks can make use of the random walk theory. In order to do this, this research suggests a technique for identifying fake news based on mixed GNN for sustained VSN. This technique is referred to as Hybrid DNN. It merges recursive neural network and convolutional neural network to lexical semantics based on GNN. Better detection results can be obtained using this mixed detection approach. With the use of GAN, a complicated tensor representation may be produced for each word in a phrase.

Then, using RNN, the serial links between groups of words are displayed and recognized. Finally, a sentence can be translated into a recognition outcome that denotes whether a part of broadcast is real or false.

The following aspects can be summed up as the main contributions of this paper:

- A quick false broadcast recognition approach based on diversified GAN is provided.
- The semantic expression of brief news is enhanced through the integration of the GAN model with CNN and RNN.

To assess the efficacy and efficiency of the Hybrid DNN, a series of comparison experiments are conducted.

The main architecture of the recommended Hybrid DNN (Fig. 2) is the first part which consists of word extraction using the lexical analysis of the dataset, in which all the verses are graded in order one after another. Only traffic related information and social event related information are extracted from the dataset and complied together in the central repository. The purpose of the central repository is to optimize the search and detection of fake news quicker. The central part of Fig. 2 is the GAN arrangement, which can be employed to categorize the dataset attributes into real and fakebased on

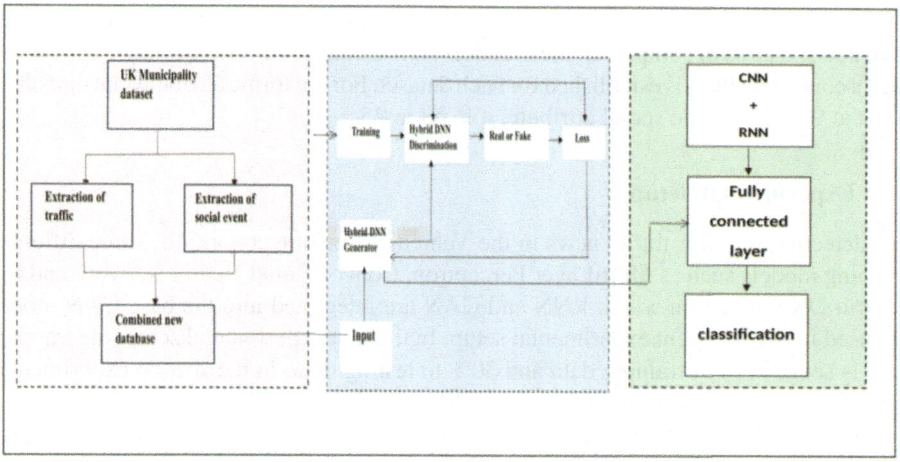

Fig. 2. Data classification in Hybrid DNN

the generator functions fed by first part which involves lexical semantic analysis. The reason behind combining CNN with the RNN model is that CNN and RNN model use convolutional layers and recurrent layers to learn from the training data.

3.2 Datasets

VSN is still not a developed and common media in today's world. Regarding VSN, there is no publicly accessible dataset. In this section, the simulative scenes are set up using a small dataset connected to traffic statistics, news, and social event information in order to guarantee generality and representativeness of samples. The United Kingdom Municipality's road traffic and social event information real dataset is publicly available with 50000 recordsets. The essential attributes in the dataset *info* and *event* are considered for analysis. The info attribute is related to the traffic related to the information that passes between the VSN devices that are moving towards the school and indoor stadium.The event attribute has the details that are related to the social events that are happening in the public places.The values related to the both the attributes are listed in the below Table 1.

Table 1. Traffic and Social event related words extracted from dataset

Traffic related info	Social Events related event
Traffic jam	congress
Road work	function
Bumper-to-bumper	meeting
Alternate route	party
Turn lane	get together

The length of various news stories varies widely, which makes it difficult to identify fake news. It aims to compile all the broadcast into a single, uniform quantity. To address this, a consistent limit is established for each dataset. For the traffic attribute, the threshold is set at 90, and for the social attribute, it is set at 25.

3.3 Experimental Setup

For detecting the fake traffic news in the Vehicular Social Network, the three different learning models such as MultiLayer Perceptron, Convolutional Neural Network and the Hybrid DNN model in which RNN and GAN are integrated into the base CNN model are used in two different experimental setup. In the first experimental setup, the training data is set to 70% of training data and 30% to testing data. In the second experimental setup the training data is set to 80% and 20% to testing data.

By manually selecting search terms from the UK Municipality Dataset, the dataset for categorizing traffic microblogs into those that are relevant to traffic and those that are irrelevant to traffic was compiled. The adopted key words are listed in Table 1. Lastly, by searching the UK Municipality Dataset using the sample key phrases listed in Table 1, over 50,000 candidate microblogs are gathered. The microblogs are then manually classified into one of two categories, namely traffic appropriate or traffic irrelevant. The remaining collection of candidate microblogs, which consists of all candidate ones without the traffic applicable entry, is divided into 6,000 randomly selected microblogs. Of the candidate microblogs, 5,000 are assigned the traffic relevant class. Five thousand traffic-relevant microblogs and six thousand traffic-irrelevant ones make up the final dataset used to train classifiers.

4 Results and Analysis

Four measures that are frequently used to evaluate classification outcomes are chosen as the metrics in this study because fake news detection is a binary classification problem. The above indicated four evaluation measures, Precision, Recall, F-score, and Accuracy, are the focus of the major experimental findings. When a positive sample is accurately predicted, the situation is referred to as True Positive (TP), when a positive sample is accurately predicted but is actually false, as False Positive (FN), when a negative sample is accurately predicted but is actually true, as True Negative (TN), and when a negative sample is accurately predicted but is actually false, as False Negative (FP). The formula for the four metrics is given as below:

$$P = \frac{No.(TP)}{No.(TP) + No.(FP)} \tag{5}$$

$$R = \frac{No.(TP)}{No.(TP) + No.(FN)} \tag{6}$$

$$F = \frac{2*P*R}{P + R} \tag{7}$$

$$P = \frac{No.(TP) + No.(TN)}{No.(TP) + No.(FP) + No.(TN) + No.(FN)} \tag{8}$$

Several common methods are chosen as the standard approaches for comparison in order to relate the performance of the Hybrid DNN and current approaches. To create a more focused comparison, the three approaches were chosen based on the concept of this work. From the figures given below, the three different algorithms are compared under the various performances.The graphical representation of the two different tests done by changing the training data set to 70% and 80% respectively are also presented (Figs. 3 and 4). It is evident that the hybrid DNN is outperforming the other two approaches.

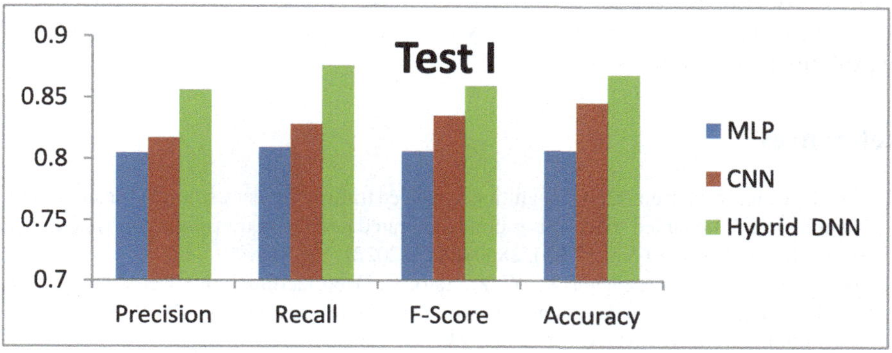

Fig. 3. Performance Metrics for Test I scenario

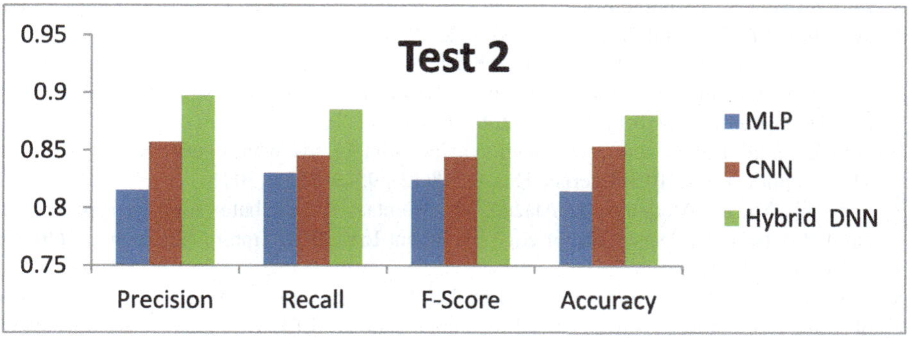

Fig. 4. Performance Metrics for Test II scenario.

5 Conclusion

The identification of contents is one of several significant problems that the prevalence of VSN continues to face. This paper makes an effort to investigate this issue from a deep learning standpoint and suggests the Hybrid DNN model. Everyone is aware that spotting fake news is a type of cognitive activity for people. Between machine intelligence-based recognition and human intelligence-based recognition in VSN, there is still some room for improvement. It should be highlighted that in order to guarantee the quickness of real-time discovery, the suggested Hybrid DNN must disregard numerous social network aspects. To create better feature spaces, some attributes, such as user summaries or context characteristics, are also helpful. Additionally, because the training is still focused on disconnected samples, it may not be as useful in actual VSN circumstances. Although VSN is still in expansion, it is far from being able to satisfy all of the demands of automobile users.

In the forthcoming work, nodes clustering, social degree of the nodes will be considered for the formation of VSN and also images extracted from Road side Units may be used for further increasing the performance. Still the best locations for the implementation of the central repository and security of the vehicular social nodes (VSN) have to be explored in the future work.

References

1. Tan, L., et al.: Speech emotion recognition enhanced traffic efficiency solution for autonomous vehicles in a 5G-enabled space–air–ground integrated intelligent transportation system. IEEE Trans. Intell. Transport Syst. **23**(3), 2830–2842 (2022)
2. Yu, K., Tan, L., Lin, L., Cheng, X., Yi, Z., Sato, T.: Deep-learning-empowered breast cancer auxiliary diagnosis for 5GB remote e-health. IEEE Wireless Commun. **28**(3), 54–61 (2021). https://doi.org/10.1109/MWC.001.2000374
3. Meng, D., et al.: A data-driven intelligent planning model for UAVs routing networks in mobile Internet of Things. Comput. Commun. **179**, 231–241 (2021)
4. Yu, K., Lin, L., Alazab, M., Tan, L., Gu, B.: "Deep learning-based traffic safety solution for a mixture of autonomous and manual vehicles in a 5G-enabled intelligent transportation system. IEEE Trans. Intell. Transport Syst. **22**(7), 4337–4347 (2020)
5. Guo, Z., Yu, K., Jolfaei, A., Ding, F., Zhang, N.: Fuz-Spam: label smoothing-based fuzzy detection of spammers in internet of things. IEEE Trans. Fuzzy Syst. **30**(11), 4543–4554 (2022). https://doi.org/10.1109/TFUZZ.2021.3130311
6. Guo, Z., et al.: Robust spammer detection using collaborative neural network in Internet-of-Things applications. IEEE Internet Things J. **8**(12), 9549–9558 (2021)
7. Feng, C., Yu, K., Aloqaily, M., Alazab, M., Mumtaz, S.: Attribute based encryption with parallel outsourced decryption for edge intelligent IoV. IEEE Trans. Veh. Technol. **69**(11), 13784–13795 (2020)
8. Guo, Z., Yu, K., Li, Y., Srivastava, G., Lin, J.C.-W.: Deep learning embedded social Internet of Things for ambiguity-aware social recommendations. IEEE Trans. Netw. Sci. Eng. **9**(3), 1067–1081 (2022). https://doi.org/10.1109/TNSE.2021.3049262
9. Cui, L., Wu, J., Pi, D., Zhang, P., Kennedy, P.: Dual implicit mining based latent friend recommendation. IEEE Trans. Syst. Man Cybern.-Syst. **50**(5), 1663–1678 (2020)
10. Lin, L., Chen, Z.: Social rumor detection based on multilayer transformer encoding blocks. Concurrency Comput. **33**(6), e6083 (2021). https://doi.org/10.1002/cpe.6083
11. Guo, Z., Yu, K., Jolfaei, A., Bashir, A.K., Almagrabi, A.O., Kumar, N.: Fuzzy detection system for rumors through explainable adaptive learning. IEEE Trans. Fuzzy Syst. **29**(12), 3650–3664 (2021)
12. Ma, T., Zhou, H., Tian, Y., Al-Nabhan, N.: A novel rumor detection algorithm based on entity recognition, sentence reconfiguration, and ordinary differential equation network. Neuro Comput. **447**, 224–234 (2021)
13. Chen, X., Zhou, F., Zhang, F., Bonsangue, M.: Catch me if you can: a participant-level rumor detection framework via fine-grained user representation learning. Inform. Process. Manage. **58**(5), 102678 (2021)

14. Tu, K., Chen, C., Hou, C., Yuan, J., Li, J., Yuan, X.: Rumor2vec: a rumor detection framework with joint text and propagation structure representation learning. Inform. Sci. **560**, 137–151 (2021)
15. Lao, Shi, C., Yang, Y.: Rumor detection with field of linear and non-linear propagation. In: Proceedings of the Web Conference (WWW), Ljubljana, Slovenia, pp. 3178–3187 (2021)
16. Wu, Z., Pi, D., Chen, J., Xie, M., Cao, J.: Rumor detection based on propagation graph neural network with attention mechanism. Expert Syst. Appl. **158**, 113595 (2020). https://doi.org/10.1016/j.eswa.2020.113595
17. Xu, F., Sheng, V.S., Wang, M.: Near real-time topic-driven rumor detection in source microblogs. Knowl.-Based Syst. **207**, 10639 (2020)
18. Neisari, A., Rueda, L., Saad, S.: Spam review detection using self-organizing maps and convolutional neural networks. Comput. Secur. **106**, 102274 (2021). https://doi.org/10.1016/j.cose.2021.102274

A Comparative Study of Mapping Names for Automating Attendance of Online Classes Using Machine Learning Models

Dillip Rout[1(✉)], Bholanath Roy[2], Ayesha Bee[2], and Deepshikha Routray[3]

[1] C.V. Raman Global University, Bhubaneswar, India
dillip.rout.iitb@gmail.com
[2] Maulana Azad National Institute of Technology, Bhopal, India
[3] S.C.S. Autonomous College, Puri, India

Abstract. A comparison of models for automating the attendance of online classes is presented in this research. The inputs to the models are the registration list and the participation list for certain days. The registration list contains name, registration number, class, section, contact number, *etc.* The participants list in general consists of the names of the attendees which are extracted from the platform of the online class. The aim is to match the names in the participation list to the names in the registration list appropriately. This problem leads to approximate string matching since most participants join online classes using unofficial email addresses, and thus the names mentioned in those emails may not exactly match the names on the registration list. The reasons for mismatch include substitution of part of the name by abbreviation, transposition, white space, *etc.* In this context, three models, namely, Longest Common Subsequence (LCS), Levenshtein, and Fuzzy Token Set (FTS) are applied to map the names appropriately. The models are tested on five real sets in the dataset containing 53 observations in total. The efficacy of these models is judged through descriptive statistics and ANOVA test. The median success matching rate of all models except FTS is at least 80%. Moreover, ANOVA test and post-hoc analyses verify that FTS has the lowest performance. The suggested approach of this paper is useful for a variety of online events, including webinars, workshops, meetings, *etc.*

Keywords: approximate string matching · Levenshtein distance · least common subsequence · Fuzzy Token Set · ANOVA

1 Introduction

In recent years, the shift towards online learning has become increasingly prevalent in academic institutions and industries [23,24]. The rise in popularity of online learning is due to its flexibility and convenience for learners [2,4]. It is evident that regular attendance spurs better understanding and results [10]. Hence,

A. Bajaj et al. (Eds.): IBICA 2023, LNNS 1229, pp. 124–133, 2025.
https://doi.org/10.1007/978-3-031-78940-3_12

the need to streamline administrative tasks such as attendance management has become apparently vital. Thus, online attendance has been the subject of numerous academic research. However, taking attendance during the online class is tedious, time-consuming, and a distraction for the instructor [2,23]. Therefore, automating online class attendance is a solution that can offer numerous benefits to both educators and students. Further, such a study provides practitioners and policy-makers with valuable information to enhance the quality of education.

The major inputs to automating the attendance process contain a list of registered learners and a link to the online class [20]. The list of registered learners includes information such as name, registration number (unique), class, section, *etc.*. The attendance is decided based on the identification number (e.g., registration number). Then, the names of the participants are extracted from the online class platform. These extracted names have to be matched to the names in the registration list which will eventually mapped to the corresponding registration number for marking the attendance. Now, the extracted names of the participants may vary from the names in the registration list due to a couple of reasons. For instance, the registration list is an official document whereas the extracted names are from emails which may be unofficial. Generally, people use a shorter version of the name in the email setup. Further, the ordering of the components in the name (first, middle, last) may cause mismatches. Moreover, people like to omit middle names in their unofficial documents like email, which leads to disparity. Thus, the automation of online class attendance is transformed into an approximate string matching problem [8,9,16,25,26]. Also, it needs attention and the methodology should be capable of tackling the aforementioned issues for maximizing the matching.

This paper aims to devise a methodology to automate the attendance of online classes efficiently. The challenge here is to match the names appropriately without having errors. In this regard, several Machine Learning (ML) models are applied to map those names for various batches of students over different days of the coursework duration. The particular objectives of the proposed study are listed below.

1. List out the reasons for the failure in matching, e.g., under which conditions the failures happen. Categorize the reasons for failures.
2. Match the unsuccessful names to approximate names in the registration list and find the respective registration numbers.
3. Find the significance of the difference among the results of the used models.

In this context, the failures of direct string matching are analyzed to find the pattern of mismatches. Further, LCS, Levenshtein, and FTS models are applied to match the names appropriately. A dataset containing five sets and 53 observations is used to test these models. Moreover, the performance of these models is judged through descriptive statistics and ANOVA test.

The remainder of the article is organized as follows. The next section contains a review of related literature. Then, the details of the input dataset and the proposed methodology are discussed in Sect. 3. Next, the results and analysis are carried out in Sect. 4. Lastly, the concluding remarks are presented in Sect. 5.

2 Literature Review

A survey of the relevant articles on the topic is conducted and the highlights are presented in the following two phases. The articles discussing the value of online education are covered first. Then, the algorithms that aid in locating the corresponding textual material are explained later.

The online class has become an integrated part of academics and training especially post Covid-19 [3,10,23,24]. A study indicates that student performance is directly proportional to consistent attendance in online classes [10,24]. Moreover, the positive effects of motivation, course design, and technological support on online class attendance are explored qualitatively in [3].

According to the survey, the primary obstacles to attending online classes are technological difficulties, a lack of enthusiasm, difficulties with time management, and less association with classmates and instructors [23]. A qualitative survey on focus groups through interviews finds that the flexibility of remote learning, the value of instructor presence, and the significance of peer interactions all influence students' involvement in online classes [2]. Hence, frequent interaction with students, unambiguous expectations for attendance, and the application of interactive technologies are the strategies for boosting attendance in online classes [7,29]. In this regard, automated attendance systems can save significant time for instructors by eliminating the tedious and time-consuming manual attendance tracking, educators can allocate more time to instructional activities and student engagement [19,23]. A web-based attendance management system is proposed to track student activity and store attendance records efficiently and accurately[27]. It also helps educators identify patterns of non-participation or irregular attendance and intervene early to provide support and guidance to struggling students. Thus, it is recommended that clear attendance guidelines should be established by institutions, along with expectations that are communicated to students and the use of interactive technologies to increase involvement [4].

The automation of online class attendance breaks down to string matching for successful marking of the attendance. Hence, the literature is surveyed related to string matching, classification, and string searching algorithms, including topics such as string distance, suffix trees, and approximate string matching that can be possibly utilized to address online class name matching [8,9,22,25]. A family of string classifiers is introduced in [8] based on local relatedness measurements which focus on finding optimal classifiers for sets of strings. Similarly, various topics related to combinatorial pattern matching of strings are discussed in [14]. Furthermore, a method for semantic matching of natural language web queries is presented in [11].

One of the prominent approximate string-matching algorithms based on dynamic programming is LCS [1]. Some of its variants are also used like run-length-encoded, uncompressed string match, constraints satisfying string match, *etc.*, [13,21]. Another popular algorithm used for approximation of string match is fuzzy logic [12,22,30]. In order to improve the fuzzy matching a couple of hybrid approaches combine it with Convolutional Neural Network (CNN), natural language processing, and information retrieval [9,18]. Similarly, Levenshtein distance metric is used for measuring the difference between two sequences, often used in applications like spell checkers [6]. [17] explores different classifications of sets of strings by weighting the edit operations in the Levenshtein distance. Likewise, [15] presents a modified Cross-Language Levenshtein Distance algorithm for matching names across different writing scripts and cultures.

Online learning produces results that are comparable to or even superior to those of conventional face-to-face learning according to the literature. It is essential to automate the attendance of online classes to save time and shift the attention of the instructor to focus on engaging the students. A handful of articles are found that describe the importance and essential aids to improve online class attendance. However, a negligible number of articles are found that discuss end-to-end attendance marking for online classes. Moreover, the issues faced while recording the attendance are not explained in detail. Further, the description of methodologies or algorithms for automating the attendance is not focused in the literature.

3 Methodology

The foundation of the methodology consists of three basic parts. The first part describes the case study, *i.e.*, the input files. The second part has the method that captures the attendance. The third part is the methodology consisting of a comparison of three models for matching the names. All these parts are discussed as follows.

3.1 Case Study

Table 1 contains five sets of input data covering various subjects that are undertaken for this study [20]. The subject may be the same but the batches vary. The total number of observations over all the days is 53 including all the sets. Notably, the number of students per subject and the number of attendances taken vary.

Table 1. Summary of input files

Set	Total Students	Total Days#	Subject	Program	Semester
SET-1	63	11	Android App Development	B.Tech.	V
SET-2	99	17	Android App Development	B.Tech.	IV
SET-3	55	15	Database Management Systems	B.Tech.	VI
SET-4	60	6	Agriculture Informatics	B.Sc.	III
SET-5	56	4	Machine Learning	B.Tech.	II

Total number of days the attendance is recorded.

3.2 Capturing Attendance

The method of collecting the attendance of online classes is through the extension of the web browser. In particular, Chrome web browser is used for this study. Then, an extension named Google Meet Attendance Tracker is added to Chrome. So, the extension has to be opened whenever the meeting has to be conducted. Then, the instructor has to mark (just a click) the attendance during the class. Here, it is preferred to mark the attendance multiple times since the students may go in and out of the class due to Internet fluctuations. In general, attendance is marked four times in a class, and a student is tagged as present if he/she has been marked at least once during the class.

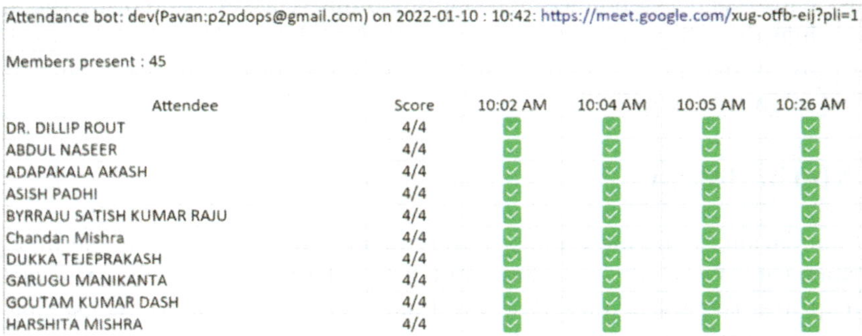

Attendance bot: dev(Pavan:p2pdops@gmail.com) on 2022-01-10 : 10:42: https://meet.google.com/xug-otfb-eij?pli=1

Members present : 45

Attendee	Score	10:02 AM	10:04 AM	10:05 AM	10:26 AM
DR. DILLIP ROUT	4/4	✓	✓	✓	✓
ABDUL NASEER	4/4	✓	✓	✓	✓
ADAPAKALA AKASH	4/4	✓	✓	✓	✓
ASISH PADHI	4/4	✓	✓	✓	✓
BYRRAJU SATISH KUMAR RAJU	4/4	✓	✓	✓	✓
Chandan Mishra	4/4	✓	✓	✓	✓
DUKKA TEJEPRAKASH	4/4	✓	✓	✓	✓
GARUGU MANIKANTA	4/4	✓	✓	✓	✓
GOUTAM KUMAR DASH	4/4	✓	✓	✓	✓
HARSHITA MISHRA	4/4	✓	✓	✓	✓

Fig. 1. Snapshot of Capturing Attendance

A snapshot of attendance capturing is given in Fig. 1. In order to preserve the symbols, this snapshot is presented in a picture although it is a table. It can be observed that there are four timestamps which are taken manually. Note that the instructor's name comes at the beginning of the list leaving apart the spaces after the header. These things need to be carefully executed during the preprocessing.

3.3 Models for Matching

The workflow of the proposed methodology is shown in Fig. 2. The major stages of processing and modeling are explained as follows.

The preprocessing includes cleansing the captured attendance (Fig. 1) to prepare a participant list. Particularly, the unnecessary rows and columns are removed. Then, only the names are collected to the participant list excluding the instructor's name.

Three models, namely, LCS, Levenshtein, and FTS are used in this study to match the names of the participant list to the names in the registration list. The background of the corresponding algorithms for each model is explained first and then the modeling part is described.

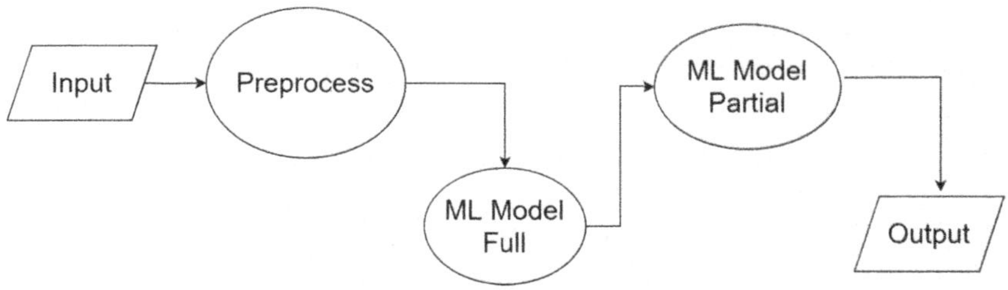

Fig. 2. Flow diagram of mapping names using ML model

The basic principles of the applied algorithms are described as follows. The LCS matches two strings approximately by finding the longest common subsequence allowing skips in between using a dynamic programming approach [16,26]. Levenshtein algorithm is based on edit distance where matching of the two strings occurs with minimum insertion, deletion, and replacement of the characters [5,16,26]. FTS operates on quantification of the overlapping between the token sets of the original strings by using Jaccard and cosine similarity [28].

The models from these algorithms use a certain threshold of matching. There are two variants used for each of the models, which are used in a pipeline (Fig. 2). The first variant is a strict or full version whereas the second variant is partial with an error tolerance threshold. The threshold for full and partial models of LCS and FTS are 100% and 90% respectively. On the other hand, these are 0 and 1 edit distance for Levenshtein model. Furthermore, the total number of matches is accounted including both these variants. Mention that simple string matching through character by character is also applied before applying these models. Finally, the results of these models are compared, and ANOVA test is conducted to check the significance of differences among the outputs.

4 Results and Analysis

The output is generated for the program by using Python programming language, and the analysis is done using JASP tool. For simplicity, the results are demonstrated in two parts, namely, failure analysis and success analysis. As per the name, the failure analysis reports all the errors and their statistics. Similarly, the success analysis describes the rate of success and its statistics. The remainder of this section contains the specific analysis of failures and successes in the respective subsections.

4.1 Failure Analysis

Analysis of the failures in matching reveals that the reasons for these can be broadly categorized into four types as shown in Table 2. For instance, the presence of white spaces between the first name and the last name. It is not feasible to capture in simple string matching and is difficult in the case of the applied models too. Furthermore, the most critical errors are the transposition of the names and abbreviated words coming as prefixes or postfixes.

Table 2. Summary of identified failure types

Type	Cause	Description
I	White space	Space missing between any of the words
II	Transposition	the first name and last are swapped positions
III	Abbreviation	Some of the middle names are abbreviated
IV	Spelling	Spelling of actual name or its components differ

The percentage of errors of every model is compared in Fig. 3. It is observed that the majority of the error percentage is falling below 20% for all the models. Precisely, the number error percentage below 20% varies across models from 33 to 42 out of 53 observations. Overall, it is observed that the error percentage follows an approximately negative exponential distribution, especially for FTS model. On the other hand, the LCS gives the highest matches with the least error rate.

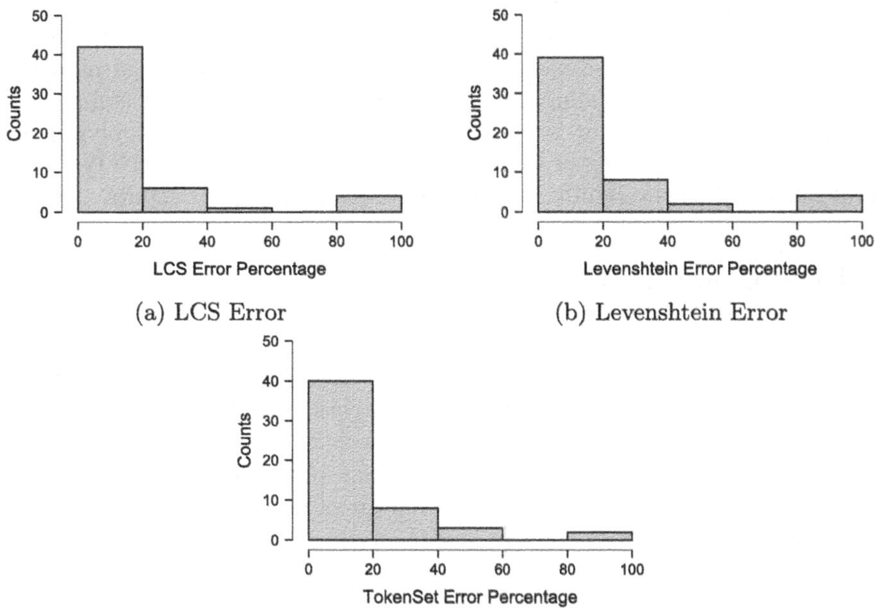

(a) LCS Error

(b) Levenshtein Error

(c) Fuzzy Token Set Error

Fig. 3. Error Distribution in Percentage

4.2 Success Analysis

The success percentage of all the models is listed and analyzed. The list contains all the observations for all the models. The analysis of success percentage is described in two parts. Firstly, the descriptive statistics of this output are compared. Secondly, the significance of differences among the results from the three models is discussed through ANOVA.

The descriptive statistics of the success percentage for all the models are shown in Table 3. It is observed that none of the outputs is following normal distribution. In particular, FTS model has right-skewed data whereas all other models have left-skewed results. The same can be confirmed from the P-value of Shapiro-Wilk test which is less than 0.001 for all the models. Furthermore, the success rate is looking good for LCS whereas FTS model seems to have the worst success rate over all the observations. The reason for LCS to be considered as good as the values of mean, median, mode, and the 25th percentile; which is at least 80%. Also, the Inter-Quartile Range (IQR) is low in comparison to others, which means that the values are compact and high. On the other hand, the values of mean, median, mode, and the 75th percentile are at most 38 for FTS. Again, its IQR and Coefficient of variation are higher than LCS and Levenshtein models. Thus, it has the lowest performance among all the models.

Table 3. Descriptive statistics for success percentage among all models

	LCS	Levenshtein	FTS
Mode	94.000	83.000	17.000
Median	93.000	83.000	17.000
Mean	81.962	78.094	30.038
Coefficient of variation	0.308	0.320	0.985
Std. Error of Skewness	0.327	0.327	0.327
P-value of Shapiro-Wilk	<.001	<.001	<.001
25th percentile	80.000	78.000	10.000
75th percentile	95.000	94.000	38.000

ANOVA test is conducted to know if the results of the models are significantly different. P-value is 1.31e−20 which is significantly less than the threshold $\alpha = 0.05$. Also, F-statistic (62.274) is considerably higher than F-critical (3.054) value. These two evidences lead to reject the null hypothesis which states that there is no significant difference among the groups. Rather, it is proven that the results of different groups do not follow the same trend. Thus, Post-hoc test is conducted on the success percentage, and the result is shown in Table 4. It is clear that FTS model has inferior performance than any of the models.

Table 4. Result of Tukey-Kramer post-hoc test for success percentage among all models

Comparison	Absolute Mean Difference	Q Critical	Significance
LCS vs Levenshtein	3.868	12.536	NO
LCS vs Fuzzy Token Set	51.925	12.536	YES
Levenshtein vs Fuzzy Token Set	48.057	12.536	YES

5 Conclusions

Automation of online class attendance is discussed in this article. A list of barriers found in matching the extracted names from the attendance into the actual names in the registration list. Broadly, the presence of white space between the parts of names, the transposition of first name and last name, usage of abbreviation in between the parts of the names, and spelling errors are the major issues while doing such mapping.

The analysis of the result shows that LCS model has better performance than Levestein and FTS models. Its mean success rate is approximately 80%. On the other hand, FTS model has the worst performance as per ANOVA and post-hoc test.

In the future, the authors would like to increase the scaling of the dataset by applying the models to a dataset like events. Further, it is essential to have a piece-wise analysis of the models. In addition, the type of failures must be counted per model by investigating the failure list. Lastly, an integrated model needs to be developed to reduce the error and have the best possible matching.

References

1. Arindam Kotal, K.K.S.: A new algorithm to find longest common sub-sequence's (2013)
2. Brown, L., Davis, C.: Student perspectives on online class attendance: a qualitative study. Int. J. Educ. Technol. **25**(4), 123–140 (2018)
3. Brown, L., Davis, M.: Exploring factors influencing online class attendance: a qualitative study. J. Dist. Educ. **15**(2), 78–95 (2017)
4. Butler, M., Morgan, M.J., Sheard, J., Simon, Falkner, K., Weerasinghe, A.: Initiatives to increase engagement in first-year ICT (2015)
5. Chen, Y., Wu, Y.: On the string matching with k mismatches. Theory Comput. Sci. **726**, 5–29 (2018)
6. Miller, F.P., Vandome, A., McBrewster, J.: Levenshtein distance: information theory, computer science, string (computer science), String metric, Damerau? Levenshtein distance, Spell checker, Hamming distance (2009)
7. Garcia, M., Martinez, S.: Strategies to promote online class attendance: a systematic review. J. Online Learn. **30**, 56–73 (2019)
8. Higa, Y., Inenaga, S., Bannai, H., Takeda, M.: A new family of string classifiers based on local relatedness, pp. 114–124. Springer, Heidelberg (2006)

9. Horng-Jyh, P.W., Jin-Cheon, N., Soo-Guan, C.K.: A hybrid approach to fuzzy name search incorporating language-based and text-based principles. J. Inf. Sci. **33**(1), 3–19 (2007)
10. Johnson, R., Williams, B.: Exploring the impact of online class attendance on academic performance. J. Dist. Educ. **15**(3), 78–95 (2017)
11. Karam, N., Benbernou, S., Hacid, M.S., Schneider, M.: Semantic matching of natural language web queries, pp. 416–429. Springer, Heidelberg (2004)
12. Kostanyan, A.: Fuzzy string matching with finite automat. In: 2017 Computer Science and Information Technologies (CSIT). IEEE (2017)
13. Liu, J., Wang, Y., Lee, R.: Finding a longest common subsequence between a run-length-encoded string and an uncompressed string. J. Complexity **24**(2), 173–184 (2008)
14. Crochemore, M., Gusfield, D.: Combinatorial Pattern Matching: 5th Annual Symposium, CPM 94 Asilomar, CA, USA, June 5-8, Proceedings (1994)
15. Medhat, D., Hassan, A., Salama, C.: A hybrid cross-language name matching technique using novel modified Levenshtein Distance. In: 2015 Tenth International Conference on Computer Engineering and Systems (ICCES). IEEE (2015)
16. Navarro, G.: Approximate string matching. In: Encyclopedia of Algorithms, pp. 1–5. Springer, Boston (2014)
17. Pearson, D.W., Janodet, J.C.: String distances and uniformities, pp. 333–339. Springer, Heidelberg (2009)
18. Pikies, M., Ali, J.: Analysis and safety engineering of fuzzy string matching algorithms. ISA Trans. **113**, 1–8 (2021)
19. Rahim, S.K.N.A., et al.: Automated attendance management and alert system (2018)
20. Rout, D.: Online attendance (2023). Title of the publication associated with this dataset: Repository of Online Class attendance
21. Sankoff, D.: Matching sequences under deletion/insertion constraints. Proc. Natl. Acade. Sci. **69**(1), 4–6 (1972)
22. Schneider, M., Bunke, H., Kandel, A.: Using fuzzy logic to match strings in documents. Int. J. Intell. Syst. **16**(5), 609–619 (2001)
23. Smith, J., Johnson, A.: Barriers to online class attendance: a systematic review. J. Educ. Technol. **20**(2), 45–62 (2015)
24. Smith, J., Johnson, A.: The impact of online class attendance on student performance. J. Educ. Res. **20**(3), 45–62 (2015)
25. Stephen, G.A.: String searching algorithms (1994)
26. Ukkonen, E.: Algorithms for approximate string matching. Inf. Contr. **64**(1–3), 100–118 (1985)
27. Upadhyaya, A., Vyas, S.: Algorithmic approach for automating attendance authentication in online classes (2023)
28. Wang, J., Li, G., Feng, J.: Extending string similarity join to tolerant fuzzy token matching. ACM Trans. Database Syst. **39**(1), 1–45 (2014)
29. Weijers, R.J., Ganushchak, L.Y., Ouwehand, K., de Koning, B.B.: "I'll be there": improving online class attendance with a commitment nudge during COVID-19 (2022). https://doi.org/10.1080/01973533.2021.2023534
30. Wu, W.Y.: A method for fuzzy string matching. In: 2016 International Computer Symposium (ICS). IEEE (2016)

Transforming Agriculture with AI and Machine Learning: A Review of Agri-Health and Crop Protection in the Agri 5.0 Era

Nisha Devi[1(✉)] and Shilpa Gupta[2]

[1] Maharaja Agrasen University, Baddi, H.P., India
er.nishasharma89@gmail.com
[2] Chandigarh University, Mohali, Punjab, India

Abstract. Crop protection and Agri-health are essential for the viability and survival of agriculture in the current period of climate change. The last ten years have witnessed the implementation of automated, intelligent, and accurate operations in line with precision agriculture health and digital Agri 1.0 to Agri 4.0 models. Crop protection has improved over time with the application of technology. The current work focuses on Agri5.0, the subsequent evolution that will facilitate a new era of smart crop management with autonomous operations, managed decision-making processes, and a decreasing amount of human labour aided by effective Machine Learning (ML) algorithms and Artificial Intelligence (AI) systems. This study's findings suggest that a new era of intelligent agriculture management, with a particular emphasis on agricultural automation, will usher in under the banner of Agri 5.0. According to the research presented in this paper, Agri 5.0 is the most recent iteration of precision agriculture, with the goal of enhancing crop management through the application of cutting-edge technology like artificial intelligence, machine learning, and big data analytics.

Keywords: Agri1.0 · Agri4.0 · Agri5.0 · Crop protection · Agri-health · sustainability · climate change · technology · intelligent crop management · controlled decision-making processes · autonomous operation · precision agriculture

1 Introduction

In order to gather information about crops and soil conditions, precision agriculture uses technologies like sensors, GPS, and drones. After that, this data is examined by AI and ML algorithms to reveal insights regarding how to maximize agricultural yields while consuming the fewest resources possible. For instance, farmers may precisely time and dose the application of water to their crops by employing sensors to track soil moisture levels. This minimizes water waste and boosts agricultural yields. For agriculture to continue to exist and be sustainable, crop protection and Agri-health are also crucial tasks. It is more crucial than ever to safeguard crops from challenges including

A. Bajaj et al. (Eds.): IBICA 2023, LNNS 1229, pp. 134–143, 2025.
https://doi.org/10.1007/978-3-031-78940-3_13

climate change, the development of pests and diseases, and other dangers. Crop management systems powered by AI and ML can help farmers identify possible hazards more quickly and precisely, lowering the need for toxic chemicals and enhancing the general health of crops. In result, farmers must use precision agriculture and AI and ML-based crop management systems to meet the many demands of modern agriculture, including increasing output and profitability, creating nutritious foods, and reducing resource consumption and environmental effect. Farmers may contribute to ensuring the ongoing survival and sustainability of agriculture in the face of climate change and other difficulties by implementing these cutting-edge strategies, techniques, and technology.

1.1 A Technological Evolution of Agriculture from Agri 1.0 to Agri 5.0

Crop protection and agricultural health have long been challenging fields for research and technology. Humans have created innovative techniques and practices throughout history to safeguard their crops. Agri1.0 represents the prehistoric era when farming was primarily done by labour and how it was carried out in the traditional way [1]. In Agri 2.0 to tackle the common crop pests, synthetic pesticides and specialized equipment were used. At that point, agriculture advanced economically, trying to produce more food at a lower cost, moving in the direction of a more industrialized agriculture [2]. Agri 3.0 is based on technique using cutting-edge technology and data-based modeling as the primary instruments for decision-making and managing agricultural systems. Precision agriculture was developed as a result of this revolutionary idea. Telematics, GNSS (global navigation satellite systems), Device Guidance, and sensor technology were used to improve Agri protection tasks, environmental impacts, lower the cost of pesticides, and raise grain quality [3]. Digital technologies are used by Agri 4.0 to make the agriculture industry more intelligent, efficient, and sustainable. Agriculture-related technologies have advanced to find more effective farming methods and to increase sustainability. This includes all digitalization and automation processes in business and daily life, including huge amounts of data, robotics, the Internet of Things (IoT), augmented and virtual reality (AR), and Artificial Intelligence (AI). The next phase of this evolution, known as Agri 5.0, in which the new generation of smart agriculture management with supervised decision-making approaches, autonomous activities with less human effort replaced by cutting-edge robotics, artificial intelligence (AI), and efficient machine learning (ML) algorithms. [4].

1.2 Agri-Health and Crop Protection with Technological Evolution of Agriculture

The necessity to meet the extraordinary challenges that contemporary agribusiness will encounter during the next few decades is one of the main forces behind this new age. The first of these difficulties is how climate change will affect agricultural systems. Due to extreme heat and water scarcities or surpluses brought on by heavy rainfall in vast productive areas, climate change affects agricultural operations and results in unpredictable crop seasons. These circumstances invariably cause crop damage, a rise in dangerous pests, a deterioration in crop quality, and a threat to food safety [5–7]. For both farmers and researchers, the impacts of climate change on agriculture are a serious worry. A few instances of extreme weather events that can be disastrous for crops are heat waves, floods, and drought resulting in lower yields and lower food quality. Additionally, climate change might make it easier for pests and illnesses to spread, which would hurt crops and make food security less secure. Researchers are looking into novel crop protection and Agri-health technologies and methods to solve these issues. As they can aid farmers and academics in better understanding the intricate interactions between crops, pests, and the environment, AI and ML are becoming more and more crucial to this endeavour. AI and ML can assist in identifying patterns and trends by analyzing vast amounts of data from sensors, satellites, and other sources that would be challenging or impossible to identify using conventional approaches.

The requirement to enhance agricultural output in order to fulfill the demands of an expanding human and animal population is second biggest difficulties that contemporary agriculture will confront in the ensuing decades. The need to protect food security by decreasing the use of agrochemicals and imposing rigorous regulations throughout the entire agricultural supply chain [8] only makes this problem more difficult. The agricultural industry is under increasing pressure from the expanding world population to produce more food while minimizing its negative environmental effects. This is a complicated problem that calls for a multifaceted solution. Reducing the use of agrochemicals, such as pesticides and fertilizers, which may effects the environment, food and human health, which is a crucial goal.

2 Agri 5.0(Agriculture 5.0)

Food systems must now strike a compromise between the need to produce enough affordable, nutrient-dense food and the equally important objective of stopping the destruction of ecosystems [9]. The last ten years have seen the adoption of automated, intelligent, and precise operations in line with the precision agriculture health and digital Agri4.0 models, as crop protection has consistently reinvented itself via the use of technology. The current rise of AI in the primary, industrial, and service sectors can be attributed to developments in machine learning. Although artificial intelligence (AI) refers to the study and creation of computers with the capacity to observe, reason, learn, adapt, make decisions, and act rationally in order to accomplish tasks within a specific environment, this is still true.

3 Machine Learning in Agri-Health and Crop Protection

Machine learning algorithms are gaining popularity across many industries as a tool to decrease the workload and resources needed from humans as well as to speed up the production of results. The literature search focused on studies applying ML algorithms within the context of Agri- health and crop protection.

3.1 Literature Survey

Plant protection pesticides have been used extensively to protect and control diseases and insects due to its high efficacy, speed, low cost and affordability, improving the utilization of pesticides Ml algorithms are used. Due to the varied and nonlinear effects of spray parameters and spatial parameters on droplet size, the polynomial regression has a poor fitting impact. In this situation, using machine learning techniques, the problem can be transformed to High-dimensional linearity derived from low-dimensional nonlinearity. Here multi-layer perceptron (MLP), extreme learning machine (ELM), decision tree (DT), support vector regression (SVR), and radial basis function neural network (RBFNN) algorithms of ML were used [10].

The main variables in the research sector include the detection of multiple diseases, deficiencies, and factors affecting crops' productivity. The three rice diseases studied by the author are brown spot, rice blast, and bacterial blight. Here, transfer learning methods were employed instead of traditional machine learning methods. This discovery could be utilised to help farmers identify rice disease early so that timely remedial action can be implemented [11]. Utilising the MC-CARS-SPA-RFR algorithm, improved the spectral reaction properties, removed the defining bands more thoroughly and effectively, significantly increased the ability to detect powdery mildew, and has a promising future for use in the targeted prevention and management of wheat powdery mildew [12]. To increase nutrients concentration in harvested plant organs the following implementation of machine learning techniques were used. The use of N, P, K, and Cu in leaves is positively connected with a extended development of the plant stage till grain filling, which may result in an increased concentration of protein, P, K, and Cu in grains. A quantitative trait loci (QTI, QTL-4H-1) was employed, and it had a beneficial effect on concentration of these nutrients existed in the leaves [13].

To improve greenhouse agriculture requirements like crop production, irrigation control system and soil moisture consistency the intelligent irrigation system were used. In order to improve the needs for greenhouse agriculture, a RSCS (Remote Sensing Assisted Control System) were proposed which gives the high precision ratio, performance ration, data transmission rate, agriculture production ratio, irrigation control ratio,lowest moisture ratio and CO_2 emission ratio [14]. To optimize Nitrogen management system the deep reinforcement learning and crop simulations were used with DSSAT (Decision Support System for Agro technology Transfer) in which soft actor-critic algorithms, deep Q-network and the soil and Gym-DSSAT interface were proposed [15]. New deep learning methods, incorporating the technique of deep semantic segmentation networks, are efficient ways to find pertinent contextual elements and improve image categorization outcomes. Transfer learning occurs in two steps which involves domain adaptation, where a source domain (an object) is changed to a target domain (a plant), and the other involves category adaptation, where a major category is changed to a minor category. In this case, transfer learning and segmentation of entire leaf areas were proposed using FCNs (fully convolutional networks) [16]. Three-stage multi-branch self-correcting trait estimation network (TMSCNet) for CPA that can offer an empirical foundation for real-time crop protection monitoring. Additionally, seed morphology study is essential for producing higher-yielding and higher-quality crop varieties as well as for comprehending the taxonomic relationships between diverse plant families and genera [17].

Broad-scale applications of ML and satellite imagery include evaluating bacterial blight control in coffee cultivation utilising multiple ecological elements like Landsat-8 surface relief morphometry and reflectance values, VIs, and utilising hydrological (RF) characteristics, SVM, and Naive Bayes [18]. The ambiguity of differentiating food affected by various biotic (such as bacterial and fungal infections) and abiotic (such as water status) factors that exhibit similar symptoms on spectral features in olive and almond trees was also reduced by the use of SVM with a Gaussian kernel and RF techniques [19], using SDA, FLDA, and k-NN algorithms, early stage anthracnose crown rot disease (caused by the fungus Colletotrichum) was classified in the strawberry crop [20], Cotton plant spectral modelling for autumn armyworm (Spodoptera frugiperda) protection using RF, XGBoost, Naive Bayes, LoR, SVM, MLP, and k-NN algorithms [21], Xanthomonas vesicatoria bacteria, as well as late blight and target fungi, were detected in tomato leaves using PCA and k-NN algorithms [22], DT and MLP combat aurel wilt disease against N and Fe nutrient shortages in avocado leaves [23], Sugar beetroot leaves were examined for the presence of the fungus Rhizoctonia root and crown rot (Rhizoctonia solani) using PLS, RF, k-NN, and SVM [24].

A Numerous machine learning algorithms are employed in the field of crop protection to find hazardous pests and plant diseases. Table 1 lists a few machine learning (ML) algorithms that can be utilised to safeguard crops and create the ideal agriculture management system (Table 2).

4 Crop Protection Emerging Technologies Towards Agri 5.0

To achieve precision agriculture there are two basic goals like one is early detection of crop pests and other is to design autonomous decision making system. Before a disease, weed, or epidemic causes too much harm, the former will enable the adoption of more effective prevention strategies at the ideal moment. Given the detrimental effects that the current climate change scenario is having on agricultural systems as a result of the spread of invasive or newly developed pests, it is especially important to build data-driven early detectors. In order to accomplish this, the adoption of Ag5.0 technologies will make it easier to combine data from various tools and sources (such as climate data, remote sensing, sensors, agriculture management systems, etc.) and evaluate the severity of the pests, which will enhance early warning systems and diagnostic software. The latter aims to develop robust autonomous machines that are able to perform the three essential steps of precision crop protection simultaneously in real time namely identifying crop diseases, invasive plants, or plagues at various spatial and temporal scales, examining crop and pest information, and deciding whether to apply a customised site-specific management customised to each crop-pest scenario [19, 20]. Agriculture disciplines are entering the digital age thanks to a variety of innovative electronic gadgets, multi-core embedded systems, IoT ecosystems, smart sensor innovations, and architectures for specialist graphics processing. The purpose of these devices is to gather and use crop data [21].

Technology convergence is making it possible to transform conventional agricultural sensors into advanced sensors with integrated AI computation, or Sensors with a specific chip included at the same circuit that can handle Machine Learning tasks and even do object perception and analysis simultaneously. These AI-sensors would make it easier to identify, target, and control many pest challenges in the present (such as plant species recognition) and create optimised prescriptions for crop protection. The extensive adaptation and evolution of machine learning algorithms used in ML-optimized

Table 1. ML algorithms in the domain of crop protection

References	Objectives	Disease/Result obtained	ML Algorithms
[10]	Improving the utilization of pesticides using plant protection spray	-----	MLP, ELM, DT, SVR, RGFNN
[11]	brown spot, rice blast, and bacterial blight were studied	rice diseases/---	transfer learning methods
[12]	prevention and management of wheat powdery mildew	Wheat fungus/---	PLSR,SVM,RF
[13]	To increase nutrients concentration in harvested plant organs	--/ increased concentration of protein, P, K, and Cu in grains	QTL (quantitative trait loci) (QTL-4H-1)
[14]	improve the needs for greenhouse agriculture	--/ Lower the moisture content ratio and CO_2 emission ratio	RSCS
[15]	optimize Nitrogen management system	--/ Increase soil fertility and crop production	soft actor-critic algorithms, deep Q-network and Gym-DSSAT interface
[16]	Semantic Plant Segmentation	--/ increase crop production	FCN
[17]	designed a novel self-correcting structure	--/ producing higher-yielding and higher-quality crop varieties	TMSCNet
[18]	Detection and progress of bacterial blight in coffee	Coffee bacteria /--	RF, SVM and Naive Bayes
[19]	Detection of bacteria and fungus in the spieces	Almond and Olivia - bacteria /fungus /--	SVM, RF
[20]	Detection of anthracnose crown rot disease in strawberries	Strawberry fungus /--	FDA, SDA, k-NN
[21]	Detection of fall armyworm in cotton plants	Cotton/ fall armyworm /--	SVM, MLP, and k-NN
[22]	To examine bacterial spots and fungal disease in tomato	Tomato bacteria/--	PCA, k-NN

(*continued*)

Table 1. (*continued*)

References	Objectives	Disease/Result obtained	ML Algorithms
[23]	N and Fe deficiency and early and late detection of laurel wilt in avocado	Avocado fungus/ nutrient efficiency/--	DT, MLP
[24]	Examine rhizoctonia root and crown rot in leaves	X sugar beet/ fungal	PLS, RF, k-NN, linear SVM, radial SVM

Table 2. Results and Findings

Reference	Parameter used	Result and findings
[10]	VMD and RS	SVR had best result for VMD i.e. $R_c = 0.9974$, $R_p = 0.9929$ MLP had best result for RS i.e. $R_c = 0.9504$, $R_p = 0.9537$
[11]	InceptionResNetV2(ML) and XceptionNet(Deep learning)	InceptionResNetV2 gives best result then XceptionNet
[12]	RMSE, MAE	For OR R^2 increased by 14.39%, for OR-RFR R^2 increased by 14.39%, and the R^2 of RMSE and MAE decreased by 23.9 and 27.87%
[13]	High-throughput hyperspectral imaging	Concentration of N, P and K in leaf reaches the coefficients of determination to 0.90, 0.75 and 0.89
[14]	precision ratio, performance ratio, data transmission rate, agriculture production ratio and irrigation control ratio	Attains the increased precision ratio = 95.1%, performance ratio = 96.35%, data transmission rate = 92.3%, agriculture production ratio = 94.2% irrigation control ratio = 94.7% Decreased moisture content ratio by 18.7%, and CO2 emission ratio by 21.5%
[16]	F-measure criterion	F-measure criterion was 0.953 for the A2 dataset, and it was 0.355 higher than direct adaptation, and 0.527 higher than non adaptation

(*continued*)

Table 2. *(continued)*

Reference	Parameter used	Result and findings
[17]	TMSCNet to detect weight, dry weight, height, diameter and leaf area of lettuce, NRMSE, NMSE	R^2 values are 0.9514, 0.9696, 0.9129, 0.8481, and 0.9495, NRMSE values are15.63, 11.80, 11.40, 10.18, and 14.65% and NMSE = 0.0826
[20]	SDA, FDA and k-NN	Mean classification accuracy was 71.3%, 70.5%, and 73.6% for SDA, FDA, and k-NN
[21]	F- measure	Healthy plant For OLI, F-Measure = 0.865, For MSI, F-Measure = 0.886 For damaged plant F-measure = 0.912

Abbreviations used: VMD (volume median diameter), RS (relative span), SVR(support vector regression), MLP (multi-layer perceptron), R^2 (coefficient of determination) RMSE (root mean square error), MAE (mean absolute error), OR (original reflectivity), RFR (random forest regression), RSCS (Remote Sensing Assisted Control System), TMSCNet (three-stage multi-branch self-correcting trait estimation network), NRMSE (normalized root mean square error), NMSE (normalized mean squared error), SDA (discriminant analysis), FDA (Fisher discriminant analysis), OLI (Landsat-8 sensor), MSI (Sentinel-2 sensor).

gateways, which enable efficient execution of equipment tasks with limited resources. Ag5.0's usage of ML-Optimized Gateways will enable the optimisation of edge computing hardware, lower latency, and boost privacy, all of which will lead to the development of models that are more reliable and secure. To achieve comprehensive approach to crop management, precision crop protection is moving towards a system approach with numerous connected techniques. In this approach proximal, and remote sensing are essential tools to achieve every abiotic and biotic factor which can affect the crop and food health. To connect devices (such as devices, computer processors, and actuators) and transfer information collected by sensors, telecommunications are necessary in this framework. This networking environment provides value to the activities of analysing data, pest estimation, decision-making, and farm management.

5 Conclusion

Global demand for agricultural products is expected to increase as a result of urbanization, dietary changes, population expansion, and rising incomes. The focus and degree of automation of future Agri 5.0 approaches to precision crop protection will largely determine how well they succeed in replacing current Agri 4.0 approaches. The exciting new era commencing Agri 5.0 of intelligent crop management has been covered in the paper, Using regulated decision-making procedures, autonomous operations, and gradually reduced human effort supported by AI systems and effective Machine Learning (ML) algorithms. These technologies have been utilized, as we have seen, to raise crop yields, make the best use of available resources, and safeguard crops from pests and disease. Increased reliance on AI and ML in agriculture management may have advantages, but there may also be disadvantages that need to be taken into account. It is critical to carry on with the responsible and sustainable research and development of these technologies with an emphasis on maximizing their advantages while lowering their hazards. Overall, we have high hopes for intelligent crop management's future and believe that it will increase global food security and sustainability. For further research in this area we propose that as much of the current research on AI and ML in agriculture has focused on staple crops such as wheat, rice, and corn; however, there is potential to expand these technologies to other crops and regions, including those that are currently underserved by modern agricultural practices.

References

1. Happe, K., Balmann, A., & Kellermann, K.: The Agricultural Policy Simulator (Agripolis) An Agent-Based Model To Study Structural Change In Agriculture (Version 1.0) (No. 918-2016-72618) (2004).
2. Ruane, A.C., et al.: Biophysical and economic implications for agriculture of +1.5 and +2.0 C global warming using AgMIP Coordinated Global and Regional Assessments. Clim. Res. **76**(1), 17–39 (2018)
3. Rahmann, G., et al.: Organic agriculture 3.0 is innovation with research. Org. Agric. **7**, 169–197 (2017)
4. Javaid, M., Haleem, A., Singh, R.P., Suman, R.: Enhancing smart farming through the applications of Agriculture 4.0 technologies. Int. J. Intell. Netw. **3**, 150–164 (2022)
5. Hoegh-Guldberg, O., et al.: The human imperative of stabilizing global climate change at 1.5 °C. Science **365**, eaaw6459 (2019)
6. Mulla, S., Singh, S.K., Singh, K.K., Praveen, B.: Climate change and agriculture: a review of crop models. Global Climate Change Environ. Policy Agric. Perspect. 423–435 (2020)
7. Falkland, T., White, I.: Freshwater availability under climate change. In: Kumar, L. (ed.) Climate change and impacts in the pacific. SC, pp. 403–448. Springer, Cham (2020)
8. van Dijk, M., et al.: Stakeholder-designed scenarios for global food security assessments. Glob. Food Sec. **24**, 100352 (2020)
9. Fraser, E.D., Campbell, M.: Agriculture 5.0: reconciling production with planetary health. One Earth **1**(3), 278–280 (2019)
10. He, Y., Wu, J., Fu, H., Sun, Z., Fang, H., Wang, W.: Quantitative analysis of droplet size distribution in plant protection spray based on machine learning method. Water **14**(2), 175 (2022)
11. Sharma, M., Kumar, C.J., Deka, A.: Early diagnosis of rice plant disease using machine learning techniques. Archiv. Phytopathol. Plant Protect. **55**(3), 259–283 (2022)
12. Feng, Z.H., et al.: Hyperspectral monitoring of powdery mildew disease severity in wheat based on machine learning. Front. Plant Sci. **13**, 828454 (2022)
13. Grieco, M., et al.: Dynamics and genetic regulation of leaf nutrient concentration in barley based on hyperspectral imaging and machine learning. Plant Sci. **315**, 111123 (2022)
14. Zhou, Y., Xia, Q., Zhang, Z., Quan, M., Li, H.: Artificial intelligence and machine learning for the green development of agriculture in the emerging manufacturing industry in the IoT platform. Acta Agriculturae Scandinavica, Section B—Soil & Plant Science **72**(1), 284–299 (2022)

15. Wu, J., Tao, R., Zhao, P., Martin, N.F., Hovakimyan, N.: Optimizing nitrogen management with deep reinforcement learning and crop simulations. In: Proceedings of the IEEE/CVF Conference on Computer Vision and Pattern Recognition, pp. 1712–1720 (2022)
16. Sakurai, S., Uchiyama, H., Shimada, A., Arita, D., Taniguchi, R.I.: Two-step transfer learning for semantic plant segmentation. In: ICPRAM, pp. 332–339 (2018)
17. Zhang, Q., Zhang, X., Wu, Y., Li, X.: TMSCNet: A three-stage multi-branch self-correcting trait estimation network for RGB and depth images of lettuce. Front. Plant Sci. **13**, 982562 (2022)
18. De Carvalho Alves, M., Pozza, E.A., Sanches, L., Belan, L.L., de Oliveira Freitas, M.L.: Insights for improving bacterial blight management in coffee field using spatial big data and machine learning. Trop. Plant Pathol. 1–22 (2022).
19. Zarco-Tejada, P.J., et al.: Divergent abiotic spectral pathways unravel pathogen stress signals across species. Nat. Commun. **12**(1), 6088 (2021)
20. Lu, J., Ehsani, R., Shi, Y., Abdulridha, J., de Castro, A.I., Xu, Y.: Field detection of anthracnose crown rot in strawberry using spectroscopy technology. Comput. Electron. Agricult. **135**, 289–299 (2017)
21. Ramos, A.P.M., et al.: Detecting the attack of the fall armyworm (Spodoptera frugiperda) in cotton plants with machine learning and spectral measurements. Precision Agric. **23**(2), 470–491 (2022)
22. Lu, J., Ehsani, R., Shi, Y., de Castro, A.I., Wang, S.: Detection of multi-tomato leaf diseases (late blight, target and bacterial spots) in different stages by using a spectral-based sensor. Sci. Rep. **8**(1), 2793 (2018)
23. Abdulridha, J., Ampatzidis, Y., Ehsani, R., de Castro, A.I.: Evaluating the performance of spectral features and multivariate analysis tools to detect laurel wilt disease and nutritional deficiency in avocado. Comput. Electron. Agric. **155**, 203–211 (2018)
24. Barreto, A., Paulus, S., Varrelmann, M., Mahlein, A.K.: Hyperspectral imaging of symptoms induced by Rhizoctonia solani in sugar beet: Comparison of input data and different machine learning algorithms. J. Plant Dis. Prot. **127**(4), 441–451 (2020)

NBA 2K20 Player Rating Predictions Using Machine Learning and Ensemble Learning Approaches

Selim Buyrukoğlu[1], Mohammed Rashad Baker[2](✉), Kamal H. Jihad[3], Taha Etem[1], and Gonca Buyrukoğlu[4]

[1] Faculty of Engineering, Computer Engineering, Çankırı Karatekin University, Çankırı, Turkey
[2] Department of Software, College of Computer Science and Information Technology, University of Kirkuk, Kirkuk, Iraq
mohammed.rashad@uokirkuk.edu.iq
[3] College of Science, University of Kirkuk, Kirkuk, Iraq
[4] Faculty of Science, Department of Statistics, Çankırı Karatekin University, Çankırı, Turkey

Abstract. The sports industry has witnessed significant growth, capturing the attention of a diverse audience. Within this burgeoning industry, athletes hold a substantial share, and their performances and ratings assume paramount significance, particularly in the determination of their remuneration. This study delves into the realm of sports video games, focusing on the NBA 2K20 official video game, employing regression analysis and various Machine Learning (ML) algorithms, including k-Nearest Neighbors (Knn), Lasso, and Extreme Gradient Boosting (XGBoost), to evaluate player ratings. Moreover, the study investigates the impact of ensemble methods on model performance. In this study, we applied ML algorithms within the domain of sports video games. This endeavour not only yielded valuable insights into the gaming experience but also highlighted the potential for data-driven decision-making within the sports industry. This study shows that Knn consistently exhibits superior predictive accuracy in both paradigms. While Boosting showed improvement, Bagging emerged as particularly prominent, engendering incremental enhancements in MAE and R2 values across all models, highlighting its effectiveness in ameliorating predictive precision.

Keywords: Machine Learning · Bagging · Boosting · NBA 2K20 · Video Games

1 Introduction

In sports science, analysis of players' data is important for managers in order to correctly manage players [1]. Managers observed players in order to obtain detailed information about them during the training in the past. Then, they made a plan for each player in terms of correctly managing his/her improvement for the next/following training. However, people making observations during the data collection process may make mistakes [2]. In other words, the data obtained by the managers as a result of observation and the

A. Bajaj et al. (Eds.): IBICA 2023, LNNS 1229, pp. 144–153, 2025.
https://doi.org/10.1007/978-3-031-78940-3_14

analysis of these data may not be at the desired level. In this sense, players may not manage correctly in terms of their improvement, and so it can be considered an important problem in sports science. On the other hand, the use of Artificial Intelligence (AI), Machine Learning (ML), and Deep Learning (DL) algorithms in terms of analysing data about players' performance has significantly increased with the development of technology especially in the last decade [3].

Data analysis have been carried out employing ML algorithms in nearly all sport science including basketball, football, volleyball, handball, etc. It should be also highlighted that ML learning algorithms mostly preferred to analyse online game data besides a variety of sports sciences.

Qader et al. [4], conducted a study wherein they utilized Artificial Neural Networks (ANN) to train the background of the DAMA game. They implemented the NEAT algorithm, coupled with extensive training time involving ANN, which yielded promising results. This utilization of ML algorithms empowers managers by allowing them to focus on detailed player data and a plethora of features.

The interdisciplinary nature of sports science, which includes diverse areas like basketball and football can be examined as a research topic. For instance, one study applied machine learning methods to FIFA19 soccer video game for forecasting player ratings, using dataset from Kaggle. The researchers tested different feature selection techniques like ReliefF, Gain Ratio, Chi-Square, and Information Gain for better performance. Buyrukoğlu and Savaş conducted a machine learning approach to predict FIFA19 player ratings. Their work focuses on how machine learning can be applied to predict performance metrics for athletes and sports video game characters [5]. Notably, the researchers employed a sophisticated stacked ensemble model, incorporating Level 0 models, including Deep Neural Network (DNN), Random Forest (RF), and Gradient Boosting (GB), and a Level 1 model, Logistic Regression (LR). Impressively, by employing the Chi-Square feature selection method in conjunction with the proposed stacked model, a commendable accuracy rate of 83.9% was achieved. While football undoubtedly enjoys widespread recognition as one of the most prominent subjects within the realm of sports sciences, it is essential to acknowledge that basketball holds a similar stature, as recognized by sports scientists [6, 7]. Consequently, the subsequent paragraph will expound upon existing scholarly contributions that delve into the examination of basketball-related data through the prism of ML algorithms.

In a study by Zaid and Rajendran [8], the accuracy of ML algorithms in predicting adult income was explored. The study employed the "Adult income dataset," a publicly available resource on Kaggle. The Decision Tree (DT) and Gaussian Naïve Bayes (GNB) algorithms were the principal ML approaches examined. Notably, the DT model exhibited a higher accuracy rate at approximately 84%, outperforming the GNB model, which achieved a 79% accuracy rate. However, a noteworthy limitation of this study is that it predicted income class rather than actual salary.

Asadi and Tasdemir [9], aimed to predict the market value of football players using ML techniques. The dataset utilized in this investigation was sourced from the publicly available FIFA20 video game dataset on Kaggle. The RF regression model yielded the highest R2 score of 95%, while other regression methods, including LR, Multiple Linear Regression (MLR), and RT, were also employed.

Jain et al. [10], conducted a study focused on predicting the salaries of NBA players using ML methodologies. They compiled a private dataset by collecting data from the 2020-2021 season from ESPN and NBA official websites. Among various regression models, the RF regressor achieved the highest R2 result of 88.18%, surpassing Multiple Linear Regression (MLR), DT, and Extreme Gradient Boosting (XGB).

Li [11], proposed a study to create a salary regression model for NBA players. This study primarily relied on empirical economic models to analyze detailed parameters related to NBA player salaries. However, a notable limitation of this study is the absence of advanced systems such as ML or DL.

Horvat et al. [12], applied ML algorithms to predict the outcomes of NBA games. This extensive study analyzed data from 6567 NBA basketball games spanning five seasons, from 2013/2014 through 2017/2018. Data collection was facilitated by a web scraping program to obtain statistics from the official NBA website, and analysis was conducted using the Basketball Coach Assistant (BCA) information system. Feature selection was manually performed to optimize results, with K-Nearest Neighbors (KNN) achieving the highest accuracy of 57.9% compared to LR, Naïve Bayes (NB), DT, Multilayer Perceptron Layer (MLP), RF, and LogitBoost.

Schanilec et al. [13], conducted a study to predict MVP candidates for an NBA season using public NBA data and four distinct ML algorithms: Support Vector Machine (SVM), Naïve Bayes (NB), Random Forest (RF), and Gradient Tree Boosting (GTB). All of the ML models attained accuracy scores exceeding 99%.

Thabtah et al. [14], aimed to predict NBA game results through ML methods using a public dataset available on Kaggle, which contained player statistics spanning 30 years. Feature selection techniques, including Multiple Regression, Correlation Feature Set (CFS), and RIPPER algorithm, were applied. Three different models, NB, Artificial Neural Network (ANN), and Logistical Model Tree (LMT), achieved accuracy rates of 80%, 80%, and 83%, respectively, compared to other feature sets.

Zuo et al. [15], proposed a model for predicting player ratings based on a study that utilized publicly available datasets from Amazon reviews. This model employed a correlation-based regression approach to evaluate prices in categories such as pet supplies, compact discs, office products, and video games.

Nor et al. [16], conducted a study to predict user ratings in video games, focusing on visualization and prediction processes using web analytics. Notably, this study exclusively utilized the RF classification method, achieving an accuracy of 96.12%.

Karaçam et al. [17], explored the mental state and emotion recognition of referees in basketball games. Data collection was conducted using the Warwick-Edinburgh Mental Well-Being Scale. However, a significant limitation of this study is its reliance on the authors' subjective opinions, with no application of data processing methods such as ML.

The aforementioned studies provided convincing scores based on their purposes. However, to the best of the authors' knowledge, researchers did not focus on comparing the performance of the bagging and boosting ensemble approach in the prediction of basketball players' overall ranking grades. Thus, the objectives of this study are listed below with bullet points according to the limitations of the studies.

- To determine the optimal dataset by applying data preprocessing techniques.

- To make a prediction of basketball players' overall ranking score by employing Knn, Lasso, XGBoost, Bagging, and Boosting Ensemble algorithms.
- To compare the performance of the employed Bagging and Boosting Ensemble approaches.

This study contributes the determination of the player's marketing value correctly based on these objectives.

The structure of this manuscript is organized as follows: The proposed basketball players' overall ranking prediction approach is presented in Sect. 2. Detailed information about the results obtained from employed ML algorithms and a discussion of the results are available in Sect. 3. Conclusion and further studies are shown in the final section.

2 Materials and Method

In this section we provide a concise overview of our research framework, as illustrated in Fig. 1. We delve into the components, techniques, and processes involved in developing and assessing the transparency of our ML model. Additionally, we emphasize the critical role of tools, data sources, and analytical methods in laying the groundwork for presenting our findings and results in subsequent sections.

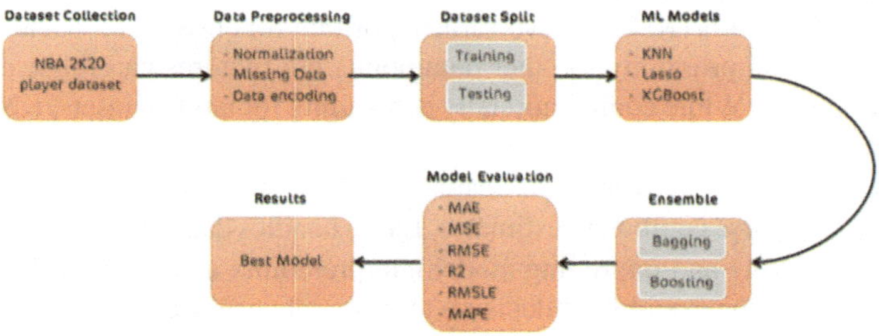

Fig. 1. The proposed ML structure

2.1 Data Collection

The dataset was used from Kaggle, a highly trusted data repository known for its reliability. Specifically, the dataset used is the "NBA 2k20 player dataset", which includes comprehensive information on 429 current NBA players, including their full names, countries of origin, and nationalities [18]. Kaggle is a trusted database provider that strict data verification procedures to ensure the reliability of the dataset for data mining applications. Some of the features like player names, nationality, and game statistics in NBA 2K20 dataset provide a solid foundation for investigating player attributes and performance metrics in a predictive model. Selection of the data is an import metric for research. Results will be better with a well-prepared dataset [19].

2.2 Data Preprocessing

The data preprocessing has some important data preparation procedures. Firstly, normalizing numerical variables to standardize scales is the most important part of the preprocessing. Imputing missing values by taking means, and encoding categorical features numerically are the other important specifications of a dataset. These steps improve data quality. Preprocessing procedures make the dataset suitable for later analysis and modeling. This process is especially important for features with different units or scales and for dealing with missing real-world data [20].

2.3 ML Models

Machine learning algorithms specialize in modeling relationships between variables to predict continuous outcomes. However, each algorithm has unique strengths and limitations. Researchers must carefully consider the right model for a given problem and dataset. As a result, advancing and extending ML to make it applicable across diverse industries and domains is crucial [21].

Classical statistical methods rely on assumptions like normality and equal variance that are often violated in practice. Machine learning algorithms present an alternative approach in these situations. Machine learning methods focus on making optimal predictions from the available data without losing information or imposing restrictions.

K-nearest neighbors (KNN) is a versatile algorithm for classification and regression. It predicts based on proximity to other data points but requires careful selection of K and distance metric. While computationally demanding for large datasets, KNN remains valuable for its simplicity and interpretability [22].

Lasso regression is known for its feature selection and regularization capabilities. By encouraging sparse coefficient estimates, Lasso handles datasets with many features, preventing overfitting and improving model interpretability across domains [23].

XGBoost uses an ensemble of decision trees to enhance accuracy. It leverages gradient descent optimization for speed, regularization to control overfitting, and provides feature importance scores for interpretability. Highly effective at capturing complex data patterns, handling missing values, and diverse data types [24].

2.4 Ensemble Methods

Ensemble methods like Boosting and Bagging are crucial machine learning techniques, improving predictive performance and model robustness [25]. Bagging trains models on random data subsets then combines predictions, reducing variance and overfitting [26]. Boosting iteratively trains base learners to enhance better models for improving accuracy of the system [27]. These methods leads to highly accurate ensemble classifiers with reduced bias. Ensemble methods exemplifies the "wisdom of crowds" concept, with combined models often outpacing individual ones, making them vital in machine learning.

2.5 Evaluation Metrics

Assessing performance of a regression model is essential part of the predictive applications., Mean Absolute Error (MAE), Mean Squared Error (MSE), Root Mean Squared Error (RMSE), R-squared (R^2), Root Mean Squared Logarithmic Error (RMSLE), and Mean Absolute Percentage Error (MAPE) are some of the crucial metrics of a regression model [28]. MAE measures average absolute differences between predictions and actuals, indicating accuracy. MSE and RMSE account for squared and squared-rooted differences, emphasizing outlier impact. R^2 quantifies the explained variance in the target variable, while RMSLE is useful for logarithmic transformations and underestimation sensitivity. MAPE calculates prediction errors as a percentage of actual values, offering accuracy insights. These metrics collectively guide model selection for specific tasks and datasets, providing a comprehensive performance assessment.

3 Results and Discussion

In this section, we present the results of suggested ML models, both individually and in ensemble configurations. The performance metrics, including MAE, MSE, RMSE, R^2, RMSLE, and MAPE, are tabulated for each model and ensemble method. Out dataset was used 70:30 for training and testing consequently. The following subsections provide detailed performance results and are referred to in the subsequent discussion.

3.1 Performance of Individual ML Models

In Table 1, an extensive evaluation of the performance of the suggested three distinct ML models is presented. The discernible trend within the results is the superior performance of the Knn model across multiple performance metrics. Specifically, Knn attains the lowest MAE at 3.3017, signifying its proficiency in generating predictions characterized by minimal average absolute errors. Furthermore, Knn prominently exhibits the highest R^2 value, registering at 0.5378. This R^2 value signifies an admirable model fit to the data, denoting a considerable extent of variance explained by the model. Nevertheless, it is noteworthy that Knn does display a relatively elevated RMSLE of 0.0544, indicating a potential variability in its performance when applied to different data instances.

Lasso and XGBoost, while providing competitive results, show slightly higher errors and lower R^2 values compared to Knn. This implies that Knn may be better suited for this specific prediction task based on individual model performance.

Table 1. Performance of Individual ML Models

ML Model	MAE	MSE	RMSE	R^2	RMSLE	MAPE
Knn	3.3017	20.9553	4.4475	0.5378	0.0544	0.0412
Lasso	4.1044	25.9807	5.0099	0.4166	0.0628	0.0528
XGBoost	4.3781	30.4268	5.4491	0.3041	0.0688	0.0570

3.2 ML Model Performance Using Bagging Ensemble

Table 2 displays the outcomes when the identical ML models are employed in conjunction with the Bagging ensemble technique. Bagging aims to reduce variance by aggregating multiple instances of a model trained on different subsets of the data. Knn continues to perform better in this ensemble model, with a minimal MAE of 3.2760 and a maximum R^2 score of 0.5398 between the models. It's interesting to see that using bagging has slightly improved each model's performance, especially when it comes to raising R^2 values and decreasing MAE.

These findings imply that Bagging is useful in raising the models' prediction accuracy, with Knn continuing to be the best option.

Table 2. ML Model Performance Using Bagging Ensemble

ML Model	MAE	MSE	RMSE	R^2	RMSLE	MAPE
Knn	3.2760	20.9054	4.4344	0.5398	0.0542	0.0408
Lasso	4.1031	26.0101	5.0127	0.4160	0.0628	0.0528
XGBoost	4.3579	29.9281	5.3983	0.3181	0.0682	0.0567

3.3 ML Model Performance Using Boosting Ensemble

Table 3 showcases the results obtained when employing the identical ML models in conjunction with the Boosting ensemble methodology.

Unlike Bagging, Boosting focuses on reducing bias and improving model accuracy by iteratively giving more weight to misclassified instances. In this scenario, Knn maintains its lead in terms of MAE of 3.2751 and R^2 of 0.5374 compared to Lasso and XGBoost. However, it is noteworthy that the Boosting ensemble meth-od does not exhibit the same level of improvement observed with Bagging.

These results indicate that while Boosting can enhance the performance of the models, its impact is less pronounced than that of Bagging in the context of our prediction task. Knn still stands out as the most favorable model in terms of pre-dictive accuracy.

Table 3. ML Model Performance Using Boosting Ensemble

ML Model	MAE	MSE	RMSE	R2	RMSLE	MAPE
Knn	3.2751	21.0004	4.4404	0.5374	0.0548	0.0413
Lasso	4.1221	26.6630	5.0836	0.3964	0.0641	0.0535
XGBoost	4.3039	29.2126	5.3317	0.3350	0.0673	0.0560

The results presented in this section illustrate the performance of individual machine learning models as well as ensemble methods that combine multiple models. The k-nearest neighbors algorithm consistently exhibits superior predictive power, whether used by itself or within an ensemble method. This indicates that KNN is a viable and dependable choice for the particular prediction problem under investigation. When selecting the final model to deploy, computational efficiency and interpretability are critical additional factors to consider alongside accuracy. To further boost the predictive capabilities of the models, future research could explore tuning the models' hyperparameters and testing other ensemble approaches to amalgamate the strengths of multiple models. Moreover, domain-specific insights and feature engineering may provide opportunities to improve model performance and gain a deeper understanding of the underlying data patterns.

4 Conclusion and Future Works

In this manuscript, insights into the gaming experience and player performance in NBA 2K20 were sought by predicting basketball players' overall ranking scores in the study. Utilizing Knn, Lasso, XGBoost, Bagging, and Boosting Ensemble algorithms, patterns and relationships within the dataset were aimed to be uncovered, thereby offering specific insights into players' in-game performance.

Moreover, this study delved into the impact of employing ensemble methodologies on model performance. In both ensemble paradigms, Knn consistently exhibited superior predictive accuracy. Notably, Bagging engendered incremental enhancements in MAE and R2 values across all models.

While Boosting did elicit some degree of improvement in model performance, it fell short of achieving the commensurate level of enhancement observed with Bagging. These findings underscore the efficacy of ensemble techniques, with Bagging particularly prominent, in ameliorating the predictive precision of the models for the specified prediction task. Nonetheless, Knn consistently emerged as the most judicious choice, whether as an individual model or as an integral component within ensemble frameworks.

The research advocates for additional exploration of alternative ensemble methods, refinement of hyperparameters, and the application of domain specific expertise to enhance model performance and gain deeper insights. The conclusion emphasizes the importance for practitioners to weigh interpretability and computational efficiency alongside predictive accuracy when selecting models for real-world applications.

The study recommends applying ensemble techniques, particularly Bagging, in real-world sports for improved player rating predictions. This enhances accuracy in strategic decisions about team composition, player recruitment, and performance evaluation. Prioritizing interpretability and computational efficiency ensures practicality in dynamic sports environments, enhancing decision-making.

Future research should explore alternative ensemble techniques and fine-tune hyperparameters for better model performance. Domain-specific expertise and advanced feature engineering can provide deeper insights into data patterns and improve predictions. When selecting a model for real-world use, consider not only predictive accuracy but also interpretability and computational efficiency.

References

1. O'Gorman, J., Partington, M., Potrac, P., Nelson, L.: Translation, intensification and fabrication: professional football academy coaches' enactment of the Elite Player Performance Plan. Sport Educ. Soc. **26**(3), 309–325 (2021). https://doi.org/10.1080/13573322.2020.1726313
2. Easton, K.L., McComish, J.F., Greenberg, R.: Avoiding common pitfalls in qualitative data collection and transcription. Qual Health Res **10**(5), 703–707 (2000). https://doi.org/10.1177/104973200129118651
3. Tyagi, A.: P.C.-R. A. on M. Learning, and undefined 2022, "Artificial intelligence and machine learning algorithms. igi-global.com. https://www.igi-global.com/chapter/artificial-intelligence-and-machine-learning-algorithms/307466. Accessed: 29 Sep 2023
4. Qader, B.A., Jihad, K.H., Baker, M.R.: Evolving and training of neural network to play DAMA board game using NEAT algorithm. Informatica (Slovenia) **46**(5), 29–37 (2022). https://doi.org/10.31449/inf.v46i5.3897
5. Buyrukoğlu, S., Savaş, S.: Stacked-based ensemble machine learning model for positioning footballer. Arab J. Sci. Eng. **48**(2), 1371–1383 (2023). https://doi.org/10.1007/S13369-022-06857-8/TABLES/6
6. Cai, W., Yu, D., Wu, Z., Du, X., Zhou, T.: A hybrid ensemble learning framework for basketball outcomes prediction. Phys. A: Stat. Mech. Appl. **528**, 121461 (2019). https://doi.org/10.1016/J.PHYSA.2019.121461
7. Rodrigues, F., Pinto, Â.: Prediction of football match results with machine learning. Procedia Comput. Sci. **204**, 463–470 (2022). https://doi.org/10.1016/J.PROCS.2022.08.057
8. Zaid, M., Rajendran, T.: Higher classification accuracy of income class using decision tree algorithm over naive bayes algorithm. Adv. Parallel Comput. **41**, 555–561 (2022). https://doi.org/10.3233/APC220079
9. Al-Asadi, M.A., Tasdemir, S.: Predict the value of football players using FIFA video game data and machine learning techniques. IEEE Access **10**, 22631–22645 (2022). https://doi.org/10.1109/ACCESS.2022.3154767
10. Jain, A., Jain, S., Pancinovia, N.M., George, J.P.: A non-linear approach to predict the salary of NBA athletes using machine learning technique. In: 2022 International Conference on Trends in Quantum Computing and Emerging Business Technologies, TQCEBT 2022, Institute of Electrical and Electronics Engineers Inc. (2022). https://doi.org/10.1109/TQCEBT54229.2022.10041664.
11. Li, N.: The determinants of the salary in NBA and the overpayment in the year. Clemson University (2014). https://tigerprints.clemson.edu/all_theses/2037/. Accessed 27 Sep 2023
12. Horvat, T., Job, J., Logozar, R., Livada, Č: A data-driven machine learning algorithm for predicting the outcomes of NBA games. Symmetry **15**(4), 798 (2023). https://doi.org/10.3390/sym15040798
13. Schanilec, M., Smith, R., Harun, I., Ali, M., Schoh, M., Mahmoud, M.: Classification of NBA MVP candidates using machine learning. In: The 2022 World Congress in Computer Science, Las Vegas, Nevada (2022). https://www.researchgate.net/publication/360756477_Classification_of_NBA_MVP_Candidates_Using_Machine_Learning. Accessed 27 Sep 2023
14. Thabtah, F., Zhang, L., Abdelhamid, N.: NBA game result prediction using feature analysis and machine learning. Ann. Data Sci. **6**(1), 103–116 (2019). https://doi.org/10.1007/S40745-018-00189-X/TABLES/4
15. Zuo, X., Liang, S., Yuan, X., Yu, S., Yang, B.: Dynamic item feature modeling for rating prediction in recommender systems. Neurocomputing **549**, 126412 (2023). https://doi.org/10.1016/J.NEUCOM.2023.126412
16. Nor, M.H.M., Liong, C.Y., Gunasekaran, H.: Visualisation and prediction of user ratings on video games using web analytics. In: AIP Conference Proceedings, vol. 2500, no. 1 (2023), https://doi.org/10.1063/5.0128976/2875193.

17. Karaçam, A., Orhan, B.E., Özdemir, A.S., Sabuncu, A.A., Mereuță, C.: Examination of the relationship between basketball referee's mental well-being and emotional regulation levels. Heliyon **9**(6), e16768 (2023). https://doi.org/10.1016/J.HELIYON.2023.E16768
18. Isaienkov, K.: NBA 2k20 player dataset," Kaggle Dataset.
19. Shi, Z., Li, M., Wang, M., Shen, J., Chen, W., Luo, X.: NPIPVis: a visualization system involving NBA visual analysis and integrated learning model prediction. Virtual Reality Intell. Hardware **4**(5), 444–458 (2022). https://doi.org/10.1016/j.vrih.2022.08.008
20. Bouke, M.A., Abdullah, A.: An empirical study of pattern leakage impact during data preprocessing on machine learning-based intrusion detection models reliability. Expert Syst. Appl. **230**, 120715 (2023). https://doi.org/10.1016/j.eswa.2023.120715
21. Baker, M.R., et al.: Uncertainty management in electricity demand forecasting with machine learning and ensemble learning: case studies of COVID-19 in the US metropolitans. Eng Appl Artif Intell **123**, 106350 (2023). https://doi.org/10.1016/j.engappai.2023.106350
22. Sumayli, A.: Development of advanced machine learning models for optimization of methyl ester biofuel production from papaya oil: Gaussian process regression (GPR), multilayer perceptron (MLP), and K-nearest neighbor (KNN) regression models. Arabian J. Chem. **16**(7), 104833 (2023). https://doi.org/10.1016/j.arabjc.2023.104833
23. Wada, T.: Out-of-sample forecasting of foreign exchange rates: the band spectral regression and LASSO. J. Int. Money Financ. **128**, 102719 (2022). https://doi.org/10.1016/j.jimonfin.2022.102719
24. Bansal, N., Singh, D., Kumar, M.: Computation of energy across the type-C piano key weir using gene expression programming and extreme gradient boosting (XGBoost) algorithm. Energy Reports **9**, 310–321 (2023). https://doi.org/10.1016/j.egyr.2023.04.003
25. Hu, X., Mei, H., Zhang, H., Li, Y., Li, M.: Performance evaluation of ensemble learning techniques for landslide susceptibility mapping at the Jinping county, Southwest China. Natural Hazards **105**(2), 1663–1689 (2021). https://doi.org/10.1007/s11069-020-04371-4
26. Ganaie, M.A., Hu, M., Malik, A.K., Tanveer, M., Suganthan, P.N.: Ensemble deep learning: a review. Eng. Appl. Artif. Intell. **115**, 105151 (2022). https://doi.org/10.1016/j.engappai.2022.105151
27. Yuan, X., Xie, L., Abouelenien, M.: A regularized ensemble framework of deep learning for cancer detection from multi-class, imbalanced training data. Pattern Recognit. **77**, 160–172 (2018). https://doi.org/10.1016/j.patcog.2017.12.017
28. Baker, M.R., Mahmood, Z.N., Shaker, E.H.: Ensemble learning with supervised machine learning models to predict credit card fraud transactions. Revue d'Intelligence Artificielle **36**(4), 509–518 (2022)

Crime Analytics on Location Based Borough Prediction Using Deep Learning

Jibin Joseph, P. V. Anusree, and K. Asha(✉)

Department of Advanced Computing, School of Information Technology, St. Joseph's University, Bangalore, Karnataka, India
ashak@sju.edu.in

Abstract. Crime remains a persistent challenge within urban landscapes, necessitating innovative approaches for prevention and the improvement of public safety. Through the provision of advanced technologies, law enforcement organisations will be able to allocate resources, make data-driven, and informed decisions on crime prevention. The empirical study also sheds light on the effects of regional and contextual changes on crime prediction by revealing the performance differences of different models across cities. Focusing on the vibrant dynamics of two major cities, the empirical study offers an in-depth exploration of crime classification and prediction. Twelve distinct models were employed to classify crimes based on location and associated factors, which will help to analyse historical crime data and identify patterns and trends. The accuracy of each model was assessed to discern the effectiveness in predicting the occurrence of crime in different places. In the context of New York City, the models achieved remarkable accuracy with the highest reaching 99.94% meanwhile in San Francisco achieved an accuracy of 94.27%. These findings offer a valuable resource for law enforcement organisations, urban planners, and policymakers looking to improve public safety and security strategies, as well as for continuing discussions on crime analysis and predictive policing.

Keywords: Machine learning · legal Analytics · smart policing · crime analysis · deep learning · predictive modelling

1 Introduction

In today's world, data analysis has made significant advancements with the increasing use of technology. Crime can include various acts such as murder, theft, assault, etc. that have a negative influence on community safety and well-being. Traditional crime analytics methods often suffer from issues like data inaccuracy and incompleteness. Thus, effective crime prediction and classification are essential components of any comprehensive urban security strategy. As urban areas expand and develop, there is an increasing demand for sophisticated tools and approaches to predict and address criminal activities [21]. In order to improve public safety, accurate crime classification and prediction can support proactive law enforcement measures, resource allocation, and targeted intervention strategies. The relevance of regional and contextual elements in influencing

A. Bajaj et al. (Eds.): IBICA 2023, LNNS 1229, pp. 154–163, 2025.
https://doi.org/10.1007/978-3-031-78940-3_15

crime dynamics is further highlighted by an awareness of the variances in model performance between cities. Due to inherent connections between geographic regions and crime activities, the crime rate in different regions is often correlated. Such correlation brings challenges and opportunities to employ deep learning to learn features from historical data for accurate prediction of future crime numbers for each neighborhood [25]. The research aims to determine how well the models can identify crime patterns in various metropolitan contexts by utilizing a wide variety of machine and deep learning approaches.

The relevance of research extends to broader landscape on crime analytics, which has witnessed a surge in various crime trends on rise of gun violence, increase in hate crimes, growing threat of domestic terrorism, and rise of cybercrime. By aligning the research with the latest developments and challenges in crime analytics, the aim to underscore the relevance and urgency of adopting advanced predictive tools to enhance public safety and security. The incorporation of real-world incidents emphasizes the practical implications of work in mitigating the impact of evolving crime patterns.

2 Legal Analytics and Strategic Case Assessment

Predictive Analytics and Artificial Intelligence (AI) are swiftly advancing technologies that have been making substantial progress across a wide range of sectors, including the realm of criminal justice [1]. The number of crimes committed is said to be the product of the number of people committing crimes and the average frequency at which they commit crimes [23, 24]. Crime has been a constant and problematic issue that causes socioeconomic disparity for society as a whole [3]. When there are not enough police officers to enforce the law, civilians can risk getting injured, and the properties can become easy targets for criminals. The motivation and benefit behind predicting future crimes allows a city to better prepare for the future.

2.1 How Important Is Legal Analytics?

Over the previous year, 75% of survey participants whose law firms had access to legal analytics reported a rise in utilization within the organization, while 92% of the respondents themselves plan to use the technology more frequently in future. The majority of users perceive the technology as a valuable tool for augmenting knowledge, competitiveness, and overall operational efficiency.

Figure 1 depict the usage level recommendations of legal analytics and associated technology in real time. Legal data analytics are objective data points that legal teams use to inform strategic decision-making.

3 Literature Survey

In recent years, there has been an increase in using data analytics and various methods, including machine learning and deep learning with respect to the problems associated with crime analysis, prediction, and prevention. In a recent study, SAT SCAN and K-means clustering were used to evaluate historical crime data from Chicago [2]. A similar

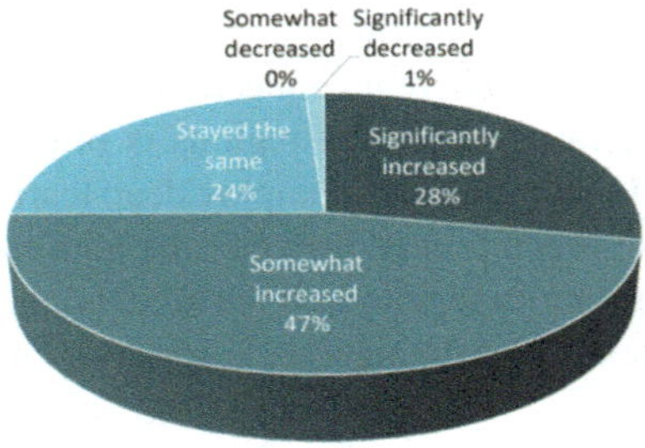

Fig. 1. Use of Legal Analytics

study was also done in the banking industry used K-Means clustering and J48 Prediction Trees [3]. DBSCAN outperformed the other methods when it came to predicting crime with aggregative clustering and clustering algorithms like K-Means [4]. By using the K-Nearest Neighbor Algorithm, a crime prediction's accuracy score increased due to a significant reduction in Mean Absolute Error and Root Mean Square Error [5]. The application of machine learning techniques has greatly aided in the analysis and prediction of crime [6]. In a related study, the San Francisco crime dataset and historical crime data, together with non-historical factors, were used to forecast crime hotspots using the Random Forest method [7]. Using the Random Forest technique, which was more accurate than Linear Regression, crime rates in Bangladesh were predicted [8]. Another study accurately classified crimes into violent and peaceful groups using the Gradient Boosting Decision Tree method [9]. A different approach to the study showed how Naive Bayesian can generate outcomes with very high accuracy rates [10]. A study to look at crime patterns in North Carolina mostly used univariate and bivariate exploratory analysis [11]. It has designed to hasten the resolution of crimes and offer precise forecasts of crime patterns in various states [12]. A complete approach to determining the perpetrator's identity was conducted using machine learning techniques [13]. Random Tree had the highest classification accuracy when a variety of ensemble learning models were used [14]. Numerous studies merged various data sets to provide better results using spatiotemporal data, auto regression methods, and the ARIMA model for statistical analysis, crime can be predicted with exceptional accuracy in a range of contexts [15]. Urban crime was predicted using spatial and temporal correlations [16]. The importance of spatiotemporal patterns in urban data has described and transfer learning techniques to improve crime prediction models across multiple boroughs [17].

A study gave a description of the Crime Prevention Decision Support System gathering data on index offences, locations, and preventative strategies [18]. A novel sentiment analysis and meteorological data-based crime prediction system were presented in order to improve forecast accuracy [19]. Another study focuses on observing, evaluating, and forecasting criminal occurrence trends in order to improve the efficiency of investigations [20]. The categorization and forecasting of crimes have also considerably benefited from the application of various deep learning techniques [21]. Another study found significant improvements in prediction accuracy for the Chicago and Los Angeles crime datasets compared to earlier studies utilizing machine learning and deep learning methodologies [22].

4 Methodology

The analysis of crime patterns in San Francisco and New York City reveals potential insights. Despite having a far lower population than New York, San Francisco experiences more crime occurrences. The number of offenses recorded on various days of the week in both cities varies noticeably. The reason for the elevated crime rate in San Francisco could be attributed to several factors, including higher population density, the presence of a larger homeless population, or its proximity to other major cities. The San Francisco dataset encompasses incident reports filed from January 1, 2018 onwards. The dataset classifies the reports into distinct categories based on the reporting method and the nature of the incident. The NYPD dataset comprises the records of arrests made in New York City, spanning from the year 2006 till the end of last year. The dataset encompasses a rich array of details, including the type of crime, the precise location, and the timestamp of the arrest.

Fig. 2. Monthly Crime Trends in New York and San Francisco

Figure 2 shows the crime rates in both New York and San Francisco have shown a decreasing trend. The crime rates in New York have declined during the course of the whole time period, dropping by more than 30% over the course of five years. The crime rates in San Francisco have also been decreasing over time. Crime rates in each location have been affected differently by the COVID-19 outbreak.

The crime rates in San Francisco remained almost constant during the pandemic, whereas in New York, the rates initially dropped but then increased again. The crime rates in San Francisco have shown a rise in a few of the months in 2019 and 2020. However, the overall crime rates have been declining, with a fall of almost 20% over the course of five years.

4.1 Preparation of Train and Test Data

For the groundwork of train and test data, NYPD and SFPD used many features. Both datasets were split into training and testing sets using the 'train_test_split' function from the 'sklearn. Model_selection' module. Categorical labels were converted to one-hot encoded vectors using 'tf.keras.utils.to_categorical.' Feature scaling or normalization was applied using 'MinMaxScaler' to ensure consistent and effective model training. The classification model, comprising dense layers were created using TesorFlow andKeras.

5 Performance Analysis

The research employed a diverse set of twelve algorithms to evaluate the performance in predicting outcomes for both San Francisco (SFPD) and New York datasets (NYPD). Structured data utilised convolutional neural networks (CNNs), leveraging the effectiveness in tasks involving structured information. When handling sequential data, recurrent neural networks (RNNs) played a vital role, further improved by the Long Short-Term Memory (LSTM) architecture to handle sequences effectively. To further enhance predictive accuracy ensemble models were used. These ensembles incorporating neural networks or utilising gradient boosting methods like CatBoost and LightGBM, consistently improved the predictive models' performance. In the context of New York City, the models achieved remarkable accuracy, with the highest reaching 99.94% using an ensemble approach combining CatBoost and LightGBM classifiers. Meanwhile, in San Francisco, the best-performing model, a Bidirectional Long Short-Term Memory (LSTM) Recurrent Neural Network (RNN), achieved an accuracy of 94.27%. The details are given in Table 1.

Table 1. Performance Analysis Using Different Models

Model	Accuracy (%)		Precision		Recall		F1 score	
	NYPD	SFPD	NYPD	SFPD	NYPD	SFPD	NYPD	SFPD
Feedforward Neural Network Classifier	99.64	92.48	0.998	0.92	0.998	0.92	0.998	0.92
Convolutional Neural Network (CNN)	99.29	93.94	0.994	0.92	0.996	0.88	0.995	0.88
Recurrent Neural Network (RNN)	99.05	93.46	0.994	0.91	0.992	0.87	0.993	0.88
Long Short-Term Memory (LSTM)	99.63	93.66	0.998	0.91	0.998	0.86	0.998	0.86
Wide and Deep Learning Model	99	94.32	0.992	0.91	0.992	0.88	0.992	0.89
Ensemble Model of Neural Networks	99.5	92.72	0.996	0.84	0.996	0.85	0.996	0.84
Enhanced Recurrent Neural Network (RNN)	99.54	89.16	0.996	0.81	0.998	0.80	0.997	0.80

(continued)

Table 1. (*continued*)

Model	Accuracy (%)		Precision		Recall		F1 score	
	NYPD	SFPD	NYPD	SFPD	NYPD	SFPD	NYPD	SFPD
Recurrent Neural Network (RNN) with Long Short-Term Memory (LSTM)	99.68	93.83	0.996	0.91	0.996	0.88	0.996	0.89
Bidirectional Long Short-Term Memory (LSTM) Recurrent Neural Network (RNN)	99.88	94.27	1.0	0.91	1.0	0.88	1.0	0.89
Tabnet	94.49	90.81	0.948	0.82	0.948	0.83	0.946	0.82
Ensemble model combining CatBoost and LightGBM	99.94	94	0.996	0.94	0.996	0.89	0.996	0.90
Ensemble Model with Random Forest, Gradient Boosting, and Support Vector Machines (SVM) Classifiers	99	94	1.0	0.94	1.0	0.89	1.0	0.90

Figures 3 and 4 depict the in-depth analysis of various models applied to two data sets, combining a neural network and two gradient-boosting models, CatBoost and LightGBM. In the case of the New York dataset, several models, including FNN, CNN, RNN, LSTM, RNN, RNN with LSTM, and various ensemble models, scored with a high accuracy of 99%. However, the Tabnet model achieved comparatively lower accuracy. The research is distinctive in its focused examination of the two urban areas, allowing for a deeper analysis of the individual characteristics. Importantly, research delves into the real-world implications and offers policy recommendations for comprehensive crime reduction.

Figures 5 and 6 describe an ensemble model combining a neural network of two gradient-boosting models, CatBoost and LightGBM, on the San Francisco dataset. SFPD achieved a strong accuracy score of 94% with the bidirectional long short-term memory (LSTM) model and recurrent neural network (RNN) model, along with the LSTM-based neural network models.

The Ensemble Model with Random Forest, Gradient Boosting, and Support Vector Machines (SVM) Classifiers also performed exceptionally well, achieving a high accuracy score of 94%.

In contrast, the Enhanced Recurrent Neural Network (RNN) model and the Deep Forward Neural Network had slightly lower accuracy scores of 89% and 88% for San Francisco data, respectively.

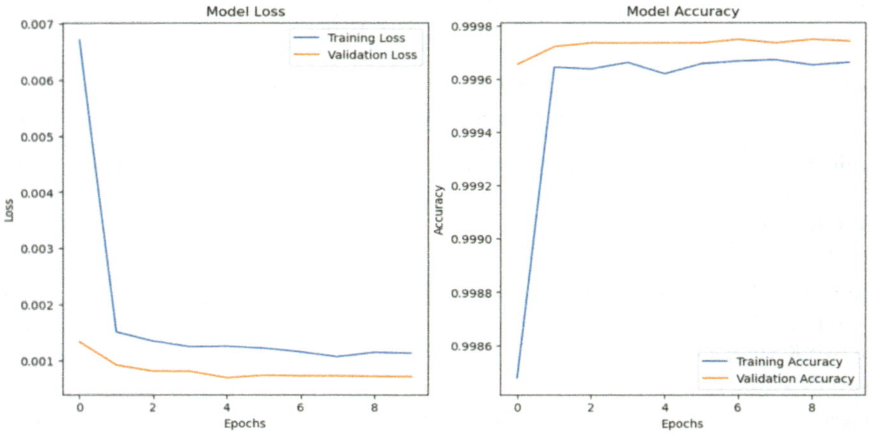

Fig. 3. Ensemble model on NYPD

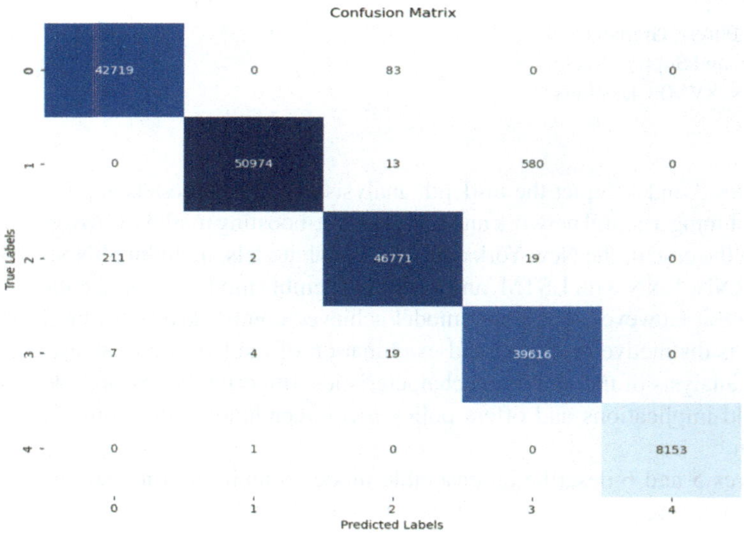

Fig. 4. Ensemble model on SFPD

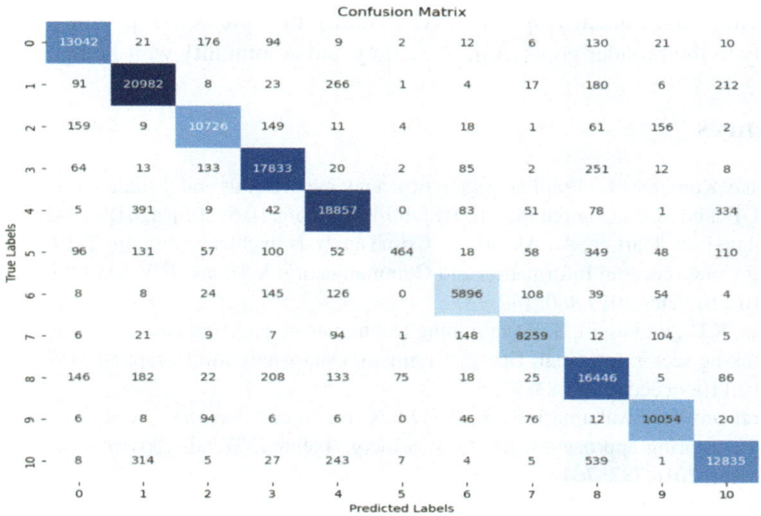

Fig. 5. Confusion matrix on NYPD dataset

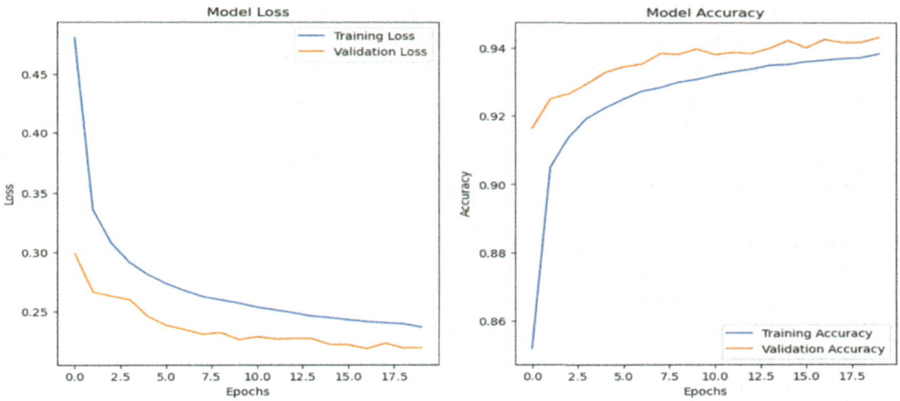

Fig. 6. Confusion matrix on SFPD dataset

6 Conclusion

The pursuit of enhancing urban security and public safety through the lens of legal analytics and smart policing has led the journey to explore the potential of deep learning models in crime analytics and prediction. Law enforcement can harness the power of ensemble models to make informed decisions. While research offers significant strides in crime prediction, there remain avenues for future exploration and refinement. Explainable ensemble models, cross-city model generalization, dynamic adaptation to temporal shifts, ethical implications and bias mitigation are the different areas that can be worked in future. The identified future directions aim to propel the field toward increased accuracy, transparency, and ethical responsibility, ensuring that predictive policing contributes positively to the broader goals of public safety and community well-being.

References

1. Vaibhav Kumar, M.L.: Predictive analytics: a review of trends and techniques. Int. J. Comput. Appl. Foundations Comput. Sci. (2018). https://doi.org/10.5120/ijca2018917434.
2. Alqahtani, A., Garima, A., Alaiad, A.: Crime analysis in chicago city. In: 2019 10th International Conference on Information and Communication Systems (ICICS) (2019). https://doi.org/10.1109/iacs.2019.8809142
3. Lekha, K.C., Prakasam, S.: Data mining techniques in detecting and predicting cybercrimes in banking sector. Int. Conf. Energy Commun. Data Anal. Soft Comput. (2017). https://doi.org/10.1109/icecds.2017.8389725
4. Sivaranjani, S., Sivakumari, S., Aasha, M.: Crime prediction and forecasting in Tamilnadu using clustering approaches. Int. Conf. Emerg. Technol. Trends (2016). https://doi.org/10.1109/icett.2016.7873764
5. Kumar, A., Verma, A., Shinde, G., Sukhdeve, Y., Lal, N.: Crime prediction using k-nearest neighboring algorithm. Int. Conf. Emerg. Trends Inform. Technol. Eng. (2020). https://doi.org/10.1109/ic-etite47903.2020.155
6. Kanimozhi, N., Keerthana, N.V., Pavithra, G.S., Ranjitha, G., Yuvarani, S.: CRIME type and occurrence prediction using machine learning algorithm. Int. Conf., Artific. Intell. Smart Syst. (2021). https://doi.org/10.1109/icais50930.2021.9395953
7. Yao, S., Wei, M., Yan, L., Wang, C., Dong, X., Liu, F., Xiong, Y.: Prediction of crime hotspots based on spatial factors of random forest. In: 15th International Conference on Computer Science and Education (2020). https://doi.org/10.1109/iccse49874.2020.9201899
8. Raza, D.M., Victor, D.B.: Data mining and region prediction based on crime using random forest. Int. Conf. Artific. Intell. Smart Syst. (2021). https://doi.org/10.1109/icais50930.2021.9395989
9. Khan, M., Ali, A., Alharbi, Y.: Predicting and preventing crime: a crime prediction model using san francisco crime data by classification techniques. Complexity (2022). https://doi.org/10.1155/2022/4830411
10. Babakura, A., Sulaiman, M.N., Yusuf, M.A.: Improved method of classification algorithms for crime prediction. Int. Symp. Biometrics Secur. Technol. (2014). https://doi.org/10.1109/isbast.2014.7013130
11. Shukla, A., Katal, A., Raghuvanshi, S., Sharma, S.: Criminal combat: crime analysis and prediction using machine learning. Int. Conf. Intell. Technol. (2021). https://doi.org/10.1109/cnit51480.2021.9498397
12. Yadav, S., Timbadia, M., Yadav, A., Vishwakarma, R., Yadav, N.: Crime pattern detection, analysis & prediction. In: International Conference of Electronics, Communication and Aerospace Technology. https://doi.org/10.1109/iceca.2017.8203676
13. Mary Shermila, A., Bellarmine, A.B., Santiago, N.: Crime data analysis and prediction of perpetrator identity using machine learning approach. In: 2nd International Conference on Trends in Electronics and Informatics (ICOEI) (2018). https://doi.org/10.1109/icoei.2018.8553904
14. Almaw, A., Kadam, K.: Crime Data Analysis and Prediction Using Ensemble Learning. Second Int. Conf. Intell. Comput. Control Syst. (2018). https://doi.org/10.1109/iccons.2018.8663186
15. Yadav, R., Kumari, S.S.: Crime prediction using auto regression techniques for time series data. In: 3rd International Conference and Workshops on Recent Advances and Innovations in Engineering (2018). https://doi.org/10.1109/icraie.2018.8710407

16. Qian, Y., Pan, L., Wu, P., Xia, Z.: GeST: A grid embedding based spatio-temporal correlation model for crime prediction. In: IEEE Fifth International Conference on Data Science in Cyberspace (2020). https://doi.org/10.1109/dsc50466.2020.00009
17. Zhao, X., Tang, J.: Exploring transfer learning for crime prediction. IEEE Int. Conf. Data Min. (2017). https://doi.org/10.1109/icdmw.2017.165
18. Retnowardhani, A., Triana, Y.S.: Classify interval range of crime forecasting for crime prevention decision making. In: 11th International Conference on Knowledge, Information and Creativity Support Systems (2016). https://doi.org/10.1109/kicss.2016.7951409
19. Chen, X., Cho, Y., Jang, S.Y.: Crime prediction using Twitter sentiment and weather. In: 2015 Systems and Information Engineering Design Symposium (2015). https://doi.org/10.1109/sieds.2015.7117012
20. Joshi, N., Shaikh, S., Tripathy, A., Sen, S., Janson, J., Karnik, S., Varghese, B.: Crime anatomization using QGIS. In: IEEE 5th International Conference for Convergence in Technology (2019). https://doi.org/10.1109/i2ct45611.2019.9033910
21. Han, X., Hu, X., Wu, H., Shen, B., Wu, J.: Risk prediction of theft crimes in urban communities: an integrated model of LSTM and ST-GCN. IEEE Access **8**, 217222–217230 (2020). https://doi.org/10.1109/access.2020.3041924
22. Safat, W., Asghar, S., Gillani, S.A.: Empirical analysis for crime prediction and forecasting using machine learning and deep learning techniques. IEEE Access **9**, 70080–70094 (2021). https://doi.org/10.1109/access.2021.3078117
23. Greenburg, D.: Time series analysis of crime rates. J. Quant. Criminol. **17**(4), 291–327 (2001)
24. Lobonţ, O.-R., Nicolescu, A.-C., Moldovan, N.-C., Kuloğlu, A.: The effect of socioeconomic factors on crime rates in romania: a macro-level analysis. Econ. Res.-Ekonomska Istraživanja **30**(1), 91–111 (2017)
25. Shukla, A., Katal, A., Raghuvanshi, S., Sharma, S.: Criminal combat: crime analysis and prediction using machine learning. In: 2021 International Conference on Intelligent Technologies (CONIT). https://doi.org/10.1109/conit51480.2021.9498397.

Education Revolution: LLM's in Online Learning and Student Engagement

P. Karthikeyan(✉), K. K. Nithish, H. Janani, P. Reshma, V. Vaishnavi, and N. Prakash

Department of Management Studies, Kongu Engineering College, Perundurai, Erode, Tamil Nadu 638 060, India
ptp.karthi@gmail.com, {nithishkk.22mba,jananih.22mba, reshmap.22mba,vaishnavi.mba,np.mba}@kongu.edu

Abstract. Large Language Models (LLMs) have revolutionized contemporary education, reshaping online learning and student engagement. This research paper explores the multifaceted impact of LLMs in education. Powered by Natural Language Processing (NLP), LLMs provide personalized, interactive, and adaptive learning experiences, catering to diverse student needs. The paper highlights the benefits of LLM integration in online education, including enhanced content delivery, automated assessment, and the creation of student-centric learning environments. Ethical concerns and challenges related to bias, data privacy, and algorithmic transparency are also investigated. The collaborative role of educators and LLMs is emphasized, underlining the importance of human-AI partnerships in education. The paper offers methods for measuring the impact of LLMs on education and explores future trends in LLM-powered education. The paper provides a global perspective on LLM adoption and utilization across different countries and regions. This research serves as a valuable resource for educators, policymakers, and stakeholders, driving the ongoing educational revolution facilitated by LLMs.

Keywords: Large Language Models · Online Learning · Student Engagement · Natural Language Processing · Education Revolution

1 Introduction

The convergence of education and technology has historically marked a significant driver of transformative change. As the world increasingly turns to digital solutions for learning, the emergence of Large Language Models (LLMs) has become a pivotal mile- stone in this ongoing evolution. LLMs, fueled by the remarkable capabilities of Natural Language Processing (NLP), are reshaping the educational landscape. They infuse fresh vitality into online learning and redefine the way students engage with educational content. This research paper embarks on a journey to unveil the profound impact of LLMs on the field of education. As exemplified by models like GPT-3 and its counterparts, LLMs demonstrate remarkable language prowess, positioning them at the forefront of the intersection between artificial intelligence and education. Their role is poised to revolutionize the acquisition of knowledge, enabling personalized learning experiences.

A. Bajaj et al. (Eds.): IBICA 2023, LNNS 1229, pp. 164–174, 2025.
https://doi.org/10.1007/978-3-031-78940-3_16

At the core of this transformation lies the promise of personalized learning, where the educational journey is finely tuned to cater to the unique requirements and preferences of each student. LLMs have exhibited exceptional prowess in comprehending context, generating text that is human-like, and fostering interactive experiences. They present educators and learners with innovative tools to collaboratively design an educational path that transcends the limitations of one-size-fits-all approaches. This educational revolution is not devoid of its challenges. The ethical considerations surrounding LLMs loom large, necessitating rigorous examination. Questions pertaining to algorithmic bias, data privacy, and the responsible use of artificial intelligence in education demand thoughtful exploration. As we march forward, it is imperative that we navigate this complex landscape with a clear understanding of the ethical imperatives that underpin the integration of LLMs in education.

1.1 Scope

1. **Educational Context:** it is situated within the broader context of education, with a specific emphasis on online learning. Online learning has become increasingly prevalent, and to study aims to explore how LLMs are shaping this educational landscape.
2. **LLMs in Education**: it will delve into the utilization of Large Language Models (LLMs) in educational settings. These models have advanced natural language processing capabilities, and to investigate how they are being harnessed to enhance various facets of education.
3. **Student Engagement:** This encompasses not only traditional notions of engagement but also includes aspects such as student motivation, active participation, interaction with learning materials, and overall satisfaction with the learning experience.
4. **Ethical Considerations:** Ethical considerations are a crucial part of this research. to examine the ethical dimensions of using LLMs in education. This involves scrutinizing issues related to fairness, bias, privacy, and transparency in the deployment of LLMs in educational contexts.
5. **Collaboration with Educator.**: to explore the collaborative relationship between human educators and LLMs. This encompasses how educators leverage LLMs as tools to augment their teaching methods and enhance student learning outcomes.

1.1.1 Objectives

- Investigate the Benefits of LLMs in Online Education
- Explore Ethical Considerations and Challenges
- Examine the Collaborative Relationship with Educators
- Present Case Studies and Best Practices
- Develop Methodologies for Measurement

2 Literature Review

Joseph Chervenak et al., 2023-the performance of the 'Chat GPT' language model in responding to clinical questions about fertility. They compare Chat GPT's responses to established clinical sources, including infertility FAQs, fertility knowledge surveys, and a reproductive medicine committee opinion. The study finds that ChatGPT can provide relevant and meaningful responses, but it has limitations such as the inability to consistently cite sources and the risk of generating false information, which may restrict its clinical utility. The study highlights the potential of language models like ChatGPT in delivering personalized healthcare information but emphasizes the importance of careful usage and ongoing evaluation.

Qingyao Ai et al., 2023 – this paper explore the convergence of Information Retrieval (IR) and Large Language Models (LLMs) like GPT-3. They discuss how LLMs have transformed the IR field by improving text understanding, generation, and knowledge inference. The report emphasizes the synergy between IR models, LLMs, and human users in forming a new technical paradigm for information seeking. They also highlight challenges related to computational costs, domain-specific limitations, trustworthiness, content generation, and more. The report offers insights into the impact of LLMs on IR research and outlines future research directions and challenges.

Jansen, B.J., Jung, S., Salminen, J. (2023) – This paper discusses the potential of using large language models (LLMs) like ChatGPT in survey research. It explores how LLMs can help generate survey responses, addressing some challenges in question wording and response bias. However, it notes that LLMs may not solve all survey research issues, especially those related to sampling or nonresponse bias. Ethical concerns, such as informed consent and privacy, are also discussed. The paper suggests using LLMs alongside other methods to ensure accurate and valid survey results and emphasizes the need for a thoughtful approach to LLM development and use.

Adrian de Wynter et al., 2023 – This paper presents an empirical evaluation of outputs generated by nine large language models (LLMs). The study examines the presence of memorized text, factual errors, logical fallacies, and discourse quality in LLM-generated content. The authors found that LLMs often produce text with memorized content, and outputs with more memorized content tend to have better discourse quality. However, factual errors and logical fallacies were also observed in LLM-generated text. The study discusses mitigation strategies to reduce memorization in LLM outputs. It highlights the importance of evaluating the reliability and quality of LLM-generated content, particularly in applications requiring human-like text generation.

Chaitali Diwan et al., 2021 – This paper focus on addressing the challenge of retaining learner engagement in online education. They propose an AI-driven approach to generate narrative fragments that enhance learning pathways, fostering interactive and engaging learning experiences. Their work emphasizes adaptability, domain-agnosticism, and the potential for transforming online education.

3 Large Language Models (LLM'S) in Online Learning

Large Language Models (LLMs) are advanced artificial intelligence models, typically based on deep learning neural networks, designed to understand, generate, and manipulate human language. In the context of online learning, LLMs are employed as powerful tools to enhance various aspects of the educational experience. These models have been trained on massive amounts of text data and are capable of processing, generating, and interpreting natural language text, enabling them to offer significant benefits to online education.

Adaptive Learning Paths: LLMs are at the core of adaptive learning systems. These systems use LLM-driven algorithms to assess a student's performance and adapt the learning materials accordingly. If a student is struggling with a particular concept, the LLM can provide additional resources or exercises to reinforce understanding, offering a personalized learning path for each student. Translation and Multilingual Support: LLMs can break down language barriers in online education. They are employed to translate course materials and instructions into multiple languages, making online learning accessible to students worldwide. This capability is particularly valuable for global educational platforms. Large Language Models (LLMs) have ushered in a new era of possibilities in education, offering a suite of versatile functions that benefit both educators and students. One of their standout features is content summarization. LLMs excel at distilling lengthy texts and academic papers into concise summaries, providing educators and students with an efficient way to grasp key concepts in complex topics. This capability not only saves time but also aids comprehension, making learning more accessible and manageable.

LLMs also play a vital role in maintaining academic integrity through plagiarism detection. By comparing student assignments to an extensive database of academic and online sources, they help ensure that submitted work is original and adheres to academic standards. This is particularly crucial in online learning environments, where the temptation to plagiarize may be higher. LLMs with Natural Language Processing (NLP) capabilities are utilized to analyze text-based interactions in online discussions and assignments. They can gauge the sentiment and emotions expressed by students, providing educators with valuable insights into student engagement and well-being. This in- formation allows for more personalized support and intervention when necessary.

4 Benefits of LLM'S in Online Learning

The incorporation of Large Language Models (LLMs) into online learning holds several compelling advantages. LLMs enable personalized learning experiences by tailoring educational content to individual student needs, enhancing engagement and com- prehension. They streamline content creation for educators, offering efficient tools to generate materials like course content and assessments. LLM-powered chatbots and virtual tutors provide continuous student support, answering queries and providing guidance 24/7. These models improve accessibility through multilingual support and assistive technologies, making online learning inclusive for diverse audiences. Auto- mated grading and feedback, content summarization, and content recommendation systems simplify the learning process, while plagiarism detection ensures academic integrity. LLMs contribute to language learning by offering pronunciation feedback and grammar correction, and their scalability enables global reach. In essence, LLMs elevate the quality and accessibility of online education, benefiting both learners and educators.

5 Challenges and Ethical Considerations

LLM-powered online learning encompasses issues like data privacy and security, algorithmic bias, transparency, overreliance on automation, the digital divide, ethical content generation, the potential loss of academic skills, teacher training needs, ethical behaviour modelling, and plagiarism prevention.

These challenges require proactive measures, policies, and ethical frameworks to ensure responsible and equitable use of LLMs in education. Incorporating Large Language Models (LLMs) into educational settings presents a host of opportunities and challenges. While these powerful AI-driven systems can enhance online learning experiences, they also raise pressing concerns that must be systematically addressed.

Data privacy and security are paramount, as educational institutions collect, store, and transmit vast amounts of student data. Robust measures must be in place to protect this sensitive information from breaches and unauthorized access. Moreover, the use of LLMs can inadvertently perpetuate biases present in the data they are trained on, leading to unfair assessments, content generation, or recommendations. Continual efforts to identify and mitigate algorithmic bias are essential, emphasizing fairness, equity, and diversity in LLM utilization. Transparency and accountability pose another challenge, as LLMs often operate as "black-box" models, making it difficult to understand their decision-making processes. Clear policies must be established to address these concerns, especially when LLMs are employed for critical tasks like grading or admissions. Ethical content generation is a priority, with educational institutions needing to com- bat the potential misuse of LLMs for unethical purposes. Additionally, there is a risk of students over-relying on LLMs for tasks like spelling and grammar correction, potentially eroding essential language and writing skills. Thus, it is imperative that institutions emphasize skill development alongside LLM use to maintain a balance between technology integration and preserving critical academic abilities. By comprehensively addressing these challenges, educational institutions can harness the benefits of LLMs while responsibly navigating their impact on online learning environments.

6 Student-Centric Learning with LLM'S:

Student-centric learning, enriched by the incorporation of Large Language Models (LLMs), redefines the educational landscape by placing individual students at the heart of their learning journey. LLMs employ advanced data analysis to provide an educational experience tailored to each student's unique needs. They generate personalized content; adapt assessments in real-time, and offer immediate, detailed feedback, allowing students to progress at their own pace. Moreover, LLMs enable 24/7 support through virtual tutors and chatbots, ensuring that students have access to assistance whenever they require it. Language support, content summarization, and recommendation systems further enhance the learning process, making education more accessible and engaging for diverse learners. By utilizing LLMs, educators can harness the power of data and artificial intelligence to create a student-centric educational environment that fosters deeper comprehension, encourages self-directed learning, and accommodates the diverse needs and backgrounds of students.

7 Teacher- LLM'S Collaboration:

The collaboration between teachers and Large Language Models (LLMs) represents a transformative partnership that can elevate the quality of education. LLMs, equipped with advanced natural language processing capabilities, assist teachers in various facets of their educational roles. They aid in content creation, enabling teachers to generate high-quality materials and enhance existing resources. Together with LLMs, educators can chart personalized learning paths for students, leveraging data analysis to tailor educational experiences. Moreover, LLMs streamline assessment processes, automating grading and offering preliminary feedback, allowing teachers to allocate more time to personalized instruction. These models provide data-driven insights, enabling educators to make informed decisions and support struggling students effectively. In language courses, LLMs serve as virtual language assistants, offering pronunciation feed- back and language exercises. They also assist in curriculum development, professional development, and time management, freeing educators to focus on meaningful interactions with students. Importantly, this collaboration promotes inclusivity by creating accessible content and fosters global collaboration by transcending language and cultural barriers. By combining the expertise of teachers with the capabilities of LLMs, education becomes a dynamic, personalized, and inclusive endeavor, benefiting learners and educators alike.

8 Student Engagement

Student engagement in the context of Large Language Models (LLMs) refers to the active participation, motivation, and interaction of students in online learning facilitated by LLMs. It involves the use of LLM-generated content, personalized learning experiences, immediate feedback, and 24/7 support to keep students involved and motivated in their educational journey. LLMs contribute to engagement by adapting content, offering interactivity, and providing tailored resources, ultimately enhancing the overall learning experience for students. LLMs have the capacity to analyze individual student data, thereby enabling the creation of personalized learning pathways. Recognizing that one-size-fits-all approaches are often ineffective in education, LLMs can tailor the curriculum to meet the unique needs and goals of each student. When learners perceive that their education aligns with their aspirations and caters to their individual strengths and weaknesses, they are more likely to engage actively with the material, finding it directly relevant to their personal and academic development.

Immediate feedback is a hallmark of LLMs. They excel in providing students with instant, constructive responses, akin to having a virtual tutor available around the clock. This immediate feedback is invaluable for student engagement, as it helps learners gauge their progress and identify areas for improvement. It also contributes to a sense of self-directed learning, where students can take ownership of their education and make real-time adjustments, keeping them motivated and on track. LLMs offer 24/7 support through virtual tutors and chatbots. This constant availability ensures that students can seek assistance whenever they encounter challenges or have questions, fostering a sense of confidence and promoting active participation in the learning process. Knowing that help is readily accessible removes barriers to engagement and encourages students to overcome hurdles proactively.

Adaptive learning is another dimension where LLMs contribute to engagement. They can dynamically adjust the difficulty of content based on individual student performance. This adaptive approach ensures that students are neither overwhelmed by overly challenging material nor bored by content that is too simple. Maintaining an optimal level of challenge is crucial for keeping students engaged and motivated, as they are consistently operating at a level that is both comfortable and stimulating for their learning journey. The integration of LLMs in education brings numerous ad- vantages, including active student participation and personalized learning pathways, immediate feedback, 24/7 support, and adaptive learning experiences. These benefits Collectively create an educational environment that is more engaging, tailored to individual needs, and conducive to meaningful and effective learning (Fig. 1).

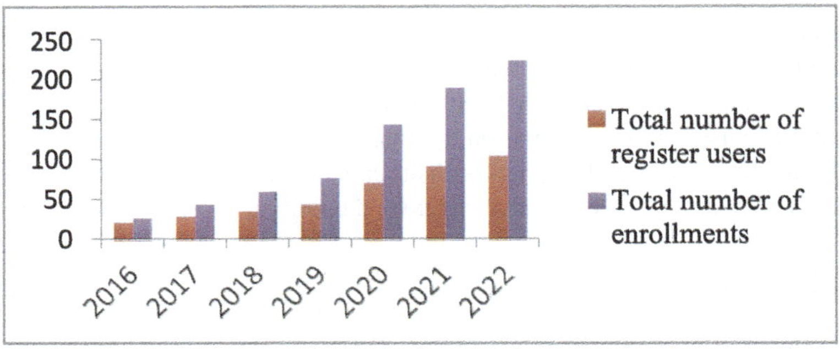

Fig. 1. Online learning platform that are total number of register users and total number of enrolments

For this analysis we have choose two primary aspects of online learning platform that are total number of register users and total number of enrolments. Register users are those who make enrolment in the learning platform and continue learning and make use of the platform. Enrolment users are those who just make enrolments and do not make use of the opportunity. These data where sourced from world economic forum, irrespective of specific online platform this data covers the entire learning platform on average. From this chart we could know that both the aspect gets increased year by year. There is a gradual decrease in the registered user when compare to the enrolment. It is almost 80% of the people register and use in 2016 and it decreases to only 48.6% register from the enrolment in 2021.This shows that people who enroll to online course are less interested to register and continuing studies in recent days. Also, it shows the sudden increase in the year 2020 this is because of the covid pandemic lockdown. The use of online platform increased in that period.

9 Student Acceptance and Satisfaction

Student acceptance and satisfaction are paramount considerations when implementing Large Language Models (LLMs) in online learning environments. To ensure the seamless integration and positive impact of LLMs, it's crucial to address both user acceptance and satisfaction. Student acceptance hinges on factors like usability, perceived utility, training, and adaptability of LLM-powered tools. Creating user-friendly inter- faces, offering relevant training, and demonstrating the practical value of LLMs in im- proving learning experiences can foster acceptance. Additionally, gathering student feedback and incorporating it into LLM enhancements can further enhance acceptance levels.

Student satisfaction, on the other hand, is deeply influenced by the quality of LLM-generated learning materials, the level of interaction and engagement facilitated by LLM-powered features, the timeliness of feedback, and the extent of personalization. High-quality, contextually relevant content, and timely feedback contribute to student satisfaction. LLMs that encourage active engagement and foster inclusivity among diverse student populations also lead to greater satisfaction. Moreover, ensuring transparent and ethical use of LLMs, coupled with robust technical support, can enhance the overall satisfaction of students engaging with LLM-powered online learning platforms. By prioritizing both acceptance and satisfaction, educational institutions can create a more enriching and effective learning environment for their student.

10 Future Trends

The future of education with Large Language Models (LLMs) is poised to usher in transformative changes. One of the foremost trends is the refinement of AI-driven personalization, where LLMs will become even more adept at tailoring learning experiences to individual student needs. This means more adaptive content, assessments, and pathways that cater to diverse learning styles and paces. Alongside this, ethical considerations will take center stage, with a heightened focus on transparency, fairness, and equity in AI-powered educational content and decision-making. LLMs will also merge with immersive technologies like Virtual Reality (VR) and Augmented Reality (AR), creating captivating learning environments that blur the lines between the physical and digital realms. Innovative assessment methods, enhanced data privacy measures, and the transition of LLMs into lifelong learning companions further characterize the future landscape of education. As global collaboration becomes more accessible through language translation and cultural adaptation, students will have access to high-quality, localized content. Interdisciplinary learning, AI-enhanced teacher training, and community building will foster holistic education experiences. Collectively, these trends signal a future where LLMs will continue to play a central role in shaping innovative, personalized, and ethically sound learning journeys for learners across the globe (Fig. 2).

Most of the people get used to online learning because of its convenience and time saving. The future enrolment of students in the online platform likely to be increased to 1.032 billion in the year of 2027.

11 Conclusion

In Large Language Models (LLMs) have ushered in a transformative era in education, revolutionizing online learning and student engagement. LLMs, driven by Natural Language Processing (NLP) capabilities, offer personalized, interactive, and adaptive learning experiences, catering to diverse student needs. They have enriched education in

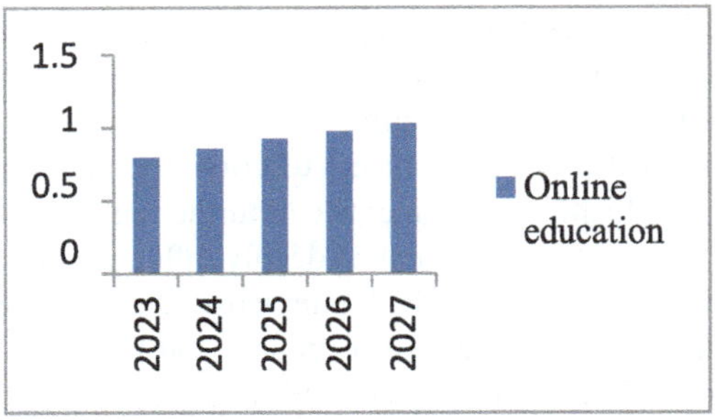

Fig. 2. Online learning because of its convenience and time saving.

numerous ways, from adaptive learning paths to virtual tutors, translation support, and content summarization. However, their integration into education comes with challenges and ethical considerations, including issues related to data privacy, bias, transparency, and overreliance on automation.

The collaboration between educators and LLMs represents a promising partnership that enhances the quality of education. LLMs assist teachers in content creation, personalized learning, assessment automation, and data-driven decision-making. They create inclusive and globally collaborative learning environments, transcending language, and cultural barriers. Student engagement is at the core of LLM-powered education, with LLMs actively involving students through interactive content, personalized learning, immediate feedback, and 24/7 support. This engagement fosters a more meaningful and effective learning experience. Future trends in LLM-powered education include advanced personalization, ethical considerations, integration with immersive technologies, innovative assessment methods, and global collaboration. These trends signal a future where LLMs will continue to shape innovative, personalized, and ethically re- sponsible learning journeys for students worldwide. As the educational landscape evolves, responsible and inclusive use of LLMs will remain paramount to ensure equitable access to quality education for all.

11.1 Summary of Key Findings

The research on Large Language Models (LLMs) in online learning reveals a transformative landscape where education becomes personalized and accessible. LLMs empower student-centric learning by tailoring content and pathways to individual needs, enhancing engagement. They provide 24/7 support through virtual tutors and chatbots, ensuring students have guidance whenever they need it. However, this integration raises ethical considerations, including algorithmic bias and data privacy, which demand careful. Moreover, collaboration between educators and LLMs augments the quality of education, with LLMs assisting in content creation, assessment automation, and data-driven decision-making. Looking ahead, future trends include advanced personalization, ethical focus, immersive technology integration, and global collaboration, signifying a promising future where LLMs continue to shape innovative, personalized, and ethically responsible learning journeys for students worldwide.

Acknowledgement. The author P. Karthikeyan would like to thank the Indian Council of Social Science Research, New Delhi, India for its financial support to ICSSR Research Project (Major)-F.No. Gen-18/2021-22/ICSSR/RP dated 14.03.2022.

References

de Wynter, A., Wang, X., Sokolov, A., Gu, Q., Chen, S.-Q.: An evaluation on large language model outputs: discourse and memorization. Nat. Lang. Process. J. **4**, 100024 (2023). www.elsevier.com/locate/nlp

Li, L., Zhang, R., Piper, A.M.: Predictors of student engagement and perceived learning in emergency online education amidst COVID-19: a community of inquiry perspective. Comput. Hum. Behav. Report. **12**, 100326 (2023). https://doi.org/10.1016/j.chbr.2023.100326

Afzal, F., Crawford, L.: Student's perception of engagement in online project management education and its impact on performance: the mediating role of self-motivation. Proj. Leadersh. Soc. **3**, 100057 (2022). www.sciencedirect.com/journal/project-eadership-and-society

B.J. Jansen, Jung, S., Salminen, J.: Employing large language models in survey research. Nat. Lang. Process. J. **4**, 100020 (2023). www.elsevier.com/locate/nlp

Hazzam, J., Wilkins, S.: The influences of lecturer charismatic leadership and technology use on student online engagement, learning performance, and satisfaction. Comput. Educ. **200**, 104809 (2023). www.elsevier.com/locate/compedu

Åkesson, J., Årzén, K.-E., Gäfvert, M., Bergdahl, T., Tummescheit, H.: Modeling and optimization with Optimica and JModelica.org—Languages and tools for solving large-scale dynamic optimization problems. Comput. Chem. Eng. **34**, 1737–1749 (2010). www.elsevier.com/locate/compchemeng

Dahalan @ Omar, N., Hassana, H., Atan, H.: Student engagement in online learning: learners attitude toward e-mentoring. In: The 3rd International Conference on e-Learning ICEL2011, 23–24 Nov 2011, Bandung, Indonesia. Procedia – Social and Behavioral Sciences, vol. 67, pp. 464–475 (2012). www.sciencedirect.com

Ertesvåg, S.K., Vaaland, G.S., Lerkkanen, M.-K.: Enhancing upper secondary students' engagement and learning through the INTERACT online, video-based teacher coaching intervention: Protocol for a mixed-methods cluster randomized controlled trial and process evaluation. Int. J. Educ. Res. **114**, 102013 (2022). www.elsevier.com/locate/ijedures

Azila-Gbettor, E.M., Abiemo, M.K., Glate, S.N.: University support and online learning engagement during the Covid-19 period: the role of student vitality. Heliyon **9**, e12832 (2023). www.cell.com/heliyon

Digital Health Cooperative Research Centre: Attention is not all you need: the complicated case of ethically using large language models in healthcare and medicine. eBioMedicine **90**, 104512 (2023).s https://doi.org/10.1016/j.ebiom.2023.104512

Aguinis, H., Villamor, I., Ramani, R.: MTurk research: review and recommendations. J. Manag. **47**(4), 823–837 (2020). https://doi.org/10.1177/0149206320969787

Ouyang, L., Wu, J., Jiang, X., Almeida, D., Wainwright, C.L., Mishkin, P.: Training language models to follow instructions with human feedback (2022). https://cdn.openai.com/papers/Training_language_models_to_follow_instructions_with_human_feedback.pdf

Peng, B., et al.: Check your facts and try again: Improving large language models with external knowledge and automated feedback 2023.. https://doi.org/10.48550/arXiv.2302.12813

Jin, L., Chen, Y., Wang, T., Hui, P., Vasilakos, A.V.: Understanding user behavior in online social networks: a survey. IEEE Commun. Mag. **51**(9), 144–150 (2013). https://doi.org/10.1109/MCOM.2013.6588663

Akyol, Z., Garrison, D.R.: Understanding cognitive presence in an online and blended community of inquiry: assessing outcomes and processes for deep approaches to learning. Br. J. Edu. Technol. **42**(2), 233–250 (2011). https://doi.org/10.1111/j.1467-8535.2009.01029.x

Andel, S.A., de Vreede, T., Spector, P.E., Padmanabhan, B., Singh, V.K., de Vreede, G.-J.: Do social features help in video-centric online learning platforms? A social presence perspective. Comput. Hum. Behav. **113**, 106505 (2020). https://doi.org/10.1016/j.chb.2020.106505

Brandt, N.D., Lechner, C.M., Tetzner, J., Rammstedt, B.: Personality, cognitive ability, and academic performance: differential associations across school subjects and school tracks. J. Pers. **88**(2), 249–265 (2020). https://doi.org/10.1111/jopy.12482

Tashakkori, A., Teddlie, C.: Issues and dilemmas in teaching research methods courses in social and behavioural sciences: US perspective. Int. J. Soc. Res. Methodol. **6**(1), 61–77 (2003). https://doi.org/10.1080/13645570305055

Taylor, G., et al.: A self-determination theory approach to predicting school achievement over time: the unique role of intrinsic motivation. Contemp. Educ. Psychol. **39**(4), 342–358 (2014). https://doi.org/10.1016/j.cedpsych.2014.08.002

A Study on the Impact of Large Language Models on Customer Satisfaction

P. Karthikeyan(✉), A. Fahima, B. S. Aishwaryaa Lakshmi, V. Kishore Kumar, V. Vaishnavi, and N. Prakash

Department of Management Studies, Kongu Engineering College, Perundurai, Erode, Tamil Nadu 638 060, India
ptp.karthi@gmail.com, {fahimaa.22mba,aishwaryaalakshmibs.22mba, kishorekumarv.22mba,vaishnavi.mba,np.mba}@kongu.edu

Abstract. The utilization of Expansive Dialect Models (LLMs) has been a game-changer within the field of client fulfillment. This paper investigates the multi-faceted relationship between LLMs and client fulfillment, shedding light on their affect, challenges, and openings. It draws on a comprehensive writing review and experimental information to supply experiences for businesses and policymakers. The discoveries uncover that LLMs can improve client fulfillment through personalized intelligent and productive bolster frameworks, but they too raise moral concerns. This paper emphasizes the noteworthiness of dependable LLM sending in client benefit and talks about future inquire about bearings.

1 Introduction

In the dynamic landscape of contemporary business, the quest for customer satisfaction stands as a cornerstone of success. As organizations strive to meet and exceed customer expectations, they navigate an intricate terrain where technology, human interaction, and evolving consumer preferences converge. Within this evolving paradigm, the emergence of Large Language Models (LLMs) has marked a pivotal juncture. These powerful and versatile AI-driven models, including but not limited to OpenAI's GPT-3, have redefined the boundaries of natural language processing, promising new dimensions in how companies engage with and serve their clientele. Customer satisfaction, long acknowledged as a paramount metric of business performance, is subject to continuous transformation. As LLMs permeate the domains of customer support, marketing, and sales, they bring forth unprecedented opportunities for enhancing the customer experience. These models can process and generate humanlike text, understand context, and personalize interactions, thereby potentially revolutionizing the dynamics of customer engagement. At the same time, the integration of LLMs into customer service strategies raises complex and multifaceted questions, ranging from ethical considerations to practical challenges, such as bias and transparency.

A. Bajaj et al. (Eds.): IBICA 2023, LNNS 1229, pp. 175–182, 2025.
https://doi.org/10.1007/978-3-031-78940-3_17

1.1 Scope

1. **Industry Relevance:** LLMs are relevant across various industries, including ecommerce, healthcare, financial services, and hospitality. The scope extends to understanding how LLMs influence customer satisfaction within these distinct sectors.
2. **Impact on Customer Support:** LLMs play a pivotal role in improving customer support through chatbots, virtual assistants, and automated helpdesk solutions. Research can focus on the effectiveness of LLMs in enhancing response times, issue resolution, and overall customer satisfaction in this context.
3. **Personalization and Customer Experience:** LLMs enable personalized interactions with customers. Research can investigate how personalized responses and content generated by LLMs influence customer experience and satisfaction.
4. **Automation and Human Interaction**: The scope also encompasses examining the balance between automation and human interaction in customer service. How LLMs complement or potentially replace human agents is a key aspect of this scope.
5. **Ethical Considerations:** The ethical implications of using LLMs in customer service are of paramount importance. Research can explore the ethical considerations surrounding data privacy, fairness, transparency, and the responsible use of LLMs to ensure customer satisfaction.
6. **Customer Feedback Analysis:** LLMs can be employed to analyze customer feedback, reviews, and social media sentiment. Research can focus on how LLMs extract insights from unstructured data to improve products, services, and customer satisfaction.
7. **Language and Cultural Sensitivity:** Understanding how LLMs handle diverse languages and cultures in customer interactions is a crucial aspect of the scope. Research can investigate how these models maintain sensitivity and adaptability across linguistic and cultural boundaries.

2 Objective

These think about has two fundamental objectives. To begin with, it points to investigate and analyze the effect of LLM on client fulfillment in several businesses and applications. Moment, it looks for to recognize the challenges and openings related with actualizing an LLM to make strides client fulfillment. Accomplishing these objectives is critical for both companies needing to progress client involvement and for a broader understanding of the part of LLM in client fulfillment.

3 Large Language Models in Customer Satisfaction

1. **Natural Language Understanding**: Large language models can understand and respond to natural language queries and feedback from customers more effectively. This can lead to more meaningful interactions and better customer support, whether through chatbots, virtual assistants, or email responses.
2. **Personalization**: These models can analyze customer data and provide personalized recommendations, product suggestions, or content, leading to a more tailored and satisfying customer experience.

3. **Chatbots and Virtual Assistants:** AI-powered chatbots and virtual assistants powered by large language models can provide immediate responses to customer inquiries, troubleshoot issues, and even assist in making purchases. This can improve response times and customer service availability.
4. **Customer Feedback Analysis:** Huge dialect models can analyze client criticism from different sources, such as studies, social media, and surveys. This makes a difference businesses superior get it client assumptions, recognize issues, and make data-driven enhancements.
5. **Translation and Multilingual Support**: Large language models can facilitate multilingual support, enabling businesses to reach a broader customer base and provide better service to non-English-speaking customers.
6. **Knowledge Base and FAQs:** Companies can use these models to create and maintain extensive knowledge bases and FAQs that are easily searchable and understandable by customers.
7. **Voice Assistants:** Large language models can power voice assistants that provide hands free customer support. Customers can get assistance through devices like smartphones and smart speakers, enhancing their convenience.
8. Predictive Analytics: Machine learning models can use customer data to make predictions about future customer behavior, allowing companies to proactively address potential issues and offer targeted promotions.

4 Review of Literature

Wulf and Meierhofer (2023) explained that LLMs have the one of a kind capacity to get it and react to complex client inquiries, adjust to different client inclinations, and provide personalized and relevantly significant back at a standardized quality level that's free of an person client benefit agent's information base. Additionally, they offer a versatile arrangement that can handle tall volumes of client intelligent, decreasing the require for broad human assets. By leveraging LLMs, businesses cannot as it were progress their client bolster but too realize noteworthy fetched reserve funds, making them an priceless resource in today's competitive trade environment.

Leissman et al. (2023) discovered that PBPM offers unused observing strategies for administrations, such as strategies for result forecast. Moving absent from the immaculate control flow-based viewpoint in prepare mining has not as it were cleared the way for a more all-encompassing investigation of commerce forms but has opened a assortment of openings for PBPM and, in turn, benefit analytics.

Leong et al. (2023) proposed leveraging conventional travel CRM criticism to create and send a transit-topic-aware huge dialect show (LLM) able of classifying open-ended content criticism to significant transit-specific themes. To begin with, we utilize semi supervised learning to design a preparing dataset of 11 wide travel themes recognized in a corpus of 6 a long time of client criticism given to the Washington Metropolitan Range Travel Specialist (WMATA).

Freire et al. (2023) explored the openings, dangers, and client acknowledgment of LLM powered CAs in two manufacturing plant settings: material and cleanser generation. A few openings and dangers are recognized through a writing audit, proof-of-concept usage, and center bunch sessions. Production line agents raise concerns with respect to information security, security, and the unwavering quality of LLMs in high-stake situations.

Kim et al. (2023) discovered that Interface creators can apply our system to deliver interfacing that empower end-users to make varieties of these objects, combine and recombine them into modern arrangements, and compare them in parallel to effectively emphasize and explore with LLMs.

Mensouri et al. (2023) centers on opinion investigation utilizing four machine learning models, specifically repetitive neural systems (LSTM, GRU, Bi-LSTM), and a pre trained dialect show (Roberta). The models are prepared to classify client audits on online stages as positive or negative.

Guan et al. (2020) found out that more than 12,000 customer service agents use ICS Assist to handle an average of more than 230,000 cases per day. Testing shows that ICS Assist completely beats the conventional manual strategy and increases deal recognition rate, deal closing rate, typical service time, customer satisfaction rate, and commercial space recovery up to 16%, 25%, 6%, 14% and 17% individually, compared to state-of the-art strategies.

5 Benefits of LLM in Customer Satisfaction

1. **24/7 Availability:** LLM-powered chatbots and virtual assistants can provide round-the clock support, ensuring that customers can get assistance at any time. This availability enhances customer satisfaction, particularly for those in different time zones or with urgent inquiries.
2. **Improved Response Times:** LLMs can instantly respond to customer inquiries, reducing the need for customers to wait in queues or for email responses. Faster response times lead to higher customer satisfaction.
3. **Consistency:** LLMs can maintain a consistent tone, messaging, and level of knowledge in interactions with customers. This ensures that every customer receives the same quality of service, contributing to a positive customer experience.
4. **Personalization:** LLMs can analyze customer data and provide personalized recommendations or responses. This personalization enhances the customer experience and can lead to higher satisfaction and loyalty.
5. **Multilingual Support:** Large language models can provide support in multiple languages, making businesses more accessible to a global customer base and ensuring that non-English-speaking customers receive quality support.
6. **Efficient Issue Resolution:** LLMs can troubleshoot common issues and provide step-by step solutions, leading to faster and more efficient issue resolution. This decreases customer frustration and increases satisfaction.
7. **Reduced Human Error:** Automation reduces the chances of human errors in customer interactions, providing a higher level of accuracy and consistency in responses.

As per the Primary data collected from 60 people we had visualized their responses according to their preference level for AI – powered Customer Support tools (Fig. 1).

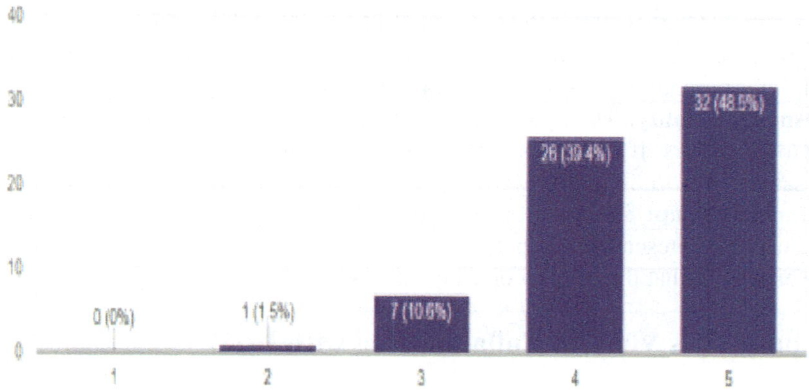

Fig. 1. Preference level for AI – powered Customer Support tools

6 Challenges of LLM on Customer Satisfaction

1. **Lack of Empathy and Human Touch:** LLMs can be perceived as impersonal, as they lack the ability to genuinely empathize with customers. Empathy is often crucial in understanding and addressing emotional or sensitive customer issues.
2. **Misunderstandings and Miscommunications:** LLMs may misinterpret customer queries or provide inaccurate responses, leading to customer frustration and dissatisfaction, particularly when dealing with complex or nuanced problems.
3. **Over-Reliance on Automation**: Overreliance on LLMs can lead to customer frustration, especially if they prefer human interactions or when the issue at hand requires a personalized, human touch.
4. **Technical Issues and Downtime:** Technical problems, such as system outages or software glitches, can disrupt LLM-powered customer support, leaving customers without assistance and negatively affecting satisfaction.
5. **Complex or Specialized Queries:** LLMs may struggle with highly technical or industry specific queries, as they lack the domain knowledge and expertise to provide accurate responses in these cases.
6. **Inability to Handle Emotional or Crisis Situations:** LLMs cannot effectively deal with highly emotional or crisis situations where a human touch, empathy, and immediate understanding are crucial.
7. **Brand Image and Trust:** LLMs may not fully capture a company's brand identity or values, potentially affecting customer trust and how they perceive the company.
8. **Integration Challenges: Integrating** LLMs into existing customer service processes and systems can be technically challenging and require substantial investment. As per the secondary data collected from 60 people, the below chart shows the customer's overall experience while getting their services through AI powered customer service

7 Customer Focused Views on LLM

1. **Convenience and Accessibility:** Many customers appreciate the convenience and accessibility that LLMs offer. They can get quick answers to their questions and resolve issues without the need to wait on hold or during non-business hours.

2. **Resolution Quality:** The effectiveness of LLMs in resolving customer issues plays a significant role in customer satisfaction. If LLMs consistently provide accurate and helpful information, customers are more likely to have a positive view.
3. **Response Quality:** The accuracy and relevance of LLM-generated responses matter to customers. If responses are consistently off-topic or unhelpful, it can lead to dissatisfaction.
4. **Human Backup:** Some customers appreciate the option to escalate from an LLM to a human representative, especially when dealing with complex issues. They value the assurance that they can receive human assistance when needed.

8 Company's View on Collaborating with LLM

1. **Competitive Advantage:** Companies may see LLMs as a means to gain a competitive edge by offering advanced features like multilingual support, 24/7 availability, and improved response times. This can set them apart from competitors and attract more customers.
2. **Content Creation:** LLMs are seen as a valuable tool for generating high-quality content, such as product descriptions, marketing materials, and blog posts. This can reduce the time and effort required for content creation and help maintain an engaging online presence.
3. **Innovation and Differentiation:** Collaboration with LLMs is viewed as a way to innovate and differentiate a company in its industry. It allows businesses to explore creative applications, such as interactive experiences and AI-powered products.

9 Future Trends of LLM on Customer Satisfaction

LLMs will get to be indeed more proficient at giving personalized suggestions, item proposals, and reactions based on person client information and behavior. Clients will connect with LLMs through voice, content, and visual interfacing, giving a more immersive and helpful encounter. LLMs will ended up more capable in giving exact and relevantly significant back in different dialects. This will upgrade openness for a worldwide client base and contribute to made strides client fulfillment. Future LLMs will be planned to way better get it and react to the passionate setting of client intelligent. They will give sympathetic and sincerely brilliantly reactions, especially in delicate or emergency circumstances. Companies will center on making a agreeable collaboration between people and LLMs. LLMs will help human client back agents, advertising real-time recommendations, mechanizing schedule assignments, and making strides the generally proficiency of back groups. There will be a more noteworthy accentuation on making LLMs more straightforward and reasonable. Clients will have the capacity to get it why LLMs create particular reactions and how they utilize client information, which is able cultivate believe. LLMs will be utilized for real-time dialect interpretation and localization, empowering businesses to supply moment back to clients in their local dialects, no matter where they are within the world. LLMs will proceed to produce high-quality substance, counting video scripts, social media posts, and personalized bulletins, improving the client engagement and brand involvement. Supportability will be a developing concern. LLMs will be utilized for prescient client benefit, expecting issues or needs based on authentic information and making proactive suggestions or help. Integration of LLMs with the Web of Things (IoT) gadgets will empower clients to control and oversee their IoT items through voice commands and common dialect intelligent, moving forward client encounters. LLMs will play a bigger part in analyzing client input from different sources, counting social media and surveys, empowering businesses to pick up more profound experiences into client opinions and inclinations.

10 Conclusion

The study on the impact of Large Language Models (LLMs) on customer satisfaction reveals a dynamic and multifaceted relationship between these advanced AI-driven models and customer experiences. LLMs, such as Open AI's GPT-3, have significantly influenced various industries, including e-commerce, healthcare, financial services, and hospitality. The paper highlights the potential benefits of LLMs in customer support, marketing, and sales, emphasizing their ability to provide personalized and efficient interactions. Along with the opportunities, the study acknowledges the ethical considerations and challenges associated with the deployment of LLMs in customer service. These challenges include issues related to empathy, misunderstandings, over-reliance on automation, technical problems, and the potential impact on brand image and trust. The study underscores the transformative potential of LLMs in reshaping customer satisfaction but emphasizes the need for careful consideration of ethical implications, addressing challenges, and fostering a harmonious collaboration between LLMs and human agents in customer service.

References

Agarwal, M., Raju, V., Dey, R., Nanda, I.: Descriptive research on AI-based tools to aid personalized customer service: Case of ChatGPT. J. Reprodu. Res. **1**(1), 140–146 (2022)

Carrasco, R.A., Muñoz-Leiva, F., Sánchez-Fernández, J., Liébana-Cabanillas, F.J.: A model for the integration of e-financial services questionnaires with SERVQUAL scales under fuzzy linguistic modeling. Expert Syst. Appl. **39**(14), 11535–11547 (2012)

Fu, M., et al.: ICS-Assist: Intelligent customer inquiry resolution recommendation in online customer service for large E-commerce businesses. In: Service-Oriented Computing: 18th International Conference, ICSOC 2020, Dubai, United Arab Emirates, December 14–17, 2020, Proceedings 18, pp. 370–385. Springer International Publishing (2020)

Kernan Freire, S., Foosherian, M., Wang, C., Niforatos, E.: Harnessing large language models for cognitive assistants in factories. In: Proceedings of the 5th International Conference on Conversational User Interfaces, pp. 1–6 (2023)

Kim, T.S., Lee, Y., Chang, M., Kim, J.: Cells, Generators, and Lenses: Design Framework for Object-Oriented Interaction with Large Language Models. In: The 36th Annual ACM Symposium on User Interface Software and Technology (UIST'23) (2023)

Lakhani, A.: Enhancing Customer Service with ChatGPT Transforming the Way Businesses Interact with Customers (2023)

Leong, M., et al.: MetRoBERTa: Leveraging Traditional Customer Relationship Management Data to Develop a Transit-Topic-Aware Language Model (2023). arXiv preprint arXiv:2308.05012

Liessmann, A., Sukhareva, M., Zilker, S., Matzner, M., Weinzierl, S.: Predicting Customer Satisfaction in Service Processes Using Multilingual Large Language Models (2023)

Liu, S., Peng, C., Wang, C., Chen, X., Song, S.: IcsBERTs: optimizing pre trained language models in intelligent customer service. Procedia Computer Science **222**, 127–136 (2023)

Mensouri, D., Azmani, A., Azmani, M.: Combining Roberta Pre-Trained Language Model and NMF Topic Modeling Technique to Learn from Customer Reviews Analysis. Int. J. Intel. Sys. Appl. Eng. **11**(1), 39–49 (2023)

Nissa, N.K., Yulianti, E.: Multi-label text classification of Indonesian customer reviews using bidirectional encoder representations from transformers language model. Int. J. Electr. Comp. Eng. **13**(5), 2088–8708 (2023)

Ouyang, L., et al.: Training language models to follow instructions with human feedback. Adv. Neural. Inf. Process. Syst. **35**, 27730–27744 (2022)

Paek, T., Chickering, D.M.: Improving command and control speech recognition on mobile devices: using predictive user models for language modeling. User Modelling and User-Adapted Interaction **17**, 93–117 (2007)

Wulf, J., Meierhofer, J.: Pioneering a new era of automated customer service with large language models (LLMs). Medium (2023)

Privacy Preserving Advanced Persistent Threat Detection Using Fed-Adv-LSTM

Lokmane Heraguemi(✉), Abdelaziz Amara Korba, and Nacira Ghoualmi-Zine

Networks and Systems Laboratory (LRS), Badji Mokhtar Annaba University, Annaba, Algeria
heraguemilokman@gmail.com, lokmane.heraguemi@univ-annaba.org

Abstract. Advanced Persistent Threats (APTs) present an ever-evolving challenge to computer systems and networks due to their stealthy and persistent nature. Traditional APT detection methods, like signature-based and anomaly-based approaches, struggle against these sophisticated threats, aggravated by the rapid evolution of APT techniques and the emergence of adversarial attacks. This study explores the application of Federated Learning (FL) to enhance APT detection. FL, a decentralized machine learning approach, enables collaborative model training across multiple devices or organizations while preserving data privacy and security. Leveraging diverse data sources enhances detection model accuracy and robustness. FL also adapts effectively to the evolving APT landscape, offering a promising tool for real-time detection. Using the DAPT 2020 dataset, this research compares Support Vector Machine (SVM) and Stacked Autoencoder with Long Short-Term Memory (LSTM) models to FL models, including investigating adversarial attacks. Results show FL significantly improves APT detection, with precision, recall, and accuracy exceeding 99%, highlighting its potential in addressing APT challenges and emphasizing privacy-preserving approaches in cybersecurity.

Keywords: Advanced persistent threats · Federated learning · Privacy preserving · Adversarial training

1 Introduction

Advanced persistent threats (APTs) pose a significant threat to computer systems and networks. APTs are characterized by their ability to stealthily infiltrate a targeted system, remain undetected for an extended period, and exfiltrate data or carry out malicious activities [1,2]. These attacks are often perpetrated by well-funded and sophisticated threat actors, including state-sponsored groups, criminal organizations, and hacktivists. APTs employ various techniques to evade detection, such as leveraging zero-day vulnerabilities, using custom malware or hacking tools, and tactics like spear-phishing or social engineering [3]. Once inside a system, they can steal sensitive data, install backdoors, and conduct reconnaissance for future attacks. The devastating impact of APTs includes

A. Bajaj et al. (Eds.): IBICA 2023, LNNS 1229, pp. 183–194, 2025.
https://doi.org/10.1007/978-3-031-78940-3_18

substantial financial losses, reputational harm, and legal obligations [4]. Identifying and mitigating APTs can be challenging, requiring sophisticated security tools and knowledgeable employees. Efficient APT detection techniques remain a major challenge for the cybersecurity industry, given the evolving nature of these threats [1,2,4]. Therefore, addressing this issue is crucial to safeguarding computer systems and networks against these persistent and sophisticated adversaries.

While the impact of Advanced Persistent Threats (APTs) on computer systems is well-recognized, detecting them remains a formidable challenge [5]. Traditional detection methods, including signature-based, anomaly-based, and behavior-based approaches, often prove insufficient against APTs, leading to false positives or negatives. The evolving attack techniques, coupled with the increasing use of artificial intelligence and machine learning in malware creation, exacerbate this issue [6]. Hence, machine learning techniques are increasingly vital for enhancing the accuracy and effectiveness of APT detection systems. These models adapt to evolving attack techniques and excel at pattern and anomaly recognition in large datasets [6].

While machine learning and deep learning exhibit promise in APT detection, they face significant limitations in practice, including the scarcity and complexity of suitable training data for APTs [7]. Additionally, these models are vulnerable to adversarial attacks, enabling attackers to manipulate input data and evade detection or trigger false alarms [8]. Moreover, concerns around data privacy and security arise when using centralized models. Federated Learning (FL) addresses these limitations by enabling collaborative model training across decentralized devices or organizations, preserving data privacy and security while only exchanging model updates instead of raw data [9]. FL leverages collective intelligence from multiple sources, enhancing model robustness and accuracy, offering a promising real-time solution for APT detection, and overcoming the constraints of traditional machine learning and deep learning approaches [9].

In order to address the limitations of traditional detection methods, the present study aims to investigate the use of FL for detecting APTs on the DAPT 2020 dataset, with the following objectives:

- To perform an initial data analysis on the DAPT 2020 dataset, extracting valuable insights into APT characteristics, and utilizing this knowledge to shape the development of our detection models.
- To compare the effectiveness of three different algorithms SVM, Stacked autoencoder with LSTM, and FL with SEA-LSTM as a local model.
- To analyze the impact of adversarial training on the proposed approache.
- To provide recommendations for future research on the use of FL for detecting APTs.

2 Background and Related Works

2.1 APT Detection

APTs are a type of cyber-attack that are characterized by their persistence, stealth, and sophistication. APTs are designed to infiltrate a targeted organization's network and remain undetected for long periods of time, allowing the attackers to collect sensitive data, steal intellectual property, and conduct other malicious activities. APTs pose a serious threat to organizations due to their ability to cause significant damage and harm over a prolonged period of time, leading to financial losses, damage to reputation, and potential legal liabilities. Additionally, APTs are becoming increasingly sophisticated and difficult to detect, making them a significant challenge for organizations to defend against. As a result, APT detection is a critical area of research and development in the field of cybersecurity.

2.2 Current Approaches for APT Detection

Various methods are employed for APT detection, including signature-based, anomaly-based, and machine learning-based approaches [1,2]. Signature-based detection relies on known patterns to identify threats swiftly but struggles with novel APTs. Anomaly-based methods can spot unknown threats but may yield false positives and require fine-tuning. Machine learning-based approaches excel at identifying unknown threats and adapting to new data but demand ample resources and are susceptible to adversarial attacks. Despite extensive research, key challenges persist. Limited data availability for training APT detection models hampers development and testing. Ensuring privacy while analyzing network data, especially in sensitive sectors, remains critical. Additionally, addressing adversarial attacks is vital, requiring robust and adaptive APT detection models. The ongoing pursuit of improved APT detection techniques underscores the need for comprehensive, efficient, and privacy-preserving solutions.

2.3 Federated Learning

Federated learning (FL) is a novel approach that can mitigate the limitations of traditional APT detection methods. FL entails training machine learning models on decentralized data sources, eliminating the need for centralized data storage [10]. By leveraging this distributed data, FL can address the issues of evasion and high false positive/negative rates associated with traditional methods. Furthermore, FL enables collaborative model training across multiple devices or organizations, enhancing model accuracy and robustness. It also maintains data privacy and security by exchanging only model updates, not raw data [11]. FL holds promise as an effective APT detection solution while preserving data privacy and addressing traditional method limitations.

As a result of its demonstrated effectiveness in preserving data privacy, FL has been widely adopted in various domains including IoT [12–14], and [15] healthcare [16,17], and [18], energy [19], smartphones [20] and [21], biology [22], vehicles [23] and [24], economy [25], and robotics [26].

Federated Learning (FL) operates through two stages: global model training and local model training. In the global model training stage, the central server dispatches a predefined model to each device [27]. These devices then train the model on their local data, and the updated model weights are sent back to the central server for aggregation. This process repeats in multiple rounds until the global model converges. The local model training stage involves fine-tuning the global model on each device's local data to create device-specific models. These device-specific models are then sent to the central server, where they are aggregated to create an improved global model for the next round of training [28]. There are different FL approaches for APT detection, including cross-silo and cross-device FL. Cross-silo FL involves training a model using data from multiple organizations or "silos", with each organization training the model on its local data [29]. Periodic aggregation and combination of model parameters create a global model that helps enhance the accuracy of APT detection models while preserving data privacy [30].

- Cross-silo FL combines data from various organizations, with each entity training the model on its local data. Periodic parameter aggregation creates a global model [29] and [30], enhancing APT detection while preserving data privacy.
- Cross-device FL trains a model using data from multiple devices within an organization, with each device training on local data. Periodic parameter aggregation creates a global model [31] and [32], enhancing APT detection. This method improves accuracy with a diverse dataset but may demand more computational resources and face practical challenges due to device heterogeneity and network constraints.

Additional FL approaches for APT detection are federated transfer learning [33], fine-tuning pre-trained models on local data, and federated ensemble learning, where multiple models are combined for improved accuracy [34]. While exploring solutions for detecting APTs that specifically target organizations and nations, Cross-silo FL emerges as a suitable approach for addressing these threats.

2.4 Federated Learning for Intrusion and APT Detection

Federated learning for APT detection has several advantages over traditional methods [35]. It maintains data privacy, reducing the risk of breaches and privacy violations. By training on diverse datasets from various devices and organizations, it creates more accurate and robust APT detection models. Moreover, FL enhances efficiency by reducing network bandwidth requirements and the need for centralized infrastructure, ultimately improving scalability and lowering costs. In summary, FL can significantly improve the accuracy and effectiveness of APT detection systems while safeguarding sensitive data [36].

In [37], the Deepfed framework employs FL for intrusion detection in cyber-physical systems (CPSs). It combines CNNs and GRUs to detect threats and utilizes the Paillier cryptosystem for model security. The model excels in precision, recall, and F-score.

In [38], the FedAGRU framework leverages FL to classify network traffic efficiently. It integrates GRU models and an attention mechanism into the Federated Averaging (FedAvg) algorithm. FedAGRU employs an improved federated averaging algorithm to detect poisoning attacks, reducing communication overhead and optimizing global model efficiency.

In [39], the authors proposed a real-time anomaly detection system for detecting intrusions in IoT networks. The system is based on Gated Recurrent Units (GRUs) models and utilizes decentralized device data for proactive recognition of intrusions. The proposed intrusion detection system (IDS) employs FL and only shares the learned weights with the central server, thereby preserving the integrity of data stored on local IoT devices. According to the evaluation results, the proposed method for intrusion detection outperforms non-FL versions of intrusion detection algorithms.

In [40], authors present an innovative intrusion detection technique for the Internet of Things (IoT) using FL. It enables model training without exposing sensitive data. The method combines local and global models, enhancing intrusion detection accuracy by training individual models on each device and aggregating them into a global model. An ensemble approach is suggested for improved accuracy. While experiments on real-world data show high accuracy and privacy preservation, centralized setups outperform federated settings, as the authors note.

In [41], APTPMFL, designed for 5G-enabled IoT scenarios using FL, detects multiple APT attack patterns. The well-trained model anticipates future APT attacks in 5G-enabled IoT environments, demonstrating efficiency with acceptable accuracy and low false positives for subsequent APT activities.

In [30], the authors proposed a system, called FLAPT, leverages FL to maintain a distributed model across multiple clients for the purpose of detecting APT attacks. The FLAPT framework involves three key phases: first, training the model using FL techniques; next, detecting attacks using the trained model; and finally, reconstructing the APT scenario to gain a better understanding of the attack. According to the experimental findings, an accuracy rate of 96.7% was achieved with negative rate lower than the compared algorithms.

In [42], a FL-based approach for proactive attack detection to enhance cyber threat hunting was introduced. This approach allows the collection of valuable information about potential cyber threats before they cause harm, leveraging insights from diverse threat environments while ensuring privacy. However, the framework has limitations, such as the cost and delays associated with collecting flow data in large-scale SDN networks. The implementation of the OpenCTI system was also constrained by time and environmental limitations.

Table 1 presents a comparative analysis of studies employing FL for APT detection.

Table 1. comparative analysis of federated learning approaches in APT detection

Paper	[30]	[41]	[42]		
Dataset	UNSW-NB15 Public	Multiple APT attacks Private	NF-UQ-NIDS public		
Full APT attack	⊠	Not known	⊠		
FL approach	Cross- silo	Cross- device	Cross- silo		
Local model	CNN-LSTM	GRU	GRU	LSTM	CNN
Metrics	Acc 96.7%	F1-score 88.7%	Acc 70 %	Acc 55 %	Acc 97%
Adversarial training	⊠	⊠	⊠		

While numerous articles highlight the promising results of FL, a critical gap exists regarding the evaluation of its susceptibility to adversarial attacks. Many authors fail to address the impact of such attacks on the performance of FL models. This knowledge gap extends beyond the articles we've cited to encompass other related researches as well. Furthermore, proposed solutions to counter adversarial attacks in the context of FL are limited. The existing literature predominantly focuses on specific types of attacks, such as data poisoning or model inversion attacks, without fully considering a wide range of adversarial attacks. As the threat landscape continues to evolve, it becomes imperative for researchers to comprehensively explore and address the vulnerabilities of FL methods. Robust defenses must be devised to safeguard the integrity and reliability of FL in real-world scenarios.

3 The DAPT 2020 Dataset

The deficiencies of current intrusion datasets, as emphasized by [43], lie in their failure to encompass the attributes of APTs such as persistence, adaptability, stealth, and complexity. In their work, the authors present an overview of the development and compilation process of DAPT 2020, a dataset specifically designed to incorporate comprehensive APT characteristics. This dataset encompasses representative attack instances from various stages of an APT, including reconnaissance, foothold establishment, lateral movement, and data exfiltration.

3.1 Data Distribution

The data is meticulously organized, categorized by VM type (public and private) and divided into five distinct days. Public VMs include Mutillidae, DVWA, and Metasploitable, with private VMs hosting services like Samba, a WordPress website, FTP, MySQL, and Nexus. This comprehensive organization provides valuable insights into VM-based cyber threats.

3.2 Scenarios to Simulate APT Attacks

Several techniques have been used. These techniques can be categorized according to the phase of the attack:

Reconnaissance	Network scan
	Application scan
	Account Bruteforce
Establish foothold	CSRF
	SQL injection
	Malware download
	Backdoor
	Reverse shell
	Command injection
Lateral movement	Internal scanning
	Account discovery
	Password dumping
	Credential theft
	Creation of user accounts
	Privilege escalation
Data exfiltration	Data theft

Therefore, the dataset encompasses four out of the five stages typically observed in an attack scenario. Consequently, the dataset comprises five distinct stages, including four attack stages and a normal state. . The dataset initially contains 87 attributes, and among these attributes, 12 are found to be unnecessary as they only contain zeros. Therefore, a total of 75 attributes will be used, of which two of them are labels (Activity/Stage).

4 Apt Detection Methodology

4.1 The Models

In this study, we leverage a quartet of models, employing two for comparative analysis and dedicating the other two to our primary research focus. The first two models utilized in this study consist of a Stacked Autoencoder with LSTM architecture and a Support Vector Machine (SVM) classifier. The first model employs a Stacked Autoencoder, which is designed to automatically learn hierarchical representations of the input data. The output of the Stacked Autoencoder serves as the input to the LSTM network, which is particularly effective at capturing sequential dependencies in the data.

On the other hand, the second model in this study is a standalone SVM classifier. SVMs are powerful machine learning algorithms that aim to find an optimal hyperplane to separate different classes of data. They are known for their ability to handle high-dimensional data and are commonly used in classification tasks.

By using the Stacked Autoencoder with LSTM architecture, the first model combines the benefits of both unsupervised learning (Stacked Autoencoder) and sequential modeling (LSTM) to extract meaningful features and capture temporal dependencies. The second model, SVM, serves as a separate classifier to make predictions based on the extracted features.

4.2 Methodology Description

These following steps are employed to ensure enhanced outcomes in the feature selection and classification process for the two models: Fed-LSTM and Fed-Adv-LSTM.

- Feature Selection: The approach begins by identifying important features through correlation analysis between numeric columns and a target variable. Relevant features are selected based on a specified correlation threshold.
- Data Preprocessing: Prior to model training, data preprocessing is conducted. This includes feature normalization using Min-Max scaling, reshaping the data to accommodate LSTM, and one-hot encoding of target labels for multiclass classification.
- Autoencoder Training: An autoencoder model is trained to extract essential features from the preprocessed data. The autoencoder architecture consists of LSTM layers, and training employs mean squared error as the loss function.
- FL Setup: The approach establishes a FL framework where a global model is trained collaboratively across multiple clients. Each client is responsible for a subset of the data for localized training.
- Client-Side Training: For each client, a local model is created, initialized with the global model's weights, and trained on the client's data. These local models are based on LSTM neural networks.
- Weight Averaging: Following local training, the approach aggregates the weights of all local models to compute updated global model weights. This step facilitates the sharing of knowledge acquired by individual clients with the global model.
- Model Evaluation: After each communication round, the approach assesses the performance of the global model using metrics such as accuracy, loss, precision, and confusion matrices. This evaluation enables tracking of model convergence and effectiveness.

Both approaches are nearly identical, with the sole distinction being in their approach to adversarial training. While they share a common structure for feature selection, data preprocessing, and FL, the second approach introduces unique elements into the adversarial training component.

4.3 Evaluation

The obtained results provide an evaluation of the model's performance using various well-established metrics, including precision, recall, F1 score, false positive rate (FPR), and true negative rate (TNR). These metrics are widely used in machine learning and can help assess the effectiveness of a classification model Table 2 reveals that both FL (Fed-LSTM) and adversarial training (Fed-Adv-LSTM) significantly improve model performance. They achieve high precision, recall, and accuracy, showcasing their effectiveness in Apt detection. Notably, adversarial training enhances model robustness without sacrificing accuracy, demonstrating its value in maintaining performance.

Table 2. Performance Metrics of the APT detection models

Metrics	SAE-LSTM	SVM	Fed-LSTM	Fed-Adv-LSTM
Precision	86.4%	95.9%	99.7%	99.3%
Recall	97.8%	95.1%	100%	99.5%
F1-score	91.2%	95.5%	99.8%	99.4%
Accuracy	95.3%	93.4%	99.7%	99.1%
TNR	94.5%	88.6%	99.2%	97.7%
FPR	5.5%	11.4%	0,8%	2,3%

5 Conclusion and Future Prospects

In conclusion, our study delves into the intricate realm of Advanced Persistent Threats (APTs), highlighting their persistent, stealthy, and sophisticated nature. Conventional detection methods often fall short due to evasion tactics and an inability to adapt to evolving threats. To overcome these challenges, we explored Federated Learning (FL) as an innovative approach for APT detection. Our evaluation on the DAPT 2020 dataset demonstrates the efficacy of FL, combined with Stacked Autoencoder and LSTM models. This approach achieved outstanding precision, recall, and accuracy, surpassing 99%. Adversarial training further bolstered model robustness without compromising accuracy.

As the threat landscape evolves, our research underscores the pivotal role of advanced detection mechanisms like FL in countering APTs. Future research should prioritize comprehensive strategies against adversarial attacks and explore FL's applicability in real-world APT scenarios. The results highlight FL's significant improvement in APT detection, emphasizing precision, recall, and accuracy.

In summary, our findings emphasize the importance of FL in enhancing APT detection capabilities, providing a foundation for future research to build upon.

References

1. Alshamrani, A., Myneni, S., Chowdhary, A., Huang, D.: A survey on advanced persistent threats: techniques, solutions, challenges, and research opportunities. IEEE Commun. Surv. Tutor. **21**(2), 1851–1877 (2019). https://doi.org/10.1109/COMST.2019.2891891
2. Maxwell, J.C.: A Treatise on Electricity and Magnetism, vol. 2, 3rd edn., pp. 68–73. Clarendon, Oxford (1892)
3. Mei, Y., Han, W., Li, S., Wu, X.: A survey of advanced persistent threats attack and defense. In: 2021 IEEE Sixth International Conference on Data Science in Cyberspace (DSC), Shenzhen, China, pp. 608–613. IEEE (2021). https://doi.org/10.1109/DSC53577.2021.00096
4. Adelaiye, O.I., Showole, A., Faki, S.A.: Evaluating advanced persistent threats mitigation effects: a review. Int. J. Inf. Secur. Sci. **7**(4), 159–171 (2018)

5. Xing, K., Li, A., Jiang, R., Jia, Y.: A review of APT attack detection methods and defense strategies. In: 2020 IEEE Fifth International Conference on Data Science in Cyberspace (DSC), Hong Kong, Hong Kong, pp. 67–70. IEEE (2020). https://doi.org/10.1109/DSC50466.2020.00018
6. Kaloudi, N., Li, J.: The AI-based cyber threat landscape: a survey. ACM Comput. Surv. **53**(1), 1–34 (2021). https://doi.org/10.1145/3372823
7. Chen, Z., et al.: Machine learning-enabled IoT security: open issues and challenges under advanced persistent threats. ACM Comput. Surv. https://arxiv.org/abs/2204.03433
8. Ren, K., Zheng, T., Qin, Z., Liu, X.: Adversarial attacks and defenses in deep learning. Engineering **6**(3), 346–360 (2020). https://doi.org/10.1016/j.eng.2019.12.012
9. Campos, E.M., et al.: Evaluating federated learning for intrusion detection in Internet of Things: review and challenges. Comput. Netw. **203**, 108661 (2022). https://doi.org/10.1016/j.comnet.2021.108661
10. Konecn'y, J., McMahan, H.B., Ramage, D., Richtárik, P.: Federated optimization: distributed machine learning for on-device intelligence (2016). arXiv:1610.02527
11. McMahan, H.B., Moore, E., Ramage, D., Hampson, S., Arcas, B.A.: Communication-efficient learning of deep networks from decentralized data. arXiv (2023). http://arxiv.org/abs/1602.05629. Accessed 08 Oct 2023
12. Rahman, S.A., Tout, H., Talhi, C., Mourad, A.: Internet of Things intrusion detection: centralized, on-device, or federated learning? IEEE Netw. **34**(6), 310–317 (2020). https://doi.org/10.1109/MNET.011.2000286
13. Abdul Rahman, S., Mourad, A., El Barachi, M., Orabi, W.A.: A novel on-demand vehicular sensing framework for traffic condition monitoring. Veh. Commun. **12**, 165–178 (2018). https://doi.org/10.1016/j.vehcom.2018.03.001
14. Ren, J., Wang, H., Hou, T., Zheng, S., Tang, C.: Federated learning-based computation offloading optimization in edge computing-supported Internet of Things. IEEE Access **7**, 69194–69201 (2019). https://doi.org/10.1109/ACCESS.2019.2919736
15. Liu, Y., et al.: Deep anomaly detection for time-series data in industrial IoT: a communication-efficient on-device federated learning approach. IEEE Internet Things J. **8**(8), 6348–6358 (2021)
16. Liu, D., Miller, T., Sayeed, R., Mandl, K.D.: FADL: federated-autonomous deep learning for distributed electronic health record. arXiv (2018). http://arxiv.org/abs/1811.11400. Accessed 08 Oct 2023
17. Huang, L., Shea, A.L., Qian, H., Masurkar, A., Deng, H., Liu, D.: Patient clustering improves efficiency of federated machine learning to predict mortality and hospital stay time using distributed electronic medical records. J. Biomed. Inform. **99**, 103291 (2019). https://doi.org/10.1016/j.jbi.2019.103291
18. Dayan, I., et al.: Federated learning for predicting clinical outcomes in patients with COVID-19. Nat. Med. **27**(10), 1735–1743 (2021). https://doi.org/10.1038/s41591-021-01506-3
19. Zhang, X., Fang, F., Wang, J.: Probabilistic solar irradiation forecasting based on variational Bayesian inference with secure federated learning. IEEE Trans. Ind. Inform. **17**(11), 7849–7859 (2021). https://doi.org/10.1109/TII.2020.3035807
20. Yang, T., et al.: Applied federated learning: improving google keyboard query suggestions. arXiv (2018). http://arxiv.org/abs/1812.02903. Accessed 08 Oct 2023
21. Hard, A., et al.: Federated learning for mobile keyboard prediction. arXiv (2019). http://arxiv.org/abs/1811.03604. Accessed 08 Oct 2023

22. Huang, Y., et al.: A high-precision method for 100-day-old classification of chickens in edge computing scenarios based on federated computing. Animals **12**(24), 3450 (2022). https://doi.org/10.3390/ani12243450
23. Mourad, A., Tout, H., Wahab, O.A., Otrok, H., Dbouk, T.: Ad hoc vehicular fog enabling cooperative low-latency intrusion detection. IEEE Internet Things J. **8**(2), 829–843 (2021). https://doi.org/10.1109/JIOT.2020.3008488
24. Abdul Rahman, S., Mourad, A., El Barachi, M.: An infrastructure-assisted crowdsensing approach for on-demand traffic condition estimation. IEEE Access **7**, 163323–163340 (2019). https://doi.org/10.1109/ACCESS.2019.2953002
25. Schreyer, M., Sattarov, T., Borth, D.: Federated and privacy-preserving learning of accounting data in financial statement audits. In: Proceedings of the Third ACM International Conference on AI in Finance, pp. 105–113. ACM, New York (2022). https://doi.org/10.1145/3533271.3561674.
26. Liu, B., Wang, L., Liu, M.: Lifelong federated reinforcement learning: a learning architecture for navigation in cloud robotic systems. IEEE Robot. Autom. Lett. **4**(4), 4555–4562 (2019)
27. Lavaur, L., Pahl, M.-O., Busnel, Y., Autrel, F.: The evolution of federated learning-based intrusion detection and mitigation: a survey. IEEE Trans. Netw. Serv. Manag. **19**(3), 2309–2332 (2022). https://doi.org/10.1109/TNSM.2022.3177512
28. Blanco-Justicia, A., Domingo-Ferrer, J., Martínez, S., Sánchez, D., Flanagan, A., Tan, K.E.: Achieving security and privacy in federated learning systems: survey, research challenges and future directions. Eng. Appl. Artif. Intell. **106**, 104468 (2021). https://doi.org/10.1016/j.engappai.2021.104468
29. Sharma, P.K., Park, J.H., Cho, K.: Blockchain and federated learning-based distributed computing defence framework for sustainable society. Sustain. Cities Soc. **59**, 102220 (2020). https://doi.org/10.1016/j.scs.2020.102220
30. Li, Z., Chen, J., Zhang, J., Cheng, X., Chen, B.: Detecting advanced persistent threat in edge computing via federated learning. In: Yu, S., Mueller, P., Qian, J. (eds.) SPDE 2020. CCIS, vol. 1268, pp. 518–532. Springer, Singapore (2020). https://doi.org/10.1007/978-981-15-9129-7_36
31. Lu, Y., Huang, X., Dai, Y., Maharjan, S., Zhang, Y.: Blockchain and federated learning for privacy-preserved data sharing in industrial IoT. IEEE Trans. Ind. Inform. **16**(6), 4177–4186 (2020). https://doi.org/10.1109/TII.2019.2942190
32. Kim, H., Park, J., Bennis, M., Kim, S.-L.: Blockchained on-device federated learning. IEEE Commun. Lett. **24**(6), 1279–1283 (2020). https://doi.org/10.1109/LCOMM.2019.2921755
33. Fan, Y., Li, Y., Zhan, M., Cui, H., Zhang, Y.: IoT defender: a federated transfer learning intrusion detection framework for 5G IoT. In: 2020 IEEE 14th International Conference on Big Data Science and Engineering (BigDataSE), Guangzhou, China, pp. 88–95 (2020). https://doi.org/10.1109/BigDataSE50710.2020.00020
34. Liu, Y., Kang, Y., Xing, C., Chen, T., Yang, Q.: A secure federated transfer learning framework. IEEE Intell. Syst. **35**(4), 70–82 (2020). https://doi.org/10.1109/MIS.2020.2988525
35. Xia, Q., Ye, W., Tao, Z., Wu, J., Li, Q.: A survey of federated learning for edge computing: research problems and solutions. High-Confid. Comput. **1**(1), 100008 (2021). https://doi.org/10.1016/j.hcc.2021.100008
36. Feki, I., Ammar, S., Kessentini, Y., Muhammad, K.: Federated learning for COVID-19 screening from Chest X-ray images. Appl. Soft Comput. **106**, 107330 (2021). https://doi.org/10.1016/j.asoc.2021.107330

37. Li, B., Wu, Y., Song, J., Lu, R., Li, T., Zhao, L.: DeepFed: federated deep learning for intrusion detection in industrial cyber-physical systems. IEEE Trans. Ind. Inform. **17**(8), 5615–5624 (2021). https://doi.org/10.1109/TII.2020.3023430
38. Chen, Z., Lv, N., Liu, P., Fang, Y., Chen, K., Pan, W.: Intrusion detection for wireless edge networks based on federated learning. IEEE Access **8**, 217463–217472 (2020). https://doi.org/10.1109/ACCESS.2020.3041793
39. Mothukuri, V., Khare, P., Parizi, R.M., Pouriyeh, S., Dehghantanha, A., Srivastava, G.: Federated-learning-based anomaly detection for IoT security attacks. IEEE Internet Things J. **9**(4), 2545–2554 (2022). https://doi.org/10.1109/JIOT.2021.3077803
40. Chatterjee, S., Hanawal, M.K.: Federated learning for intrusion detection in IoT security: a hybrid ensemble approach. arXiv (2023). https://arxiv.org/abs/2106.15349v1. Accessed 08 Oct 2023
41. Cheng, X., Luo, Q., Pan, Y., Li, Z., Zhang, J., Chen, B.: Predicting the APT for cyber situation comprehension in 5G-enabled IoT scenarios based on differentially private federated learning. Secur. Commun. Netw. **2021**, 8814068 (2021). https://doi.org/10.1155/2021/8814068
42. Thi, H.T., Hoang Son, N.D., Duy, P.T., Pham, V.-H.: Federated learning-based cyber threat hunting for APT attack detection in SDN-enabled networks. In: 21st International Symposium on Communications and Information Technologies (ISCIT), Xi'an, China, pp. 1–6 (2022). https://doi.org/10.1109/ISCIT55906.2022.9931222
43. Myneni, S., et al.: DAPT 2020 - constructing a benchmark dataset for advanced persistent threats. In: Wang, G., Ciptadi, A., Ahmadzadeh, A. (eds.) MLHat 2020. CCIS, vol. 1271, pp. 138–163. Springer, Cham (2020). https://doi.org/10.1007/978-3-030-59621-7_8

A Personal Health Assistant: Chatbot Design for Exercise and Nutrition Support

Garima Chandel[1(✉)], Sachin Yadav[2], Gauri Katiyar[3], Saweta Verma[4], and Setu Garg[5]

[1] Department of Electronics and Communication Engineering, Chandigarh University, Mohali, India
chandelgarima5@gmail.com
[2] Department of Engineering Technology and Industrial Professions, University of North Alabama, Alabama, USA
[3] Department of Electrical and Electronics Engineering, Galgotia College of Engineering and Technology, Greater Noida, India
[4] Department of Computer Science, Chandigarh University, Mohali, India
[5] Department of Electronics and Communication Engineering, I.T.S. Engineering College, Greater Noida, India

Abstract. Lack of physical activity in daily life can result in poor health and increase your risk of developing several diseases, including obesity and type-2 diabetes, which are currently the most prevalent among people. Even if someone starts an exercise program, they may not receive the proper instruction because diet and exercise go hand in hand for a healthy lifestyle. As a result, they may not see the desired results and stop. To address this, this work proposes the design of a chatbot using machine learning to provide suitable advice to make life easy. For this, an extensive study has been done and developed the user-friendly chatbot.

Keywords: Natural Language Processing · exercise · chatbot health · machine learning · personal assistant

1 Introduction

After the recent advancement of artificial intelligence and machine learning, the boom has been observed in the making of chatbots as each and every company start making its assistance for their device for amazon it was Alexa for Google assistance and the most common among them was the Siri by apple but upon that, they had very limited features cause they can only do what you say them to and that too in a limited number like they can't suggest you anything they can just find a webpage related to what you have asked them but can't suggest you something of their own [1, 2].

Nowadays machine learning algorithm is used in each and every field big tech companies like Google, Microsoft, and Instagram also uses them to show you what you like or what you liked the most, and then comes the chatbot boom after the launch of a CHATGPT which is currently the most advanced version of a chatbot in current

A. Bajaj et al. (Eds.): IBICA 2023, LNNS 1229, pp. 195–202, 2025.
https://doi.org/10.1007/978-3-031-78940-3_19

world scenario that can be built and chatbots work on the very basic principle of natural language processing(NLP) in which first the chatbot is given a user input and then it just understands it and gives you your desired output [4]. Each and every task people is nowadays done with their help of them so we thought why can't we make a chatbot that will give you exercise and nutrition tips alongside some motivation there are many limitations to it because a chatbot is not physically present it can't see whether you are losing weight or even cannot see is your technique correct while performing a specific task for which you will still need a physical trainer cause in some field humans can't replace chatbots and the other problem that we need to tackle is that many chatbots have a high dropout rate like 1/3rd of the people interacting with the chatbot leaves it after the first message so we need to first make our chatbot more interesting and human friendly to retain a large number of audience [5].

For this, we need to ask specific questions to them about their journey and what they have done what are their level like are they beginner or intermediate, or advanced alongside that we need to take care of them for someone starting a new in this field should be tackled with a lot of complex words so that he just doesn't drop out even before starting and they alongside need a daily dose of motivation until it becomes a habit for them and which will be a challenging task to perform [6].

The scope of this paper will deal with all of these problems and will try to overcome them and help people with their meal recommendations alongside giving them motivation so they don't their hope and stick to their goal in the long run as we believe this type of system will help us to make our user more engaging with the Chabot as it would be able to connect with them and thus they would be listening to it more and would be doing what the chatbot recommends them doing to stay fit and also because of this the exercise can reach to more people and they can spread the knowledge of it finally overall improving the health of the world and because of that decreasing the world diseases to an extent.

The paper is organized as follows: Sect. 2 reviews about the related work in this field of study. Section 3 describes the methodology used in present study. In Sect. 4 Results and Discussion, and the summary, comparison, and conclusions are given in Sect. 5.

2 Background

This study examines the difficulties and restrictions related to the use of chatbots to suggest meals in the context of a healthy diet. The purpose of employing chatbot systems to promote healthy eating habits is examined in this research, along with its history and drivers. Technical, theoretical, behavioural, and societal issues are all included in the discussion of the present chatbot application problems. The article suggests a pipeline for developers to follow in order to construct strong chatbot systems as a means of overcoming these difficulties. Health, recommender systems, conversational agents, human-computer interface (HCI), behaviour modification, and artificial intelligence (AI) are some of the terms connected to this study [7].

A personalized agent-based chatbot platform for nutritional counseling is presented in the present research. To help users maintain a healthy lifestyle, the system makes use of ubiquitous technology and offers dynamic, multi-modal, and adapted interactions. It provides dietary advice, monitors caloric intake, and encourages wholesome actions. The current state of agent-based chatbots, meal-tracking systems, and food recommendation systems is covered in the article. A user research and performance evaluation are offered, along with a full description of the chatbot platform's architecture and implementation. The suggested approach tackles issues with user autonomy, privacy, and customization in nutrition coaching systems [8].

The study offers a chatbot system for healthcare that uses artificial intelligence (AI) to offer medical support. To save healthcare expenses and increase accessibility to medical information, the chatbot intends to detect disorders and provide basic information prior to contacting a doctor. When a user submits a query, the system analyses it using natural language processing techniques including n-gram, TF-IDF, and cosine similarity, then retrieves pertinent replies from a knowledge database. When the chatbot is unable to answer the question satisfactorily, a professional program takes over. By facilitating productive and economical interactions between users and a virtual medical assistant, the suggested technology illustrates its potential to improve healthcare [9].

This paper examines how the growing acceptance of wearable technology has given people the means to use embedded sensors to actively track their physical activity. Existing fitness apps, however, frequently need constant user interaction and are ineffective at capturing the attention of time-constrained or less motivated people. We suggest PRO-Fit, a personalized fitness assistant architecture that makes use of machine learning and recommendation algorithms, in order to solve this. PRO-Fit uses accelerometer data to track and identify user activity, syncs with the user's calendar and suggests tailored workouts based on the user's prior behaviors, preferences, physical condition, and availability. This framework seeks to give people with different schedules and degrees of commitment a stimulating and organized approach to exercise activities. Keywords: personalized assistant, activity monitoring, categorization, and wearable technology [10].

The goal of this research paper is to construct a chatbot for the health care application that uses machine learning and natural language processing to diagnose ailments and suggest therapies based on user complaints. By allowing users to engage with it and enter their symptoms, the chatbot offers an alternative to traditional hospital visits. The chatbot then recognizes the symptoms, uses a K-nearest neighbour algorithm to forecast the condition, and suggests appropriate treatments. The system attempts to increase user knowledge of personal health issues and motivate them to take proactive steps to keep themselves in good health. The chatbot reduces the need for lengthy healthcare appointments by providing a free and open platform. Putting this system into place can greatly improve healthcare by increasing people's awareness of their health state and encouraging [11].

This paper explored how intelligent personal assistants might enhance individual health care. A smart personal assistant application is suggested that offers medical appointment scheduling using machine learning and cloud services. Users of the app may schedule appointments based on doctor availability anytime, anyplace, and identify local hospitals. By providing prompt service, this increases patient satisfaction and time savings. It is simple to use and has speech recognition as well as being available in many different languages.

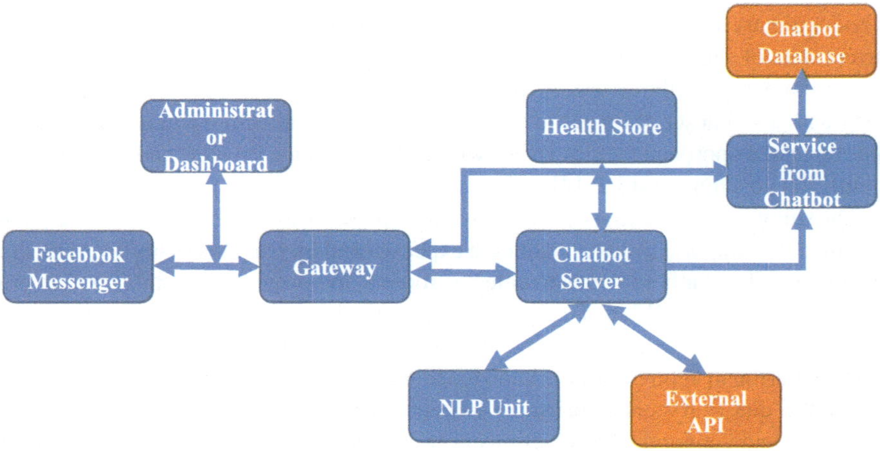

Fig. 1. Proposed Chatbot System Architecture

3 Methodology

The proposed model of chatbot architecture is shown in Fig. 1. This model works as follows: API Gateway forwards incoming requests from various customers to the appropriate service. The chatbot server receives chat messages from various clients and responds to them. Predicting the intents and entities from a given text is the responsibility of the NLU engine. Additionally, the chatbot server serves as an action server that makes requests to external APIs to carry out a certain action. All messages in the system are logged by the chatlog service. The store oversees keeping track of the messages for each user. The steps in developing the system are described below:

1) Planning and designing: To create a chatbot, we must first choose the platform on which we will build it. There are currently many platforms available to build a chatbot of your own. Here, we will choose Chatfuel as it is the most popular among students because it is affordable, simple to use, and programmable. After opting to do so, I'll be working with someone who has a working grasp of computers and chatbots to create a flowchart outlining our whole process for developing and deploying our chat bot.
2) User input handling: After we have started building our chatbot, we must first determine what services it offers and what are the most fundamental or frequent queries that users will have when they first interact with the bot because first impressions count for a lot when it comes to whether they will stick around or leave. To accomplish this, we must create a dataset that contains all the relevant data. So that we can give the best feedback to the user.
3) Blocks: For example, if someone asked for guidance on exercise planning, we would be able to respond with advice on that; it wouldn't be like they were asking about exercising and receiving advice on nutrition, or the other way around. To do this, we must first create various datasets or blocks that contain all the necessary knowledge about each relevant subject. For example, the nutrition section must include all the

fundamentals of portion control, calorie counting, and hunger management. Similarly, the exercise section must first classify the user as either a beginner, intermediate or advanced in the relevant field before offering suggestions tailored to their needs. Additionally, a dataset of motivation so that they can be motivated towards their goal.

4) Integration: As the demand will increase the datasets should also be increased cause the users will become more diverse and thus they would be requiring different types of advice for nutrition if someone is a vegetarian then he would be needing a different diet rather than someone who is a nonvegetarian alongside that the region will also affect their died like in some regions some foods are more common then on others so for that reasons we must collect different raw data from around the world in order to reach to a wider number of audience in the world alongside the exercise section should also be improved cause some people prepare calisthenics over gym some prefer running and yoga so they must also be evaluated according to their need and according to the availability of the types of equipment in their reason.
5) Testing: We must also get feedback from the user who will use the chatbot because they will be able to provide more inside information, and then we must work on what they have said. After all of this, the chatbot should be tested thoroughly and all bugs should be removed promptly so that people are able to use it without running into any issues while interacting with the bot.
6) Deployment: After completing all the testing and gathering user feedback, we must deploy our chatbot after removing it from its beta version, but even then, we must keep improving it and add additional suggestions more databases so that it would even reach more wide audience and be able to help as many people as possible. In addition, we can also deploy our chatbot on a variety of pages like on Facebook as well as on Instagram, which will help it to reach a larger number of users.

4 Result and Discussion

The goal of this study was to develop a chatbot using the chatfuel tool that deals with nutrition advice, exercise planning, along with motivating users on a regular basis so that they are engaged. The design and creation of the chatbot involved several steps, such as first planning our model, then deciding what would be the user-given input, and finally creating a dataset that includes all the input that a user can provide to the chatbot. The chatbot was successfully deployed on several platforms, including Facebook and Instagram, after various data sets were created in response to the customer's needs, requirements, and type of assistance he needs from the chatbot.

These data sets also included knowledge of nutrition and exercise. his demonstrates the growth of a health and nutrition chatbot utilizing Chatfuel, which genuinely aids the public by comprehending their inputs, providing pertinent output based on their meal-planning recommendations, to further improve its capabilities and user needs, and responding to their inquiries. It's important to note that the chatbot's effectiveness may increase in the future after the deployment of more datasets and the uploading of a more sophisticated machine learning algorithm. The chatbot may also be integrated with more extra features, such as fitness tracking APIs or customized experience. The chatbot demonstrates the technology's promise in health and wellness by providing users with a more practical and accessible platform to assist and support them as they

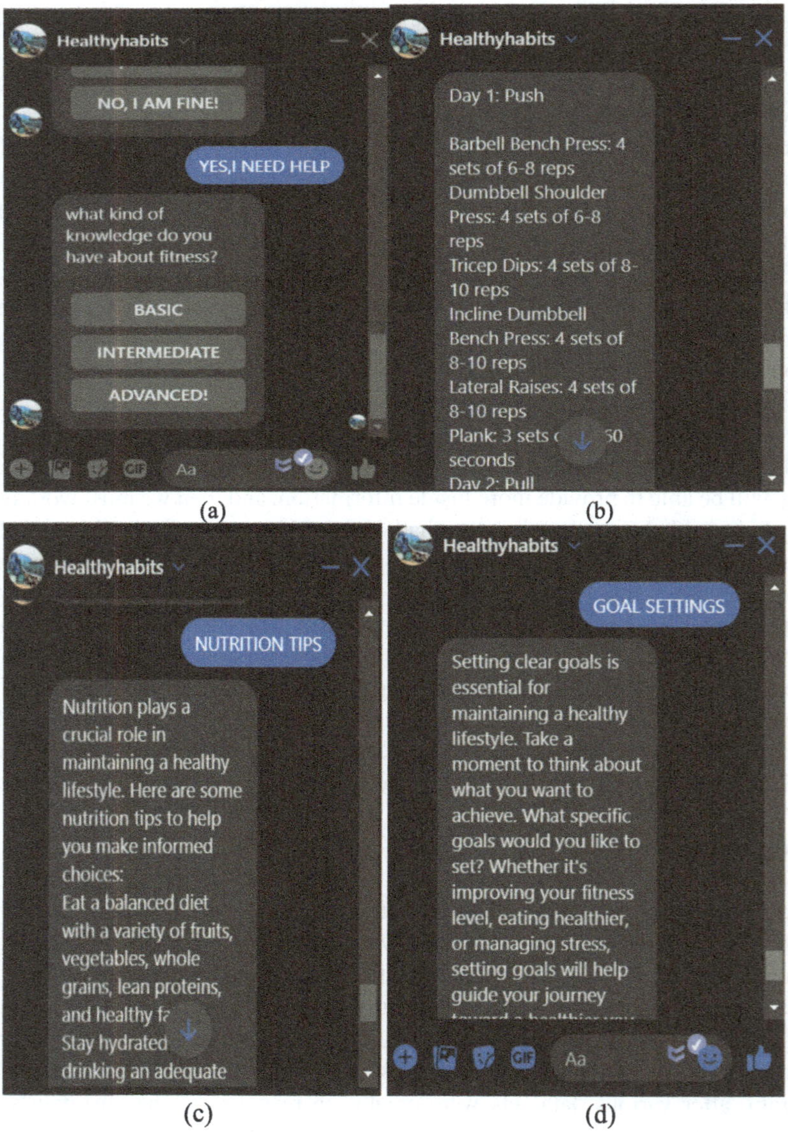

Fig. 2. Screenshots of designed chatbot output (a) Screenshots of Welcome note from chatbot (b) Screenshot of Chatbot reply to client (c) Screenshot of designed chatbot reply to client (d) Screenshot of designed chatbot reply.

adopt a better lifestyle. Figures 2 shows the sample results of developed chatbot which is replying to the query of a client in a proper way. Figure 2(a) shows a conversation about fitness knowledge levels, Figure 2(b) lists a workout routine, Figure 2(c) provides nutrition tips and Figure 2(d) discusses goal setting for a healthy lifestyle, encouraging clear objectives. Table 1 compares results of proposed methods with available studies in the same field.

Table 1. Comparison of proposed personal assistant with available ones

Author, year	Focus/Findings
Ahmed Fadill [7], 2016	Chatbot that can provide personalized exercise and nutrition tips, while also keeping users engaged and motivated. It suggests principles for programmers and emphasises the necessity of theoretical underpinnings, behaviour modification strategies, and overcoming technological constraints
David Cala aresi et al. [8], 2021	It includes the most recent developments in meal monitoring, food recommendation, and agent-based chatbots. Meal monitoring, meal suggestion, restaurant search, statistics, wearable sensor integration, periodic behaviours, and data are all included in the design
Lekha Athota et al. [9], 2020	The paper discusses the development of a medical chatbot using Artificial Intelligence that can diagnose diseases and provide basic details to reduce healthcare costs and improve accessibility to medical knowledge. The chatbot uses N-gram, TF-IDF, and cosine similarity to identify sentence keywords and make query decisions. The system also includes an expert program to handle questions not understood by the bot
Saumil Dharia et al. [10], 2018	This research has aimed to advance the capabilities of deep ocean electric manipulators and their control systems, which can be used in a variety of applications such as underwater exploration and maintenance
Rohit Binu Mathew [11] , 2019	The study focuses on creating a chatbot application that utilises machine learning and natural language processing to diagnose illnesses and suggest treatments based on user symptoms. By encouraging routine medical checkups and increasing personal health awareness, the suggested approach seeks to offer an alternative to conventional hospital visits
Arul Srinivasan et al. [3], 2018	The system architecture includes Google Cloud Platform, Dialogflow, and Heroku Cloud App Hosting. The proposed system provides quick confirmation of appointments and can be accessed via voice recognition. The application, called MedX AI Clara, offers an interface for finding doctors, selecting appointment dates, and choosing locations
Prissadang Suta et al. [12], 2020	The evolution of chatbots and their potential to enhance customer service and expedite processes are discussed in this article. It emphasizes the demand for chatbots that can have independent, conversational interactions with humans
This work	The creation of a chatbot that provides fitness and dietary advice is discussed in this study. With the help of the chatbot, users will receive appropriate suggestions on how to increase their daily physical activity. To keep a big audience interested, the chatbot is made to be user-friendly and entertaining. It addresses subjects including fitness scheduling, dietary guidance, and inspiration. The chatbot has been successfully used on website

5 Conclusion

The present study focused that the field of artificial intelligence is a journey. Machine learning has enormous potential and is shaping the way we live either its lifestyle or shopping or any other kind of stuff that we used to do in our day to day task and in this project, we have created a chatbot using a most widely used platform named chatbot which actually helped us to create a chatbot which provides exercise tips and nutrition pieces of advice to people according to their need and the bot has an enormous potential in the upcoming future because nowadays more and more people are coming and relying on chatbot for the smaller task also so this bot will help them in their fitness journey and will make them fit by not making fitness complicated and keeping it as simple as possible so that they don't drop out from their journey while also having a dose motivation in between their journey.

References

1. Zierau, N., Engel, C., Söllner, M., Leimeister, J.M.: Trust in smart personal assistants: A systematic literature review and development of a research agenda. In: International Conference on Wirtschaftsinformatik (WI). Potsdam, Germany (2020)
2. Canbek, N.G., Mutlu, M.E.: On the track of artificial intelligence: Learning with intelligent personal assistants. Journal of Human Sciences **13**(1), 592–601 (2016)
3. Srinivasan, A., Madheswari, A.N.: The role of smart personal assistant for improving personal healthcare. Int. J. Adv. Eng. Manage. Sci. **4**(11), 268274 (2018)
4. Edu, J.S., Such, J.M., Suarez-Tangil, G.: Smart home personal assistants: a security and privacy review. ACM Computing Surveys (CSUR) **53**(6), 1–36 (2020)
5. Devi, S., Merchant, Z.A.M., Siddiqui, M.S., Lobo, M.: Artificial intelligence based personal assistant. Asian Journal For Convergence In Technology (AJCT) (2019). ISSN-2350-1146
6. Kepuska, V., Bohouta, G.: Next-generation of virtual personal assistants (microsoft cortana, apple siri, amazon alexa and google home). In: 2018 IEEE 8th annual computing and communication workshop and conference (CCWC), pp. 99–103. IEEE (2018)
7. Fadhil, A.: Can a chatbot determine my diet?: Addressing challenges of chatbot application for meal recommendation (2018). arXiv preprint arXiv:1802.09100
8. Calvaresi, D., Eggenschwiler, S., Calbimonte, J.P., Manzo, G., Schumacher, M.: A personalized agent-based chatbot for nutritional coaching. In: IEEE/WIC/ACM International Conference on Web Intelligence and Intelligent Agent Technology, pp. 682–687 (2021)
9. Athota, L., Shukla, V.K., Pandey, N., Rana, A.: Chatbot for healthcare system using artificial intelligence. In: 2020 8th International conference on reliability, infocom technologies and optimization (trends and future directions) (ICRITO), pp. 619–622. IEEE (2020)
10. Dharia, S., et al.: PRO-Fit: A personalized fitness assistant framework. In: SEKE, pp. 386–389 (2016)
11. Mathew, R.B., Varghese, S., Joy, S.E., Alex, S.S.: Chatbot for disease prediction and treatment recommendation using machine learning. In: 2019 3rd International Conference on Trends in Electronics and Informatics (ICOEI), pp. 851–856. IEEE (2019)
12. Suta, P., Lan, X., Wu, B., Mongkolnam, P., Chan, J.H.: An overview of machine learning in chatbots. Int. J. Mechan. Eng. Robot. Res. **9**(4), 502–510 (2020)

Digital Banking Transformation: LLMs in Customer Service and Fraud Detection

R. Manoj(✉), K. Boobesh, A. Peter Rajan, and T. P. Saravanan

Department of Management Studies, Kongu Engineering College, Perundurai, Erode, Tamil Nadu 638 060, India
{manojr.22mba,boobeshk.22mba,peterrajana.22mba, tpsaravanan.mba}@kongu.edu

Abstract. The financial services industry is undergoing a profound transformation known as digital banking, marked by the integration of advanced technologies that reshape the customer experience and fortify fraud detection measures. At the forefront of this evolution are large language models (LLMs), leveraging artificial intelligence and natural language processing capabilities to enhance customer service interactions and bolster fraud detection mechanisms. This research paper offers a comprehensive exploration of the intersection between digital banking transformation and the integration of LLMs in customer service and fraud detection within the banking sector. The paper commences with an examination of the digital banking landscape, tracing its evolution and expanding eco- system of digital services.

The research paper concludes by outlining future trends and prospects in digital banking transformation, emphasizing the significance of LLMs in shaping the future of banking services. It provides practical recommendations for financial institutions contemplating the integration of LLMs and underscores the need for responsible and ethical AI adoption in the financial sector. This study aims to contribute to a deeper understanding of the transformative potential of LLMs in banking and their implications for customer service excellence and fraud prevention.

1 Introduction

The financial services industry has embarked on a transformative journey in response to the inexorable tide of digitalization. This paradigm shift, known as "digital banking transformation," is revolutionizing the banking landscape, fundamentally altering how financial institutions interact with their clientele and combat financial malfeasance. In this epochal transformation, the integration of large language models (LLMs) emerges as a catalytic force, reshaping the manner in which banks provide customer service and detect fraudulent activities.

Digital banking transformation is the result of a confluence of factors. As technology rapidly evolves and consumer expectations evolve with it, financial institutions are compelled to reassess their strategies and operations. From the proliferation of mobile banking apps to the advent of online customer support, digitalization has ushered in

A. Bajaj et al. (Eds.): IBICA 2023, LNNS 1229, pp. 203–211, 2025.
https://doi.org/10.1007/978-3-031-78940-3_20

an era of enhanced customer experiences, streamlined processes, and fortified security measures.

Large language models, underpinned by advanced artificial intelligence and natural language processing capabilities, have ascended as formidable tools poised to revolutionize the banking sector. With their capacity to comprehend and generate human-like text, LLMs are adept at engaging customers in conversation, addressing inquiries, and resolving issues. Moreover, their analytical prowess extends to the realm of fraud detection, where they can sift through vast datasets and identify suspicious patterns more expeditiously and accurately than conventional methods. This research paper endeavors to conduct an exhaustive exploration of the fusion between digital banking transformation and the assimilation of large language models in two pivotal domains: customer service and fraud detection. It seeks to unveil the implications, challenges, and advantages of incorporating LLMs in these facets of modern banking. Through the examination of real-world case studies and industry practices, this paper aims to offer valuable insights to practitioners and scholars interested in the convergence of artificial intelligence and the financial sector.

2 Literature Review

In this research paper, the authors introduce FLAIR (Finding Large Language Model Authenticity via a Single Inquiry and Response), a framework designed to detect conversational bots in online interactions, particularly in chat scenarios. The framework utilizes a set of carefully crafted questions that exploit the differences in how bots and humans process and generate language. These questions are divided into categories, some challenging for bots but easy for humans, and vice versa. FLAIR aims to address concerns about the misuse of large language models for malicious purposes by providing a method for online service providers to distinguish between bots and real users effectively. This framework offers a promising alternative to traditional CAPTCHAs in both detections.

This research paper investigates the impact of Large Language Models (LLMs), with a focus on ChatGPT, on academic integrity in computing education. The study assesses the effectiveness of eight publicly available LLM-generated text detectors by comparing 124 human-written submissions with 40 ChatGPT-generated submissions, along with additional submissions that underwent paraphrasing. The evaluation includes measures of accuracy, false positives, and resilience. The results indicate that CopyLeaks is the most accurate LLM-generated text detector, GPTKit excels in reducing false positives, and GLTR is the most resilient detector. The study highlights the need for improved LLM detectors to maintain academic integrity and discusses the usability of these tools in educational settings.

This study explores the impact of Large Language Models (LLMs), particularly ChatGPT, on medical imaging stakeholders, encompassing businesses, insurance entities, governments, research institutions, and hospitals. LLMs demonstrate the potential to enhance the skills of radiologists, streamline clinical workflows, and improve patient-physician communication in healthcare settings. The paper introduces the BIGR-H framework, providing a structured approach to examine these interactions. Additionally, it delves into use cases and implications, highlighting opportunities and challenges such as ethics, privacy, bias, and regulation in the context of AI-enabled healthcare.

The paper discusses the development of HuntGPT, an intrusion detection dashboard that combines machine learning-based anomaly detection, Explainable AI (XAI) techniques, and Large Language Models (LLMs), specifically ChatGPT. The system employs a Random Forest classifier for network traffic analysis, trained with the KDD99 dataset, and integrates XAI frameworks like SHAP and LIME. HuntGPT, when combined with the GPT-3.5 Turbo conversational agent, aims to provide easily understandable threat alerts. The paper evaluates the system's technical accuracy using Certified Information Security Manager (CISM) Practice Exams and assesses response quality across various metrics, highlighting the potential of LLMs and XAI in intrusion detection systems.

This paper discusses the transformation of the banking industry from traditional brick-and-mortar institutions to the world of online and virtual banking, emphasizing the benefits and challenges of internet banking. It highlights the security concerns, including internet banking fraud and various types of cybercrimes such as identity theft and phishing. The paper underlines the impact of these issues on consumer trust and financial institutions' credibility, as well as the rise of fraud in the Indian banking sector. Additionally, it touches upon the existing legal situation related to internet banking in India and the need for clear regulations to address these concerns and protect consumers.

3 Research Problem

The rapid digital banking transformation, characterized by the integration of large language models (LLMs) in customer service and fraud detection, presents a complex challenge in the banking sector. While the potential benefits of LLMs are evident, there is a pressing need to address the following research problem. How can financial institutions effectively leverage large language models (LLMs) to enhance customer service and combat financial fraud while addressing ethical considerations, regulatory compliance, and ensuring optimal operational outcomes in the context of digital banking transformation. This research problem encompasses several key dimensions, including the utilization of LLMs to improve customer service experiences, their effectiveness in fraud detection, the ethical implications of AI adoption, regulatory compliance, and the overall operational impact on financial institutions. Solving this problem requires a comprehensive examination of the multifaceted challenges and opportunities associated with the integration of LLMs in the rapidly evolving digital banking landscape.

4 Objectives

1. **To Examine the Impact of Digital Banking Transformation**: Investigate the extent to which digital banking transformation has influenced the operations and customer interactions of financial institutions, with a focus on technological advancements.
2. **To Assess the Role of Large Language Models in Customer Service**: Analyze the applications and effectiveness of large language models in improving customer service within the banking sector, including their capacity to engage in natural language conversations, address inquiries, and provide personalized support.

3. **To Evaluate Large Language Models in Fraud Detection**: Evaluate the efficacy of large language models in enhancing fraud detection and prevention mechanisms in banks, including their ability to analyze complex datasets and identify fraudulent activities in real-time.
4. **To Explore Ethical and Regulatory Considerations**: Examine the ethical implications and regulatory challenges associated with the deployment of large language models in banking, including issues related to data privacy, algorithmic bias, and compliance with industry regulations.
5. **To Provide Insights into Best Practices**: Offer practical recommendations and best practices for financial institutions looking to integrate large language models into their customer service and fraud detection processes, while ensuring responsible and ethical AI adoption.

5 Large Language Models (LLMs) in Banking

Large Language Models (LLMs) represent a groundbreaking development in the field of artificial intelligence and natural language processing. In the context of banking and finance, their potential impact is far-reaching and transformative. This section delves into the core aspects of LLMs within the banking sector, providing an in-depth exploration of their capabilities, evolution, and relevance. LLMs, such as GPT-3 and BERT, are sophisticated AI systems that excel in understanding and generating human-like text. Their capabilities encompass natural language understanding, sentiment analysis, and text generation, making them versatile tools for communication and data analysis. These models have undergone a remarkable evolution, from their nascent stages to becoming pivotal components of various industries. The banking sector, in particular, has recognized the immense potential of LLMs.

In banking and finance, LLMs find relevance in a multitude of applications. They enable personalized customer interactions, automate routine inquiries, and provide 24/7 support through chatbots and virtual assistants. Moreover, LLMs play a vital role in fraud detection and prevention by analyzing vast datasets and identifying suspicious patterns with unprecedented accuracy. We will also address the challenges and ethical considerations associated with LLM adoption in the financial industry, paving the way for an in-depth understanding of the transformative power and ethical responsibilities tied to the integration of LLMs in banking and finance.

6 Digital Banking Transformation

The financial industry is currently undergoing a monumental shift known as "digital banking transformation." This transformation marks a fundamental reimagining of how financial services are delivered, consumed, and experienced in the digital age. At its core, it is a response to changing customer expectations, fueled by technological advancements that have reshaped the very essence of banking. The digital banking landscape has evolved from its humble beginnings to a sprawling ecosystem encompassing mobile banking apps, online transactions, and a wide array of digital financial services. It has become synonymous with convenience, accessibility, and immediacy, fundamentally altering the relationship between customers and their financial institutions.

The technological advancements have underpinned this transformation. The proliferation of high-speed internet access, the ubiquity of smartphones, and the advent of cloud computing have paved the way for seamless digital experiences. Additionally, the integration of big data analytics, artificial intelligence (AI), and blockchain technology has revolutionized banking operations, enabling real-time data analysis, personalized services, and enhanced security measures. As we delve deeper into this research, we aim to dissect the various facets of digital banking transformation, exploring not only the technological advancements but also the challenges, opportunities, and future trends that shape the banking industry in this digital era.

7 Large Language Models in Customer Service

The integration of Large Language Models (LLMs) into customer service represents a pivotal transformation within the banking sector. In this section, we embark on a comprehensive exploration of the wide-ranging applications, compelling case studies, benefits, challenges, and ethical dimensions associated with the utilization of LLMs in the domain of customer service. LLMs, exemplified by influential models like GPT-3 and BERT, have redefined the parameters of customer interactions within the banking industry. These advanced AI systems possess a unique ability to comprehend and generate human-like text, enabling a spectrum of applications that transcend conventional customer service capabilities. From automating responses to addressing complex queries with natural language understanding and delivering highly personalized experiences, LLMs have become the cornerstone of modern customer support in financial institutions.

To bring the transformative potential of LLMs to life, this section will showcase a curated selection of case studies and exemplary instances from banks and financial institutions. These real-world examples will illustrate how LLMs have not only streamlined operations, improved response times, and accuracy but have also elevated customer experiences, fostering loyalty and satisfaction. The integration of LLMs in customer service is not devoid of challenges. This section will delve into the potential obstacles and intricacies, including concerns related to data privacy, accuracy, and the evolving customer expectations, which must be navigated for successful LLM implementation. Ethical considerations loom large in this narrative. We will explore the ethical dimensions, encompassing the imperative for algorithmic fairness, the mitigation of bias, and the crucial need for transparency in AI-driven customer interactions. Responsible AI practices will emerge as a central theme in this discourse.

8 Fraud Detection and Prevention

The integration of Large Language Models (LLMs) marks a profound transformation in the realm of fraud detection and prevention within the banking sector. This section embarks on an in-depth exploration of the pivotal role played by LLMs in fortifying the defenses of financial institutions against fraudulent activities. It offers a comprehensive analysis of the applications, real-world examples, effectiveness, challenges, and ethical considerations inherent in the deployment of LLMs for fraud prevention. LLMs have emerged as potent tools in the fight against financial fraud, capable of processing and comprehending vast volumes of financial data with remarkable precision. Their natural language understanding capabilities empower banks to identify irregularities, flag suspicious transactions, and detect fraud schemes that may otherwise elude traditional methods. From identity theft to transaction fraud, LLMs have redefined the efficacy and agility of fraud detection.

The transformative potential of LLMs comes to life through a curated selection of real-world case studies. These cases spotlight instances where banks have successfully employed LLMs to combat fraud, illustrating tangible results and showcasing how these models have thwarted fraudulent activities, safeguarding the financial interests of both institutions and customers. Effectiveness is a key criterion in evaluating LLMs' contribution to fraud prevention. This section will delve into empirical evidence and statistical insights that validate the efficacy of LLMs in identifying and mitigating a spectrum of fraudulent activities. These data-driven assessments underscore the pivotal role of LLMs in the battle against financial malfeasance. Ethical considerations are paramount, and this section will explore the ethical dimensions of LLMs in fraud prevention, emphasizing the need for responsible AI practices. Privacy concerns, fairness, transparency, and the imperative for unbiased fraud detection will be integral parts of this discourse.

9 Challenges and Ethical Considerations

The incorporation of Large Language Models (LLMs) into the banking sector, while promising transformational benefits, is not without its challenges and ethical concerns. This section delves into the multifaceted issues that surround the use of LLMs in the financial industry, shedding light on privacy concerns, biases in AI algorithms, regulatory and compliance challenges, transparency, and mitigation strategies. Privacy concerns stand as a paramount challenge when deploying LLMs in banking operations. These models often require access to extensive customer data for effective performance, raising questions about data privacy, consent, and security. The collection and processing of personal information by LLMs must adhere to rigorous data protection standards, necessitating robust safeguards to prevent data breaches and unauthorized access. Bias in AI algorithms is another formidable hurdle, particularly relevant in LLM-driven banking. The biases present in training data can result in discriminatory outcomes, impacting customer experiences and decisions made by financial institutions. Addressing bias requires continuous monitoring, rigorous data curation, and algorithmic fairness assessments to ensure equitable treatment for all customers.

Regulatory and compliance challenges loom large in the adoption of LLMs. Banks must navigate a complex landscape of existing regulations and standards while adapting to the novel capabilities of LLMs. This entails ensuring that LLM-driven processes adhere to data protection laws, consumer rights, and industry-specific compliance requirements, necessitating transparency, accountability, and a thorough understanding of evolving regulatory frameworks. Transparency and explain ability are vital aspects of mitigating challenges and ethical concerns. LLM-generated responses and decisions must be interpretable by both customers and regulators. Achieving transparency requires developing mechanisms for explaining LLM-driven decisions, providing insights into the reasoning behind recommendations, and ensuring that customers can understand and challenge those decisions when necessary.

10 Future Trends and Implications

The future of digital banking is poised for transformation, with Large Language Models (LLMs) at the forefront of this evolution. In this section, we gaze into the crystal ball, offering predictions for the role of LLMs in the financial sector and the implications for banks, customers, and the industry as a whole.

Predictions for the Future of LLMs in Digital Banking encompass a diverse array of possibilities. We anticipate LLMs evolving into even more sophisticated and context aware entities, capable of seamlessly understanding and addressing customer inquiries, providing personalized financial advice, and streamlining complex banking operations. The integration of LLMs into voice-activated virtual assistants may become ubiquitous, offering customers a conversational and intuitive banking experience. Moreover, LLMs are likely to play a pivotal role in automating back-office functions, enhancing compliance procedures, and accelerating decision-making processes within banks. Emerging Trends and Innovations in LLM Technology paint a picture of continual advancement. Multimodal LLMs, capable of processing text, images, and speech, may become mainstream, opening new avenues for richer customer interactions and fraud detection. We foresee improvements in natural language understanding, allowing LLMs to grasp nuanced customer queries and provide more accurate responses. Additionally, federated learning and privacy-preserving techniques may enhance data security in LLM-powered banking applications.

Anticipated Challenges and Opportunities for Banks lie ahead. While LLMs promise efficiency and convenience, banks must grapple with challenges related to data privacy, algorithmic transparency, and ethical use. However, these challenges also present opportunities for banks to lead in responsible AI adoption. Banks that successfully navigate these complexities stand to gain a competitive edge by offering customers innovative, trustworthy, and secure banking experiences. As LLMs continue to evolve, so too will the ethical and regulatory considerations. We delve into how these considerations may evolve, emphasizing the need for adaptive regulations and ethical guidelines that balance innovation with safeguarding customer rights and data. The implications for banking strategy are profound. Banks must be prepared to embrace change, invest in LLM technology, and align their strategies with the evolving landscape. Those that proactively harness LLM capabilities, while prioritizing ethical and regulatory compliance, are poised to excel in the ever-evolving terrain of digital banking transformation.

11 Conclusion

In the digital banking transformation, the integration of Large Language Models (LLMs) into the realms of customer service and fraud detection stands as a beacon of innovation and change. This research paper has delved deep into the pivotal role that LLMs play in reshaping the financial landscape. As we conclude this exploration, several key takeaways emerge, underscoring the significance and implications of this transformative convergence. First and foremost, LLMs offer unparalleled potential to elevate customer service in the banking sector. With their natural language processing capabilities, LLMs can engage customers in human-like conversations, provide real-time assistance, and enhance overall experiences. The case studies and examples presented throughout this paper have demonstrated how banks are leveraging LLMs to meet the evolving expectations of their tech-savvy clientele. LLMs have emerged as formidable allies in the fight against financial fraud.

This transformative journey is not without its challenges and ethical considerations. Privacy concerns, bias in AI algorithms, regulatory complexities, and the need for transparency have been highlighted as crucial areas of concern. Navigating these challenges while maximizing the benefits of LLM technology will be a defining task for banks in the digital era. In the face of these challenges and opportunities, the banking sector must embrace adaptability, ethical responsibility, and a commitment to customer-centricity. Proactive strategies that encompass robust data governance, bias mitigation, compliance adherence, and customer consent mechanisms will be the guiding compass for successful LLM adoption. The integration of LLMs into digital banking represents a pivotal juncture in the financial industry's evolution. As we peer into the future, the promise of enhanced customer service, advanced fraud detection, and efficient banking operations beckons. With diligence, foresight, and unwavering ethical principles, banks can harness the power of LLMs to not only meet the demands of the digital age but also redefine the banking experience for generations to come.

References

Bakhtin, A., et al.: Real or fake? learning to discriminate machine from human generated text. arXiv preprint arXiv:1906.03351 (2019)

Brown, T., et al.: Language models are few-shot learners. Advances in Neural Information Processing Systems **33**, 1877–1901 (2020)

Chew, M., Baird, H.S.: Baffletext: a human interactive proof. In: Document Recognitionand Retrieval X, volume 5010, pp. 305–316. SPIE (2003)

Chiang, W.-L., et al.: Vicuna:An open-source chatbot impressing gpt-4 with 90%* chatgpt quality (2023). URL https://vicuna.lmsys.org

Chowdhery, A., et al.: Palm:Scaling language modeling with pathways. arXiv preprint arXiv:2204.02311 (2022)

Devlin, J., Chang, M.-W., Lee, K., Toutanova, K.: Bert: Pre-training of deep bidirectional transformers for language understanding. arXiv preprint arXiv:1810.04805 (2018)

Elazar, Y., et al.: Measuring and improving consistency in pretrained language models. Transactions of the Association for Computational Linguistics **9**, 1012–1031 (2021)

Gao, H., Liu, H., Yao, D., Liu, X., Aickelin, U.: An audio captcha to distinguish humans from computers. In: 2010 Third International Symposium on Electronic Commerce and Security, pp. 265–269. IEEE (2010)

Gossweiler, R., Kamvar, M., Baluja, S.: What's up captcha? a captcha based on image orientation. In: Proceedings of the 18th international conference on World wide web, pp. 841–850 (2009)

Kirchenbauer, J., et al.: A watermark for large language models. arXiv preprint arXiv:2301.10226 (2023)

Kluever, K.A.: Evaluating the Usability and Security of a Video CAPTCHA. Rochester Institute of Technology (2008)

Lewis, M., et al.: Bart: Denoising sequence-to-sequence pre-training for natural language generation, translation, and comprehension. arXiv preprint arXiv:1910.13461 (2019)

Albreiki, B., Zaki, N., Alashwal, H.: A systematic literature review of student' performance prediction using machine learning techniques. Education Sciences **11**, 9 (2021). https://doi.org/10.3390/educsci11090552

Manuscript submitted to ACM Detecting LLM-Generated Text in Computing Education Chandan Kumar Behera and D Lalitha Bhaskari. Different obfuscation techniques for code protection. Procedia Computer Science **70**, 757–763 (2015)

Blanchard, J., et al.: Stop reinventing the wheel! promoting community software in computing education. In: 2022 Working Group Reports on Innovation andTechnology in Computer Science Education (Dublin, Ireland). Ass. Comp. Machin. 261–292 (2022). https://doi.org/10.1145/3571785.3574129

Bucci, E.M.: Automatic detection of image manipulations in the biomedical literature. Cell Death & Disease **9** (2018). https://doi.org/10.1038/s41419-018-0430-3

Budi, S., et al.: IBAtS - Image Based Attendance System: A Low-Cost Solution to Record Student Attendance in a Classroom. In: 2018 IEEE International Symposium on Multimedia (ISM), pp. 259–266 (2018). https://doi.org/10.1109/ISM.2018.00037

Implementing a Deep Learning Approach for Forecasting Student Acedemic Performance

Shazia Asif[1], Manjula Jain[2], Meenakshi[3], Khaled A. Z. Alyamani[4], Sami A. Morsi[4,5], and Abhishek Raghuvanshi[6(✉)]

[1] Higher Colleges of Technology, Dubai, UAE
[2] Teerthanker Mahaveer University, Moradabad, UP, India
[3] Apeejay Stya University Sohna, Gurgaon, Haryana, India
[4] Abqaiq Branch, King Faisal University, Al-Ahsa 31982, Hofuf P.O. Box 4000, Saudi Arabia
{kalyamani,Smorsi}@kfu.edu.sa
[5] AL-Neelain University, Khartoum, Sudan
[6] Computer Science and Engineering, Mahakal Institute of Technology, Ujjain, India
Abhishek14482@gmail.com

Abstract. In the field of educational statistics, one of the most important tasks is student classification for the purposes of performance prediction or the detection of potential dropouts. It is essential for educational institutions to be able to make accurate projections of their students' future achievements and success. Because it serves as the measure by which education systems globally are evaluated, the accomplishment of students is extremely important to those who work in education. In addition, kids who stop attending school contribute to a reduction in this efficiency; therefore, identifying these pupils as early as possible is essential. As a result, arranging educational data in the appropriate categories is of the utmost importance. The recent advances in technology have made it possible to use a broad variety of classification strategies to educational records in a way that is both effective and efficient. It is necessary to investigate a greater number of techniques and methods suitable for this sort of activity. The amount of data that is being gathered by educational institutions is growing at a rapid rate, and so is the breadth of topics that are being covered. Investigating the usefulness of unconventional methods is an interesting endeavor. This article presents a deep auto encoder based technique for forecasting students academic performance. Experimental results have shown that the accuracy of autoencoder in predicting academic performance is 98.5%.

Keywords: Deep learning · Forecasting · Student Performance · Autoencoder · Accuracy · Specificity

1 Introduction

In this day and age of data mining and knowledge discovery, the field of educational data mining has advanced to a higher level of study [1]. Discovering meaningful knowledge and mining patterns that can be applied in the educational domain or area, ranging from elementary school to higher education, is the primary focus here. EDM allows for

A. Bajaj et al. (Eds.): IBICA 2023, LNNS 1229, pp. 212–219, 2025.
https://doi.org/10.1007/978-3-031-78940-3_21

the completion of a wide variety of tasks, some of which include the classification of students, the creation of association rules and decision trees, and the analysis of student data and information [2].

Because of this, EDM is becoming a more popular choice for academic studies. The education industry is responsible for both the collection and production of massive amounts of data relating to students and other parties involved in the education systems. This kind of information is now very easily retrievable in digital form because to the widespread adoption of ERP systems in the academic sector [3]. There are a variety of applications for the data that was gathered by educationalists and Educational System Management [4]. The information that was collected reveals Big Data qualities such as volume, velocity, diversity, honesty, and value. These characteristics may be seen in the information. Every one of DM's strategies can be helpful in the context of a classroom. Procedures such as character analysis, discrimination analysis, association analysis, classification analysis, clustering analysis, and outlier analysis are essential to data mining. All of these approaches can be helpfully applied in various aspects of academic study. One method that might be utilized is called "characterization," and it's employed to find out important information about the children. We are engaging in discriminating against students whenever we create comparisons between them based on artificial standards of evaluation. Data pertaining to education can also benefit from methods of association and categorization, among other similar approaches. When we talk about educational data mining, what we really mean is just applying these methodologies to educational data in some way [5].

In the realm of data mining techniques, classification is a method of classification and prediction that is utilized frequently and is known for its high level of effectiveness. The processes of Data Mining and Knowledge Discovery make use of a large number of different classification algorithms. Educational systems are continually embracing new methodologies and technology in order to improve the quality of the data they collect from students. In order to manage the ever-increasing volume of educational data, there is an immediate need for algorithms that can be scaled across multiple nodes [6]. This article presents a deep auto encoder based technique for forecasting student's academic performance. Experimental results have shown that the accuracy of deep auto encoder in predicting academic performance.

2 Literature Review

Zafar, Junaid, and Faisal studied ML systems that may predict grades in various courses. They classified data using RBM, collaborative filtering, regression, and matrix factorization. The Pakistani Information Technology University (ITU) data was analyzed using these methods. They evaluated ITU Bachelor's Program students. The RBM technique was the best of the ML algorithms used [7]. Kim Byung-Hak and colleagues recommend utilizing Deep Learning to predict student achievement. They collected data from Udacity Nanodegree Program students. This research uses the deep learning-based GritNet algorithm. This method outperforms Logistic Regression [8].

A machine learning strategy was given by Jie, Kyeong, and Schaar (2017) to predict degree program student success. This estimate considers past and present performance.

A data-driven latent component model will be utilized to create many base predictors and an effective basis prediction in the proposed system. The architecture is bilayered. This study shows that the recommended technique outperforms the benchmarks [9]. Murat examined student performance prediction machine learning methods. This study used linear regression (LR), decision trees (DT), and naïve bayes (NB) on two datasets. NB, with 98% accuracy, is better for the first dataset, and DT, with 78% accuracy, for the second [10].

Singh and Singh predicted engineering students' academic success by discipline using machine learning (ML). This research evaluates subjects from the previous academic year to predict students' current class performance. The Naive Bayes algorithm and C4.5 Decision Tree classifier were used to classify data in this investigation. The data show that C4.5 DT outperforms NB in accuracy [11]. In 2018, Bendengnu Ksung and Prabhu suggest using a Deep Neural Network model to predict student academic performance by category. In the study, a Deep Neural network was used with Decision Tree (J48), Naive Bayes, and ANN to compare accuracy. This model outperformed other ML Algorithms with 84.3% accuracy [12].

The model that was developed by Balachew and Gobena makes use of ML techniques to make predictions about how well students will do in the future. The information on the students in this study was obtained from a database that was developed at Wolkite University. In order to build their models, they make use of a Neural Network, a Naive Bayes classifier, and a Support Vector Machine (SMO) each of which belongs to the field of machine learning. Comparisons are made between the results obtained from each prediction model, and the students' overall performance is evaluated [13].

A Neural Network is the strategy that Agrawal and Mavani suggest as the best way to predict the academic success of students. The program uses the data from sixty different engineering students as input. The academic output of a student in the past might provide insight into their current level of academic accomplishment. In this research, a neural network methodology is contrasted with the NB technique. The outcome suggests that a more comprehensive dataset will produce superior outcomes when using the Neural Network approach [14].

Ali, Helali, and Mohamad identify the impacts on students' academic achievement and course performance by using approaches from the field of data mining. The dataset was contributed by the Community College, Computer Science, and Business Administration Departments of Narjan University, which is located in Saudi Arabia. The purpose of this study is to identify the characteristics that contribute to the overall performance of students by employing a prediction model that makes use of clustering and classification techniques [15].

Deep learning and the Tensorflow Artificial Intelligence Engine were utilized in the research that Fok et al. conducted in order to develop their models of growth forecasting for pupils. They used the information from two thousand students, with seventy-five percent of it going toward training and twenty-five percent toward testing. In the piece, the writers played around with a variety of settings, including the hidden layer depth, the hidden node depth, the iteration count, and the learning rate, and then compared the results. The results of this study demonstrated that deep learning has the capacity to effectively predict the outcomes for students. Accuracy ranged from 80% to 91% of the time [16].

3 Methodology

This section presents a deep learning based model for forecasting student's academic performance. Classification model is trained and tested. Autoencoder, Deep Belief Network (DBN), and Naïve Bayes algorithm are used to classify and predict student's academic performance.

Naive Bayes [17] classifiers are statistical classifiers that are based on Bayes theory with the assumption of class conditional independence. This assumption states that the impact of a feature's value on the class supplied is unrelated to the feature values of other features. Naive Bayes classifiers are often used in machine learning and data mining applications. It does this by increasing categorisation with the individual probabilities of each attribute-value combination, which is a probabilistic strategy. It is possible for the SL system to be used to successfully teach a wide variety of difficult real-world issues, particularly those that are amenable to computer-based diagnostic procedures. We do not need to create a whole covariance matrix in order to compute the variances of the characteristics for each class since they are conditionally independent of one another (Fig. 1).

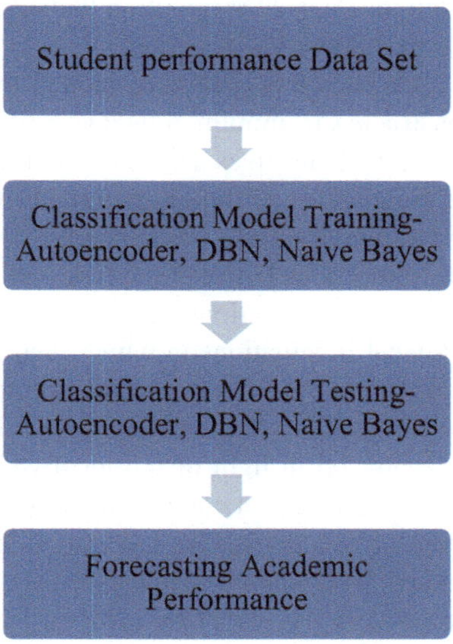

Fig. 1. Deep learning based model for forecasting student's academic performance

In the Naive Bayes classifier, P represents the probability, C' stands for the class variable, F stands for the feature variables, and the denominator does not depend on C' in any way. A Bayesian network, sometimes called a belief network, is a graphical representation (statistical model) that makes use of a Directed Acyclic Graph to characterise a collection of random variables and their conditional relationships. This kind of network is also known as a belief network (DAG). To give you an example, it may show the probabilistic correlations between different diseases and the symptoms of such diseases. In order to determine the chance that a disease is present, a network is fed a list of symptoms, and then it does the calculation. Bayesian networks are directed acyclic graphs that include hypotheses expressed at their nodes. Bayesian networks contain latent variables, observable quantities, and unknown parameters. Edges link nodes that are not directly connected to variables that are conditionally independent of one another. Conditional dependencies are represented by edges, which connect nodes that are not directly connected to variables. Each node is linked to a probability function that, given a certain set of inputs, calculates the likelihood of the occurrence of its parent variables.

The Naive Bayes method is a basic approach for choosing problem occurrence class labels from feature value vectors. This method is used in the process of creating classifier models. These classifiers are not trained with just one approach, but rather utilising a broad variety of techniques that all have a base in common. If we just consider the class variable, it is not feasible to decide which quality is more important than the others. Classifiers based on Naive Bayes may be educated via the process of supervised learning for certain sorts of probability models. Dealing with the naive bayes model does not need the use of bayesian probability or any other bayesian methods; instead, the maximum likelihood technique for parameter approximation for naive bayes models may be used. Naive Bayes classifiers are supervised learning algorithms that make use of the Bayes theorem in combination with the "naive" assumption that all pairs of attributes may be considered independent of one another.

Deep belief networks, often known as DBNs, are models that are both generative and probabilistic. They are constructed from numerous layers of covert levels that contain building blocks. The latent units themselves are binary, but the input and output latent units can be either real or binary. One of the characteristics that sets DBN apart from other networks is that the units in one layer are connected with the units in the layers that are adjacent to it. The connections between the first two levels of a DBN are done in a directed manner, whereas the connections between the deep layers are done in an undirected manner. DBNs excel in situations in which there is only a limited amount of training data available. There is not enough information to initialize the weights in an appropriate manner. Because of this, the effectiveness of the model can be reduced. The obtained weights are evaluated in light of the ideal weights that may have been produced. As a direct consequence of this, it is not difficult to accomplish either improved modeling or accelerated training of models. The layers of a DBN are built with the help of something called Restricted Boltzmann Machines (RBM). The training of the RBM enables the training of the network on a layer-by-layer basis [18]. In the DBM model, the network is trained in a manner that is less altruistic and more greedy.

A synthetic neural network that contains three or more processing layers is referred to as an autoencoder [19]. Due to the fact that the output units are connected to the input units, this is an effective instrument for unsupervised learning. An autoencoder's principal objective is to lower the dimensionality of a dataset by way of the process of learning a representation (encoding) for that dataset. The second sort of auto encoder, known as stacked denoising auto encoders, is utilized to remove noise from the signal after it has been corrupted in some way. A good representation is one that consistently generates accurate results despite the fact that the input is imprecise. When there is such a good representation available, it is significantly simpler to retrieve the authentic, unaltered input. Training the autoencoder network is necessary in order to obtain accurate outputs. After being cleaned up, the data is then used for additional analysis with a supervised machine learning method such as SVM, LR, or a Softmax Layer. Two frequent applications for autoencoders are reducing the dimensions of data and retrieving information from it. The utilization of stacked autoencoders is extremely beneficial to the process of semantic hashing, which is a method for the retrieval of information.

4 Result Analysis

The data gathering from the college is utilized in the experimental study that is being conducted [20]. In total, there are 285 distinct samples contained within this data collection. This data set is distinguished from other collections of the same sort by a total of seventeen characteristics that may be found within it. Performance is shown in Fig. 2.

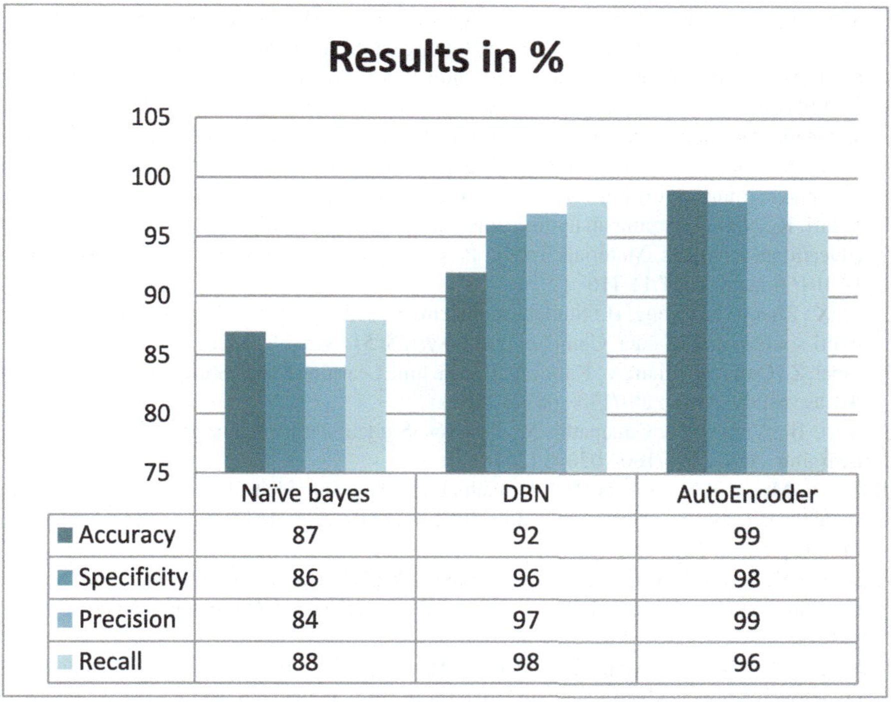

Fig. 2. Result Comparison of Classifiers for Forecasting Academic Performance

5 Conclusion

Student classification for performance prediction or dropout detection is a major educational statistics task. Educational institutions need to reliably forecast students' future success. Because student performance is used to compare all educational systems, educators value it. Early identification of at-risk kids is important since dropping out lowers efficiency. Academic information must be properly catalogued. Recent technical advances have made classification algorithms for educational data efficient and effective. This task requires additional investigation on different ways. In today's classrooms, information collection and topic variety have increased dramatically. It's intriguing to consider alternative methods. This article describes a deep auto encoder grade prediction algorithm. Autoencoder predicts academic performance with 98.5% accuracy, according to experiments.

References

1. Manjarres, A.V., Sandoval, L.G.M., Suárez, M.J.S.: Data mining techniques applied in educational environments: literature review. Digit. Educ. Rev. **33**, 235–266 (2018). https://doi.org/10.1344/der.2018.33.235-266
2. Salloum, S.A., Alshurideh, M., Elnagar, A., Shaalan, K.: Mining in educational data: Review and future directions. In: Advances in Intelligent Systems and Computing, pp. 92–102. Springer International Publishing (2020). https://doi.org/10.1007/978-3-030-44289-7_9
3. Senthil Kumar, R., Jithin Kumar, K.P.: Analysis of student performance based on classification and mapreduce approach in bigdata. Int. J. Pure and Applied Math. **118**(14), 141–148 (2018)
4. Yağcı, M.: Educational data mining: Prediction of students' academic performance using machine learning algorithms. Smart Learn. Environ. **9**, 1–19 (2022)
5. Veluri, R., et al.: Learning analytics using deep learning techniques for efficiently managing educational institutes. Materials Today: Proceedings **51**, 2317–2320 (2022). https://doi.org/10.1016/j.matpr.2021.11.416
6. Li, X., Zhang, Y., Cheng, H.: Student achievement prediction using deep neural network from multi-source campus data. Complex Intell. Syst. **8**, 5143–5156 (2022)
7. Iqbal, Z., Qadir, J., Mian, A., Kamiran, F.: Machine Learning Based Student Grade Prediction: A Case Study. ArXiv, abs/1708.08744 (2017)
8. Kim, B., Vizitei, E., Ganapathi, V.: GritNet: Student Performance Prediction with Deep Learning. ArXiv, abs/1804.07405 (2018)
9. Xu, J., Moon, K.H., Schaar, M.: A machine learning approach for tracking and predicting student performance in degree programs. IEEE Journal of Selected Topics in Signal Processing **11**, 742–753 (2017)
10. Sungar V., Shinde P., Rupnar, M.: Predicting Student's Performance using Machine Learning, Communications on Applied Electronics (CAE), Foundation of Computer Science FCS **7**(11) (2017)
11. Singh, M., Singh, J., Mahavidyalaya, S., Mandal, V.P.: Machine learning techniques for prediction of subject scores: a comparative study. Int. J. Comp. Sci. Netw. **2**(4), 77–80 (2013)
12. Bendangnuksung, Prabu, D.: Students' performance prediction using deep neural network. Int. J. Applied Eng. Res. **13**(2), 1171–1176 (2018)
13. Belachew, E.B., Gobena, F.A.: Student performance prediction model using machine learning approach: the case of wolkite university. Int. J. Adv. Res. Comp. Sci. Softw. Eng. **7**(2), 46–50 (2017)

14. Agrawal, H., Mavani, H.: Student performance prediction using machine learning. Int. J. Eng. Res. Technol. **4**(3), 111–113 (2015)
15. MohamedAly, W., Hegazy, O., Rashad, H.: Automated student advisory using machine learning. Int. J. Comp. Appl. **81**, 19–24 (2013)
16. Fok, W., He, Y., Law, K., Cheung, K.H., Ho, Y.Y.: Prediction model for students' future development by deep learning and tensorflow artificial intelligence engine. 4th International Conference on Information Management (ICIM), pp. 103–106 (2018)
17. Sivasakthi, M., Padmanabhan, K.R.A.: Prediction of students programming performance using naïve Bayesian and decision tree. In: Ranganathan, G., Fernando, X., Piramuthu, S. (eds.) Soft Computing for Security Applications. Advances in Intelligent Systems and Computing, vol 1428. Springer, Singapore (2023). https://doi.org/10.1007/978-981-19-3590-9_8
18. Neha, K., Kumar, R., Jahangeer Sidiq, S., Zaman, M.: Deep neural networks predicting student performance. In: Saraswat, M., Chowdhury, C., Kumar Mandal, C., Gandomi, A.H. (eds.) Proceedings of International Conference on Data Science and Applications. Lecture Notes in Networks and Systems, vol 552. Springer, Singapore (2023). https://doi.org/10.1007/978-981-19-6634-7_6
19. Rahul, Katarya, R.: Deep auto encoder based on a transient search capsule network for student performance prediction. Multimed Tools Appl. **82**, 23427–23451 (2023). https://doi.org/10.1007/s11042-022-14083-5
20. http://archive.ics.uci.edu/ml/datasets/university

On Hyperspectral Image Classification and Deep Learning

K. Prema and A. V. Sriharsha(✉)

Mohan Babu University, Tirupathi, India
avsreeharsha@gmail.com

Abstract. Hyperspectral imaging is a technique for capturing images of objects in multiple dimensions. It combines spectroscopic and imaging technology to produce multidimensional images. Hyperspectral imaging (HSI) allows us to examine any object's internal and external characteristics. Every characteristic of an object has a distinctive spectral signature that is created as a result of variations in the reflectance or emittance of the object's material. Hyperspectral imaging (HSI), which is non-destructive, is now being used in the food production, medical diagnosis, precision agriculture, pharmaceutical, recycling, and environmental monitoring industries. The various HSI classification techniques based on conventional, deep learning, and pre-trained classifiers will be reviewed.

Keywords: Medical imaging analysis · medical hyperspectral imaging · deep learning

1 Introduction

The electromagnetic spectrum, a limited part of our vision, contains spectral responses that distinguish objects. However, multispectral imaging sensors can capture images in infrared and visible segments, enabling material identification based on unique spectral signatures [1]. Since the late 1960s, multispectral imaging technology has been utilized for remote sensing in satellite and airborne systems [4] for environmental and land observation. The advent of precision medicine and artificial intelligence offers opportunities for hyperspectral medical research, particularly in hyperspectral imaging (HSI). HSI, based on multispectral remote sensing technology, offers a multiple channels, wide spectral range and high resolution for continuous spectral range imaging. HSI Images utilize a fundamental technique of classification based on multiple classifiers, categorizing them into Knowledge-based, Sub-Pixel, Per-field, Contextual, or Per-Pixel categories based on pixel data. HSI technology is increasingly used in biomedicine to classify cervical cancer, detect skin tumors, and understand oxygen transport mechanisms in vascular tumors. Support Vector Machines (SVM) [7] have been used to segment multispectral images of bone marrow cells with an accuracy of 94%, identify abnormal blood vessel parts, and distinguish between arteries and veins in pigs. These developments provide more reliable, rich, and useful information than other imaging techniques.

A. Bajaj et al. (Eds.): IBICA 2023, LNNS 1229, pp. 220–228, 2025.
https://doi.org/10.1007/978-3-031-78940-3_22

MRI and other imaging techniques have significantly impacted medical imaging, particularly in surgical procedures and diagnosing diseases. These methods have paved the way for the creation of more sophisticated examination techniques, like computed tomography (CT) scans, which can now evaluate head or brain-related injuries. Hyperspectral imaging (HSI) is a medical technique that creates a reflectance spectrum out of all the pixels in an image by aligning images in specific spectral bands.

Image enhancement is a crucial aspect of hyperspectral imaging [2], utilized in various fields such as remote sensing, agriculture, food safety assessment, disaster monitoring, feature extraction, classification, object detection, and art conservation. Through the transmission of light through tissue samples, medical imaging is a vital tool used in the diagnosis and detection of various types of cancers. Future breast cancer research may benefit from the success Medical Hyperspectral Imaging (MHSI) has had in separating tumours from healthy tissues in a rat breast tumour model. MHSI will significantly simplify the diagnosis of gastric cancer, a leading cause of cancer-related deaths globally. MHSI applications in medical diagnoses have led to the development of multiple algorithms for accurately classifying cancerous tissues [5], promoting various techniques for processing and analysis.

2 Related Work

An imaging spectrometer is the main piece of hardware used by HSI [5] to obtain image data from the region of interest using continuous narrowband data. A vital tool in medical diagnosis is HSI data, which is produced by spectroscopy, electronics, fine instrumentation, and computer hardware. Since the 1990s, it has been incorporated into medical diagnosis, offering image and spectral information, a breakthrough for surgeons, and improving the overall performance of imaging spectrometers. A light source, optical imaging system, spectroscopic system, detector, sample platform, and computer system make up HSI systems. Fluorescence is excited by ultraviolet light in these systems. Through the use of a lens, the HSI collects light that is then divided into distinct spectral lengths by means of prisms or diffraction gratings, recording the wavelengths and their locations.

2.1 Data Acquisition

A new method called hyperspectral imaging reconstructs reflection spectra using a small range of nearby wavelengths to produce three-dimensional hypercube data. By giving diagnostic details on tissue physiology, appearance, and composition, this information facilitates non-invasive biopsies, histopathological and fluorometric analysis, and the understanding of disease biology. For use in biomedicine, various spectral imaging systems, including handheld hyperspectral imagers, have been researched. A continuous portion of the spectrum is covered by the spectral imaging and spectroscopy technique known as hyperspectral imaging, refer Fig. 1. It offers continuous scanning imaging, enabling the collection of each pixel's reflectance, absorption, or fluorescence spectra. Compared to other techniques, this one offers richer spectral bands and greater resolution.

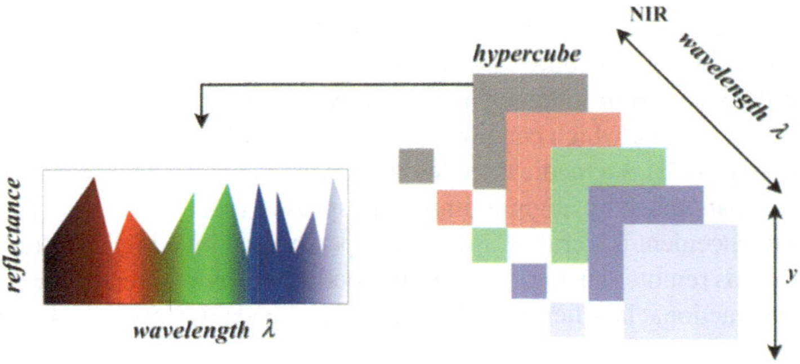

Fig. 1. Spectral data cube.

The HSI system works by using a light source to radiate spatial information, which then travels through a lens and a slit before being directed by dispersion devices onto a detector. A mechanical push sweep collects adjacent images from each row of sample space data, treating each row as a two-dimensional image. This results in a hypercube with two spatial dimensions and one spectral dimension.

Hyperspectral systems are classified into four methods: whiskbroom, push broom, staring, and snapshot, based on their acquisition methods for spectral and spatial information.

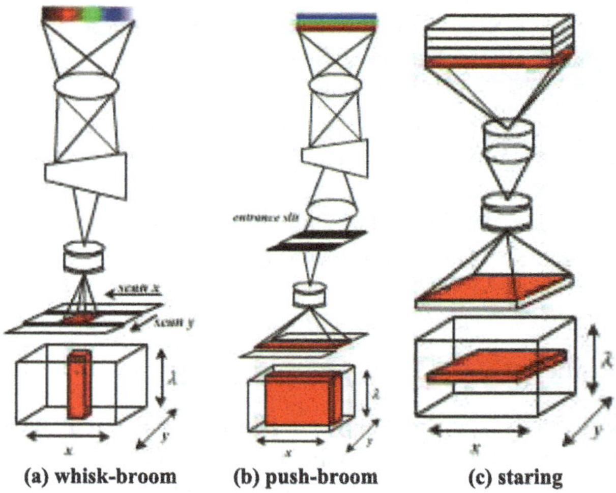

Fig. 2. Basic Spectral Imaging methodologies

In whiskbroom imaging systems, a single point is swept along the spatial and spectral dimensions using a revolving scanning device. A spectral picture data cube is created when signals are processed by a front optical system, spectrometer, and swivel scanning mirror.

In Medical biology tissue detection uses a line-wise-scan method referred to as the push broom imaging system, to produce high-resolution images with a resolution spectra of 5 nm. As depicted in the Fig. 2; A CMOS-based methods are used to produce data cube of size $640 \times 480 \times 100$ by the camera with hyper-spectra in push broom methodology with an acquisition time of one-tenth of a minute. The diameter of the camera, which is mounted on a mobile medical system, is determined by the objective lens.

A 2-D image consisting of single-band representations is captured by the staring-type of imaging system, that uses data cube spectra which were created by means of sweeping the spectra across various bands of wavelengths. A focusing optical system is used after filters like LCTF and AOTF have been used to scan the spectrum. Real-time spectral information is revealed using this technique, which is essential for targeting and focalization.

The imaging systems that take a frame of eye snap shot, without considering spatial parameters or spectra with a single exposure (at presto) on to the grab zone. Although this mode's resolution is constrained, it can make up for it by expanding the sampling area. Both remapped and dispersed images can be used.

2.2 Medical Hyper Spectral Imaging

Hyperspectral images, particularly medical ones, are crucial for medical diagnosis and treatment. They form a hypercube because they have hundreds of spectral bands and visible spectra.

Their high dimensionality, however, can result in dimensional catastrophe and data redundancy. Band selection can help with this problem; six approaches are compared, and deep learning architectures and image preprocessing operations are discussed.

2.2.1 Preprocessing with Spectra

Despite various hindrances such as noise in the instrument, redundancy of bands within images and high-dimensionality due to resolutions, preprocessing algorithms are needed after acquisition of hyperspectral data to remove extra noise. In hyperspectral radiation observations, the Standard Normal Variate (SNV) normalisation technique frequently used in resolving scatter effects, reduce particle inhomogeneity and resolve NIR diffuse reflectance spectrum effects. Smoothing filters, which preserve data characteristics and achieve smooth denoising of the original spectrum, can be used to remove spectral noise from HS sensor data. The choice of waveband is essential for minimising hyper-spectral data redundancy. Ranking, search, clustering, sparsity, embedded learning, and hybrid scheme-based hyperspectral bands are six of the types used. Adaptive subspace partitioning techniques enhance precision and effectiveness.

2.2.2 Dimensionality Reduction

Feature downscaling transforms hyperspectral data into lower dimensions, improving interpretability without losing information. Principal component analysis reduces dimensionality, retaining maximum information and eliminating redundant noise.

3 Applications

Advancements in instrumentation engineering and computer technology have improved food assessment, enabling faster and more accurate evaluations. Near Infrared (NIR) spectroscopy overcomes limitations, while hyperspectral imagery is ideal for assessing food quality and safety. CT and MRI are commonly used for clinical analysis, with MRI yielding better results. Modern spectral imaging techniques, including hyperspectral image analysis, provide real-time biomarker information and image guided surgery.

The world's crop production needs to double by 2050 due to population growth, but yields aren't increasing fast enough. Increasing crop production directly reduces global poverty and undernourishment. Advancements in remote sensing hyperspectral images enable cost-effective crop stress evaluation. Hyperspectral remote sensing technology has been widely used in water resource management, allowing accurate estimation of parameters. Recent advancements in remote sensing technology have improved early flood warning systems, reducing detection time and enhancing response time.

Hyperspectral remote sensing offers reliable information on water quality parameters like chlorophyll, turbidity, and oxygen demand. It aids in evaluating wetland parameters, classifying lakes, and analyzing algal blooms, enhancing ecosystem quality. Forensic document experts traditionally utilized chemical solution-based methods to analyze historic documents, but these methods were time-consuming, sensitive to temperature changes, and irreversible in damage.

3.1 Datasets

Due to privacy or medical ethics concerns, the majority of authors avoid making public experimental data and codes [3]. For algorithm and model research, some institutions do provide pertinent datasets, such as hyperspectral image data from animal tissues. When possible, public datasets are listed (Table 1).

Table 1. Various types of hyperspectral datasets and their descriptions

Datasets	Description
HSI Human Brain Database	On the basis of these datasets, many frameworks have been proposed, and it is common for hyperspectral public datasets.
The Hyperspectral SRS and Fluorescence data	Provide access to hyperspectral SRS and organelle fluorescence images utilized in training deep-learning prediction models using U-within-U-Net.
Microscopic Hyperspectral Choledoch Dataset	The links provide access to a dataset for microscopy hyperspectral and color images of cholangiocarcinoma, but due to upload space limitations, only part of the data has been uploaded. The original files are available at http://bio-hsi.ecnu.edu.cn/.
Multispectral Imaging Dataset of Colorectal tissue	The links to the dataset allows download of images of four classes of abnormalities, including benign, normal, and cancerous, represented by infra-red and visual spectrum bands.

3.2 Classification

Medical highlight images (MHSI) are increasingly used for medical diagnostic applications, particularly in cancer detection and cell classification. Machine learning (ML) employs supervised and unsupervised techniques for learning, extraction, detection and recognition tasks using data and statistical models. Deep Learning improves learning capabilities by automatically growing network layers and figuring out weight parameters while continuously learning features from various data sets, as a neural network approach [7]. HSI is a network model used in image processing for hyperspectral image classification, combining multidimensional data with spatial information, requiring downscaling for optimal results. Researchers are concentrating more on extracting information on spatial and spectral features as a dual branch structure in deep learning. This method enriches information and improves classification accuracy by integrating spatial and spectral features. Deep learning has been used in various fields, including MHSI, but its maturation in MHSI has been slower.

3.3 Detection

Hyperspectral images are more potential and drive wide scope for development, but in medicine, the hyperspectral images are less frequently used for detection [7, 8] and extraction during conventional clinical procedures. Detection and Extraction in conventional pathological imaging is typically used later, by the experts as an essential activity in the diagnosis in clinical medicine.

The support vector machine (SVM) performs classification on the original data sets that are represented in raw as wavelets [9], where features are speculative, that are composed of average spectra. The SVM classifies better on such images with clear spectra, when employed for detecting cancerous and healthy tissues. A tumour mask is created to discriminate between diseased and healthy tissues and testing shows enhanced discriminating of overlapping spectra using wavelet feature classification.

Medical Hyperspectral Imaging (MHSI) uses a convolutional neural network to classify medical hyperspectral images for tumor detection. Purely spectral and structural methods have been outperformed by two-stream convolutional models in the detection of tongue squamous cell carcinoma. Beng et al. created a method to give spectral dimension as input stacked into a DenseNet2D-MS network to order detect in-vivo pharyngeal cancers. Halicek et al. looked into the ability of hyperspectral images to detect cancer using an Inception-v4 convolutional neural network architecture with an algorithm consisting of mapping as gradient-like activation. Using stacked visible and near-infrared light, a single-stream U-net architecture was used in surgery that takes references from hyper-spectral imaging in real-time segmentations of deployments. Deep learning [10–13] for semantic segmentation in tumour detection on HSI data is an example of the value of NIR spectroscopy for tumour capture. Detection of breast cancer cells using a microscope facilitating the hyperspectral imaging associated with the deep learning software for the study of digital pathology, the applications are two additional diseases for which MHSI has been used in the detection process. In order to detect features beneath the skin's surface, both manual and AI methods such as local feature detection and deep learning for skin feature extraction and detection are employed, demonstrating the system's capacity to track skin features and localise them more effectively than local manual features.

3.4 Medical Applications

A research method that combines spectral and spatial features to extract rich spectral information and texture structure has been developed as a result of the increase in resolution of hyperspectral medical images. This raises classification precision. The ability to detect cancer using optical imaging is made possible by changes in cell morphology and absorption, scattering, and fluorescence characteristics. For conditions like stomach, breast, cervical, skin-like diseases, and head and neck conditions [14, 15], HSI can provide broad-area tissue images, improving diagnostic accuracy. Table 2 lists various approaches for using hyperspectral images in medicine along with their successes.

Table 2. Various methods of medical hyperspectral image applications

Applications	Different Methods		Different Achievements	
	Machine Learning	Deep learning	Accuracy	Sensitivity
Stomach cancer	Spectral Angle Mapper	3DCNNN with Radial Basis Function, Multilayer Perceptron	93%	97.19%
Bran cancer	Spectral Angle Mapper	U-Net, 2DCNN, 1 DCNN	94%	77%
Head and Neck cancer	SVM, Random Forests, k-NN	Regression, DCNN	upto 93.5%	69% to 71%
Skin cancer	Spectral Angle Mapper, k-means	CNN		87% to 93%
Eye diseases	Spectral Angle Mapper	CNN	70% aribtrary	80% arbitrary
Coloncancer	MLP	3D + 2D CNN	70% aribtrary	86% to 88%

4 Conclusions

The literature review identifies common machine learning and deep learning models that are successfully applied and widely used in the medical hyperspectral field. The classification and detection of spatial and spectral features in hyperspectral images uses these models, especially 2D, 3D, and 2D CNN combined with 3D CNN. The full convolutional network of the classical U-Net is frequently used for more effective models in image segmentation tasks. To enhance feature extraction and performance, network models make use of a variety of modules, including Inception multiscale processing, 3D attention, transformer, Gabor filter, DWT, and DCC. These directions for research can improve model performance. The difference between predicted and true values is calculated using loss functions like cross-entropy loss, SoftMax loss, R-square, RMSE, and MSE in order to assess how well the model fits the data.

References

1. Tong, Q., Zhang, B., Zheng, L.: Hyperspectral Remote Sensing: Principle, Technology and Application, 1st ed., pp. 38–65. Higher Education Press, Beijing, China (2006)
2. Shan, L.: Sub-Pixel Small Target Detection in Hyperspectral Images. Xi'an University of Electronic Science and Technology, Xi'an, China (2015)
3. Sree, S.R., Agarwal, R., Markkandan, S., et al.: Integrated learning algorithms-based epileptologist assistive tool for seizure detection and prediction. Soft. Comput. (2023). https://doi.org/10.1007/s00500-023-07913-7
4. Radhakrishnan, P., Anbarasi, A., Srujan Raju, K., Sai Thrinath, B.V.: Detection of colon cancer using image processing. Cybern. Syst. (2023). https://doi.org/10.1080/01969722.2023.2175131
5. Srujan Raju, K., et al.: Statistical Evaluation of Network Packets in an Intrusion Detection Mechanism Using ML and DL Techniques. Cybernetics and Systems (2023). https://doi.org/10.1080/01969722.2023.2175137
6. Gadepaka, L., Lingabathina, V.: Correlation Analysis of Cognitive Regions in Automated Anatomical Labeling Atlas Using LSTM. In: Kumar, A., Ghinea, G., Merugu, S., Hashimoto, T. (eds.) Proceedings of the International Conference on Cognitive and Intelligent Computing. Cognitive Science and Technology. Springer, Singapore (2023). https://doi.org/10.1007/978-981-19-2358-6_52
7. Lakshmi Narayanan, S., et al.: Design and implementation of cyber threat intelligence data mining model. In: Satapathy, S.C., Lin, J.CW., Wee, L.K., Bhateja, V., Rajesh, T.M. (eds.) Computer Communication, Networking and IoT. Lecture Notes in Networks and Systems, vol 459. Springer, Singapore (2023). https://doi.org/10.1007/978-981-19-1976-3_22
8. Reddy Madhavi, K., Madhavi, G., Rupa Devi, B., Kora, P.: Detection of pneumonia using deep transfer learning architectures. Int. J. Adv. Trends in Comp. Sci. Eng. **9**(5), 8934–8937 (2020). ISSN 2278-3091, https://doi.org/10.30534/ijatcse/2020/292952020
9. Prabhakar, T., Srujan Raju, K., Reddy Madhavi, K.: Support Vector Machine Classification of Remote Sensing Images with the Wavelet-based Statistical Features. Fifth International Conference on Smart Computing and Informatics (SCI 2021), Smart Intelligent Computing and Applications, Volume 2. Smart Innovation, Systems and Technologies, vol 283. Springer, Singapore (2022)
10. Rajani, A., Kora, P., Madhavi, R., Jangaraj, A.: Quality Improvement of Retinal Optical Coherence Tomography, 1–5 (2021). https://doi.org/10.1109/INCET51464.2021.9456151
11. Reddy Madhavi, K., et al.: COVID-19 Detection using Deep Learning. 20th International Conference on Hybrid Intelligent Systems-HIS 2020, at Machine Intelligence Research (MIR) labs, pp. 1–7. USA, Springer AISC, 1375 (2020)
12. Kora, P., et al.: EEG-Based Brain-Electric Activity Detection During Meditation Using Spectral Estimation Techniques, pp. 687–693 (2021). https://doi.org/10.1007/978-981-16-1941-0_68
13. Abbagalla, S., Rupa Devi, B., Anjaiah, P., Reddy Madhavi, K.: Analysis of COVID-19-Impacted Zone Using Machine Learning Algorithms. Springer series – Lecture Notes on Data Engineering and Communication Technology **63**, 621–627 (2021)

14. Reddy, U.J., Dhanalakshmi, P., Dileep Kumar Reddy, P.: Image Segmentation Technique Using SVM Classifier for Detection of Medical Disorders. Ingénierie Des Systèmes D'information **24**(2) (2019)
15. Somasekar, J., et al.: A dataset for automatic contrast enhancement of microscopic malaria infected blood RGB images. Data in brief **27**, 104643 (2019)

A Federated Learning Methodology for Preserving Privacy in Healthcare Systems

S. Swetha and A. V. Sriharsha(✉)

Mohan Babu University, Tirupathi, India
avsreeharsha@gmail.com

Abstract. Federated learning (FL) is a distributed machine learning system that enables local data sharing for medical IoT devices and EHRs, streamlining healthcare processes and promoting user data privacy. However, FL faces privacy challenges due to model updates being shared between clients and servers, potentially breaching privacy requirements. Healthcare organizations need to address privacy issues in machine learning due to digitization of patient data. Federated learning allows medical institutions to train deep models collaboratively using homomorphic encryption. This study aims to improvise the erstwhile privacy-preserving systems with federated learning system for medical and healthcare data by implementing a realistic implementation of a preprint with a proposed encryption technique (MK-CKKS). The solution involves adjusting a basic "ring learning with error" method, forking a federated learning framework for Python, and incorporating a communication process with protocol buffers. Experimental assessments confirm the framework's strong model performance despite these adjustments.

Keywords: federated learning · privacy preserving · deep learning

1 Introduction

Federated Learning (FL) is a distributed machine learning system that eliminates data sharing by allowing participants to train local models on local data. Participants share model parameters with the federated learning server, which sends new weights back by aggregating the model weights. Due to the built-in advantages of distributed models of architectures for training and learning, federated learning has become very well-liked in the medical industry.

Generally, remote healthcare systems are designed so that all (encrypted) data exchanged between base stations and hospitals is collected and redirected by a central node, referred to as a data concentrator for the remainder of this work. Insurance firms and government agencies may find it useful to monitor public health or create statistics using the context information, or metadata, about these interactions. Even in cases when cryptography is employed, patients' private information may still be leaked to the data concentrator through metadata, which could result in data misuse such as depriving people of an insurance contract. Having many data concentrators that don't

A. Bajaj et al. (Eds.): IBICA 2023, LNNS 1229, pp. 229–239, 2025.
https://doi.org/10.1007/978-3-031-78940-3_23

communicate with one another might be a simple way to alleviate some of these problems. The implementation and upkeep of this solution would be costly. Furthermore, there is no assurance that data concentrators won't share information with one another. Using a new pseudonym for each message would be an additional way to resolve this. This strategy is insufficient to prevent adversaries from jeopardising patient privacy; if the data concentrator is aware of the location of the base station, it could expose unique patient IDs.

FL in the healthcare industry enables collaboration between various healthcare institutions and the training of ML models based on decentralised data without intervention of sharing sensitive data pertaining to patients.

It has become more widely used in hospitals and other healthcare facilities. In a study conducted by Google and Stanford Medicine, FL was used to create a predictor of hospital readmissions. Local storage was used to hold patient data from several hospitals. The distributed model of machine learning is trained using local data from various medical institutions without disclosing any private information. The model that emerged offered distinctive readmission risk evaluations while protecting patient privacy [5]. The distributed model of machine learning in face of federated learning has shown potential enormously in the drug discovery, disease diagnosis, population health management. The FL is profoundly in vogue by startups, pharmaceutical companies, academic research labs completely involved in train models for drug discovery, disease detection, and patient optimization. FL also aids in resource allocation, disease prevention, and proactive intervention with the bottom line of preserving patient data confidentiality.

Federated learning, which allows for local training on edge devices and increases data privacy, is a decentralised method of developing machine learning models. By sending the central server encrypted model updates, this strategy guarantees data security. A further benefit of federated learning is access to heterogeneous data [6], which enables models to be trained on private and sensitive data while maintaining security and privacy. Models become more generalizable as a result.

Privacy: With federated learning, training can take place locally on the edge device rather than being sent to a central server, which eliminates the risk of data breaches.

Data security is ensured by the central server receiving only the encrypted model updates. Additionally, only aggregated results can be decrypted using secure aggregation techniques like the Secure Aggregation Principle.

Access to heterogeneous data is ensured by federated learning, which is spread across numerous devices, locations, and businesses. It allows for the secure and private training of models on delicate data, such as financial or medical information. Additionally, more diverse data allows for the generalisation of models (Fig. 1).

The central server keeps a general baseline model. The client devices receive copies of this model and use them to train the models using the local data they produce. Individual gadget models grow more personalised and improve the user experience over time. In the following phase, secure aggregation techniques are used to share the updates (model parameters) from the locally trained models with the primary model hosted at the central server. In order to provide new learnings, this model mixes and averages many inputs

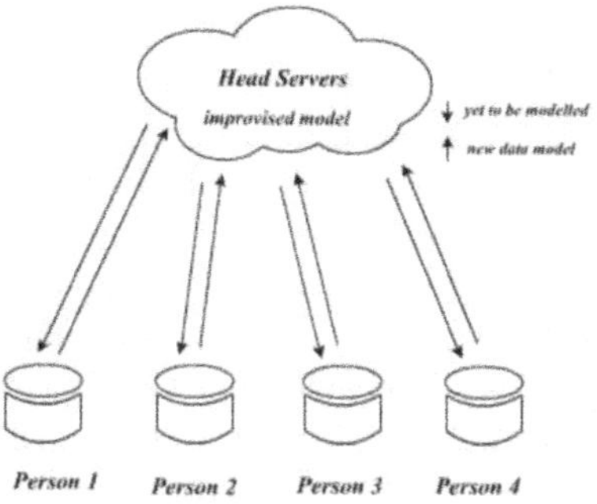

Fig. 1. A generalized federated learning framework.

[5]. Because the data are gathered from many sources, there is more potential for the model to be generalised.

The central model is shared once more with the client devices for the following iteration after being retrained on the new parameters. The models gain a variety of data points with each cycle and advance without compromising privacy.

2 Related Work

CNNs are widely used for image classification in COVID-19 lung scans [2], while privacy-preserving FL allows remote devices to work together and develop shared prediction models while preserving the privacy of local data. Through the use of several data sources, FL makes it possible to capture a wider variety of features and models, that leads to the creation of strong and reliable models. Due to its modular architecture and adjustable capabilities, with modifications [15] made for customised communication, the Python module Flower is found to be the best choice for the implementation Federated Learing. The choice of Flower is, because of its modular design and simplicity in customising it to meet different needs. The communication module is built-in in the Flower and adjustments required to provide custom conversations; are also briefly discussed in this section. Lattice-based encryption is compared to conventional encryption in order to discuss how the latter may be able to address problems brought on by the former. The cryptographic problem ensemble with the properties of Learning With Errors (LWE) is specifically contrasted with the Ring Learning With Errors (RLWE) encryption scheme, which is optimal in memory and time. The study also discusses encryption jargon like multi-key HE. The mathematical concepts are the underpinnings of the propounded multi-key-homomorphic-encryption (MKHE) scheme xMK-CKKS, which is based on the CKKS scheme, are then described. FL architecture provides some privacy protection in healthcare data, but it doesn't offer waterproof protection. Healthcare data, including digital health records, contains sensitive and confidential personal information. FL systems lack formal proof of privacy guarantees and can expose patients' private information. To improve privacy and security in healthcare, FL systems need additional mechanisms, focusing on privacy leakage from model gradients.

Li et al.'s [9] PPMA scheme addresses privacy concerns in smart grid data aggregation, while our research advances this by using advanced data compression algorithm and efficient encryption techniques and with less computational overheads.

Pu et al. [10] introduce PMAP, a lightweighted, privacy-preserving protocol for IoD environments, demonstrating flexibility in facing security assailants and better functions in terms of cost of computation, conservative energy, and overheads in communication.

Sala et al. [11] developed a re-configurable design combining PUF and TRNG on an FPGA platform, achieving results competently in TRNG and compactness throughput. Their research also proposed a lightweight, privacy-preserving multi-key cryptography for IoT security.

Sun et al. [12] proposed a modern method, MADAR, a mutual-authentication framework for privacy-preservation applied in VANs, enhancing efficiency, with inter-vehicle authentication asymmetrically with resisting computational denial-of-service attacks, targeting network and security challenges.

2.1 Convolutional Neural Network

Convolutional Neural Networks are artificial neural networks used in deep learning [8] for image classification and computer vision. They use matrix multiplication and algebraic principles to detect features in the images, reducing manual feature engineering and revolutionizing image processing and computer vision tasks. This study compares CNN-based models trained using COVID-19 [2, 3] scanning of lungs in centralized and decentralized environments. The goal is to measure performance with and without post-quantum secured updates [13] with differences between local training and FL techniques.

2.2 Privacy Concerns

In order to train models on dispersed devices or servers without sending data to a centralised place, Google introduced FL in 2016. It protects privacy, lowers communication and bandwidth needs [14], and works well in situations where network connectivity is constrained or privacy laws are strict [3, 7]. A decentralised training process is used in place of the conventional method of centralising data in FL. Data isn't transferred to a centralised server; instead, local training is carried out on each participant device using the data that is already present there. Subsequently, the local models created on every device are combined to form the global model. Each device then receives the global model back, which is utilised to improve and hone the local models. When compared to traditional centralised machine learning techniques, FL has many benefits. Privacy protection is one of its main advantages because data stays localised on individual devices and isn't shared with third parties. Additionally, FL makes it simple to train models on big, diverse datasets, which improves the models' precision and generalizability. A wider spectrum of patterns and changes can be captured by using a variety of data sources using FL, leading to the creation of more reliable and strong models.

Although FL has advantages over conventional techniques, it is vulnerable to adversarial threats such deep leaking, sybil assaults, inference attacks, model poisoning, and sybil attacks. Decentralised FL entails decentralised coordination among devices or nodes, as opposed to centralised FL, which uses a centralised server to handle the training process across several devices. Heterogeneous FL accommodates varying capabilities of devices or nodes, assigning computation and communication tasks based on their capabilities. These variants ensure consistent data exchange and global model computation.

The research aims to implement differential privacy and multi-key HE for encrypted model update aggregation using privacy-preserving FL. Centralised learning is employed to efficiently manage and control encryption keys, and a suggested strategy is presented to counteract model inversion risks (Fig. 2).

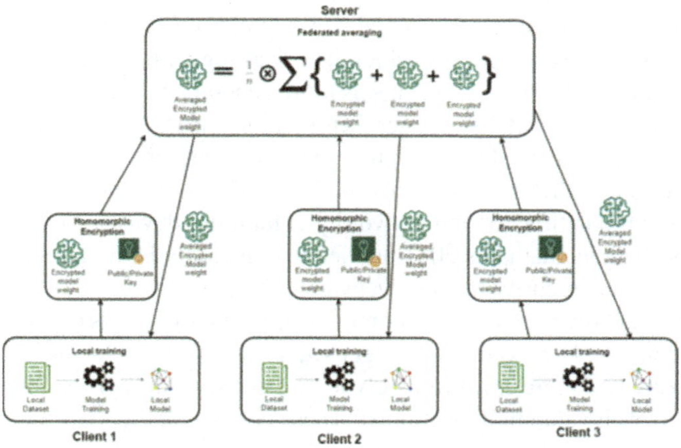

Fig. 2. A centralized federated learning framework for privacy.

2.3 Flower Framework

Flower (Flwr) is an open-source Python library for FL that facilitates seamless flow of machine learning models between client devices and servers for training and evaluation. Flower provides a unified model training, analytics, and evaluation approach across various workloads, machine learning frameworks, and programming languages, allowing developers to use their preferred tools. OpenMined's PySyft, a potential privacy-preserving FL framework, shows promise for future industry standards. The framework's modular and open architecture allows for easy subclassing of core components, including clients and servers, for tailored needs. However, Flower lacks a feature for creating custom messages between server and clients. The solution approach section will demonstrate successful modification of the code.

2.4 Homomorphic Encryption

HE encryption allows computation with encrypted data, divided into partial, somewhat, and fully categories [4]. Partially HE (PHE) allows a single operation; somewhat HE

(SHE) bounded computation of a set of operations, and fully HE (FHE) allows arbitrary addition and multiplication operations. Despite advancements in HE and FHE, they still face challenges such as increased computational overhead, data expansion, and noise management. RAD's 1978 and 2009 work has led to significant improvements, but these issues still pose challenges in practical systems. Techniques like bootstrapping, modulus reduction, and dimension reduction are used to mitigate noise accumulation. Further research is needed to improve HE efficiency and effectiveness.

2.5 Ring Learning

LWE is a quantum robust cryptography method that creates a secret key and public key using a one-dimensional matrix A and a two-dimensional matrix B, with the public key being determined by selecting s values. The learning with errors problem known as RLWE, or learning with errors over rings, is more general and is focused on polynomial rings over finite fields. Similar to integer factorization and discrete logarithm problems, it might one day serve as the basis for public-key cryptography. The solution to RLWE can be used to solve the shortest vector problem. Homomorphic encryption enables sensitive data computations on untrustworthy computing devices, as demonstrated by Brakersky and Vaikuntanathan's 2011 publication on "Fully Homomorphic Encryption from Ring-LWE and Security for Key Dependent Messages."

A collection of polynomials with a common unknown variable is known as a ring. It is necessary to convert plaintext into an integer sequence in order to secure it. The plaintext can now be expressed more easily as a polynomial by using this conversion, which allots a coefficient to each value in the plaintext. These coefficients serve as placeholders for integers in a modular arithmetic system and are normally selected using a prime number q as the modulus. This modulus enables a symmetrical and convenient representation of the data. When adding polynomials modulo q, the coefficients are added collectively, and the result is subsequently reduced by the modulus q. However, multiplying polynomials results in a higher degree polynomial, which calls for a modulus reduction to bring the polynomial down to its original degree. For modulus reduction and other operations, the symmetric value range of the modulus q is helpful. Any further noise or mistakes are generated inside this value range, often with a low standard deviation, and are frequently modelled as zero-mean Gaussian distributions.

3 Experimental Study

In a Flower library setup, server and client functions are implemented in separate scripts called [server.py] and [client.py], with the server initiating the FL process and specifying optional parameters like a strategy object. Although it offers great flexibility, the conventional Flower arrangement is limited in its ability to provide unique messages for server-client communication (Fig. 3).

3.1 Mathematical Approach

The multi-key homo-morphic encryption technique uses CKKS, where a public key and secret key are generated, where it is used for privacy preserving. The fully homo-morphic

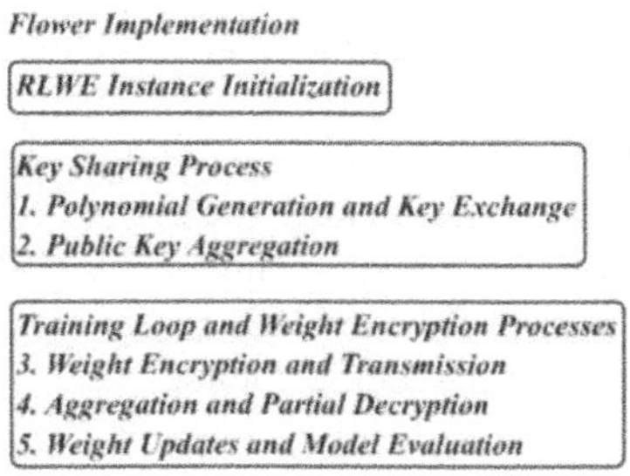

Fig. 3. A Ring Learning Without Error framework for Flower API.

encryption scheme uses analytical tools that run directly on the encrypted data while yielding the same encrypted results as similar to that of functions that run on plain-text. The CKKS method is used for private summation in the analytical functions. The HE uses this method to increase security and protect privacy on sensitive data by securely aggregating the encrypted models in the flower framework. Therefore, the CKKS scheme blended with the HE ensures the consistent encryption in federated learning, by reducing the computational loads on individual persons with collaborative approach. As a result, the cooperative encryption method employs model updates that are hidden from the server and clients. The partial encryption system also gives an extra layer of security.

The technique is annotated with the formulations; the RLWE parameters are initiated with the parameter n, the modulo of cipher text q and the three Gaussian distributions viz., χ, ψ and ϕ for key distribution, encryption noise and partial decryption noise respectively. Key generation with the polynomial a is an set containing (n, q, χ, ψ, ϕ, a), The aggregated public key $\tilde{b}$ is defined as:

$$\tilde{b} = \sum_{i=1}^{N} b_i = \sum_{i=1}^{N} (-s_i \times a + e_i) mod\ q \rightarrow \quad (1)$$

$$\text{Encryption } (\mathrm{m}_i, \tilde{b}, \ \mathrm{a}) \text{ of client } i \text{ is denoted as} ct_i = \left(c_0^{d_i}, c_1^{d_i}\right) \rightarrow \quad (2)$$

$$\text{The homo - morphic addition of } (ct_1, \ldots ct_n) \text{ is } C_{sum} = \sum_{i=1}^{N} ct_i \rightarrow \quad (3)$$

$$\text{And the decryption share } D_i \text{ in the partial decryption is } (s_i \times C_{sum_1} + e_i^*) \rightarrow \quad (4)$$

$$\text{The sum of plain texts that are recovered is denoted by} \approx \sum_{i=1}^{N} m_i \rightarrow \quad (5)$$

$$\text{Therefore, for each client } ct_i, \text{ where is the polynomial plain text to be encrypted is denoted as } ct_i = \left(c_0^i, c_1^i\right) \rightarrow \quad (6)$$

$$\text{And the partial decryption share as } D_i = (s_i \times C_{sum_1} + e_i^*) \rightarrow \quad (7)$$

After each client sends the encrypted ciphertexts, the server requests a partial decryption share based on the computed c1sum. The clients compute their partial decryption shares, indicated as "d," by a sequence of modular operations using c1sum, their private key, and an error polynomial.

The server then uses homomorphic aggregation to create the variable dsum from each of the received partial decryption shares. The client's RLWE instance performs the partial decryption process by using the equation presented in Eq. (7). The server can use this method to collectively decrypt the combined ciphertexts without being aware of the specific model modifications of any specific clients.

3.2 Datasets

In this study around 7593 images from 466 patients suffered with COVID-19, 6893 images from 604 patients with normal lung X-ray scan data sets are used from Kaggle. All images are normalized, scaled into 128 × 128 for consistent adoption into deep learning framework (Fig. 4).

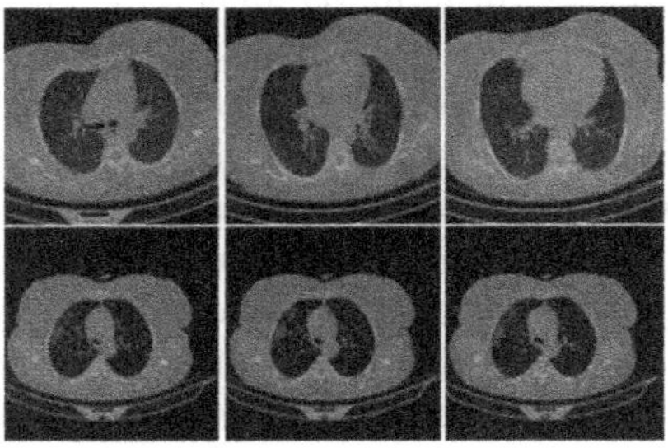

Fig. 4. X-Ray images of Lung scan.

75% of samples were allocated for training, 10% for validation, and 25% for testing, ensuring a balanced distribution and effective evaluation of the model's performance across different sets. The data segregation is shown in the following Table 1:

Table 1. Description of Dataset Segregation

Dataset	Rows	Label
Training	1500	Positive
	1500	Negative
Testing	500	Positive
	500	Negative

A five-round encrypted FL procedure was used in testing with different client counts. AWS EC2 instances were used for training and assessing the models, and a g4dn.x4large instance was used for the final testing to provide a thorough assessment of the suggested encrypted FL scheme.

4 Results and Discussions

One of the crucial performance indicators for classification is accuracy, since it quantifies how well a model can forecast instances in a dataset by calculating the proportion of instances that are correctly classified. In classification tasks, precision is a critical metric that evaluates a model's capacity to prevent false positives by calculating the ratio of true positives to false positives. Another key performance indicator in classification tasks is recall, also referred to as sensitivity or true positive rate. It measures how well a model can avoid producing false negatives by calculating the percentage of true positive instances.

Another classification performance metric is F1-score that integrates precision and recall, using a harmonic mean to assess a model's effectiveness on both positive and negative instances. Time is the length of time it takes a programme or algorithm to complete a particular task or operation. In our study, we used seconds as the unit of time (Table 2).

Table 2. Performance Measure

Metric	Value
Accuracy	0.927
Precision	0.925
Recall	0.925
F1-Score	0.925
Time	175000 ms
Memory	1820 mb

Various instance of experiments have been conducted with HE variants using the CNN framework, the quintessence of the ideal results are depicted in this section.

5 Conclusion

The HE-based FL scheme is thoroughly verified and evaluated considering the vital factors of performance such as time, memory, F1-score, precision, recall, accuracy. The scheme is tested on Amazon Web Services EC2 instance with scenarios involving up to 10 clients and compare it to plain FL and locally trained models.

Potential upgrades could further enhance the privacy, usability, and efficiency of our solution. First, privacy may be enhanced by demonstrating that all customers begin with the same initial weights and removing the requirement for full model weight disclosure in the first round. Second, depending on a variety of factors, employing Flower's built-in code for parallel processing during encrypted weight sharing may significantly reduce the overall execution time. Thirdly, a more complex model update procedure may be produced for each client by using a weighted average approach that takes the volume of training data into consideration. Finally, we might decrease the precision of the model weights in order to use smaller integer types and ease memory constraints. To ensure that memory optimisation does not obstruct the model's ability to learn, a compromise must be struck. Additional research could determine the optimal degree of precision reduction to maximise memory savings with minimal impact on model performance.

References

1. Abouelmehdi, K., Beni hssane, A., Khaloufi, H., Saadi, M.: Big data security and privacy in healthcare: a Review. Procedia Comput. Sci. **113**, 73–80 (2017)
2. Ye, J.: The Role of Health Technology and Informatics in a Global Public Health Emergency: Practices and Implications From the COVID-19 Pandemic
3. Maftouni, M., Law, A.C.C., Shen, B., Grado, Z.J.K., Zhou, Y., Yazdi, N.A.: A robust ensemble-deep learning model for COVID-19 diagnosis based on an integrated CT scan images database. In: Proceedings of the IIE Annual Conference, Online, 22–25 May 2021, Institute of Industrial and Systems Engineers (IISE): Peachtree Corners, GA, USA, pp. 632–637 (2021)
4. Wibawa, F., Catak, F.O., Sarp, S., Kuzlu, M.: BFV-based homomorphic encryption for privacy-preserving CNN models. MDPI Cryptogr. **6**, 34 (2022)
5. Prema, K., Sriharsha, A.V.: Differential privacy in big data analytics for haptic applications. Technology **8**(3), 11 (2017)
6. Reddy Madhavi, K., VijayaSambhavi, Y., Sudhakara, M., Srujan Kotagiri, R.: COVID-19 isolation monitoring system. In: Springer series – Lecture Notes on Data Engineering and Communication Technology, vol. 63, pp. 601–609 (2021)
7. Kovuri, K., Katha, C., Sriharsha, A.V., Siddardha, B., Hitesh Reddy, A.: Adaptable fog computing framework for healthcare 4.0. In: International Conference on Soft Computing and Pattern Recognition, pp. 24–32. Cham: Springer Nature Switzerland (2022)
8. Sree, S.R., Agarwal, R., Markkandan, S., et al.: Integrated learning algorithms-based epileptologist assistive tool for seizure detection and prediction. Soft. Comput. (2023). https://doi.org/10.1007/s00500-023-07913-7
9. Radhakrishnan, P., Anbarasi, A., Srujan Raju, K., Sai Thrinath, B.V.: Detection of colon cancer using image processing. Cybern. Syst. (2023). https://doi.org/10.1080/01969722.2023.2175131
10. Srujan Raju, K.: et al.: Statistical evaluation of network packets in an intrusion detection mechanism using ML and DL techniques. Cybern. Syst. (2023) https://doi.org/10.1080/01969722.2023.2175137
11. Gadepaka, L., Lingabathina, V.: Correlation analysis of cognitive regions in automated anatomical labeling atlas using LSTM. In: Kumar, A., Ghinea, G., Merugu, S., Hashimoto, T. (eds.) Proceedings of the International Conference on Cognitive and Intelligent Computing. Cognitive Science and Technology. Springer, Singapore (2023). https://doi.org/10.1007/978-981-19-2358-6_52
12. Lakshmi Narayanan, S., Shunmugavel, S., Prasanth, R., Satheesh Kumar, M., Srujan Raju, K., Suthendran, K.: Design and implementation of cyber threat intelligence data mining model. In: Satapathy, S.C., Lin, J.CW., Wee, L.K., Bhateja, V., Rajesh, T.M. (eds) Computer Communication, Networking and IoT. Lecture Notes in Networks and Systems, vol 459. Springer, Singapore (2023). https://doi.org/10.1007/978-981-19-1976-3_22
13. Madhavi, K.R., K. Suneetha, Srujan Raju, K., Kora, P., Madhavi, G., Kallam, S.: Detection of COVID-19 using X-ray Images with Fine-tuned Transfer Learning (2023)

14. Varalaxmi, G., Sahithi, R.B., Emmanuel, S., Yalamarthi, K., Swaraja, K.R.M., Sujatha. C.: Diagnosis of gastrointestinal diseases using modern CNN techniques. In: 2023 IEEE 8th International Conference for Convergence in Technology (I2CT), pp. 1–6. IEEE (2023)
15. Somasekar, J., Ramesh, G., Ramu, G., Reddy, P., Madhavi, K., Praveen, J.: Beneficial image preprocessing by contrast enhancement technique for SEM images (2023)

Machine Learning-Based Multimodal Depression Diagnosis for IT Personnel Using CCTV and Messaging Streams

B. Manjulatha[1(✉)] and Suresh Pabboju[2]

[1] OU Research Scholar and Vignana Bharathi Institute of Technology, Hyderabad, Telangana, India
manjulathakbr@gmail.com
[2] IT, CBIT, Hyderabad, India

Abstract. Stress is a prominent topic in public health and has permeated all aspects of our everyday lives. Stress has recently become a necessary component of working in the corporate world, particularly in the extremely competitive economy of today. Because of the ongoing mental and physical strain of their jobs, employees in the IT business are more likely to have a variety of health issues. Stress either causes, prolongs, or aggravates diseases. To prevent long-term health problems, it is crucial to identify stress early on and provide medicine. With the use of machine learning and image processing, this research creates non-intrusive techniques for computer-aided diagnostics to identify stress in IT workers. Machine learning models are used to assess visual data taken from continuous video surveillance in various environments, such as work, home, travel, etc., to identify different levels of stress. The creation of machine learning models to assess and identify distinct classes of stress based on multimodal text and visual feature extraction is the main emphasis of the project.

Keywords: Depression · Noninvasive · LSTM · CNN · PSS-10 Stress Scale

1 Introduction

Physical and emotional events in work environment create stress. Modern work environment has become very competitive both within and across organizations. The gap between the work expectations and the perceived ability of employee creates stress for employees. The prolonged stress affects the physical and mental well being of the employees. According to recent research on effect of stress on health, prolonged stress is found to induce abnormality in cardiac rhythm, create depression and also induce suicidal ideation. Higher stress levels can also affect the human immune system and create cancer. The stress is most prevalent among employees and recent Americal institute survey [1] has claimed that almost 80% of employees experience stress and nearly half of them need assistance to manage the work induced stress.

Currently, intrusive stress measurement with physiologists assistance is the most adopted method to measure stress. Physiologists evaluate the subject with questionnaires

A. Bajaj et al. (Eds.): IBICA 2023, LNNS 1229, pp. 240–249, 2025.
https://doi.org/10.1007/978-3-031-78940-3_24

and from the feedback stress is evaluated. But this technique is qualitative, intrusive and time consuming. Also many times, subjects does not want to share the information and this can result in erroneous results. Many works have been proposed using biochemical and physiological indicator for stress evaluation. But these methods are intrusive and costly. Many research works have been done with aim to find a non-intrusive, continuous and quantitative technique for stress evaluation.

Communication between humans is 70–90% non verbal with 55% visual and 38% vocal [1]. Thus the human communication has its significant influence on non verbal components. The non verbal behavior does have any linguistic content but they have variety of communicative behaviors like facial expression, smile, eye glance, body postures, pitch changes, speech dysfluency, loudness etc. These non verbal communication means carry more emotional information [2]. Some of the non verbal communication means and the emotion they convey in a discussion are illustrated in Table 1.

Table 1. Expression emotion correlation

Expression	Emotion
Frown	Disapproval
Smile	Agreement
Blank	Boredom

These normal verbal communication means can be used to identify if the meaning conveyed through verbal communication is authentic or fake.

This survey explores the techniques for stress monitoring and their applicability in work environment. The aim of this work is to identify the research gaps in current solution and improvements for effective non intrusive monitoring. Based on the research gaps, the directions for future research are presented.

2 Related Work

Zhang et al. (2020) [2] presented a two-level framework for detecting stress. A weighted vector is created from a combination of face, and action-level representations learned from videos. Stress assessment, however, is based solely on snapshots and does not consider changes in facial cues over time. Han et al. (2018) [3] used deep learning techniques to detect stress in speech. The speech signals are used to extract the coefficients of Mel filter banks. The features that make up these features are input into long short-term memory (LSTM) so that it can classify stress according to them. As a result of this approach, stress detection is approximately 66.4% accurate.

Yogesh et al. (2017) [4] analyzed speech signals to extract glottal and speech information. Stress is detected using the mean shift clustering method. An assessment of stress is based on wristband physiological signals, as described by Sevil et al. (2017) [5]. The amount of blood in the blood vessels and the skin temperature were significantly affected by stress. This work aimed to design a system for detecting stress based on this observation. Empatica E4 wristband signals were collected and thresholded to detect stress using the signal energy value. Giannakakis et al. (2017) [6] developed a stress detection system based on facial cues, taking into account both non-volunteer and semi-volunteer actions. Eye movements, mouth activity, head motion, and heart rate are considered to classify stress.

Pampouchidou et al. (2016) [7] investigated how stress relates to mouth activity characteristics. Using the Eigen method, mouth features are extracted using template matching. There is a higher correlation between stress, normalized opening rate per minute, and opening intensity among different mouth features. Gao et al. (2014) [8] developed a method to detect a driver's emotional state through facial expressions in real-time. We extract features from the face using discrete cosine transforms (DCT) and SIFT feature extractors. Stress is classified using an SVM classifier. Viegas et al. (2018) [9] used 17 facial action units to detect stress using binary classifiers. The facial action units were taken from each video frame captured during the person's various activities. There was a strong correlation between the facial action units and stress levels. Gavrilescu et al. (2019) [10] developed a system for detecting stress based on facial expressions. This system produced an anxiety and depression scale called the Depression anxiety stress scale (DASS).

However, there are still a number of difficulties in developing a generalised emotion recognition model. One of these difficulties is that an EEG-based emotion classifier trained on one stimulus may find it difficult to handle other stimuli. Recently, electroencephalogram (EEG)-based emotion recognition has drawn more attention in the field of human–computer interaction (HCI). This subject has received minimal attention. This problem is examined by the writers in [11], who also ascertain if feature selection may be a practical solution. Following their viewing of the video clip, 12 healthy participants experienced emotional reactions. Features from the EEG were retrieved, including power spectral density (PSD) and brain asymmetry (BAY). Then, samples taken from one video were sent to both the training and testing sets, or to the cross-stimulus conditions, where samples from the same video were only sent to one set, the training set, or the testing set, as an alternative. These conditions were used to evaluate the support vector machine (SVM) classifier. Accuracy for the within-stimulus 5-class categorization was 93.31% for PSD and 85.39% for BAY, which was a satisfactory result. Unfortunately, cross-stimulus categorization declined to low levels (46.22% and 46.2%, respectively). With the use of SVM-recursive feature elimination (RFE) to choose the most robust feature subset, the cross-stimulus classifier's mean 5-class performance was greatly enhanced, reaching 68.89% for PSD and 64.44% for BAY after testing.

A unique element for emotion identification was presented by the authors in [12] and named the subspace alignment auto-encoder (SAAE). We integrate a subspace alignment solution and an auto-encoder network into a single framework by using both nonlinear transformation and a consistency requirement. Because the aligned source domain has a distribution resembling the target domain, it can be used to align the source domain with the target domain along with its class label. From there, any supervised technique can be used to train a classifier for classification in the target domain.

Human emotions, which are a result of everyday activities, lead to psychological changes in people. In order to comprehend these modifications in behaviour patterns, ective computing research has surfaced. Since emotions play a crucial role in our everyday lives, authors in [13] used electroencephalogram (EEG) data to examine the effects of both good and negative emotions. There are three groups of emotions that have been studied: contentment, rage, and tranquilly. Ten volunteers' EEG signals are captured in real time as they view two-minute videos featuring a variety of emotions. Next, from the raw EEG, the fractal dimension characteristic was retrieved. Using a radial basis function (RBF) kernel and Support Vector Machine (SVM) with an average accuracy of 60%, the recovered features have been further categorized to identify emotional states. Empathy identification from EEG data is demonstrated using the suggested methods.

In situations like the study of human emotions, the performance of conventional artificial neural networks is inadequate. Modified ANNs that perform better than traditional systems are therefore required. To address the computational complexity and accuracy shortcomings of classic neural networks, the authors of [14] propose the Deep Kohonen Neural Network (DKNN) and the Circular Back Propagation Neural Network (CBPN). Using Electroencephalography (EEG) data, the performance of the ideas is investigated in terms of categorizing human emotions. The suggestions perform better than similar approaches, as confirmed by validation.

The authors of [15] describe an automated technique that uses electrodermal activity (EDA) data to classify emotions in toddlers. The obtained raw EDAs' time-frequency analysis offers a feature space that may be used to identify various moods. The recorded EDA signals are then subjected to the complex Morlet (C-Morlet) wavelet function in order to achieve this. The dataset utilised in this work consists of 100 infants under 30 months old's EDA recordings in addition to a collection of multimodal recordings of social and communicative behaviour. Two experts annotate the dataset in order to extract the temporal sequence related to the three primary emotions—acceptance, boredom, and joy. The children's facial expressions and the EDA time sequences are checked for synchrony during the annotating process. Using a support vector machine (SVM) classifier, many tests are carried out on the annotated EDA signals to categorise emotions. The numerical outcomes demonstrate that applying the suggested wavelet-based features significantly enhances the emotion classification performance in comparison to alternative approaches.

EEG data were utilised by the authors in [16] to categorise two emotions: pleasure and melancholy. Images of people smiling and crying are shown to participants in order to elicit these feelings. They suggest using a frequency band search technique to identify the ideal band to filter the recorded EEG data into. The author classified these two emotions using linear-SVM and common spatial patterns (CSP).

In order to evaluate the accuracy of emotion identification between the Biopac MP150 laboratory sensor and the Empatica E4 wearable sensor, the authors in [17] built machine learning models. The accuracy of data gathered utilising wearable sensors and laboratory equipment is comparable, according to the results. These findings bolster the validity of emotion identification outside of experimental settings.

The authors of [18] looked into whether using noninvasive, inexpensive EEG, EMG, and GSR sensors could accurately distinguish between high and low emotional valence and arousal. They provide the findings of a 19-person empirical investigation. Even in a cross-subject classification context, they were able to attain state-of-the-art classification performance for both valence and arousal, negating the necessity for individual model tweaking and training.

The authors of [19] made an attempt to ascertain how various emotions affected physiological signals. As stimuli, they used photos from the International Affective Picture System (IAPS) database. They have used six different physiological signals: skin conductance (SC), electrocardiogram (ECG), electromyogram (EMG), skin temperature (ST), respiration rate (RR), and arterial pulse rate (PR). They have taken into consideration five different types of emotions: high valence high arousal (HVHA), high valence low arousal (HVLA), low valence high arousal (LVHA), and neutral (Blank). Their experiment demonstrates that there are sufficient differences in physiological signal characteristics for various emotional stimuli.

In order to enhance the effectiveness of the emotion identification system, authors in [20] presented a novel adaptive technique for combining many emotional modalities. As emotional modalities, three-channel forehead biosignals were used in conjunction with peripheral physiological data (blood volume pressure, skin conductance, and interbeat intervals). Each of the experiment's twenty-five volunteers saw pre-selected video clips to provoke six fundamental emotions: anger, sorrow, fear, disgust, happiness, and surprise. At the same time, physiological signals were recorded. The recorded signals in this multimodal emotion detection system were classified into many units, each of which independently detected the emotions. The final result was then obtained by fusing the data using the adaptive weighted linear model.

Authors in [21] have used the minimal number of channels and frequency bands of the EEG data to classify emotional states in the arousal/valence dimensions. EEG signals have been broken down into matching frequency bands using discrete wavelet transforms, and various characteristics have subsequently been retrieved. From the collected characteristics, the emotional states have been identified using the support vector machine and K-nearest neighbour classifiers. The results show that the 10-channel EEG signal collected has an arousal level classification accuracy of 86.75% and a valence level classification accuracy of 84.05%. Furthermore, compared to the low-frequency bands of the EEG signal, employing the high-frequency bands—specifically, the gamma band—yields superior accuracy.

Using a Valence-Arousal emotion model, the authors in [22] investigate the derived features of the bispectrum for quantifying emotions. They arrive at a feature vector using backward sequential search. Based on the suggested characteristics, cross-validated classification accuracies of 64.84% for Low/High Arousal and 61.17% for Low/High Valence were achieved on the DEAP data set; these accuracies are close to those reported in the literature.

A technique for identifying emotions based on computing EEG data and identifying emotions from human brain activity was proposed by the authors in [23]. EEG data are obtained from the brain's scalp and used to measure how the body reacts to various stimuli related to the four main emotions found on the IAPS emotion stimuli. To identify the emotional state of the test participant, features from the EEG signals are extracted using Kernel Density Estimation (KDE) and categorised using an artificial neural network classifier.

From the survey, we could see following gaps.

- Not much work available specifically for IT environment.
- Majority of works in literature are on intrusive means using sensors and bio signals which cannot be used in IT environment during working hours. Also, the solution is costly when applied for larger employee base. There is a need for non-intrusive and low-cost stress detection mechanism using camera feeds.
- The mechanisms are snapshot based without considering the temporal and spatial correlation in behavior of the concerned employee and work environment characteristics.

Following are the objectives of the research work.

1. Design of continuous stress monitoring model for an IT professional write from his entry into office to the time of exit from the office.
2. Identification of visual features to be extracted from the CCTV camera feeds to be useful for mental state mapping.
3. Design of machine learning models for classification of different levels of stress and mental state predictions considering temporal and spatial work environment characteristics.
4. Implement a proof of concept of proposed model and measure the accuracy of stress detection. Compare the accuracy with existing works.

3 Methodology

The proposed methodology is given below (Fig. 1).

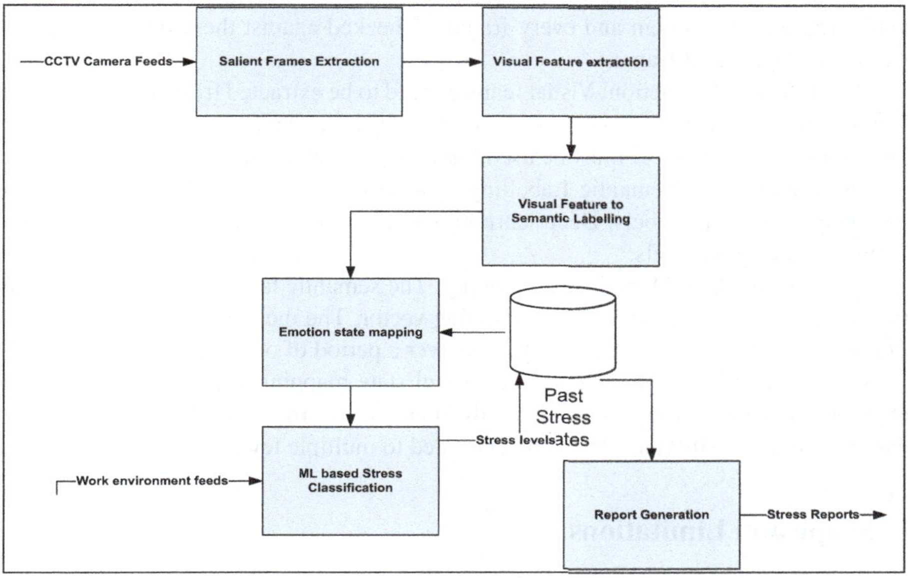

Fig. 1. Proposed Methodology

The proposed solution takes two inputs.

- Camera Feeds
- Work Environment Text Message Feeds

Camera feeds captured from CCTV are feed into the system. Salient Frames which have information need to detect stress are first extracted from the CCTV feeds. Once the salient frames are extracted, Visual features are extracted from it. The extracted visual features are translated to semantic labels like employee shouting, employee crying etc. An emotion state map is constructed for each person over a period of day and passed as input to Machine Learning (ML) based stress classification module. ML based stress classification module classifies the emotion state mapping over a period of day to different levels of stress. While classification, it also uses work environment feeds like performance appraisal given to employee, Saturday working day etc. to classify the stress with higher confidence. The stress level observed on each day is stored in a database of past stress states.

Report is generated and sent to employees based on the observation of past stress states. Algorithms are needed for following tasks in the proposed solution.

- Salient Frame extraction
- Visual Feature extraction
- Visual Feature to Semantic Labelling
- Semantic labels to Mental state mapping
- ML based stress classification

Salient Frame extraction: CCTV video feed has lot of frames and not all these frames has characteristic information needed for stress detection. From the large pool of frames, the significant frames are extracted using various decision rules. The decision rules are configured into the system and every frame is checked against these decision rules to check if its significant frame.

Visual Feature Extraction: Visual features need to be extracted from the signification frame. Before extracting the visual features, the object in the image must be segmented and each segmented object must be used for visual features extraction.

Visual Feature to Semantic Labelling: The visual features of the object must be classified to semantic labels. Deep learning models can be used to classify the visual features to semantic labels.

Semantic labels to Mental state mapping: The semantic labels over a period of day must be translated to a Mental state mapping vector. The mental state mapping vector represents the emotion changes of a person over a period of one day.

ML based stress classification: The mental state mapping vector along with work environment feed is used as input to the machine learning model for stress level classification. The stress levels can be classified to multiple levels.

4 Scope and Limitations

The research studies shows following areas under the scope for solution design.

1) Identify the best set of text features with higher correlation to PSS10 stress scale.
2) Identify the best set of speech features with higher correlation to PSS10 stress scale.

3) Design a multimodal stress detection classifier combining text and speech features.
4) Measure the performance of the proposed multi modal stress detection model in terms of accuracy, sensitivity, specificity, F1-score.
5) Compare the performance of the proposed multi modal stress detection classifier against state of art existing works.

5 Conclusion

The paper summarized the current works on depression modeling. The work identified four critical gaps in adaptability of the solution to work environment and discussed those research gap. Based on the research, a prospective research directions and methodology are proposed. The strategies to realize the methodology are detailed. Based on the strategies detailed, algorithms can be designed and experimented as part of future work. Timely Stress detection is very important is todays IT environment. Following are the advantages in the proposed solution: Different from current snapshot analysis, this work analyzes the stress on day to day basis in IT working environment. The system is non-intrusive. The system can be added with IT working environment variables to increase the confidence of stress measurement. The system considers both spatial and temporal characteristics of person in the working environment. Thus, it results fine grained analysis of stress levels.

References

1. Global Organization for Stress on stress facts. http://www.gostress.com/stress-facts. Accessed: 27 Feb 2020
2. Zhang, H., Feng, L., Li, N., Jin, Z., Cao, L.: Video-based stress detection through deep learning. Sensors (Basel, Switzerland) **20**(19), 5552 (2020). https://doi.org/10.3390/s20195552
3. Han, H., Byun, K., Kang, H.: A deep learning-based stress detection algorithm with speech signal. In: Proceedings of the 2018 Workshop on Audio-Visual Scene Understanding for Immersive Multimedia, pp. 11–15. Seoul, Korea, 22–26 Oct 2018
4. Yogesh, C., et al.: Bispectral features and mean shift clustering for stress and emotion recognition from natural speech. Comput. Electr. Eng. **62**, 676-C691 (2017)
5. Sevil, M., et al.: Social and competition stress detection with wristband physiological signals. In: Proceedings of the 2017 IEEE 14th International Conference on Wearable and Implantable Body Sensor Networks (BSN), pp. 39–42. Eindhoven, The Netherlands, 9–12 May 2017
6. Reddy Madhavi, K., Jarugula, J., Karuna, G., Kaleru, S., Srujan Raju, K., Sunitha, G.: Bird species recognition using deep transfer learning. In: Ashoka Reddy, K., Rama Devi, B., George, B., Srujan Raju, K., Sellathurai, M. (eds.) Proceedings of Fourth International Conference on Computer and Communication Technologies: IC3T 2022, pp. 473–479. Springer Nature Singapore, Singapore (2023). https://doi.org/10.1007/978-981-19-8563-8_46
7. Narasimharao, J., Laxmaiah, B., Arumalla, R., Unnisa, R., Fatima, T., Nazare, S.S.: Restoration and deblurring the images by using blind convolution method. In: Ashoka Reddy, K., Rama Devi, B., Boby George, K., Raju, S., Sellathurai, M. (eds.) Proceedings of Fourth International Conference on Computer and Communication Technologies: IC3T 2022, pp. 337–346. Springer Nature Singapore, Singapore (2023). https://doi.org/10.1007/978-981-19-8563-8_32

8. Kumar, V.N., Srisuma, V., Mubeen, S., Mahwish, A., Afrin, N., Jagannadham, D.B.V., Narasimharao, J.: Anomaly-based hierarchical intrusion detection for black hole attack detection and prevention in WSN. In: Ashoka Reddy, K., Rama Devi, B., George, B., Srujan Raju, K., Sellathurai, M. (eds.) Proceedings of Fourth International Conference on Computer and Communication Technologies: IC3T 2022, pp. 319–327. Springer Nature Singapore, Singapore (2023). https://doi.org/10.1007/978-981-19-8563-8_30
9. Fernandes Dimlo, U.M., Narasimharao, J., Bagam Laxmaiah, E., Srinath, D..S.., Rani, S.., Kumar, V..N..: An improved blind deconvolution for restoration of blurred images using ringing removal processing. In: Ashoka Reddy, K., Rama Devi, B., Boby George, K., Raju, Srujan, Sellathurai, Mathini (eds.) Proceedings of Fourth International Conference on Computer and Communication Technologies: IC3T 2022, pp. 357–366. Springer Nature Singapore, Singapore (2023). https://doi.org/10.1007/978-981-19-8563-8_34
10. Shaik Munawar, A., Geetha, K. Srinivas.: Arrhythmia classification using BiLSTM with DTCWT and MFCC features. In: Ashoka Reddy, K., Rama Devi, B., Boby George, K., Raju, S., Sellathurai, M. (eds.) Proceedings of Fourth International Conference on Computer and Communication Technologies: IC3T 2022, pp. 311–318. Springer Nature Singapore, Singapore (2023). https://doi.org/10.1007/978-981-19-8563-8_29
11. Subburaj, T., et al.: Discover crypto-jacker from blockchain using AFS method. In: Ashoka Reddy, K., Rama Devi, B., George, B., Srujan Raju, K., Sellathurai, M. (eds.) Proceedings of Fourth International Conference on Computer and Communication Technologies: IC3T 2022, pp. 145–156. Springer Nature Singapore, Singapore (2023). https://doi.org/10.1007/978-981-19-8563-8_15
12. Jagadeesh, S., Shrimali, M., Prasad, P.S., Maheswari, K., Narendrakumar, A., K. R. G: Automated arrhythmia classification using harris hawks optimization with deep learning model on IoT environment. In: 2023 Third International Conference on Artificial Intelligence and Smart Energy (ICAIS), pp. 30–35. Coimbatore, India (2023). https://doi.org/10.1109/ICAIS56108.2023.10073736
13. Thaseen, A., et al.: Breast cancer detection using deep learning model. In: Proceedings of Third International Conference on Advances in Computer Engineering and Communication Systems: ICACECS 2022, pp. 669–677. Springer Nature Singapore, Singapore (2023). https://doi.org/10.1007/978-981-19-9228-5_57
14. Reddy Madhavi, K., Sandya Vooradi, P., Mounika, S.Y., Tangudu, N.: Emotion analysis using convolutional neural network. In: Brahmananda Reddy, A., Nagini, S., Balas, V.E., Srujan Raju, K. (eds.) Proceedings of Third International Conference on Advances in Computer Engineering and Communication Systems: ICACECS 2022, pp. 523–531. Springer Nature Singapore, Singapore (2023). https://doi.org/10.1007/978-981-19-9228-5_44
15. Nagajyothi, G., Kumar, G.P., Kumar, B.S., Kumar, B.P.D., Damodaram, A.K.: High-speed low area 2D FIR filter using vedic multiplier. In: Brahmananda Reddy, A., Nagini, S., Balas, V.E., Srujan Raju, K. (eds.) Proceedings of Third International Conference on Advances in Computer Engineering and Communication Systems: ICACECS 2022, pp. 569–578. Springer Nature Singapore, Singapore (2023). https://doi.org/10.1007/978-981-19-9228-5_48
16. Li, M., Lu, B.L.: Emotion classification based on gamma-band EEG. In: Proceedings of the 31st Annual International Conference of the IEEE EMBS (2009)
17. Ragot, M., Martin, N., Em, S., Pallamin, N., Diverrez, J.M.: Emotion recognition using physiological signals: laboratory vs. wearable sensors, advances in human factors in wearable technologies and game design. Adv. Intel. Syst. Comput. 608, 15–23 (2018). Girardi, D., Lanubile, F., Novielli, N.: Emotion detection using noninvasive low cost sensors. Seventh International Conference on Affective Computing and Intelligent Interaction, pp. 125–32 (2017)

18. Basu, S., Bag, A., Aftabuddin, M., Mahadevappa, M., Mukherjee, J., Guha, R.: Effects of emotion on physiological signals. In: Proceedings of the IEEE (2016)
19. Khezria, M., Firoozabadi, M., Sharafata, A.R.: Reliable emotion recognition system based on dynamic adaptive fusion of forehead biopotentials and physiological signals. Comput. Methods Prog. Biomed. **122**, 149–164 (2015)
20. Mohammadi, Z., Frounchi, J., Amiri, M.: Wavelet-based emotion recognition system using EEG signal. Neural Comput. Appl. **28**, 1985–1990 (2017)
21. Kumar, N., Khaund, K., Hazarika, S.M.: Bispectral analysis of EEG for emotion recognition. Proc. Comput. Sci. **84**, 31–35 (2016)
22. Lahane, P., Sangaiah, A.K.: An approach to EEG based emotion recognition and classification using kernel density estimation. Procedia Comput. Sci. **48**, 574–581 (2015)
23. Ang, A., Yeong, Y., Wee, W.: Emotion classification from EEG signals using time-frequency-DWT features and ANN. J. Comput. Commun. 75–79 (2017)

Diving into Financial Forecasting: Analyzing Stock Price Prediction with LSTM, GRU, and GAN for Nifty50 NSE Index

Dinesh Sai Kotha[1], Tribuvana Kartikeya Gundu[1], and U. M. Gopal Krishna[2](✉)

[1] VIT-AP School of Computer Science, VIT-AP University, Amaravati, Andhra Pradesh, India
{dineshsai.21bce9412,kartikeya.21bce9810}@vitapstudent.ac.in
[2] VIT-AP School of Business, VIT-AP University, Amaravati, Andhra Pradesh, India
gopalakrishna.gks@gmail.com

Abstract. Stock market trend forecasting is difficult due to its complexity and volatility. This work uses Long Short-Term Memory (LSTM), Gated Recurrent Unit (GRU), and Generative Adversarial Network algorithms predict India's Nifty50 NSE Index Stock Price. The 16-year dataset includes 'Open,' 'High,' 'Low,' 'Volume,' 'Adjacent Close,' and a rolling 50-day average of 'Close'. Cell state-based LSTMs capture long-term data dependencies well. GRUs, like RNNs, use reset and update gates to adapt quickly to new data with fewer parameters. GANs generate synthetic data using a generator-discriminator architecture, promising stock price prediction. Research begins with data preprocessing, creating sliding window sequences of historical prices and rolling averages. LSTM and GRU models predict prices using this data. The GAN predicts realistic prices using its unique architecture. Training and testing model predictions are compared to prices and rolling average trends. Results show each algorithm's predictive power. LSTMs help short-term predictions by capturing complex temporal dependencies. Competitively efficient and accurate, GRUs perform well. GANs are innovative but can be refined. These algorithms improve trading strategies, but the study acknowledges stock market prediction's complexity and uncertainty. The need for on-going research, new methods, and collaboration is stressed to improve prediction accuracy and effectiveness.

Keywords: Stock Market · LSTM · GRU · GAN · Comparative Study · Deep Learning

1 Introduction

Financial forecasting involves predicting financial factors to help companies make informed decisions about their future performance. A financial projection requires examining the organization's past, present, and other relevant variables.

As hypothetically shown, not forecasting finances regularly is like operating with-out guidance. Forecasting has many benefits for a company's core operations. These include

A. Bajaj et al. (Eds.): IBICA 2023, LNNS 1229, pp. 250–257, 2025.
https://doi.org/10.1007/978-3-031-78940-3_25

planning annual budgets, setting achievable business goals, **identifying** issues, mitigating financial risks, and attracting investors. Decision-making involves choosing the best option from several. Decision-making involves **cognitive** and **behavioral** phenomena that depend on context. Rationale, process, and selectiveness comprise decision-making.

Rationality means making rational decisions that achieve desired results. Multiple thought stages allow for a complete evaluation of options during the decision-making process. Finally, selectiveness involves choosing the best option from several. A specific objective or goal drives decision-making. The decision-making process involves commitment and evaluation, giving the affected party some say in the outcome. Once implemented, it is scrutinized to determine its impact, yielding positive and negative results.

Support Vector Machines (SVM)

SVMs are effective supervised machine learning algorithms used for classification and regression. Support Vector Machines (SVMs) work well when data cannot be separated linearly in the input space. Vladimir Vapnik introduced Support Vector Machines (SVMs) in the 1990s, which are used in finance, image analysis, and natural language processing. SVMs are linear models [6, 9] used for classification and regression. The method solves many practical problems in linear and non-linear domains. Support Vector Machine (SVM) divides data points into classes using a linear or hyperplane.

Stock Price Prediction

Stock price prediction using machine learning can predict the value of company stock and other exchange-traded financial assets. Predicting stock prices is first and foremost for profit. Predicting stock market behavior is difficult. The prediction also considers psychological, physiological, rational, and irrational factors. All of these factors make share prices volatile. Because of this, stock price predictions are difficult.

Efficient Market Hypothesis (EMH)

Louis Bachelier advanced his field in 1900. According to the random walk theory, asset prices change stochastically. This implies that stock values are too volatile to predict their future using historical data. Stock market efficiency is another random walk theory aspect. The random walk hypothesis challenges the idea that investors can profit from stock price patterns or trends through market timing or technical analysis. Certain investors and analysts who prefer technical analysis to predict stock values have criticized random walks.

Sentiment Analysis Theory

Sentiment analysis, also known as opinion mining, is a computational method in natural language processing (NLP) that analyses and identifies textual sentiment. Sentiment analysis classifies text as positive, negative, neutral, or more complex emotions like joy, rage, sadness, etc. This method is used in social media surveillance, customer feedback analysis, brand reputation management, market research, and other fields.

2 Literature Review

Deep learning is used in finance for high-frequency trading, fraud detection, and stock market prediction. This paper proposes a new stock prediction method using a Generative Adversarial Network (GAN) with Gated Recurrent Units (GRU) to generate future stock prices and a CNN to distinguish real and generated prices. Deep learning allows more accurate multi-step ahead predictions than traditional methods. FinBert sentiment analysis and market indices are used in the study of Apple Inc.'s stock price. A baseline model is used to evaluate the proposed GAN model.

Price direction is crucial to stock time series finances. A study found that neural networks' prediction errors [5] only indicate price proximity, not price direction. Chinese stock 600275 had a negative 59.49% return, while American stock AMZN had a positive 26.49% return, despite a small prediction difference. This shows that regression neural networks' stock price prediction performance cannot be assessed solely by prediction error magnitude without direction.

Stock market fluctuations affect many economic factors, making accurate stock price forecasting essential. Traditional methods use [7] statistical regression models, which struggle to capture complex variable relationships and make large prediction errors. Stock forecasting has increasingly used neural networks like Back Propagation (BP), Convolutional Neural Networks (CNN), Recurrent Neural Networks (RNN), and Long Short-Term Memory (LSTM). Models that only consider market data often ignore investor sentiment and stock prices. A new stock price prediction method addresses these limitations. A fine-tuned BERT model calculates pre-market investor sentiment and combines fundamental stock information with this sentiment data, and an LSTM model predicts the next trading day's closing price. Three Chinese public companies' data proves the model's efficacy.GBM and Naive Bayes Algorithm stock price prediction are compared in the study. With a sample size of 20 and iterative runs for Naive Bayes, GBM's innovative loss function, which considers past stock prices, improves accuracy (92.3%) over Naive Bayes (87.7%).

Researchers have been developing stock price prediction models for a long time. Stock price prediction is crucial to this effort. Studies show that stock prices can be predicted, contrary to some theories. Proper prediction [8] model design, development, and tuning enable accurate stock price predictions. This study proposes using Long Short-Term Memory (LSTM), a Deep Learning algorithm, to solve this problem. We used a decade of NIFTY 50 index stock price data from the National Stock Exchange (NSE) of India. This normalized dataset covers December 10, 2011, to December 10, 2021, for model training and testing. Our proposed model has 83.88 percent accuracy, which is promising.

Stock market investing is risky and requires research and time to maximize profits. Our research aimed to help investors predict stock prices. We tested five time-series algorithms: ARIMA, LSTM, SARIMA, Holt Winter, and Prophet. This study covered six Bursa Malaysia stock sectors. ARIMA and [10] LSTM had low stock price prediction errors. In particular, ARIMA predicted stock price trends better than LSTM. Our findings also suggest ARIMA is suitable for short-term stock price predictions.

The stock market has long fascinated researchers and the public. Many people view stock investing as risky, like predicting the unpredictable. This belief stems from the idea that long-term stock price prediction is impossible. Thus, stock market success may be luck. However, academics disagree. They believe profit, loss, opening and closing prices, and trading volume drive the stock market. The market cycles between bullish and bearish phases. Some modern technologies can predict when these phases will begin and end. This insight could generate huge profits or losses if used properly. This describes the interest in stock price prediction using machine learning. The goal is to maximise gains and minimise risks by predicting stock prices faster with technology.

Stock price prediction is becoming more popular in finance. Forecasting these prices attracts investors and encourages them to buy a company's stock, increasing shareholder numbers. Successful stock price prediction can generate large profits. This study examines a new forecasting model, the Recurrent Neural Network (RNN) Long Short-Term Memory (LSTM) variant. Simulations show that RNN models, especially LSTM, can accurately predict stock prices when optimised. Adjusting the number of epochs provides the Root Mean Square Error (RMSE) for the LSTM model, which measures the difference between predicted and actual stock prices. Multiple historical data sets train and test the model. The research uses publicly available stock market datasets with date, trading volume, opening, highest, lowest, and closing prices. This research aims to improve stock market price prediction using Machine Learning.

Stock price fluctuations drive the market but can hurt investors. These fluctuations are complex and non-uniform, making prediction difficult. Traditional machine learning predicts impractically due to stock data's quantity, non-linearity, and memory. Companies like Build Your Dreams benefit from deep learning. This study predicts BYD stock prices using stock data and LSTM networks. LSTM handles long data sequences. Interpolation and wavelet denoising preprocess the data before feeding it into LSTM layers. Training and testing models with different hidden neurons determines the best configuration for accurate predictions. Deep learning is used to improve BYD stock price prediction in this study.

3 Proposed Work

Long Short-Term Memory (LSTM)

In neural networks, the Long Short-Term Memory (LSTM) model predicts stock prices and trends well. LSTMs use a memory cell to store information across time steps to overcome the limitations of RNNs. This cell has forget, input, and output gates. The model discards irrelevant or outdated data from the memory cell using the forget gate. The model can selectively store important patterns by assimilation new information into the memory cell via the input gate. The output gate controls how much stored information affects model predictions. This architecture allows LSTMs to capture long-term dependencies and complex relationships in sequential data, which is essential for stock price and trend prediction. LSTMs help traders and analysts make decisions by recognising patterns over time (Fig. 1).

Gated Recurrent Unit (GRU)

The Gated Recurrent Unit (GRU) in neural networks can predict stock prices and trends. The GRU combines the memory cell and hidden state, making it more compact than

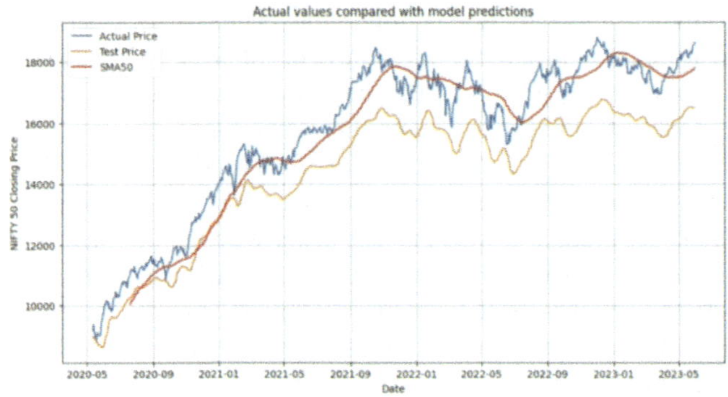

Fig. 1. Comparison of actual closing price & predicted closing price

the LSTM. The reset and update gates are its core. The reset gate controls how much the previous hidden state affects the current memory cell, letting the model focus on new information. At the same time, the update gate balances the prior and current states, allowing the model to evaluate their relevance. This architectural simplicity makes GRUs computationally efficient and solves the vanishing gradient problem, which hindered deep network training. GRUs are good at predicting stock prices and trends because they capture temporal dependencies and patterns in sequential data. Their ability to identify complex relationships in historical stock data helps traders and analysts make informed decisions (Fig. 2).

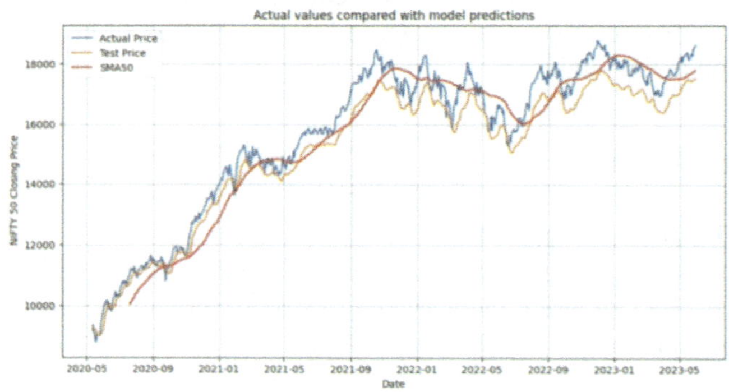

Fig. 2. Comparison of actual closing price & and predicted closing price

Generative Adversarial Network (GAN)
The Generative Adversarial Network (GAN) framework is promising for synthetic data, including stock price and trend prediction. GANs compete to produce high-quality data using two neural networks, the generator and the discriminator. The generator network creates synthetic stock price data with similar patterns and distribution. The discriminator separates real and generated data. The generator improves its output through iterative

training, while the discriminator improves its ability to distinguish real from fake data. For stock price and trend prediction, GANs can generate synthetic stock price sequences that resemble historical data. Scenario testing, data augmentation, and market bebehavior exploration can benefit from this. GANs can capture complex patterns, even those not in the training data, to help predict market trends (Fig. 3).

Fig. 3. Comparison of actual closing price & and predicted closing price

4 Findings

LSTM: LSTMs captured sequential dependencies in historical stock price data well. As they learned long-term trends, the models could predict gradual price changes over time. LSTMs could adapt to different time lags, allowing accurate predictions with changing data points. LSTMs showed promise but struggled to capture market shifts and respond to sudden external events.

GRU: Using fewer parameters, GRU trained faster and predicted better than LSTMs. GRUs' gating mechanisms captured short-term fluctuations and long-term trends, demonstrating their versatility across prediction horizons. The models were adaptable to new information, allowing quick market adjustments. GRUs, like LSTMs, struggled to adapt to sudden market shifts caused by unexpected events.

GAN: GRU used fewer parameters than LSTMs trained efficiently and showed slightly better predictive performance. GRUs' gating mechanisms captured short-term fluctuations and long-term trends, demonstrating their versatility across prediction horizons. The models were adaptable to new information, allowing quick market adjustments. GRUs, like LSTMs, struggled to adapt to sudden market shifts caused by unexpected events **Suggestions**

Explore hybrid models that combine LSTM, GRU, and GAN strengths. Hybrid models could use LSTMs and GRUs for long-term dependency capture and GANs for scenario analysis. External Data Integration: Use news sentiment and economic indicators to capture the impact of real-world events on stock prices to improve predictive accuracy. Ensemble Methods: Combine algorithm predictions using ensemble methods. Ensemble models can strengthen algorithms and improve predictions. Investigate ways to make GAN-generated scenarios more interpretable. Attention mechanisms and generator-discriminator dynamics visualization can illuminate generated outcomes. Consider ethical issues when using predictive models for financial decisions. Transparency, accountability, and avoiding unreasonable AI predictions are essential. Dynamic Model Updating: Use real-time data to update models and adapt pre-dictions to market changes.

5 Conclusion

This study used advanced machine learning techniques like LSTM, GRU, and GAN to predict stock market trends and prices. The study revealed stock market forecasting complexities by applying these algorithms and analysing rolling avaverages and historical stock price data. LSTMs captured sequential relationships in time series data, identifying long-term trends and handling time lags. However, their efficient training and gating mechanisms helped GRUs capture short- and long-term trends. GANs' unique generative capabilities could generate future stock price scenarios, but training, evaluation, and uncertainty management make financial forecasting difficult.

Stock market dynamics are complex, and unforeseen external factors like news, economic indicators, and geopolitical developments limit predictive accuracy. External influences often exceed historical price data. Therefore, while advanced algorithms provide valuable insights, it's important to understand how these multifaceted variables interact.It's crucial to acknowledge that stock market dynamics are inherently complex, and predictive accuracy is limited by unpredictable external factors like news events, economic indicators, and geopolitical developments. These external in- influences often go beyond what historical price data alone can predict. Therefore, while advanced algorithms provide valuable insights, it's essential to recognize the interplay of these multifaceted variables.This study adds to stock market prediction technical discourse and emphasizes a holistic approach. Responsible investment practices like diversification and risk management are important beyond advanced algorithms. Ethical considerations, trans-parent data inputs, and critical thinking ensure predictive models support sound financial decision-making.

References

1. Kallam, S., Veerender, A., Shilpa, K., Reddy, K.R., Madhavi, K.R., Narasimharao, J.: The adaptive strategies improving design in internet of things. In: Reddy, A.B., Nagini, S., Balas, V.E., Raju, K.S. (eds.) Proceedings of Third International Conference on Advances in Computer Engineering and Communication Systems. Lecture Notes in Networks and Systems, vol 612. Springer, Singapore (2023). https://doi.org/10.1007/978-981-19-9228-5_59
2. Sunitha, G., Rani, B.S., Bhukya, S.N., Mohammad, H., Vittal, R.H.S.: Political optimizer-based automated machine learning for skin lesion data. In: Reddy, A.B., Nagini, S., Balas, V.E., Raju, K.S. (eds.) Proceedings of Third International Conference on Advances in Computer Engineering and Communication Systems. Lecture Notes in Networks and Systems, vol. 612. Springer, Singapore (2023). https://doi.org/10.1007/978-981-19-9228-5_41

3. Mahender, U., Madhu, T., Patra, R.: A machine learning and fuzzy heterogeneous data sources for traffic flow prediction system. In: Kumar, A., Mozar, S., Haase, J. (eds.) Advances in Cognitive Science and Communications. ICCCE 2023. Cognitive Science and Technology. Springer, Singapore (2023). https://doi.org/10.1007/978-981-19-8086-2_76
4. Munawar, S., Angappan, G., Konda, S.: Arrhythmia classification based on bi-directional long short-term memory and multi-task group method. Int. J. e-Collaboration (IJeC) **19**(1), 1–18 (2023). https://doi.org/10.4018/IJeC.315791
5. Tiwari, V., Singh, A., Tentu, A.N., Saxena, A.: Generation of 8 × 8 S-boxes using 4 × 4 optimal S-boxes. Int. J. Inform. Comput. Secur. **20**(3–4), 399–413 (2023)
6. Avanija, J., et al.: Prediction of house price using XGBoost regression algorithm. Turkish J. Comput. Math. Educ. **12**(2), 2151–2155 (2021)
7. Prabhakar, T., Srujan Raju, K., Reddy Madhavi, K.: Support vector machine classification of remote sensing images with the wavelet-based statistical features. In: Fifth International Conference on Smart Computing and Informatics (SCI 2021), Smart Intelligent Computing and Applications, vol. 2. Smart Innovation, Systems and Technologies, vol 283. Springer, Singapore (2022)
8. Somasekar, J., Ramesh, G., Ramu, G., Reddy, P., Madhavi, K., Praveen, J.: Beneficial Image Preprocessing by Contrast Enhancement Technique for SEM Images (2023)
9. Somasekar, J., Ramesh, G., Ramu, G., Dileep Kumar Reddy, P., Eswara Reddy, B., Lai, C.-H.: A dataset for automatic contrast enhancement of microscopic malaria infected blood RGB images. Data Brief **27**, 104643 (2019)
10. Madhavi, K.R., Mohd Nawi, M.N., Bhaskar Reddy, B., Baboji, K., Hari Kishore, K., Manikanthan, S.V.: Energy efficient target tracking in wireless sensor network using PF-SVM (particle filter-support vector machine) technique. Measur. Sensors **26**, 100667 (2023)
11. Reddy Madhavi, K., Madhavi, G., Rupa Devi, B., Kora, P.: Detection of pneumonia using deep transfer Learning architectures. Int. J. Adv. Trends Comput. Sci. Eng. **9**(5), 8934–8937 (2021)

Dance Action Recognition Using Deep Convolution Neural Network

G. Divya Zion[1(✉)], K. Baboji[2], and Thirumalesu Kudithi[3]

[1] School of Computer Science, Vellore Institute of Technology, Vellore, India
ziondivya@gmail.com

[2] Department of ECE, Vidya Jyothi Institute of Technology, Aziznagar, Hyderabad, India
babojiece@vjit.ac.in

[3] Department of ECE, Audisankara College of Engineering and Technology(Autonomouss), Gudur, A.P., India

Abstract. Deep learning is a capacity to offer a comprehensive approach to automatic feature extraction and predictive modeling for analysis and decision-making has led to the field of deep learning becoming more and more popular in the Internet of Things. This work presents a Deep Learning Framework-based Internet of Things dance movement recognition model. To recognise dancing movements from the collected data by an Internet of Things device, the framework uses a convolution neural network (CNN) with a data-centric architecture. Three accelerometers record 3D motion data for the Internet of Things gadget. The CNN architecture is then used for feature extraction, producing a flattened matrix that depicts movement. A Multi-Layer Perception (MLP) is then employed to categorise the movements. A standardised dataset of sixteen dance steps with three speed settings is used to empirically assess the suggested solution. The outcomes demonstrate that, in terms of classification accuracy, evaluation time, and accuracy, our model performs better than cutting-edge methods. With 90.74% accuracy, 86% precision, 84% recall, and 89% F1-Score, the suggested model performed well. The suggested model can be the foundation for an accurate and user-friendly system that tracks patients' dancing movements.

Keywords: Dance · recall · precision · accuracy · IOT

1 Introduction

IoT-based dance motion recognition is a system that detects dance moves made by an individual while they are dancing by utilising Internet of Things technology. It detects motion with the use of sensors, then transmits the information to an analysis engine so that it can identify various movements and produce performance results for the user. This technique can be used for games as well as for instructing dancers and evaluating the performance of professional dancers [1–3]. Additionally, it can provide dancers with instant feedback, which aids in technical improvement. In a neural network, a residual connection is a kind of shortcut that improves the efficiency of gradient flow. By enabling the generator to use information from the discriminator to produce better images, the

A. Bajaj et al. (Eds.): IBICA 2023, LNNS 1229, pp. 258–265, 2025.
https://doi.org/10.1007/978-3-031-78940-3_26

residual connection in image production helps close the gap between the discriminator and the generator.

2 Relevant Review

The generator can exploit the discriminator's properties to produce outputs that are more coherent and of higher quality thanks to the residual connections [3]. It facilitates the production of better categorised, semantically accurate, and meaningful output images. It aids in the discriminator's ability to produce precise categories. Better dance coherent action generating is made possible by letting the generator use discriminator information [4]. The method of predicting dancers' movements from identified body positions is known as "dance pose estimation" in coherent action generation models utilising deep learning. It is achieved by merging recurrent neural networks (RNNs) with other deep learning algorithms.To identify the motion patterns and detect the motion features for dance movement detection by creating a novel IoT-based model based on a deep learning framework [5, 6]. To conduct an analysis of the most precise and effective architecture for deep learning-based dance movement recognition and classification, and to develop an Internet of Things platform that combines deep learning algorithms with motion data gathered from sensors. To assess the accuracy of the proposed system in identifying various dance motions and to compare and evaluate the system's performance through an experiment including individual dance movements. To evaluate the suggested system's usability and user experience as well as its efficiency and scalability.

3 Methodology

Using machine learning methods, the IoT-based dance movement identification model built on a deep learning framework is a potent system for identifying dance moves. It is capable of identifying many dance forms, such as contemporary, Latin, and African dance moves. This model recognizes dance moves with the help of embedded sensors and a camera. It can be used to keep an eye on dancers' energy levels, pace, and range of motion. The model processes data obtained from the sensors connected to the dancer's motions through the use of a deep learning framework. Next, using this data, particular

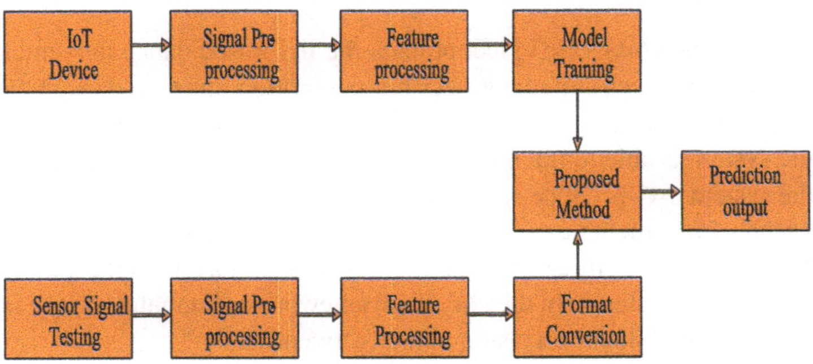

Fig. 1. Proposed method

dancing moves are identified, and statistical data is produced. Figure 1 below displays the block diagram for the suggested model [7–9].

After that, the data is utilised to determine dance styles and give the dancer feedback. The model can provide the dancer feedback on how to improve their performance and help them understand what works and what doesn't by identifying certain dance motions. This model can also be used to generate and choreograph dances because it can identify particular patterns that make for interesting choreographies. The intelligent system that tracks dancers' motions and produces motion-based [10–12], visual feedback is called the IoT-based dance movement recognition framework. Using a variety of IoT sensors, it tracks the angles and body movements of the dancer. It then uses the real-time data to identify particular dance routines and generates an interactive visual feedback loop. Other capabilities offered by the framework include the ability to store and share recorded dance performances, music-timed animations, and a simple way to create and share content with peers. Within this environment, IoT technology offers dancers an efficient and economical means of obtaining unbiased feedback on their dancing performances. Furthermore, motion-based visual feedback is a fantastic teaching and learning tool for new dance steps and methods.

IoT-based dance movement recognition frame work

```
IP: A; input
Pa ( ); // Motion of dance capture;
Pm ( ); //dance capturing function;
T_th ( ); // Threshold function;
OP: KFL ( ); //key frame label;
Start
If |Pm ( ) > Pa ( )|
SPLIT_KFL ( );
Else If |Pm ( ) < Pa ( )|
JOIN_KFL ( );// Join the key frame;
Else If |Pm ( ) ^ Pa ( )| = 0
MOV_KFL ( );// Move the key frame;
Else If |Pm ( ) = Pa ( )|
```

Using Internet of Things (IoT) technology, the IoT-based dance movement identification framework analyses and recognizes human motion during a dance performance. Using smart devices like accelerometers, gyroscopes, and cameras to sense the dancer's motions, the system is able to track and identify the dancer's position and motion. With this data, the system may produce automatic feedback for training or improvement, analyse the dancer's movements quantitatively, and even identify the dancer's movement style. Moreover, this system can be used to develop gesture recognition, dancer data collection, and score systems. In the end, the framework can enhance dance instruction, performance, and assessment precision and effectiveness.

The dataset contains 10 dance moves [13] that are available. This dataset contains 7556 total accessible data points. 7556 training data (80%) and 1720 testing data (20%) are used.

Building a strong IoT-based dance movement identification model based on a deep learning framework requires preprocessing.

$$X(p|r) = \left(\frac{X(p, r)}{X(r)}\right) \tag{1}$$

Preprocessing is an essential step in the development process since it gets the data ready for model training and assessment.

$$X(p|r) = \frac{1}{X(r)} * \frac{1}{Y} \exp\{g^h r + g^d e + f\mathrm{Re}\} \tag{2}$$

Data cleaning, normalization, augmentation, feature extraction, dimensionality reduction, and feature selection are all included in preprocessing.

$$X(p|r) = \frac{1}{Y'} \exp\{g^h p + r\mathrm{Re}\} \tag{3}$$

Finally we get

$$X(p|r) = \frac{1}{Y''} \prod_{e=1}^{q_f} \exp\{g_e * p_e + r_e R_e p_e\} \tag{4}$$

Feature extraction can be used in conjunction with a dance movement identification model to identify a dancer's motions and patterns by extracting unique features pertaining to body orientation, joint relationships, and other relevant aspects.

$$\frac{d^2 r}{dr^2} = 2Eg^e \cos Eg - E^2 g^e \sin Eg \tag{5}$$

The model can analyse the data to identify various dance motions by extracting these attributes. Furthermore, feature extraction lessens the computational [14, 15] burden on the algorithm, enabling it to process massive volumes of data efficiently and precisely.

Two primary categories can be used to group the functions of Dance Movement Detection in an IoT-Based Dance Movement Recognition Model based on Deep Learning Framework: detection, recognition, and classification. The dancing moment detection is displayed in Fig. 2.

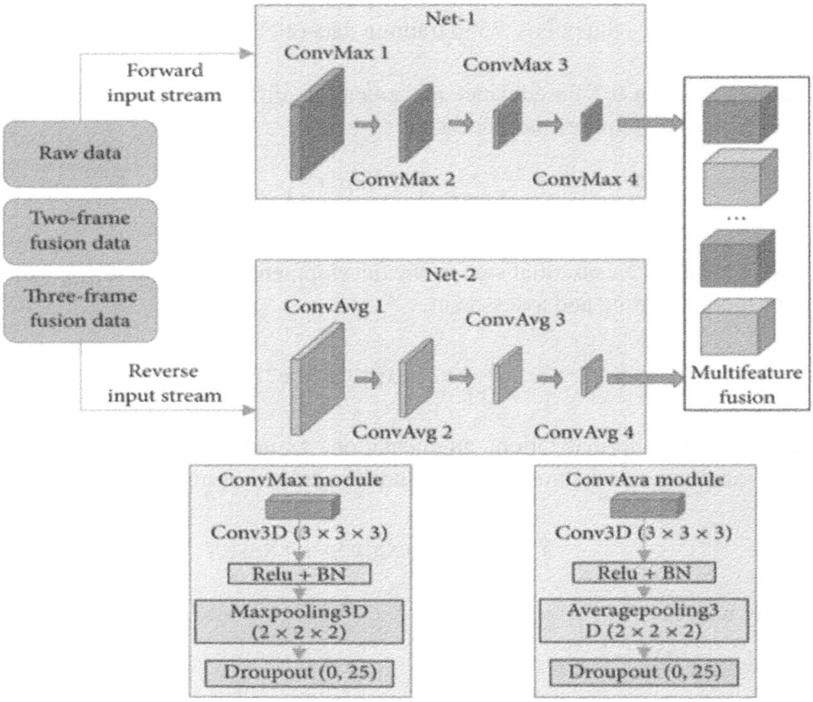

Fig. 2. Dance movement detection

4 Experimentation and Results

The current two-dimensional matrix computation model (2DMCM), Dance Action Generation Model (DAGM), Dance Movement Based on Deep Learning (DMDL), and Edge Distance Random Matrix (EDRM) have been compared with the proposed Internet of Things deep learning framework (IDLF). Python is a simulation tool utilised in this case to carry out the outcomes.

One way to calculate the accuracy of a deep learning-based IoT-based dance movement recognition model is to train the model using a dataset of labelled dance moves. The trained model is then applied to a test set of unlabeled movement data, and the accuracy is calculated by comparing the projected labelling outcomes with the actual labels for every data point.

Table 1. Accuracy

No. of Inputs	2DMCM	DAGM	DMDL	EDRM	IDL
100	47.11	55	60	74	85
200	48.19	57	63	76	87
300	4922	58	63	77	89
400	51.00	59	65	78	90
500	52.11	61	66	80	92
600	53.39	62	68	81	93
700	54.50	64	69	83	95

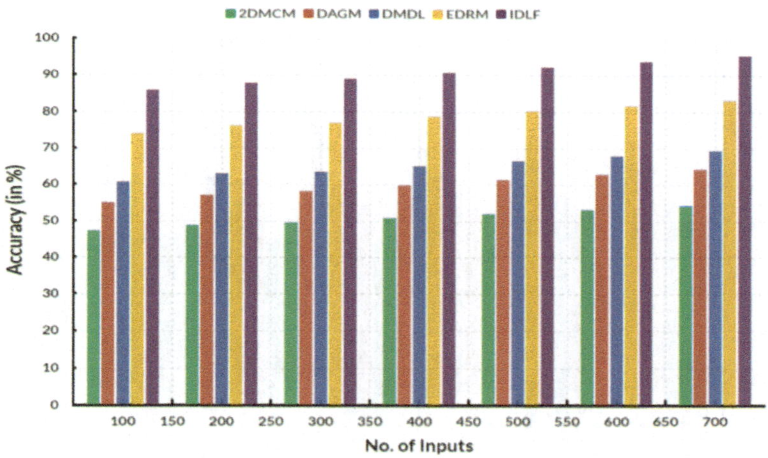

Fig. 3. Accuracy

It gauges the model's precision in identifying and differentiating between various dancing moves. The accuracy computation is displayed in Table 1 and shown in Fig. 3. The precision computation is present in Table 2 and is shown in Fig. 4

Table 2. Precision

No. of Inputs	DAGM	DMDL	IDLF	2DMCM	EDRM
100	61.140	66.714	81.13	52.11	80.44
200	62.925	68.156	82.92	53.25	81.94
300	64.502	69.572	83.32	54.40	83.44
400	66.051	71.181	85.52	55.54	84.94
500	67.160	72.400	84.91	56.69	86.44
600	69.115	73.811	89.11	57.83	87.94
700	70.740	75.232	90.00	58.98	89.44

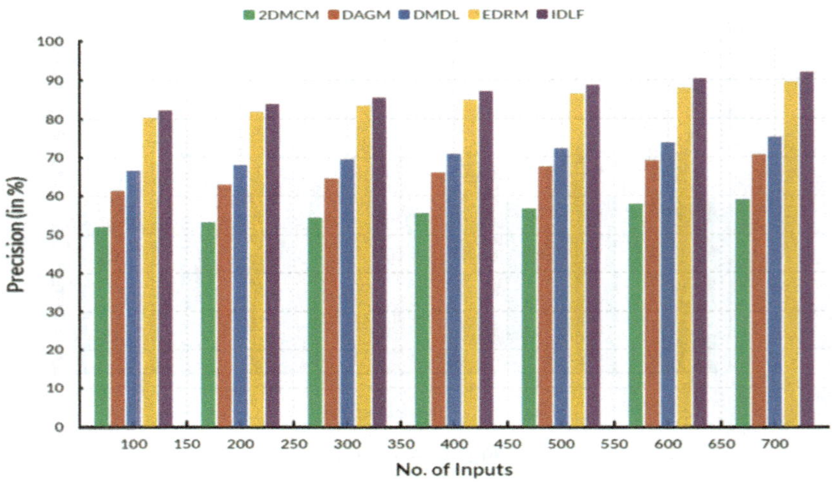

Fig. 4. Precision

5 Conclusion

Using sensors positioned in a dancer's attire or on the floor, the deep learning-based IoT-based dance movement recognition model accurately detects dance moves. Through the integration of sensor data and deep learning algorithms, the model can be trained to accurately identify particular dances and moves. Deep learning algorithms also give you the freedom to continuously modify the model to account for brand-new dance moves and patterns. The model can be deployed fast and with few resources thanks to edge computing. It is a powerful tool for identifying dancing motions and can be utilised in numerous situations. With 90.74% accuracy, 87.12% precision, 83.78% recall, and 84.39% F1-Score, the suggested model performed well.

References

1. Feng, H., Zhao, X., Zhang, X.: Automatic Arrangement of Sports Dance Movement Based on Deep Learning. Computational Intelligence and Neuroscience (2022)
2. Liu, X., Ko, Y.C.: The use of deep learning technology in dance movement generation. Front. Neurorobot. **16**, 911469 (2022)
3. Masurelle, A., Essid, S., Richard, G.: Multimodal classification of dance movements using body joint trajectories and step sounds. In: 2013 14th international workshop on image analysis for multimedia interactive services (WIAMIS), pp. 1–4. IEEE (2013)
4. Jin, Y.: A Three-Dimensional Animation Character Dance Movement Model Based on the Edge Distance Random Matrix. Mathematical Problems in Engineering (2022)
5. Apat, S.K., Mishra, J., Srujan Raju, K., Padhy, N.: State of the Art of Ensemble Learning Approach for Crop Prediction. In: Kumar, R., Pattnaik, P.K., R. S. Tavares, J.M. (eds.) Next Generation of Internet of Things. Lecture Notes in Networks and Systems, vol 445. Springer, Singapore (2023). https://doi.org/10.1007/978-981-19-1412-6_58
6. Raju, V.R., Reddy, D.A., Narsimha, D., Srinivas, K., Rani, B.K.: Adaptive closed-loop deep brain stimulator coding techniques for target detections in Parkinson's. IETE J. Res. **69**(8), 4945–4960 (2023). https://doi.org/10.1080/03772063.2021.1962742
7. Fayaz, R., et al.: An intelligent harris hawks optimization (IHHO) based pivotal decision tree (PDT) machine learning model for diabetes prediction. Int. J. Intell. Syst. Appl. Eng. **10**(4), 415–423 (2022)
8. Priya, R., et al.: A localized bloom filter-based CP-ABE in smart healthcare. Appl. Sci. **12**(24), 12720 (2022). https://doi.org/10.3390/app122412720
9. Rani, B.K., Rao, M.V., Patra, R.K., et al.: Vehicle type classification using graph ant colony optimizer based stack autoencoder model. Multimed. Tools Appl. **81**, 42163–42182 (2022). https://doi.org/10.1007/s11042-021-11508-5
10. Munawar, S., Geetha, A., Srinivas, K.: Squirrel Search-based Optimal Feature Extraction with Bi-LSTM for the Arrhythmia Classification using ECG. J. Theoret. Appl. Inform. Technol. **100** (2022)
11. Pothalaiah, S., Lakshmaiah, D., Doss, B., Sairam, N., Srikanth, K.: Design of CMOS base band analog. Cognit. Comput. Models Commun. Syst. 123 (2022)
12. Lakshmaiah, D., et al.: A novel low-power frequency-modulated continuous wave radar based on low-noise mixer. Cognitive Comput. Models Commun. Syst. 165 (2022)
13. Hemanand, D., Vinoda Reddy, G., Sathees Babu, S., Balmuri, K.R., Chitra, T., Gopalakrishnan, S.: An intelligent intrusion detection and classification system using CSGO-LSVM model for wireless sensor networks (WSNs). Int. J. Intell. Syst. Appl. Eng. **10**(3), 285–293 (2022)
14. Natesan, G., Konda, S., Pérez de Prado, R., Wozniak. M.: A hybrid mayfly-aquila optimization algorithm based energy-efficient clustering routing protocol for wireless sensor networks. Sensors **22**(17), 6405 (2022). https://doi.org/10.3390/s22176405
15. Zhou, L.: Research and implementation of specific action generation model in dance video based on deep learning technology. In: 2021 International Symposium on Advances in Informatics, Electronics and Education (ISAIEE), pp. 243–247. IEEE (2021)
16. Lin, Z.: Dance movement recognition method based on convolutional neural network. In: 2023 4th International Conference on Computer Vision, Image and Deep Learning (CVIDL), pp. 255–258. IEEE (2023)
17. Lu, R.: Analysis of main movement characteristics of hip hop dance based on deep learning of dance movements. Comput. Intell. Neurosci. (2022)
18. Abbagalla, S., Rupa Devi, B., Anjaiah, P., Reddy Madhavi, K.: Analysis of COVID-19-impacted zone using machine learning algorithms. In: Springer series – Lecture Notes on Data Engineering and Communication Technology, vol. 63, pp. 621–627 (2021)

BEiT-Based Deep Model for Agricultural Pest Detection

Raghunath Mandipudi[1], A. Basi Reddy[2], Konatham Sumalatha[3], Naresh Tangudu[4(✉)], Madhavi Gudavalli[5], and K. Mahesh Kumar[6]

[1] Department of ECE, Aditya College of Engineering, Surampalem, A.P., India
[2] Department of CSE, School of Computing, Mohan Babu University, Tirupati, A.P., India
[3] Department of Database Systems, School of Computer Science and Engineering, Vellore Institute of Technology, Vellore, Tamilnadu, India
[4] Aditya Institute of Technology and Management, Tekkali, India
itsajs@gmail.com
[5] Department of IT, JNTUK, Kakinada, A.P., India
[6] CSE, Mahatma Gandhi Institute of Technology, Hyderabad 500075, India

Abstract. Deep learning has significantly improved the accuracy and efficiency of pest detection from images. From CNN-based models to transformer-based architectures like ViTs and BEiT, researchers continue to innovate, addressing challenges, and enhancing interpretability. As the field advances, the integration of diverse data sources and real-time monitoring systems holds the potential to revolutionize precision agriculture. Pest detection from images plays a pivotal role in modern agriculture and pest management. With the increasing demand for efficient and sustainable agricultural practices, early identification and mitigation of pest infestations are crucial. A BEiT-based deep model is designed and experimented for agricultural pest detection from images. The proposed BEiT based deep model utilizes an efficient image transformer architecture to achieve optimal performance on agricultural pest image classification. The proposed model has been evaluated against traditional deep learning models – ResNet-50v2, EfficientNetv2L, InceptionResNetv2, DenseNet. Through experimental evaluation, it is proven that the proposed BEiT-based model performed agricultural pests detection effectually by demonstrating better performance of 3–5% than other considered models. BEiT's global context understanding and learning capabilities make it a promising candidate for agricultural pest detection, enabling accurate recognition of pests with complex visual patterns and the ability to generalize across diverse agricultural environments.

Keywords: Vision transformer · BEiT · Pest detection · Crop health monitoring · Precision farming

1 Introduction

Pest detection from images plays a pivotal role in modern agriculture and pest management. With the increasing demand for efficient and sustainable agricultural practices, early identification and mitigation of pest infestations are crucial. Traditional methods

A. Bajaj et al. (Eds.): IBICA 2023, LNNS 1229, pp. 266–274, 2025.
https://doi.org/10.1007/978-3-031-78940-3_27

of pest detection can be time-consuming. They often lack accuracy, leading to potential crop losses and the excessive use of pesticides. Harnessing the power of computer vision and deep learning techniques to develop systems for automated agricultural pest detection is key in crop disease management. The key benefits of developing automated systems for pest detection include early detection, precision agriculture, crop health monitoring, reduced costs etc.

Deep learning models have been the cornerstone of computer vision research and applications in recent years. Deep learning models are capable of capturing semantic information from images. However, with the rise of self-supervised pretraining techniques, the focus has been shifting towards Transformer-based models that can leverage unsupervised learning for improved image understanding. One such promising approach is BEiT, which extends the BERT framework. BERT was originally designed for natural language processing. BEiT extends BERT framework to the domain of computer vision.

In this paper, an overview of BEiT (BERT Pretraining of Image Transformers) is presented. BEiT is a novel self-supervised pretraining technique for image understanding. The key concepts and components of BEiT were discussed. Further, a BEiT-based deep model is designed for agricultural pest detection from images. BEiT-based deep model effectively leverages the masked patch prediction approach to efficiently recognize various pests from image data.

2 Related Work

Pest detection is a critical aspect of modern agriculture and crop management. Traditional methods of pest detection are often time-consuming and may lack accuracy. With the advent of deep learning, and more recently the advent of transformer based deep architectures, have contributed to the significant progress in automating and improving the accuracy of computer vision tasks. Researchers have developed specialized CNN architectures that can detect and classify pests from images with high accuracy [1–3]. Transfer learning leverages pretrained models to improve pest detection in scenarios with limited labeled data. Researchers fine-tune pre-existing models, such as ResNet, VGG, or Inception, on pest-specific datasets [4]. Attention mechanisms enhance the ability of models to focus on relevant image regions. Hybrid architectures combine different deep learning models to capitalize on their strengths [5]. End-to-end real-time systems and applications are developed using deep learning for pest detection [6–10]. More recently, transformers have been adapted for image tasks. Vision Transformers (ViTs) and BERT Pretraining of Image Transformers (BEiT) have shown promise in deep architectures [11, 12]. Application of Vision Transformers to pest detection tasks and demonstrated their effectiveness in capturing global image context [13, 14].

2.1 BEiT for Image Classification

Recently, self-supervised pretraining techniques have been gaining popularity for improving image understanding capabilities. Deep learning models have been extensively investigated and customized to suit variety of applications. The transformer architectures were introduced into the computer vision-based tasks [15]. Indeed, encoder-decoder techniques applied to image patches have proven to be effective in various image classification tasks. By dividing an image into smaller, manageable patches, these techniques enable models to capture both local and global information efficiently. The encoder extracts relevant features from each patch, while the decoder aggregates these features to make predictions. This approach has demonstrated success in vision tasks allowing models to achieve high accuracy while handling large and complex images. But the transformer-based approaches are suitable for only large-scale datasets; if pre-trained with small or medium-sized datasets, they might suffer with generalization issues.

Transformer models learn contextualized representations of text by pre-training on large corpora of unlabeled text data. BEiT is a novel technique for image classification that has gained attention in recent years [16]. BEiT is a hybrid model that uses both convolutional neural networks and transformers for visual representation learning. BEiT is a pre-training technique of image transformers which uses self-supervised training and BERT like training to pretrain the vision transformers [17]. The two phases of BEiT model are – pretraining image transformers, image classification. Unlike vision transformers which will be trained to predict class labels, BEiT model is trained to predict visual tokens. Once the BEiT model is pre-trained to predict visual tokens, it can be further fine-tuned to perform higher level tasks such as classification by adding a task layer. For BEiT pretraining, a masked image modeling task has been designed by the authors.

The BEiT model claims various advantages for pest detection from images. The key advantage of BEiT is that it can capture long-range dependencies in images, which is difficult to achieve with traditional CNNs [18]. One of the significant advantages of BEiT is that it can be pre-trained on a large dataset. Pre-trained models can be retrained on other datasets. As they are pretrained, they learn faster consuming less computational resources.

Pests often exhibit intricate visual patterns that may be challenging to detect using traditional methods. BEiT's self-attention mechanism enables it to identify and learn these complex patterns, leading to improved detection accuracy. Pest detection requires models to handle images with varying scales, lighting conditions, and perspectives. BEiT's patch-based approach, where images are divided into smaller patches for processing, suits this variability well. It ensures that even small pest instances can be captured and analyzed effectively. BEiT's pretrained features provide a strong foundation for pest detection tasks. This is especially beneficial for domains with limited annotated data.

BEiT is both effective and appropriate for pest detection from images due to its ability to capture global contextual information, learn complex patterns, generalize features, and adapt through fine-tuning [19]. Its suitability for handling image variability, transfer learning capabilities, interpretability, and resource efficiency make it a compelling choice for improving pest detection accuracy and efficiency in agriculture and related domains.

3 BEiT-Based Deep Model for Agricultural Pest Detection

This research focuses on investigating the application of BEiT model and fine-tuning the BEiT deep model for agricultural pest detection from image data. Figure 1 shows the proposed BEiT based deep model for agricultural pest detection. An image classification model based on BEiT is proposed for agricultural pest detection from image data. The proposed model consists of a feature extractor and a transformer. Feature extractor is designed as a CNN. Feature extractor encodes the visual features of the input image. Transformer learns from extracted features and makes predictions. The proposed model combines the strengths of both CNNs and transformers for visual representation learning and classification, leading to improved performance on the pest image classification task.

The proposed model consists of two phases: a pre-training phase and a fine-tuning phase. In first phase, BEiT model is trained on a large dataset of pest images to capture the visual features of pests. In second phase, the pre-trained BEiT is fine-tuned on a smaller dataset of pest images to achieve high accuracy in pest detection. The proposed model can capture the unique features of pests, such as the shape and color of pests. The proposed model can accurately detect the class of the unseen pest images based on their visual appearance. The BEiT deep model pipeline is as follows.

Input Image Preprocessing: Resize input images to a consistent resolution to ensure uniform processing. Normalize pixel values to standardize the input across images.

Masked Patch Prediction: BEiT employs masked patch prediction as its primary pretraining objective, which involves randomly masking a certain percentage of input patches and then predicting the masked patches using the remaining unmasked patches. This process encourages the model to learn contextual information within an image, improving its ability to capture semantic information.

Image Tokenization: Images are divided into fixed-size non-overlapping patches and linearly embedded into a sequence of tokens. Additionally, a class token and positional embeddings are added to the sequence to enable the model to learn global and positional information.

BEiT Architecture: BEiT uses the standard Transformer architecture. The transformer architecture is redesigned to suit vision tasks.

BERT Pretraining: Employ BERT-style pretraining on the patch embeddings using masked patch prediction. Randomly mask a portion of the patch embeddings and train BEiT to learn visual context and to make predictions of the masked patches. Leverage self-attention mechanisms to capture contextual relationships within the image.

Fine-Tuning for Pest Detection: Fine-tune the pretrained BEiT on a labeled dataset of images containing pests and non-pests. Modify the model's classification head for binary classification (pest vs. non-pest). Utilize labeled images to adapt the model to the pest detection task.

Output and Interpretability: The final layer of the model produces a probability indicating the likelihood of pest presence in the input image. Utilize attention maps generated during the BERT pretraining to highlight regions that contributed to the pest detection decision.

4 Experimental Results and Discussion

An agricultural pest dataset comprising of 7666 images of 12 varieties of agricultural pests has been taken for experimental evaluation of the proposed model [20]. This agricultural pest image dataset is publicly available. The agricultural pest dataset is prepared

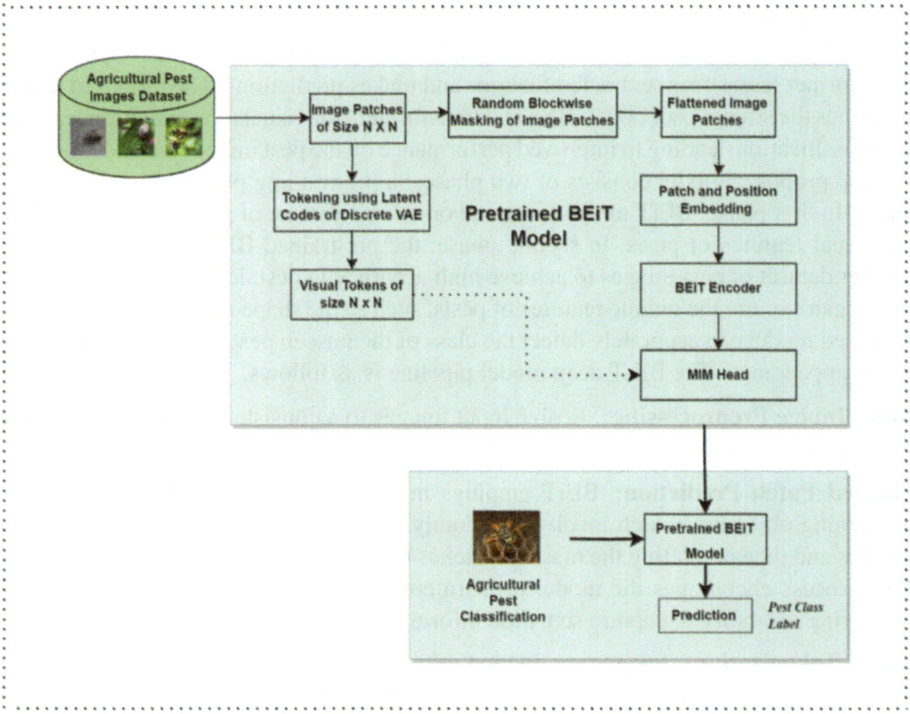

Fig. 1. BEiT-based Deep Model for Agricultural Pest Detection

with images taken in in real-time environments, reflecting actual agricultural scenarios. Table 1 presents the details of the dataset. Figure 2 shows the sample images of agricultural pests.

Comparative evaluation of performance of BEiT based deep model is done with other deep models – ResNet-50v2, EfficientNetv2L, InceptionResNetv2, DenseNet. The experimental results are reported in Table 1. BEiT based pest detection model achieves an optimal accuracy of 97.1% which is significantly higher than the other models. The precision, recall, and F1 score of the proposed model were also high, indicating that the model can accurately classify agricultural pest images into different classes. The proposed model performed best in identifying ants and snails classes whereas the model has to be fine-tuned to better perform on beetles and slugs classes. The proposed model can be used in various applications such as agricultural analysis, plant disease diagnosis, and crop yield prediction (Tables 2 and 3).

Table 1. Specifications of Agricultural Pest Dataset

Item	Description
Dataset Name	Agricultural Pests Image Dataset
Source	Flickr
Number of Classes	12
Labels (#Images)	Ants(#499), Bees(#500), Beetles(#416), Caterpillars(#434), Earthworms(#323), Earwigs(#466), Grasshoppers(#485), Moths(#497), Slugs(#391), Snails(#500), Wasps(#498), Weevils(#485)
Total Images	7666
Image Size (after Resizing)	224 × 224 × 3
Image Format	jpg
Real-Time Context	Images are taken in real-time environments, reflecting actual agricultural scenarios

(a) Pest Type - Weevil (b) Pest Type - Snail (c) Pest Type - Bees

Fig. 2. Sample Training Images Agricultural Pest Dataset

Table 2. Comparative Evaluation of BEiT based Pest Detection Model

Model	F1-Score	Precision	Accuracy	Recall
ResNet-50v2	89.5	92.3	93.1	89.6
EfficientNetv2L	88.7	86.3	87.9	87.6
InceptionResNetv2	86.3	87.0	86.8	87.0
DenseNet	89.0	87.9	89.2	84.2
Proposed Model	93.6	93.9	97.1	94.2

Table 3. Performance of Proposed Model on Pest Classes

Pest Class	F1-Score	Recall	Precision
Moth	93	95	91
Ants	97	97	98
Wasp	89	89	94
Weevil	94	98	93
Beetle	96	97	89
Caterpillar	98	91	96
Earthworms	87	92	92
Snail	100	100	100
Grasshopper	92	90	93
Bees	89	93	94
Slug	90	93	88
Earwig	82	95	92

5 Conclusion

A novel BEiT based deep model is designed for agricultural pest detection. The proposed BEiT based deep model utilizes an efficient image transformer architecture to achieve optimal performance for agricultural pest detection. By leveraging the masked patch prediction approach, BEiT demonstrated considerable performance on image dataset for agricultural pest detection. BEiT holds significant promise for a wide range of real-world vision applications. Despite advancements, challenges like dataset quality, model generalization, and interpretability remain. The incorporation of multi-modal data, including spectral and thermal imaging, could further improve pest detection accuracy. The integration of real-time monitoring and sensor networks could enhance the practicality of deep learning-based pest detection systems. While challenges and limitations remain, ongoing research and development in the field are likely to address these issues and further refine BEiT's capabilities, solidifying its position as a powerful tool for image understanding and analysis. Deep learning has significantly improved the accuracy and efficiency of pest detection from images. From CNN-based models to transformer-based architectures like ViTs and BEiT, researchers continue to innovate, addressing challenges, and enhancing interpretability. As the field advances, the integration of diverse data sources and real-time monitoring systems holds the potential to revolutionize pest detection in agriculture.

References

1. Ebrahimi, M.A., Khoshtaghaza, M.H., Minaei, S., Jamshidi, B.: Vision-based pest detection based on SVM classification method. Comput. Electron. Agric. **137**, 52–58 (2017)

2. Venkatasaichandrakanth, P., Iyapparaja, M.: Pest detection and classification in peanut crops using CNN, MFO, and EViTA algorithms. IEEE Access (2023)
3. Sunitha, G., Madhavi, K.R., Avanija, J., Reddy, S.T.K., Vittal, R.H.S.: Modeling convolutional neural network for detection of plant leaf spot diseases. In: 2022 3rd International Conference on Electronics and Sustainable Communication Systems (ICESC), pp. 1187–1192. IEEE (2022)
4. Zha, M., Qian, W., Yi, W., Hua, J.: A lightweight YOLOv4-Based forestry pest detection method using coordinate attention and feature fusion. Entropy **23**(12), 1587 (2021)
5. Prabhakar, T., Srujan Raju, K., Reddy Madhavi, K.: Support vector machine classification of remote sensing images with the wavelet-based statistical features. In: Proceedings of Fifth International Conference on Smart Computing and Informatics, pp. 603–613. Springer (2022)
6. Liu, L., et al.: PestNet: an end-to-end deep learning approach for large-scale multi-class pest detection and classification. IEEE Access **7**, 45301–45312 (2019)
7. Chen, J.W., Lin, W.J., Cheng, H.J., Hung, C.L., Lin, C.Y., Chen, S.P.: A smartphone-based application for scale pest detection using multiple-object detection methods. Electronics **10**(4), 372 (2021)
8. Sunitha, G., Sudeepthi, A., Sreedhar, B., Shaik, A.B., Farooq, C.L.: RetinaNet and vision transformer-based model for wheat head detection. In: 2023 5th International Conference on Inventive Research in Computing Applications (ICIRCA), pp. 151–156. IEEE (2023)
9. Suma, K.G., Aswini, J., Sunitha, G., Balaji, K.: AI-driven applications in high-tech agriculture. In: Handbook of Research on AI-Equipped IoT Applications in High-Tech Agriculture, pp. 23–37. IGI Global (2023)
10. Rustia, D.J.A., et al.: Automatic greenhouse insect pest detection and recognition based on a cascaded deep learning classification method. J. Appl. Entomol. **145**(3), 206–222 (2021)
11. Nennuri, R., Kumar, R.H., Prathyusha, G., Tejaswini, K., Kanishka, G., Sunitha, G.: A multi-stage deep model for crop variety and disease prediction. In: International Conference on Soft Computing and Pattern Recognition, pp. 52–59. Springer Nature Switzerland, Cham (2022)
12. Dou, Z.Y., et al.: An empirical study of training end-to-end vision-and-language transformers. In: Proceedings of the IEEE/CVF Conference on Computer Vision and Pattern Recognition, pp. 18166–18176 (2022)
13. Shereesha, M., Hemavathy, C., Teja, H., Reddy, G.M., Kumar, B.V., Sunitha, G.: Precision mango farming: using compact convolutional transformer for disease detection. In International Conference on Innovations in Bio-Inspired Computing and Applications, pp. 458–465. Springer Nature Switzerland, Cham (2022)
14. Devika, K.D., Saxena, A.: Dynamic Authentication Using Visual Cryptography. In: Satapathy, S.C., Lin, J.CW., Wee, L.K., Bhateja, V., Rajesh, T.M. (eds.) Computer Communication, Networking and IoT. Lecture Notes in Networks and Systems, vol/ 459. Springer, Singapore (2023). https://doi.org/10.1007/978-981-19-1976-3_20
15. Maithri, Y., Saxena, A.: Browser extension for digital signature. In: Satapathy, S.C., Lin, J.CW., Wee, L.K., Bhateja, V., Rajesh, T.M. (eds.) Computer Communication, Networking and IoT. Lecture Notes in Networks and Systems, vol 459. Springer, Singapore (2023). https://doi.org/10.1007/978-981-19-1976-3_21
16. Singh, A., Tiwari, V., Tentu, A.N., Saxena, A.: Securing communication in IoT environment using lightweight key generation-assisted homomorphic authenticated encryption. In: Satapathy, S.C., Lin, J.CW., Wee, L.K., Bhateja, V., Rajesh, T.M. (eds.) Computer Communication, Networking and IoT. Lecture Notes in Networks and Systems, vol 459. Springer, Singapore (2023). https://doi.org/10.1007/978-981-19-1976-3_26

17. Srujan Raju, K., Jagtap, V., Kulkarni, P., Varaprasad Rao, M. (2023). Gameplay Cognitive Decision Support Using Statistical and Non-statistical Parametric Fusion. In: Satapathy, S.C., Lin, J.CW., Wee, L.K., Bhateja, V., Rajesh, T.M. (eds.) Computer Communication, Networking and IoT. Lecture Notes in Networks and Systems, vol. 459. Springer, Singapore. https://doi.org/10.1007/978-981-19-1976-3_24
18. Kumar Apat, S., Mishra, J., Srujan Raju, K., Padhy, N.:. IoT-Assisted crop monitoring using machine learning algorithms for smart farming. In: Kumar, R., Pattnaik, P.K., R. S. Tavares, J.M. (eds.) Next Generation of Internet of Things. Lecture Notes in Networks and Systems, vol. 445. Springer, Singapore (2023). https://doi.org/10.1007/978-981-19-1412-6_1
19. Folle, L., et al.: DeepNAPSI multi-reader nail psoriasis prediction using deep learning. Sci. Rep. **13**(1), 5329 (2023)
20. Agricultural Pests Dataset: https://www.kaggle.com/datasets/gauravduttakiit/agricultural-pests-dataset. Accessed May 2023

Enhancing Image Captioning Through a Hybrid Deep Model: The ResNet-LSTM Approach

M. Lavanya[1], B. Rupa Devi[2], Sreenivasulu Gogula[3(✉)], Kuntrapakam Divya[4], Madhavi Gudavalli[5], and Shaik Mohammad Rafee[6]

[1] Department of CSE(AIML), Sri Venkateswara College of Engineering and Technology, Chittoor, A.P., India

[2] Dept of CSE, Annamacharya Institute of Technology and Sciences, Tirupati, A.P., India

[3] Department of CSE(Data Science), Vardhaman College of Engineering, Hyderabad, India
aceaimldept@gmail.com

[4] Department of CSE, School of Computing, Mohan Babu University (Erstwhile Sree Vidyanikethan Engineering College), Tirupati, A.P., India

[5] Department of IT, JNTUK, Kakinada, A.P., India

[6] Department of AI and ML, Sasi Institute of Technology and Engineering, Tadepalligudem, A.P., India

Abstract. The task of image captioning plays a crucial role in enabling machines to describe visual content accurately and coherently. Deep models have the potential to facilitate applications as a bridge between visual content and natural language. A novel hybrid deep learning model for enhancing automated image captioning using ResNet-LSTM approach has been designed in this research. The proposed hybrid model combines the power of residual networks and long short-term memory networks to generate descriptive captions for images. To evaluate the performance of proposed model, extensive experiments were conducted and compared it to several single-stage image captioning models. Notable models included VGG16, ResNet50, and M-RNN. The results, as measured by BLEU-1, BLEU-2, BLEU-3, and BLEU-4 scores, demonstrated that the proposed hybrid model consistently outperformed the other models. Specifically, the proposed model achieved higher scores across all BLEU metrics, indicating hybrid model's proficiency in generating contextually relevant captions. By integrating the feature extraction capabilities of ResNet with the sequential learning abilities of LSTM, the quality of generated captions was enhanced. These findings underscore the potential of hybrid models in bridging the gap between visual content and natural language. Deep learning is a transformative technology for developing applications to content understanding in diverse domains.

Keywords: Hybrid deep model · Image captioning · Content understanding · Accessibility solutions · Deep learning

1 Introduction

In the era of rapidly advancing technology, the fusion of computer vision and natural language processing has given rise to a transformative capability - automated image captioning. This innovative approach empowers machines to perceive visual content

A. Bajaj et al. (Eds.): IBICA 2023, LNNS 1229, pp. 275–283, 2025.
https://doi.org/10.1007/978-3-031-78940-3_28

and express it in the form of meaningful textual descriptions [1]. It not only opens up new horizons in human-computer interaction but also offers immense potential across a diverse array of applications [2].

The significance of automated image captioning is twofold. Firstly, it addresses fundamental accessibility challenges by making visual content inclusive to individuals with visual impairments. For these individuals, textual descriptions of images serve as windows into a world that would otherwise remain concealed. Secondly, it plays a pivotal role in content understanding across a multitude of domains, including healthcare, e-commerce, and content recommendation systems [3, 4]. By endowing machines with the ability to comprehend and describe visual scenes, automated image captioning transcends the boundaries of traditional computer vision.

Yet, the path to effective automated image captioning is fraught with challenges. Capturing the intricate details, contextual nuances, and salient features of images in order to describe them in human language requires the model to not only "see" but also "understand". Ambiguities, diverse image styles, and the need for coherent and contextually relevant descriptions further compound these challenges [5].

Deep learning models, particularly neural networks, are uniquely equipped to tackle the complexity of image captioning [6]. These models, inspired by the human brain's architecture, can autonomously extract hierarchical and meaningful features from images. Simultaneously, they excel in the sequential generation of text, making them ideal candidates for bridging the gap between visual perception and natural language expression.

A novel hybrid deep learning model using ResNet-LSTM approach is proposed in this research. This model integrates two powerful neural network architectures to tackle image captioning comprehensively. This dual-stage architecture harnesses the strengths of both ResNet and LSTM. The challenges of accurate image perception and context-aware text generation are both addressed by the proposed hybrid model.

2 Related Work

Image captioning represents automated generation of contextual descriptions of images. A remarkable progress in recent years in this area. Early image captioning systems often relied on hand-crafted features and rule-based methods. These approaches, while pioneering, faced limitations in handling the complexity and diversity of real-world images [7]. They struggled to capture nuanced details and context.

Deep learning has brought transformative advancements in the way machines generate descriptive and contextually relevant captions for images. The combination of CNNs and RNNs enabled end-to-end learning for image captioning. CNNs are exceptionally good at extracting hierarchical and meaningful features from images. RNNs, with their sequential and recurrent nature, are well-suited for generating text one word at a time. Models like Neural Image Caption (NIC) and Show-Observe-Tell demonstrated the power of end-to-end deep learning in generating more contextually relevant captions [8, 9]. Attention mechanisms, inspired by human visual perception, became a focal point of research [10]. Models such as "Adaptive Attention" and "Up-Down" incorporated attention to focus on relevant regions of an image while generating captions. These mechanisms improved the generation of detailed and coherent descriptions.

Research extended beyond single-modal image captioning to include multimodal approaches that incorporate additional modalities such as audio and text [11]. Furthermore, there has been a surge in multilingual image captioning models, enabling cross-lingual understanding of visual content [12–14]. The integration of pre-trained language models into image captioning pipelines has demonstrated the potential of leveraging contextual embeddings and cross-modal transfer learning. This approach enhances the richness and accuracy of generated captions.

3 Proposed Methodology

A complete and detailed architecture for the proposed hybrid deep model for automated image captioning is presented in this section. The proposed model uses a two-stage approach with ResNet for feature extraction in Stage 1 and LSTM for caption generation in Stage 2.

3.1 Stage 1: Feature Extraction with ResNet

In the first stage, a ResNet is used for feature extraction from images. ResNet is a popular choice due to its excellent performance in image-related tasks [15]. A ResNet model, which has been pre-trained is used. The layers and their configurations for Stage 1 are as follows.

Input Layer:

a) Input Size: Variable, depending on the input image dimensions (224×224 pixels for standard ImageNet pre-trained models).
b) Input Channels: 3 (for RGB images).

ResNet Base Model:

a) Base Architecture: Pre-trained ResNet.
b) Pre-trained Weights: Loaded from a pre-trained model.
c) Input: RGB Image.
d) Output: Feature Map with Spatial Dimensions ($7 \times 7 \times$).

Global Average Pooling Layer:

a) Purpose: Reduce the spatial dimensions and convert the feature map into a fixed-size feature vector.
b) Output: 2048-dimensional feature vector.

Feature Extraction Model (Stage 1 Output):

a) Output: The final output of Stage 1 is a fixed-size feature vector of dimension 2048. Output represents the high-level features extracted from the input image.

3.2 Stage 2: Caption Generation with LSTM

In the second stage, an LSTM network as the caption generation module. This LSTM-based model will take the feature vector from Stage 1 as input and generates captions word by word. The layers and their configurations for Stage 2 are as follows.

Input Layer (Stage 2):

a) Input Size: 2048 (dimension of the feature vector from Stage 1).

Word Embedding Layer:

a) Purpose: Convert words into dense vectors (embeddings) that the model can work with.
b) Vocabulary Size: The size of the vocabulary in the dataset.
c) Output: Word embeddings of a fixed dimension (300).

LSTM (Long Short-Term Memory) Cells:

a) Purpose: Generate captions word by word.
b) LSTM Units: Configurable (512 units).
c) Input: Word embeddings from the Word Embedding Layer.
d) Output: Hidden states and cell states for each time step.

Dense Layer and Softmax Activation:

a) Purpose: Predict the next word in the caption sequence.
b) Input: LSTM hidden states.
c) Output: Probability distribution over the vocabulary (a vector of vocabulary size).

Caption Generation Model (Stage 2 Output):

a) Output: The final output of Stage 2 is a sequence of words representing the image caption. This sequence is generated word by word, and the model repeats the LSTM cell for each time step until maximum caption length is reached.

The hybrid model (ResNet-LSTM) takes an input image, extracts features using Stage 1, and generates captions using Stage 2. The caption generation process continues until predefined maximum caption length is reached. This comprehensive architecture combines the strengths of ResNet for feature extraction and LSTM for sequential data generation to produce descriptive captions for input images. The hybrid approach ensures that the model first captures relevant image features and then generates contextually relevant captions (Fig. 1).

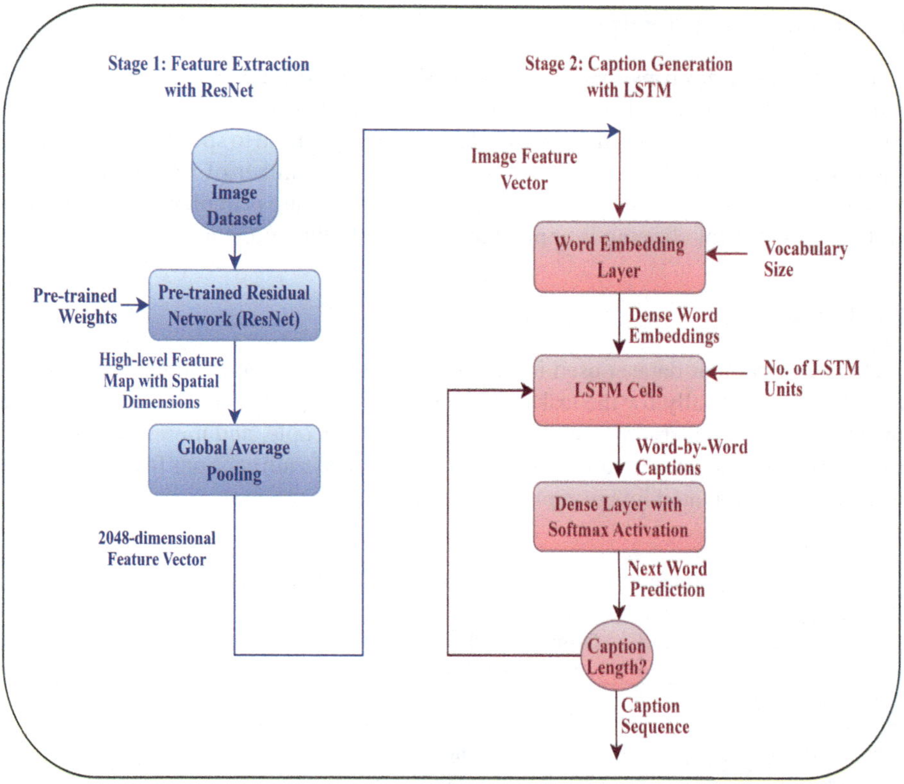

Fig. 1. Hybrid Deep Model using ResNet + LSTM Approach for Image Captioning

4 Experimental Results and Discussion

4.1 Model Configuration

The proposed hybrid model consists of feature extraction model (stage-1) and caption generation model (stage-2). The input image is passed through stage-1. Stage-1 comprises of pre-trained ResNet model with weights learned from the ImageNet dataset. Given an image as input, the ResNet model extracts high-level features and generates a feature map with spatial dimensions. Global average pooling is applied to the feature map. Global pooling reduces spatial dimensions and converts it into a fixed-size 2048-D feature vector. This feature vector serves as the input for stage-2, where caption generation using LSTM takes place. Word embedding layer understands the semantics of caption words and generates good caption sequences. It converts discrete words into dense word embeddings of a fixed dimension 300. BERT is used for word embedding layer [16]. The LSTM Cells generate captions word by word. The number of LSTM units are configured to be 512. Caption generation model outputs a caption sequence. Words in the caption sequence are predicted by the dense layer with softmax activation. Caption sequence is generated word by word. LSTM cell repeats process until maximum caption length is generated.

4.2 Training and Optimization

Categorical Cross-Entropy loss is used as loss function for sequence prediction. Adam optimizer with an adjustable learning rate is used for hyper-parameter optimization. The training strategy called as teacher forcing during training is used to speed up convergence. The batch size is configured to be 32 and 64. Epochs: Number of training epochs are configured from 30–50 based on convergence. Early stopping is implemented to prevent overfitting. Gradient clipping is applied to avoid exploding gradients.

4.3 Dataset

Flickr8k is a popular dataset used in the field of computer vision and natural language processing, specifically for the task of image captioning [17]. The dataset contains 8K images. It includes everyday scenes, animals, objects, people, and more. Each image is associated with human annotations which serve as ground truth captions for experimental evaluation of automated image captioning models.

Table 1. Sample Outputs of Proposed Hybrid Model

Image	Ground Truth Caption Sequences	Predicted Caption Sequence
	✓ Dogs playing in sand at beach ✓ Two dogs playing with each other in sand at beach ✓ Two brown dogs playing at the beach	Two dogs running on sand
	✓ Child and woman near water lake in a park at the city ✓ Two people watching duck in water ✓ Several people at park and two people near lake	Two people near water and trees

4.4 Experimental Results

Table 1 shows the sample source images, source image captions and the caption sequence predicted by the ResNet-LSTM hybrid model. The evaluation of proposed hybrid model is performed using Bilingual Evaluation Understudy (BLEU) metric. It evaluates the quality of predicted caption sequences. It measures the similarity between a predicted caption sequences and ground truth caption sequences. It computes a similarity score based on the overlap of n-grams between predicted and ground truth caption sequences (Table 2).

Table 2. Comparative Evaluation of Proposed Hybrid Model

Deep Model	BLEU-1	BLEU-2	BLEU-3	BLEU-4
VGG16	55.80	35.20	24.00	17.70
ResNet50	54.40	39.80	26.00	19.60
M-RNN	52.18	32.09	29.17	24.19
ResNet50-LSTM (Proposed Model)	57.12	42.76	37.20	29.46

The proposed hybrid model is evaluated against traditional single-stage image captioning deep models -M-RNN, VGG16, ResNet50 [18]. The VGG16 model achieves competitive results, with a relatively high BLEU-1 score, indicating that it generates captions that contain a good amount of overlap with the ground truth caption sequences. However, as the n-gram order increases, BLEU scores decrease. This scenario suggests that VGG16 struggles with generating longer and more detailed captions. The ResNet50 model performs well, especially in terms of BLEU-2, indicating that it captures meaningful word pairs in its captions. However, like VGG16, it faces challenges with higher-order n-grams. The M-RNN model has a high BLEU-1 score, but its performance decreases significantly as the n-gram order increases. This suggests that while it generates reasonable single words or short phrases, it struggles with generating coherent and matching longer phrases. Proposed model, which combines ResNet50 for image feature extraction with LSTM for caption generation, outperforms the other models across all BLEU metrics. It achieves the highest scores, indicating that it generates captions that better match the ground truth caption sequences, in terms of individual words and in terms of longer phrases.

Overall, based on these results, the proposed hybrid model using ResNet50 + LSTM approach appears to be most effective than the single-stage models for image captioning, as it consistently achieves higher BLEU scores across different n-gram orders. These results suggest that proposed hybrid model generates more accurate and contextually relevant captions for images.

5 Conclusion

In this research, a hybrid deep learning model using ResNet-LSTM approach for image captioning is investigated. The two-stage architecture is designed emphasizing the role of ResNet for feature extraction and LSTM for caption generation. The experimentation results yielded compelling evidence of the hybrid model's superiority over traditional single-stage models such as VGG16, ResNet50, and M-RNN. Proposed model consistently achieved higher BLEU scores across all n-gram orders, demonstrating its proficiency in generating descriptive and contextually relevant captions for images. This enhanced performance is attributed to the combination of powerful image feature extraction capabilities from ResNet and sequential text generation capabilities from LSTM. The success of the ResNet-LSTM approach underscores the significance of hybrid models in addressing the challenges of image captioning. Deep learning is a transformative technology for developing applications to content understanding in diverse domains. Further exploration and refinement of hybrid architectures hold the potential to push the boundaries of machine understanding of visual content, bringing us closer to more effective human-computer interaction and content interpretation. Future research may focus on generating captions that provide more fine-grained and detailed descriptions of images. Cross-modal understanding, where information from one modality enhances understanding in another, will be a critical area of exploration.

References

1. Suresh, K.R., Jarapala, A., Sudeep, P.V.: Image captioning encoder–decoder models using CNN-RNN architectures: a comparative study. Circuits Systems Signal Process **41**(10), 5719–5742 (2022)
2. Chohan, M., et al.: Image Captioning using Deep Learning: A Systematic. Image **11**(5) (2020)
3. Tiwary, T., Mahapatra, R.P.: An accurate generation of image captions for blind people using extended convolutional atom neural network. Multimedia Tools and Applications **82**(3), 3801–3830 (2023)
4. Liu, H., Wang, G., Huang, T., He, P., Skitmore, M., Luo, X.: Manifesting construction activity scenes via image captioning. Autom. Constr. **119**, 103334 (2020)
5. Karthikeyan, C.: Prediction of climate change using SVM and naïve Bayes machine learning algorithms. Turkish J. Comp. Maths. Educ. (TURCOMAT) **12**(2), 2134–2139 (2021)
6. Elhagry, A., Kadaoui, K.: A thorough review on recent deep learning methodologies for image captioning (2021). arXiv preprint arXiv:2107.13114
7. Devika, K.D., Saxena, A.: Dynamic authentication using visual cryptography. In: Satapathy, S.C., Lin, J.CW., Wee, L.K., Bhateja, V., Rajesh, T.M. (eds.) Computer Communication, Networking and IoT. Lecture Notes in Networks and Systems, vol 459. Springer, Singapore (2023). https://doi.org/10.1007/978-981-19-1976-3_20
8. Maithri, Y., Saxena, A.: Browser Extension for Digital Signature. In: Satapathy, S.C., Lin, J.CW., Wee, L.K., Bhateja, V., Rajesh, T.M. (eds.) Computer Communication, Networking and IoT. Lecture Notes in Networks and Systems, vol 459. Springer, Singapore (2023). https://doi.org/10.1007/978-981-19-1976-3_21
9. Singh, A., Tiwari, V., Tentu, A.N., Saxena, A.: Securing Communication in IoT Environment Using Lightweight Key Generation-Assisted Homomorphic Authenticated Encryption. In: Satapathy, S.C., Lin, J.CW., Wee, L.K., Bhateja, V., Rajesh, T.M. (eds.) Computer Communication, Networking and IoT. Lecture Notes in Networks and Systems, vol 459. Springer, Singapore (2023). https://doi.org/10.1007/978-981-19-1976-3_26
10. Srujan Raju, K., Jagtap, V., Kulkarni, P., Varaprasad Rao, M.: Gameplay Cognitive Decision Support Using Statistical and Non-statistical Parametric Fusion. In: Satapathy, S.C., Lin, J.CW., Wee, L.K., Bhateja, V., Rajesh, T.M. (eds.) Computer Communication, Networking and IoT. Lecture Notes in Networks and Systems, vol 459. Springer, Singapore (2023). https://doi.org/10.1007/978-981-19-1976-3_24
11. Kumar Apat, S., Mishra, J., Srujan Raju, K., Padhy, N.: IoT-assisted crop monitoring using machine learning algorithms for smart farming. In: Kumar, R., Pattnaik, P.K.R.S. Tavares, J.M. (eds.) Next Generation of Internet of Things. Lecture Notes in Networks and Systems, vol 445. Springer, Singapore (2023). https://doi.org/10.1007/978-981-19-1412-6_1
12. Muhammad Shah, F., Humaira, M., Jim, M.A.R.K., Saha Ami, A., Paul, S.: Bornon: Bengali image captioning with transformer-based deep learning approach. SN Computer Science **3**, 1–16 (2022)

13. Sunitha, G., et al.: Speech Recognition based Assistant Bee: Vision to the Impaired. In: 2022 4th International Conference on Inventive Research in Computing Applications (ICIRCA), pp. 1535–1540. IEEE (2022)
14. Kannan, K.S., Sunitha, G., Deepa, S.N., Babu, D.V., Avanija, J.: A multi-objective load balancing and power minimization in cloud using bio-inspired algorithms. Comput. Electr. Eng. **102**, 108225 (2022)
15. Sowmya, T.S., Narasimhulu, T., Sunitha, G., Manikanta, T., Venkatesh, T.: Vision Transformer based ResNet Model for Pneumonia Prediction. In: 2023 4th International Conference on Electronics and Sustainable Communication Systems (ICESC), pp. 316–321. IEEE (2023)
16. Devlin, J., Chang, M.W., Lee, K., Toutanova, K.: Bert: Pre-training of deep bidirectional transformers for language understanding (2018). arXiv preprint arXiv:1810.04805
17. Flickr8k Dataset for image captioning, https://www.kaggle.com/datasets/adityajn105/flickr8k
18. Mao, J., Xu, W., Yang, Y., Wang, J., Huang, Z., Yuille, A.: Deep captioning with multimodal recurrent neural networks (m-rnn) (2014). arXiv preprint arXiv:1412.6632

Solid Waste Classification Using Transformer Model

K. Ayyappa Swamy[1(✉)], G. M. Madhavi[2], K. Reddy Madhavi[3], Naresh Tangudu[4], C. V. Krishnaveni[5], and S. Hema Chandra[6]

[1] Department of ECE, Aditya Engineering College, Surampalem, India
ayyappa.kondru@gmail.com
[2] Amity University, Bengaluru, India
[3] School of Computing, Mohan Babu University, Tirupati, India
[4] Aditya Institute of Technology and Management, Tekkali, India
[5] SKR and SKR GCW(A), Kadapa, India
[6] Sanskrithi School of Engineering, Puttaparthi, India

Abstract. Waste classification is a critical component of modern waste management systems. The fusion of cutting-edge technology with the ever-pressing issue of waste management holds great promise for a sustainable and cleaner future. Integration of transformer models in waste management systems can transform the landscape of solid waste classification and redefine the way of managing waste streams. This research paper contributes to this evolving field by presenting a comprehensive investigation into the utility of transformer models for solid waste classification. This paper addresses the challenge of solid waste classification using advanced deep learning models and presents a transformer-based architecture specifically tailored for this task. The importance of accurate waste classification in reducing environmental impact, optimizing recycling, and complying with regulations is emphasized. In experimental comparisons against established models, including Simple CNN, ResNet101, MobileNet, and DenseNet121, the proposed transformer model showcased superior performance, achieving a remarkable accuracy of 96.1%. This performance highlights the potential of self-attention mechanisms in effectively capturing intricate waste features. While other models performed well, the Transformer model's results underscore its significance in waste classification.

Keywords: Waste management systems · Solid waste · Vision transformers · Smart city · Deep Learning

1 Introduction

This is an era characterized by rapid urbanization and increased industrialization. The management of solid waste has become a paramount concern for communities and governments worldwide [1]. Solid waste comprises of a diverse range of materials. It presents significant environmental, health, and economic challenges. Traditional waste management practices often rely on manual sorting and limited automation [2]. This process makes it labor-intensive, error-prone, and unsustainable in the face of burgeoning waste volumes.

A. Bajaj et al. (Eds.): IBICA 2023, LNNS 1229, pp. 284–291, 2025.
https://doi.org/10.1007/978-3-031-78940-3_29

Efficient waste management hinges on the accurate categorization of solid waste. Proper classification enables separation of waste items into different categories such as recyclables, non-recyclables, hazardous waste, non-hazardous waste, organic materials, inorganic materials etc. [3]. Accurate classification is crucial for optimizing recycling efforts. Thus, reducing landfill waste and minimizing environmental impact. Moreover, it is essential for complying with regulations and ensuring public safety [4, 5].

Historically, waste classification has predominantly relied on human judgment, which is subject to inconsistency and fatigue. Manual sorting, despite its longstanding presence, is not scalable to meet the growing demands of modern waste management systems [6]. These challenges have paved the way for the integration of advanced technology and artificial intelligence into waste classification processes.

In this research paper, a novel approach is presented to address the pressing issue of solid waste classification. The proposal is made by harnessing the power of transformer models. Transformer models, originally designed for natural language processing tasks, have exhibited remarkable capabilities in capturing complex patterns and relationships within non-sequential data. By adapting these models to the realm of waste classification, the aim is to revolutionize the way waste is sorted and managed.

2 Literature Review

Waste management is an interdisciplinary field [7]. It has garnered significant attention in recent years due to its environmental, economic, and societal implications. The challenge of efficiently sorting and classifying solid waste materials has prompted researchers to explore innovative AI solutions.

Traditional methods for waste classification have primarily relied on manual sorting and human expertise [8]. These techniques, although widely practiced, are labor-intensive and prone to errors. Human factor introduces subjectivity and inconsistency in waste classification processes.

Early efforts to automate waste classification incorporated machine learning algorithms [9]. Researchers applied machine learning methods to classify waste materials. The process is based on features of waste materials like color, shape, and texture [10]. These approaches showed promise in improving the efficiency of automated waste detection systems.

Accuracy of automated systems improved significantly with the advent of deep learning [11, 12]. Deep models demonstrated exceptional performance in image-based waste classification tasks [13, 14]. Notable studies employed CNNs for tasks like identifying recyclable materials from images. These successes highlighted the potential of AI-driven solutions for waste management.

Transformer models gained prominence for their ability to capture complex patterns in non-sequential data [15]. Recent research has extended the application of transformers to various domains beyond language, including computer vision and tabular data [16–18]. The adaptability of transformers suggests their suitability for tackling the intricacies of waste classification tasks. Some recent studies have explored the integration of transformer models for various classification scenarios. These works have demonstrated the potential of transformers to learn intricate relationships among data. The current research builds upon these emerging trends, focusing on the specific application of transformer models for solid waste classification.

3 Transformer Model for Solid Waste Classification

The research work undertaken in this paper is to investigate the efficiency of vision transformer models for classifying solid waste materials [18]. The transformer-based architecture for the undertaken task is presented in Fig. 1. Mathematical model of the proposed transformer-based model is presented below.

Input Data: Let N represent batch size, HxW represent image size, C represent number of channels. Let X represent the input data where $X \in R^{N \times H \times W \times C}$.

Positional Encodings: Learnable positional encodings help the model understand the spatial relationships between pixels. Positional encodings are added to X to incorporate spatial information. Let $P \in R^{H \times W \times C}$ denote the positional encodings.

Embedding Layer: This layer processes input images to extract meaningful features. The convolutional layer with ReLU activation captures local patterns. Let $E(X) \in R^{N \times H' \times W' \times C'}$ represent the output of the embedding layer. H', W', C' are the dimensions after embedding.

Transformer Encoder Blocks: The core of the transformer architecture consists of multiple encoder blocks. Let T represent this layer's function. Each encoder block consists of the following:

a) Multi-Head Self-Attention Layer: This layer allows the model to capture both local and global dependencies. Given $E(X)$, let Q, K, V be linear projections of $E(X)$. Let h be the number of attention heads. The attention weights A^i are computed for each head i. Output is concatenated and projected to obtain the output.
b) Layer Normalization (Sub-Layer 1): It is applied to output from previous layer.
c) Feed-Forward Neural Network Layer: Feed-forward network with ReLU processes output from attention layer, capturing complex feature interactions. An ReLU activation function is embedded between two linear transformations.
d) Layer Normalization (Sub-Layer 2): Layer normalization is applied to the feed-forward output.
e) Residual Connections: These are added to each sub-layer's output.

Multi-Head Self-Attention Mechanism: This mechanism computes scaled dot-product attention for each attention head i.

$$A_i(Q_i, K_i, V_i) = softmax\left(\frac{Q_i K_i^T}{\sqrt{d_k}}\right) V_i \tag{1}$$

Outputs from multiple heads are concatenated and projected to obtain the self-attention output.

Position-wise Feed-Forward Neural Network: This network applies two linear transformations to each position independently.

$$FFN(X) = ReLU(XW_1 + b_1)W_2 + b_2 \tag{2}$$

where W_1, b_1, W_2, b_2 are learnable parameters.

Global Average Pooling Layer: Pooling is applied on output of transformer blocks to obtain a fixed-size representation of the image features. Let G represent this layer's function.

Output Layer: This layer produces the final classification probabilities for each solid waste category. As the solid waste classification is a multi-class classification problem, softmax function is used at this layer. Let O represent this layer's function.

The forward pass of the model can be represented as:

$$Z_1 = E(X) \tag{3}$$

$$Z_2 = Z_1 + P \tag{4}$$

$$Z_3 = T(Z_2) \tag{5}$$

$$Z_4 = G(Z_3) \tag{6}$$

$$Output = O(Z_4) \tag{7}$$

The potential benefits of transformer-based model are manifold. A highly accurate waste classification model powered by transformers could significantly enhance the efficiency of waste management operations, reduce errors, and facilitate recycling initiatives. Furthermore, it can contribute to a cleaner environment by minimizing the improper disposal of hazardous materials.

4 Experimental Results and Evaluation

A comprehensive TrashNet dataset encompassing various waste categories is employed [19]. A transformer architecture tailored to the unique characteristics of waste classification is designed. Through rigorous experimentation and evaluation, performance of the proposed model is assessed for accurately categorizing solid waste. Table 1 presents the design of configuration of the proposed transformer-based model for solid waste classification.

Figure 2 shows the images of correctly classified and misclassified samples by the proposed transformer model. In comparison to traditional deep learning approaches (Simple CNN, ResNet101, MobileNet, DenseNet121) proposed solid waste classification transformer model demonstrates enhancement of nearly 3% in classification accuracy, showcasing its superior performance (Table 2). Traditional deep learning-based methods rely on manually engineered image features. Hence, these approaches

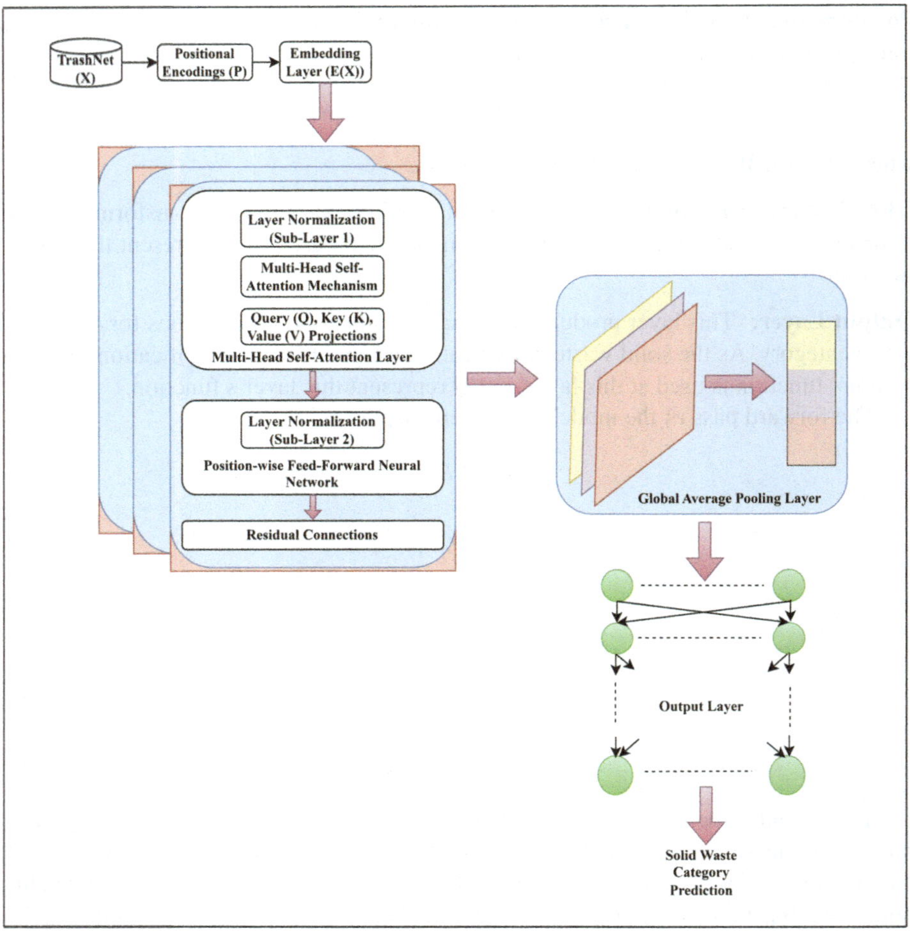

Fig. 1. Proposed Transformer-based Model

often struggle to effectively represent the intricate features found in waste images. Shallow classifiers also face challenges in efficiently realizing classification within feature spaces. Consequently, these traditional methods frequently yield suboptimal classification accuracy, limiting their practical utility in real-world waste classification scenarios.

Simple CNN model, while performing reasonably well with an accuracy of 87.65%, fell short in comparison to the Transformer model. This suggests that more complex architectures are beneficial for capturing intricate waste features. ResNet101 and DenseNet121 achieved competitive results, with accuracies of 91.57% and 93.76%, respectively. These models excel in feature extraction due to their deep architectures. MobileNet, while offering computational efficiency, exhibited a lower accuracy of 89.34%. This suggests that in tasks where feature richness is essential, more lightweight models may struggle to match the performance of larger architectures. The

Table 1. Proposed Model Configuration

Layer	Hyperparameter	Configuration
Embedding Layer	Type	2D Convolutional Layer
	No. of Filters	64
	Kernel Size	(3, 3)
	Activation Function	ReLU
Positional Encoding	Type	Learnable Positional Encoding
Transformer Encoder Blocks	No. of Blocks	6
Multi-Head Self-Attention Layer	No. of Heads	8
Feed-Forward Neural Network Layer	Hidden Units	2048
	Activation Function	ReLU
Output Layer	Type	Dense
	No. of Units	6
	Activation Function	softmax

Table 2. Performance of Deep Models on TrashNet

Model	Overall Accuracy	Precision	Recall
Simple CNN	87.65	86.2	88.5
ResNet101	91.57	89.1	90.5
MobileNet	89.34	83.2	85.8
DenseNet121	93.76	90.5	91.8
Proposed transformer model	96.10	91.7	93.2

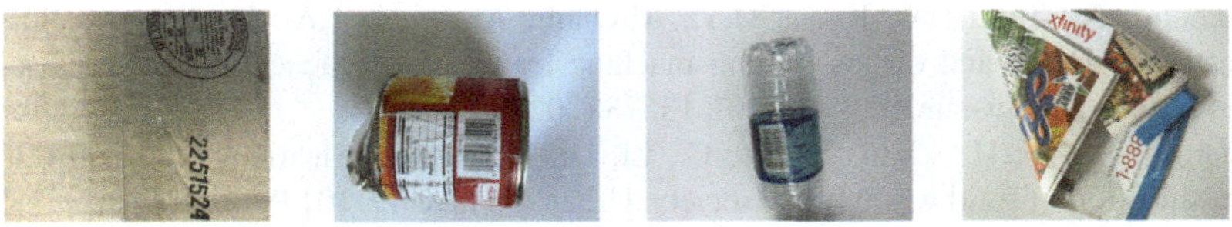

(a) Correct Classification as Paper (b) Correct Classification as Metal (c) Correct Classification as Plastic (d) Misclassification as Metal

Fig. 2. Classification Samples of the Proposed Transformer Model

Proposed Transformer Model outperformed all other models with an accuracy of 96.1%. This demonstrates the effectiveness of leveraging self-attention mechanisms for feature extraction and classification in the context of solid waste images.

5 Conclusions

Accurate solid waste classification is imperative for effective waste management, environmental sustainability, and regulatory compliance. This research introduces a transformer-based model tailored for waste classification and demonstrates its superiority over established models. The experimental results illustrate the importance of advanced deep learning architectures in capturing complex waste features, thereby improving classification accuracy. The proposed transformer model's success underscores its potential for revolutionizing waste classification and management. However, future research endeavors should focus on expanding datasets, optimizing model deployment for practical scenarios, and enhancing model interpretability.

In conclusion, the integration of advanced deep learning models in waste classification is a promising avenue for sustainable waste management, with far-reaching implications for environmental conservation and resource optimization. While AI-driven waste classification holds promise, it is not without challenges. Robustness to variations in waste appearance, scalability to handle large datasets, and interpretability of model decisions remain open research questions. Additionally, the social and ethical implications of automating waste management processes require careful consideration.

References

1. Xia, W., Jiang, Y., Chen, X., Zhao, R.: Application of machine learning algorithms in municipal solid waste management: a mini review. Waste Manage. Res. **40**(6), 609–624 (2022)
2. Guo, H.N., Wu, S.B., Tian, Y.J., Zhang, J., Liu, H.T.: Application of machine learning methods for the prediction of organic solid waste treatment and recycling processes: a review. Biores. Technol. **319**, 124114 (2021)
3. Lu, W., Chen, J.: Computer vision for solid waste sorting: a critical review of academic research. Waste Manage. **142**, 29–43 (2022)
4. Shahab, S., Anjum, M., Umar, M.S.: Deep learning applications in solid waste management: a deep literature review. Int. J. Adv. Comp. Sci. Appl. **13**(3) (2022)
5. Namoun, A., Tufail, A., Khan, M.Y., Alrehaili, A., Syed, T.A., BenRhouma, O.: Solid waste generation and disposal using machine learning approaches: a survey of solutions and challenges. Sustainability **14**(20), 13578 (2022)
6. Zhang, H., Cao, H., Zhou, Y., Gu, C., Li, D.: Hybrid deep learning model for accurate classification of solid waste in the society. Urban Climate **49**, 101485 (2023)
7. Joshi, L.M., Bharti, R.K., Singh, R.: Internet of things and machine learning-based approaches in the urban solid waste management: Trends, challenges, and future directions. Expert. Syst. **39**(5), e12865 (2022)
8. Shahab, S., Anjum, M.: Solid waste management scenario in India and illegal dump detection using deep learning: an AI approach towards the sustainable waste management. Sustainability **14**(23), 15896 (2022)
9. Charan, N.S., et al.: Solid Waste Management Using Deep Learning. In International Conference on Soft Computing and Pattern Recognition, pp. 44–51. Springer Nature Switzerland, Cham (2022)
10. Lu, W., Huo, W., Gulina, H., Pan, C.: Development of machine learning multi-city model for municipal solid waste generation prediction. Front. Environ. Sci. Eng. **16**(9), 119 (2022)

11. Murthy, D.S., et al.: An Investigative Study of Shallow, Deep and Dense Learning Models for Breast Cancer Detection based on Microcalcifications. In 2022 International Conference on Data Science, Agents & Artificial Intelligence (ICDSAAI), Vol. 1, pp. 1–6. IEEE (2022)
12. Devika, K.D., Saxena, A.: Dynamic authentication using visual cryptography. In: Satapathy, S.C., Lin, J.CW., Wee, L.K., Bhateja, V., Rajesh, T.M. (eds.) Computer Communication, Networking and IoT. Lecture Notes in Networks and Systems, vol 459. Springer, Singapore (2023). https://doi.org/10.1007/978-981-19-1976-3_20
13. Maithri, Y., Saxena, A.: Browser extension for digital signature. In: Satapathy, S.C., Lin, J.CW., Wee, L.K., Bhateja, V., Rajesh, T.M. (eds.) Computer Communication, Networking and IoT. Lecture Notes in Networks and Systems, vol 459. Springer, Singapore (2023). https://doi.org/10.1007/978-981-19-1976-3_21
14. Singh, A., Tiwari, V., Tentu, A.N., Saxena, A.: Securing Communication in IoT Environment Using Lightweight Key Generation-Assisted Homomorphic Authenticated Encryption. In: Satapathy, S.C., Lin, J.CW., Wee, L.K., Bhateja, V., Rajesh, T.M. (eds.) Computer Communication, Networking and IoT. Lecture Notes in Networks and Systems, vol 459. Springer, Singapore (2023). https://doi.org/10.1007/978-981-19-1976-3_26
15. Srujan Raju, K., Jagtap, V., Kulkarni, P., Varaprasad Rao, M.: Gameplay cognitive decision support using statistical and non-statistical parametric fusion. In: Satapathy, S.C., Lin, J.CW., Wee, L.K., Bhateja, V., Rajesh, T.M. (eds.) Computer Communication, Networking and IoT. Lecture Notes in Networks and Systems, vol 459. Springer, Singapore (2023). https://doi.org/10.1007/978-981-19-1976-3_24
16. Kumar Apat, S., Mishra, J., Srujan Raju, K., Padhy, N.: IoT-assisted crop monitoring using machine learning algorithms for smart farming. In: Kumar, R., Pattnaik, P.K., R.S. Tavares, J.M. (eds.) Next Generation of Internet of Things. Lecture Notes in Networks and Systems, vol 445. Springer, Singapore (2023). https://doi.org/10.1007/978-981-19-1412-6_1
17. Sowmya, T.S., Narasimhulu, T., Sunitha, G., Manikanta, T., Venkatesh, T.: Vision transformer based ResNet model for pneumonia prediction. In: 2023 4th International Conference on Electronics and Sustainable Communication Systems (ICESC), pp. 316–321. IEEE (2023)
18. Sunitha, G., Sudeepthi, A., Sreedhar, B., Shaik, A.B., Farooq, C.: RetinaNet and Vision Transformer-Based Model for Wheat Head Detection. In: 2023 5th International Conference on Inventive Research in Computing Applications (ICIRCA), pp. 151–156. IEEE (2023)
19. Shereesha, M., et al.: Precision mango farming: using compact convolutional transformer for disease detection. In: International Conference on Innovations in Bio-Inspired Computing and Applications, pp. 458–465. Springer Nature Switzerland, Cham (2022)

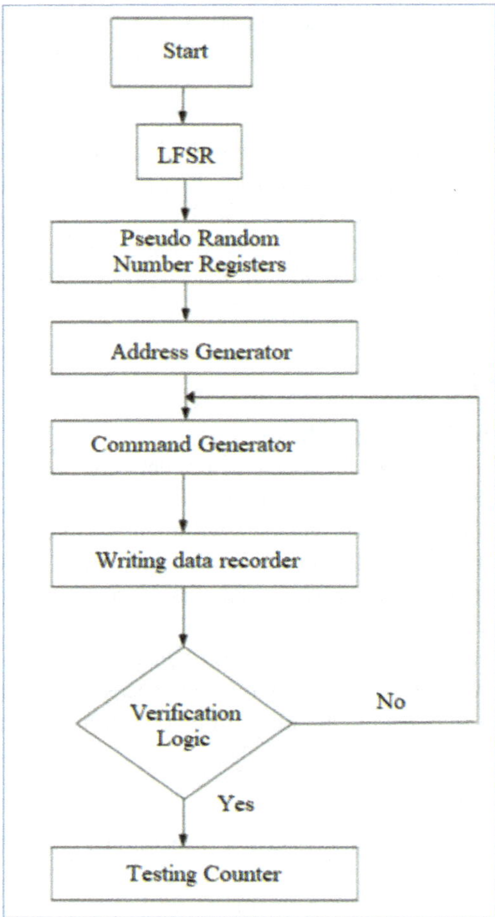

Fig. 1. Flow control diagram of Random Testing Circuits with LFSR Technique.

Table 1. Comparison of Parameters

Parameter	LFSR Testing circuit	LFSR Based Random Testing Circuit
Errors	94%	9%
Delay	93%	18%
Area	97.45%	19%
Speed	12%	97.42%

Figure 3 presents a contrast between area and speed. As this graphic illustrates, the LFSR-based random testing circuit efficiently increases speed while decreasing area.

Enhancing Image Quality: Geometric Correction with CNN + STN Model

J. Aswini[1(✉)], A. Revathi[2], Ramayanam Jhansi Rani[3], Kuntrapakam Divya[4], T. Ravi Kumar[5], and A. Basi Reddy[6]

[1] Department of AI and ML, Saveetha Engineering College, Affiliated to Anna University, Chennai, India
aswini.jayaraman@gmail.com

[2] Department of CSE(DS), Sri Venkateswara College of Engineering and Technology, Chittoor, India

[3] Department of Computer Applications, Chadalawada Ramanamma Engineering College, Tirupati, India

[4] Department of CSE, School of Computing, Mohan Babu University (Erstwhile Sree Vidyanikethan Engineering College), Tirupati, India

[5] Department of CSE, Aditya Institute of Technology and Management, Tekkali, India

[6] Department of Computer Science and Engineering, School of Computing, Mohan Babu University, Tirupati, India

Abstract. The presence of geometric distortions in real-world images is pervasive and arises from various sources, including camera misalignment, viewpoint changes, and lens imperfections. These distortions can lead to misalignment of objects, making it challenging for deep learning models to accurately identify and classify them. Geometric correction is a foundational task in computer vision, aimed at rectifying spatial distortions and misalignments within images. This paper presents a novel CNN + STN model tailored for geometric correction and explores its efficacy in rectifying spatial distortions and misalignments within images. Through extensive experimentation, CNN + STN model's performance is rigorously evaluated on synthetic dataset. Experimental results reveal its ability to enhance image quality, accurately register visual data, and improve the performance of downstream computer vision tasks. The integration of CNN + STN promises to unlock new frontiers in computer vision applications, solidifying its position as an indispensable asset in the computer vision toolkit.

Keywords: Image quality · Geometric distortions · Geometric corrections · Computer Vision · Deep learning

1 Introduction

In the realm of computer vision and image processing, accurate and reliable geometric correction of images is an essential task with profound implications for a wide range of applications [1]. Geometric distortions, such as noise, rotation, scaling, translation, and perspective variations, can degrade the quality of images and adversely affect the

A. Bajaj et al. (Eds.): IBICA 2023, LNNS 1229, pp. 292–299, 2025.
https://doi.org/10.1007/978-3-031-78940-3_30

performance of deep learning models designed for tasks such as image classification, object detection, and scene analysis [2]. Addressing these distortions is crucial to harness the full potential of computer vision systems.

The presence of geometric distortions in real-world images is pervasive and arises from various sources, including camera misalignment, viewpoint changes, and lens imperfections [3]. These distortions can lead to misalignment of objects, making it challenging for deep learning models to accurately identify and classify them. Consequently, there is a pressing need for robust solutions that not only mitigate the effects of noise and distortions but also enhance the interpretability and accuracy of computer vision systems [4].

In response to this challenge, this paper proposes an effective solution for geometric correction, centered around a Convolutional Neural Network with a Spatial Transformer Network (CNN + STN). This model is designed to automatically the geometrical distortions in images, thus facilitating superior performance of deep learning models. By integrating the STN into the architecture, proposed approach enables the network to learn and apply spatial transformations that correct geometric distortions while preserving essential features of the underlying objects.

The significance of our proposed CNN + STN model lies in its potential to substantially improve the performance of deep learning systems operating on real-world image data. Through precise geometric correction, the accuracy, robustness, and generalization capabilities of these models are enhanced. The objective is to open up new horizons for a wide array of computer vision applications.

2 Literature Review

Geometric correction is a fundamental task in computer vision, aimed at rectifying spatial distortions and misalignments within images [5]. These distortions can arise from a variety of sources, including camera perspective, lens imperfections, and physical distortions during image capture. Effective geometric correction is essential for ensuring that computer vision algorithms can accurately analyze and interpret visual data.

Geometric correction has a rich history dating back to early image processing techniques. Classical methods, such as affine transformations and homography, were employed to rectify spatial distortions [6]. These methods played a pivotal role in applications like aerial imagery analysis and cartography.

With the advent of deep learning, the field of computer vision witnessed a transformation [7–9]. Researchers began to leverage CNNs to learn geometric transformations directly from data [10, 11]. Notable contributions include the Spatial Transformer Network (STN) which allows networks to learn and apply spatial transformations such as rotation, scaling, and translation [12, 13]. This marked a significant shift toward data-driven geometric correction methods.

STNs can be applied to improve the performance of UAVs in autonomous navigation tasks. STNs could be used to adaptively adjust the camera perspective, correct for distortion, or transform images to better suit the needs of object detection, obstacle avoidance, and path planning in real-time UAV applications [14]. This research could lead to more efficient and accurate UAV navigation systems, with potential applications in fields such as agriculture, surveillance, and search and rescue operations.

Geometric correction finds applications across various domains such as Satellite and aerial imagery, medical imaging, cultural heritage preservation, robotics, augmented reality etc. [15–17]. Despite significant advancements, several challenges persist. Addressing non-rigid distortions, where objects deform or change shape, remains a research challenge. Achieving real-time geometric correction for applications like robotics and augmented reality requires efficient algorithms. Methods to train geometric correction models with limited annotated data are an ongoing focus area.

Geometric correction is a critical component of computer vision. The transition from classical methods to data-driven approaches has significantly improved the accuracy and adaptability of geometric correction [18]. Ongoing research continues to address challenges and extend the capabilities of these methods, making geometric correction an indispensable tool in the realm of computer vision.

3 CNN + STN Model for Enhancing Image Quality

Geometric correction is a crucial step in computer vision and image processing, aiming to rectify spatial misalignments and distortions within images. These distortions can arise from various sources, including camera misalignment, perspective changes, and lens aberrations [19]. The integration of deep learning models, specifically the CNN + STN architecture, has emerged as a powerful solution to address these challenges.

Architectural description of the CNN + STN model for enhancing the quality of aerial imagery is presented in this section. This description outlines the purpose of each stage and how the proposed model improves the quality of the imagery.

1. **Input:** Let I represent the input aerial image with dimensions $W \, x \, H$ pixels and C color channels $I \in R^{H \times W \times C}$.
2. **Stage 1 - Convolutional Neural Network (CNN) Feature Extraction:** The first stage of the model is the CNN feature extraction stage. This stage captures relevant visual features from the raw aerial imagery. It helps in reducing noise, enhancing distinguishable patterns, and preparing the data for subsequent geometric correction.
   a. Input: The input to this stage is the raw aerial image, which may contain geometric distortions, noise, and other artifacts.
   b. CNN Layers: The CNN consists of multiple convolutional layers. Each layer is responsible for detecting different patterns and features at various levels of abstraction. These layers apply convolutional filters, batch normalization, and ReLU activation functions. For each convolutional layer i, N_i convolutional filters of size K_i $x \, K_i$ with a stride of S_i are applied. The output of each convolutional layer is given

$$O_i = \text{ReLU}(W_i \, \text{x} \, O_{i-1} + b_i) \tag{1}$$

   where $O_{i\text{-}1}$ is output of previous layer, W_i are the learnable convolutional kernels, b_i are biases.
   c. Pooling Layers: These layers downsample the feature maps and reduce spatial dimensions. This helps in abstracting higher-level features and reducing computation.

d. Flatten Layer: Feature maps are flattened into a 1D vector F creating a representation of the extracted features.

$$F = ReLU(W_f.Flatten(O_k) + b_f) \quad (2)$$

where O_k is the output of last pooling layer, W_f is the weight matrix, b_f is the bias vector.

3. **Stage 2 - Spatial Transformer Network (STN) for Geometric Correction:** The second stage of the model is the STN. It corrects geometric distortions and aligns the imagery with real-world coordinates.
   e. Input: The input to this stage is the flattened feature vector F obtained from the CNN feature extraction stage.
   f. Localization Network (LocNet): Let W_l be the weight matrix and b_l be the bias vector. LocNet learns transformation parameters Θ, including scaling factors (s_x, s_y), a rotation angle θ, and translation offsets (t_x, t_y), which describe the geometric correction needed. The output of the LocNet is

$$\Theta = (s_x, s_y, \theta, t_x, t_y) = Activation(W_l.Flatten(I) + b_l) \quad (3)$$

   g. Grid Generator: Using the learned transformation parameters Θ, a grid is generated to define how pixels in the input feature map should be sampled. This grid accounts for scaling, rotation, and translation. The grid is represented by a set of points (x_i, y_i) that define how the pixels in the input feature map should be sampled. The grid generation can be expressed by following equations.

$$x_i = s_x.x + t_x \quad (4)$$

$$y_i = s_y.y + t_y \quad (5)$$

These equations define how the pixel coordinates in the input feature map are transformed to obtain the corrected feature map.

   h. Sampler: It uses the generated grid to intermediate feature map F. from the flattened feature vector, producing a spatially corrected feature map. The sampling operation is implemented using bilinear interpolation. The output is corrected feature map $F_{transformed}$.

$$F_{transformed}(x, y, c) = \Sigma_{ij} I(i, j, c).weight(x - i, y - j) \quad (6)$$

where $(x\text{-}i, y\text{-}j)$ are the bilinear interpolation weights based on the grid coordi nates.

4. **Objective Function:** The objective function is to optimize weight vectors and bias vectors of CNN feature extractor and the STN, as well as a geometric correction loss term.
5. **Output:** The output of the CNN + STN model is the spatially transformed feature map $F_{transformed}$ which contains features from the input feature map that have been geometrically corrected or transformed according to the learned transformation parameters.

4 Experimental Results and Evaluation

In this section, experimental methodology and results of the study are presented. The experimentation is aimed at evaluating the impact of geometric correction on the classification performance of various deep learning architectures. First, creation of a synthetic dataset with artificially introduced noise is described. Subsequently, the experimental setup, results performed on the noisy dataset is outlined. Then application of CNN + STN model for geometric correction is discussed. The experimentation is repeated on the geometrically corrected dataset and results are analyzed. Finally, comprehensive analysis of the results is provided.

4.1 Creation of Synthetic Noisy Dataset

To assess the effectiveness of geometric correction, a synthetic dataset is created containing images with geometric distortions. Two types of geometric distortions are performed on the original images - rotation, scaling. Geometric distortions are applied on each original image to generate four synthetic images. Also, original images are included in the synthetic dataset. This synthetic dataset simulates real-world scenarios where images may suffer from distortions during acquisition. Figure 1 shows sample images with and without noise.

(a) Original Image (b) Scaled and Rotated Image

Fig. 1. Sample Images With and Without Geometric Distortions

4.2 Classification on Geometrically Distorted Dataset

Classification experiments on the geometrically distorted dataset are performed using three popular deep learning architectures: ResNet101, EfficientNetB2 and DenseNet121. Each architecture was trained on the geometrically distorted dataset, and the performance was assessed in terms of classification accuracy.

4.3 Geometric Correction Using CNN + STN

To address the geometric distortions, present in the synthetic image dataset, CNN + STN model is proposed. This model incorporates a CNN which learns feature maps from noisy images and STN performs geometric correction. The STN network learns to align and register objects in the images correctly (Fig. 2 and Table 1).

Table 1. Experimental Results

Input	Model	Accuracy (%)
Image Dataset with Noise	ResNet101	85.23
	EfficientNetB2	88.65
	DenseNet121	91.49
Geometrically Corrected Image Dataset	ResNet101	89.78
	EfficientNetB2	92.86
	DenseNet121	95.62

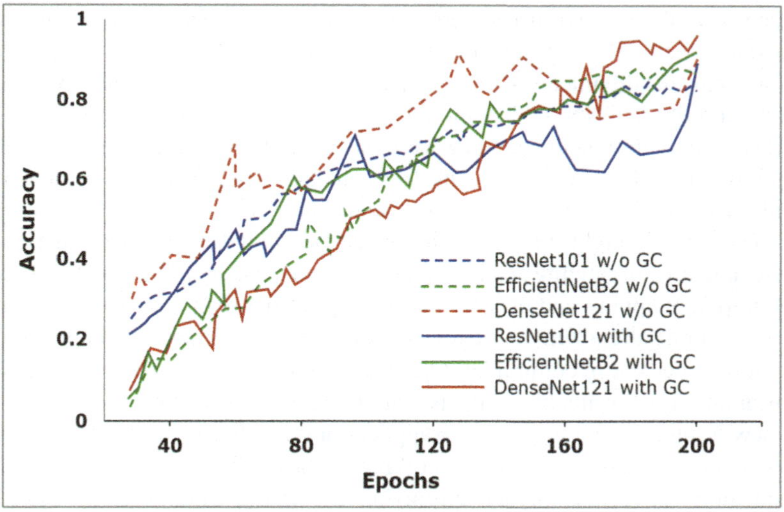

Fig. 2. Performance Evaluation of Deep Models Before and After Geometric Correction

4.4 Performance Comparison

Significant improvements in classification accuracy are observed on the geometrically corrected dataset compared to the geometrically distorted dataset for all three deep learning architectures. The geometric correction process effectively handled distortions, allowing the models to focus on relevant features, resulting in enhanced classification performance.

4.5 Impact of Geometric Correction

The geometric correction also had the added benefit of making the deep learning models more robust to variations in object orientation and position. This improved robustness is essential for real-world applications where images may exhibit geometric distortions due to camera perspective or other factors.

Experimental results demonstrated the effectiveness of geometric correction using the proposed CNN + STN model in improving the classification performance of deep learning architectures. The correction process not only enhanced accuracy but also increased the robustness of the models. This research underscores the importance of addressing geometric distortions in image datasets, particularly when deploying deep learning models for classification tasks in real-world scenarios. The findings have practical implications for various computer vision applications, including object recognition, image classification, and scene analysis.

5 Conclusions

In this study, a comprehensive exploration of the CNN + STN model's capabilities are presented in the context of geometric correction in computer vision. This research demonstrates that CNN + STN model, offers a powerful solution to rectify spatial distortions and misalignments within images. The experimental results paint a compelling picture of the CNN + STN model's effectiveness. Consistent improvements in image quality and alignment accuracy were observed through experimental evaluation of proposed model. These enhancements translate into tangible benefits for downstream computer vision tasks, where image quality and spatial registration are pivotal.

This research highlights the versatility of deep learning for enhancing image quality in diverse domains. In satellite and aerial imagery analysis, deep learning facilitates precise geographic mapping and disaster monitoring. In the realm of medical imaging, it aids in the accurate alignment of images from different modalities, enhancing diagnostic capabilities. Moreover, in cultural heritage preservation, it contributes to the restoration and conservation of invaluable artifacts. The integration of CNN + STN promises to unlock new frontiers in computer vision applications, solidifying its position as an indispensable asset in the computer vision toolkit. Yet the challenges that persist, including addressing non-rigid distortions and optimizing real-time processing are to be researched upon. Ongoing research endeavors aim to tackle these issues, driving the evolution of geometric correction techniques.

References

1. Shan, S., et al.: Distortion-corrected image reconstruction with deep learning on an MRI-Linac. Magnetic Resonance in Medicine (2023)
2. Li, X., Zhang, B., Sander, P.V., Liao, J.: Blind geometric distortion correction on images through deep learning. In: Proceedings of the IEEE/CVF Conference on Computer Vision and Pattern Recognition, pp. 4855–4864 (2019)
3. Nakamura, Y., et al.: Possibility of deep learning in medical imaging focusing improvement of computed tomography image quality. J. Comput. Assist. Tomogr. **44**(2), 161–167 (2020)
4. Charan, N.S., et al.: Solid Waste Management Using Deep Learning. In: International Conference on Soft Computing and Pattern Recognition, pp. 44–51. Springer Nature Switzerland, Cham (2022)
5. Sowmya, T.S., Narasimhulu, T., Sunitha, G., Manikanta, T., Venkatesh, T.: Vision Transformer based ResNet Model for Pneumonia Prediction. In: 2023 4th International Conference on Electronics and Sustainable Communication Systems (ICESC), pp. 316–321. IEEE (2023)

6. Shereesha, M., et al.: Precision Mango Farming: Using Compact Convolutional Transformer for Disease Detection. In: International Conference on Innovations in Bio-Inspired Computing and Applications, pp. 458–465. Springer Nature Switzerland, Cham (2022)
7. Mohan, E., Rajesh, A., Sunitha, G., Konduru, R.M., Avanija, J., Ganesh Babu, L.: A deep neural network learning-based speckle noise removal technique for enhancing the quality of synthetic-aperture radar images. Concurrency and Computation: Practice and Experience **33**(13), e6239 (2021)
8. Finnveden, L., Jansson, Y., Lindeberg, T.: Understanding when spatial transformer networks do not support invariance, and what to do about it. In: 2020 25th International Conference on Pattern Recognition (ICPR), pp. 3427–3434. IEEE (2021)
9. Chianucci, D., Savakis, A.: Unsupervised change detection using spatial transformer networks. In: 2016 IEEE Western New York Image and Signal Processing Workshop (WNYISPW), pp. 1–5. IEEE (2016)
10. Apat, S.K., Mishra, J., Srujan Raju, K., Padhy, N.: State of the art of ensemble learning approach for crop prediction. In: Kumar, R., Pattnaik, P.K.R.S., Tavares, J.M. (eds.) Next Generation of Internet of Things. Lecture Notes in Networks and Systems, vol 445. Springer, Singapore (2023). https://doi.org/10.1007/978-981-19-1412-6_58
11. Raju, V.R., Reddy, D.A., Narsimha, D., Srinivas, K., Rani, B.K.: Adaptive closed-loop deep brain stimulator coding techniques for target detections in Parkinson's. IETE J. Res. **69**(8), 4945–4960 (2023). https://doi.org/10.1080/03772063.2021.1962742
12. Fayaz, R., et al.: An Intelligent Harris Hawks Optimization (IHHO) based Pivotal Decision Tree (PDT) Machine Learning Model for Diabetes Prediction. Int. J. Intel. Sys. Appl. Eng. **10**(4), 415–423 (2022)
13. Priya Remamany, K., et al.: A Localized Bloom Filter-Based CP-ABE in Smart Healthcare. Applied Sciences **12**(24), 12720 (2022). https://doi.org/10.3390/app122412720
14. Rani, B.K., Rao, M.V., Patra, R.K., et al.: Vehicle type classification using graph ant colony optimizer based stack autoencoder model. Multimed Tools Appl **81**, 42163–42182 (2022). https://doi.org/10.1007/s11042-021-11508-5
15. Munawar, S., Geetha, A., Srinivas, K.: Squirrel Search-based Optimal Feature Extraction with Bi-LSTM for the Arrhythmia Classification using ECG. J. Theoret. Appl. Info. Technol. 100 (2022)
16. Pothalaiah, S., Lakshmaiah, D., Doss, B., Sairam, N., Srikanth, K.: Design of CMOS Base Band Analog. Cognitive Computing Models in Communication Systems, 123 (2022)
17. Lakshmaiah, D., et al.: A Novel Low-Power Frequency-Modulated Continuous Wave Radar Based on Low-Noise Mixer. Cognitive Computing Models in Communication Systems, 165 (2022)
18. Hemanand, D., et al.: An intelligent intrusion detection and classification system using CSGO-LSVM model for wireless sensor networks (WSNs). Int. J. Intel. Sys. Appl. Eng. **10**(3), 285–293 (2022)
19. Natesan, G., Konda, S., de Prado, R.,P., Wozniak, M.: A hybrid mayfly-aquila optimization algorithm based energy-efficient clustering routing protocol for wireless sensor networks. Sensors **22**(17), 6405 (2022). https://doi.org/10.3390/s22176405

An Effective Netb0-CNN Transfer Learning Model for Traffic Sign Recognition and Safety

Suraya Mubeen[1(✉)], Attuluri Uday Kiran[1], P Raveendra Babu[2], A. Pradeep Kumar[2], Madhavi Pingili[3], and Bommireddy Prasanthi[3]

[1] CMR Technical Campus, Hyderabad, Telangana, India
cmrtc.paper@gmail.com, suraya418@gmail.com
[2] CMR College of Engineering and Technology, Hyderabad, Telangana, India
{praveendrababu,drapradeepkumar}@cmrcet.ac.in
[3] CMR Engineering College, Hyderabad, Telangana, India
prashanthi.veerla@cmrec.ac.in

Abstract. For effective traffic management and road safety, traffic sign recognition is essential. Here, we provide a novel method for traffic sign recognition that makes use of the potent EfficientNetB0-CNN transfer learning model. This study fine-tunes EfficientNetB0, which is well-known for its remarkable performance in picture classification tasks, with a tailored Convolutional Neural Network (CNN) for traffic sign identification. The training dataset for the model includes a wide range of traffic signs that are subjected to different lighting situations, weather fluctuations, and road surroundings. The model can use pre-trained weights thanks to the transfer learning process, which speeds up convergence and boosts performance. The model is extensively tested and its hyper parameters are fine-tuned in order to maximize its precision and effectiveness. The outcomes reveal that the suggested EfficientNetB0-CNN model recognizes traffic signs with amazing accuracy and resilience, highlighting its potential for use in autonomous cars and traffic safety systems. The suggested method creates new opportunities for improving technologies for recognizing traffic signs, which will make transportation systems safer and more intelligent.

Keywords: Traffic Sign recognition · transfer learning · EfficientNetB0 · Convolutional Neural Network · traffic safety

1 Introduction

To achieve complete and successful traffic management, one must possess the ability to recognize traffic indicators [1]. As the number of automobiles on the road keeps rising, it is becoming more and more important to understand traffic signs and be able to detect them. Handmade characteristics and rule-based algorithms were the mainstays of the conventional traffic sign recognition approaches, both of which were frequently limited in their ability to handle complex scenarios and changes in the signs' appearance.

A. Bajaj et al. (Eds.): IBICA 2023, LNNS 1229, pp. 300–308, 2025.
https://doi.org/10.1007/978-3-031-78940-3_31

Traffic sign recognition has been impacted by the revolution in computer vision brought about by the advent of deep learning models in recent years [2]. Deep learning models, and specifically convolutional neural networks (CNNs), have demonstrated exceptional performance in numerous image recognition tasks, suggesting that they could be a viable solution for traffic sign detection and classification.

Traffic signs are essential for both guiding traffic and ensuring the safety of drivers, thus it is vital to properly identify the signs that are used on the roads. Traffic signs convey important information, such as speed restrictions, warnings, forbidden activities, and instructions [3, 4]. The topic of traffic sign identification offers a wide range of possible applications for deep learning, and these applications are quite diverse [5]. To start, it can be included into advanced driving assistance systems (ADAS) to provide alerts and warnings based on signs that are noticed, helping drivers to stay safe and follow traffic laws. Second, smart city infrastructure might incorporate the ability to recognize traffic signs. This improves the ability of the traffic management authority to keep an eye on and regulate the movement of vehicles.

One crucial area of research and development is the use of deep learning models for traffic sign identification. With the right application, deep learning may improve traffic management, increase road safety, and pave the way for the widespread adoption of intelligent transportation systems. The ability to properly recognize and comprehend traffic signs is crucial for realizing a future where vehicles are safer and more intelligent when operating on public roadways [6].

2 Literature Review

A novel Thin but Deep convolutional neural network architecture that can recognize traffic signs and consumes less energy was proposed by Wasif Arman Haque et al. [7]. Two novel lightweight networks that can achieve more accurate recognition with the same amount of trainable factors in the models were proposed by Jianming Zhang et al. [8]. Information is transferred from a larger, learnt model known as the teacher network to a smaller one known as the student network through information distillation. In order to increase the accuracy of interpreting traffic signs, the authors additionally include a new area in the teaching network that combines two sets of feature channels with dense connections. The student network consists of one fully linked layer and five convolutional layers in a straightforward end-to-end architecture. Jayant Mishra et al. [9] presented a successful technique for automatically identifying images of traffic signs. The four GTSRB, GTSDB, BTSC, and TSRD standard traffic sign picture files—which consist of various traffic sign images—are the primary sources used in this procedure. This strategy produces the greatest results when using the CNN model design. It facilitates the driver's ability to operate the vehicle safely. A new traffic sign collection featuring 105K street-level images from around the globe, covering 400 hand-labeled traffic sign classes in a variety of scenarios, climates, and lighting situations, was unveiled by Christian Ertler et al. [10]. The collection contains 52K completely captioned images. The authors also demonstrate how to augment the dataset with 53K partially tagged, semi-supervised images. This is the most extensive and most a diverse collection of traffic sign data. It consists of images from around the globe captioned with details about the kind of traffic sign they are.

The deep learning approach based on convolutional neural networks was proposed by Riadh Ayachi et al. [11]. A convolutional neural network needs to be trained with a large amount of data, as is common knowledge. The authors created a set of data for traffic sign recognition in order to address the issue. 10,500 images representing 73 distinct kinds of traffic signs are included in the package. The photos were shot under actual weather conditions on actual Chinese highways.

After some time, Cen Han et al. [12] developed a real-time technique based on an upgraded Faster-RCNN for the detection of minor traffic signs. To be more precise, the writers first determine the appearance of miniature traffic signs using a small area plan creator. The TSD literature was thoroughly analyzed by Chunsheng Liu et al. [13]. The analyzed detection methods are divided into five main categories by the authors: LIDAR-based, color-based, shape-based, color-and-shape-based, and machine-learning-based. The methods within each category are further subdivided into numerous subcategories to aid in understanding and summarizing the mechanics of various approaches.

The deconvolution region-based convolutional neural network (DR-CNN) was first described by Zhigang Liu et al. [14]. The output of the convolution layer is first modified by adding a deconvolution layer and a normalizing layer. A three-stage, real-time system for traffic sign recognition and classification that combines image segmentation, traffic sign detection, and classification based on the input picture was the focus of E. Sathishkumar et al.'s research [15, 16]. A color enhancement technique is applied to a photograph to eliminate red areas. Convolutional neural networks (CNN) are used to determine the content of the traffic signals that are received during the detection, classification, and identification processes. The idea behind this plan is to use an RFID-based system and an RF module to automatically turn green the traffic signal in an ambulance's path [17, 18]. Transmits the accident location right away to the main server in order for the ambulance to arrive at the scene in time, save lives, and identify the scene [19]. The primary server locates the closest ambulance to the scene of the accident and provides the emergency vehicle with the precise location of the incident [20, 21].

3 Proposed Model

Transfer learning is a successful technique for traffic sign recognition that gets around the problem of not having enough training data by using pre-trained deep neural networks. The process involves adapting a well-trained model—often trained on a sizable dataset like ImageNet—to the task of identifying traffic signs. After the model has been trained, this phase is performed. The traffic sign dataset is used to fine-tune the successive levels of the pre-trained model, which capture broad visual qualities in the early layers. Through the use of previously learned visual representations, transfer learning enables the model to efficiently extract important features from traffic sign photographs. The application of the transfer learning strategy makes this feasible. In the context of traffic sign recognition tasks, this approach improves the model's performance, reduces the training time, and addresses the problems caused by a dearth of available data.

EffcientNet-B0

Researchers employed by Google first demonstrated the EfficientNetB0 convolutional neural network (CNN) architecture in 2019. It is a part of the EfficientNet family, which was created to provide cutting-edge accuracy on image classification tasks while requiring less computational power in terms of processing computational data and model size.

EfficientNetB0's architecture consists of multiple layers that work together to extract information from input images and generate precise predictions. Let us examine each of these stages one by one:

The Layer of Input: Receiving the image and forwarding it to the next layer for additional processing is the responsibility of the input layer. The maximum size of the input layer is determined by the picture's dimensions, and it may vary based on the specific implementation.

Convolutional Layers: EfficientNetB0 consists of multiple convolutional layers, each of which convolves a collection of filters over the input image to extract features. These filters are trained to identify an extensive range of features and patterns at different scales. Depthwise separable convolutions are used in the convolutional layers of EfficientNetB0. The computational complexity is decreased as a result of these convolutions, which divide the channel-wise and spatial convolution operations.

Depthwise Separable Convolutions: Depthwise separable convolutions are a type of convolution that are more efficient than ordinary convolutions, and EfficientNetB0 uses them extensively. Pointwise convolution is used to integrate the output of depthwise convolutions by combining the results across channels using 1x1 filters. A single filter is independently applied to each input channel via depthwise convolution. This approach results in a decrease in the overall number of parameters as well as the computational cost.

Inverted Residual Blocks: EfficientNetB0 adopts inverted residual blocks as well; it gets this idea from MobileNetV2. These blocks use projection and expansion layers and consist of a sequence of convolutions in 1x1 and 3x3 sizes. The network may be able to gather more complex data because of the expansion layer, which also allows the network to have additional channels. The number of channels that are removed during the operation determines the size of the projection layer.

Pooling Layers: In order to minimize the spatial dimensions and downsample the feature maps, EfficientNetB0 includes pooling layers. This reduces the requirement for processing power while also helping to gather the most crucial information. The most common type of pooling operation is called max pooling, and it selects the biggest value from each pooling window.

Fully Connected Layers: At the end of the network, EfficientNetB0 frequently contains fully linked layers. After the feature maps have been flattened, these layers use them to do either regression or classification, depending on the task at hand. Depending on how it is implemented and what the intended result is, there may be changes to the total number of linked layers and the sizes of those layers.

Activation Functions: Activation functions add non-linearity to the network, allowing it to discover complex feature associations. Two activation functions that are frequently used in EfficientNetB0 are the softmax activation for multi-class classification, which creates a probability distribution across the classes, and the Rectified Linear Unit, or ReLU, which turns negative values to zero. These two activation functions set values that are negative to zero.EfficientNetB0 was designed with the overall objective of developing a system that balances model size, computing efficiency, and accuracy. It accomplishes this by utilizing effective pooling methods, inverted residual blocks, and depthwise separable convolutions. As so, it minimizes the number of parameters and computations required while achieving great accuracy. It can therefore be used in situations where resources are few.

2019 saw the initial release of the EfficientNetB0 convolutional neural network (CNN) model. It is a member of the EfficientNet family of models, which are made to achieve high accuracy with minimal resources in terms of computation power and parameters. The following are some of the EfficientNetB0 model's most noteworthy features:

Multiplier Scaling: Compound scaling is used by EfficientNetB0 to get higher performance. This suggests that in order to achieve the best possible balance between accuracy and efficiency, the model's depth, breadth, and resolution are all scaled together. Through scaling these dimensions, the model is able to achieve better representation power while requiring comparatively less computer power.

Level-wise Distinctive Convolution: EfficientNetB0 uses the depth-wise separable convolution as its basic building block. This sort of convolution factors the conventional convolution into two different types: a point-wise convolution and a depth-wise convolution. It maintains expressive capability while streamlining the computational process.'

Linear bottlenecks and Inverted Residuals: The model includes inverted residuals, which are frequently utilized in mobile designs. Inverted residuals enable an efficient information flow through the use of bottleneck layers, which reduce feature size prior to convolution application. The application of linear bottlenecks can help to some extent with the information bottleneck problem that may occur in mobile systems.

Squeeze-and-Excitation (SE) Blocks: In order to increase the network's representational power, EfficientNetB0 integrates SE blocks. SE blocks are able to determine channel relationships by using an adaptive weighting technique to each channel's properties. By using this method, the model may focus its attention on channels that offer more valuable data, which improves the model's overall performance (Fig. 1).

Efficient and Efficient Scaling of Model Parameters: The fifth characteristic of EfficientNetB0 aims to optimize the scaling of model parameters in order to achieve an efficiency and accuracy trade-off. To achieve the best feasible balance between the two conflicting objectives, an empirical search is conducted to estimate the scaling coefficients for depth, breadth, and resolution.

Getting Ready with Big Datasets: EfficientNetB0 performs better when pre-trained on large-scale image datasets like ImageNet. Pretraining allows the model to learn generalized visual representations, which can subsequently be improved by applying it to specific tasks with smaller datasets. Cutting Edge Performance On many benchmark datasets, EfficientNetB0 has demonstrated its ability to achieve state-of-the-art performance. It achieves a high degree of accuracy while concurrently using fewer parameters and less processing power when compared to other models with comparable performance. By combining SE blocks, inverted residuals, depth-wise separable convolutions, compound scaling, and efficient parameter scaling, EfficientNetB0 creates a robust and

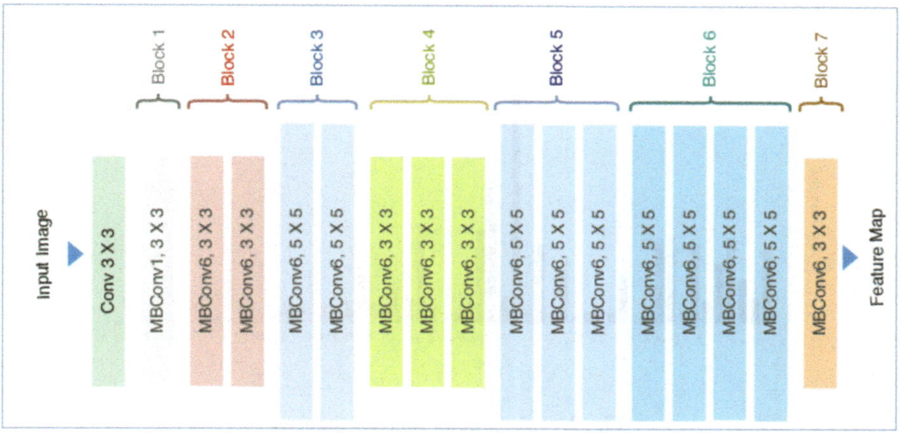

Fig. 1. Architecture of EffcientNet-B0 [17]

effective convolutional neural network model for image recognition applications. The goal of doing this is to achieve peak performance.

4 Experimental Results

This section discusses the simulation results of the suggested method. The Kaggle dataset was used. The suggested technique was used with the dataset. The collection includes several photos of traffic signs along with their descriptive text. These pictures were used to train and evaluate the suggested technique.

Labeling is the process of classifying or labeling each image in a collection based on its characteristics or content. It involves adding relevant information about the objects, traits, or concepts visible in photos, either manually or automatically. Since the labels offer a way to categorize and identify the content of the photographs, machine learning algorithms may be able to learn from the labeled data and produce predictions or classifications based on those labels. Labeling the photos in a dataset is crucial for supervised learning tasks, where the labeled data is used to train models to recognize and classify related images in the future. Figure 2 displays the labeled picture instances.

The dataset's example photos are displayed in Fig. 3. "Training accuracy" refers to how accurately the model predicts the training dataset. It assesses the model's proficiency in describing or estimating the training cases that it was exposed to during the training phase. The suggested model's test accuracy is 99.32%. Although a high training precision does not guarantee that the model will perform well on new data, it does indicate that the model matches the training data well. Validation accuracy, on the other hand, gauges how well the model's output agrees with an alternative validation dataset.

Typically, this dataset is not utilized in training. Rather, it is employed to evaluate the model's performance on newly acquired data. The model's ability to anticipate events that it has never observed before is determined by its validation accuracy. The graphical representation of training and validation accuracy is displayed in Fig. 4.

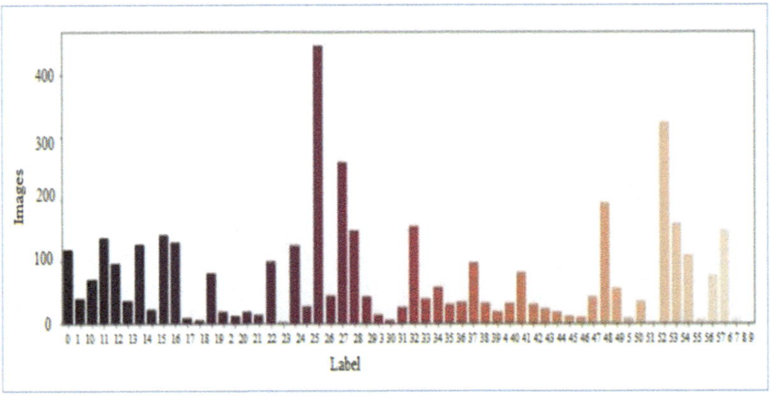

Fig. 2. Labelling of image instances

Fig. 3. Input Images

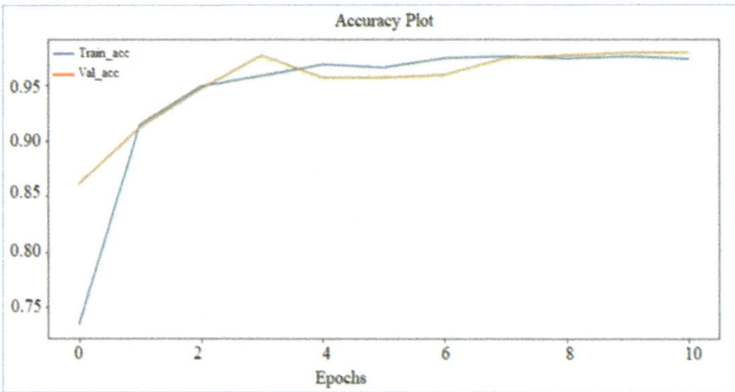

Fig. 4. Training and validation accuracy plots

5 Conclusion

Finally, this study uses the EfficientNetB0-CNN transfer learning model to provide a novel and very successful method for traffic sign recognition. With a modified CNN and the power of EfficientNetB0 as the backbone architecture, the suggested model performs remarkably well in identifying different types of traffic signs. With a great deal of trial and error, the model is improved to an exceptional 99.32% test accuracy. This astounding accuracy is evidence of the model's capacity to recognize and classify traffic signs with accuracy in a variety of real-world scenarios, such as varied lighting, weather, and road surroundings. The suggested approach's effectiveness is demonstrated by the high test accuracy of 99.32%, which suggests that traffic safety systems and autonomous cars may benefit from its use. The device can greatly improve road safety by preventing accidents and promoting effective traffic management because to its quick and precise recognition of traffic signs. With EfficientNetB0 serving as a pre-trained backbone and transfer learning applied, the model can take advantage of its extensive data set, which accelerates convergence and improves performance on the traffic sign identification job.

References

1. Srivastava, V., Mishra, S., Gupta, N.: Automatic detection and categorization of road traffic signs using a knowledge-assisted method. Procedia Computer Science **218**, 1280–1287 (2023)
2. Bi, Z., Xu, F., Shan, M., Yu, L.: YOLO-RFB: An Improved Traffic Sign Detection Model. In: International Conference on Mobile Computing, Applications, and Services, pp. 3–18. Springer International Publishing, Cham (2021)
3. Wali, S.B., et al.: Vision-based traffic sign detection and recognition systems: Current trends and challenges. Sensors **19**(9), 2093 (2019)
4. Zhu, Y., Yan, W.Q.: Traffic sign recognition based on deep learning. Multimedia Tools and Applications **81**(13), 17779–17791 (2022)
5. Tai, S.-K., Dewi, C., Chen, R.-C., Liu, Y.-T., Jiang, X., Hui, Y.: Deep learning for traffic sign recognition based on spatial pyramid pooling with scale analysis. Appl. Sci. **10**(19), 6997 (2020)
6. Alghmgham, D.A., Latif, G., Alghazo, J., Alzubaidi, L.: Autonomous traffic sign (ATSR) detection and recognition using deep CNN. Procedia Computer Science **163**, 266–274 (2019)
7. Vankdothu, R., Hameed, M.A., Ameen, A., Unnisa, R.: Brain image identification and classification on Internet of Medical Things in healthcare system using support value based deep neural network. Computers and Electrical Engineering **102**, 108196 (2022)
8. David, L.G., et al.: Tool wear monitoring using improved dragonfly optimization algorithm and deep belief network. Applied Sciences **12**(16), 8130 (2022). https://doi.org/10.3390/app12168130
9. Mehbodniya, A., Varaprasad Rao, M., David, L.G., Gerard Joe Nigel, K., Vennam, P.: Online product sentiment analysis using random evolutionary whale optimization algorithm and deep belief network. Pattern Recognition Letters, 159
10. Sunitha, D., Patra, R.K., Babu, N.V., Suresh, A., Gupta, S.C.: Twitter sentiment analysis using ensemble based deep learning model towards COVID-19 in India and European countries. Pattern Recognition Letters **158**, 164–170 (2022). ISSN 0167-8655
11. Balmuri, K.R., et al.: A long short-term memory network-based radio resource management for 5G network. Future Internet **14**(6), 184 (2022). https://doi.org/10.3390/fi14060184
12. Ahmad, S.J., Unissa, I., Shoukath Ali, M., Kumar, A.: Enhanced security to MANETs using digital codes. J. Info. Secur. Appl. **66**, 103147 (2022). ISSN 2214-2126, https://doi.org/10.1016/j.jisa.2022.103147
13. Soma, S., Sonth, M.V., Gowre, S.C.: Design of two-dimensional photonic crystal based ultra compact optical RS flip-flop. Photon Netw. Commun. **43**, 109–115 (2022). https://doi.org/10.1007/s11107-021-00955-7

14. Sudhamani, C.: Detection probability maximization through optimization of samples in cognitive radio networks. Multimed Tools Appl **81**, 12275–12285 (2022). https://doi.org/10.1007/s11042-021-11089-3
15. Chinna Rao, R., et al.: Real-Time Implementation and Testing of VoIP Vocoders with Asterisk PBX Using Wireshark Packet Analyzer. J. Interconnect. Netw. **22**(Supp. 01), 2141030 (2022)
16. Rao, M.V., Reddy, D.A., Ampavathi, A., Shaik, M.: Data mining for cyber-physical systems. Data Mining and Machine Learning Applications, pp. 235–280. https://doi.org/10.1002/9781119792529.ch10
17. Ahmed, T.: Classification and understanding of cloud structures via satellite images with EfficientUNet. SN Computer Science **3**(1) (2020)
18. Paparao Nalajala, N.M., Umarani, R.: Design of Intelligent Road Traffic Control System for Ambulance using RF and GSM Technology. Int. J. Adv. Trends in Comp. Sci. Eng. **5**(1), 70–73 (2016)
19. Madhavi, K.R., et al.: Bird species recognition using deep transfer learning. In: Reddy, K.A., Devi, B.R., George, B., Raju, K.S., Sellathurai, M. (eds.) Proceedings of Fourth International Conference on Computer and Communication Technologies. Lecture Notes in Networks and Systems, vol 606. Springer, Singapore (2023). https://doi.org/10.1007/978-981-19-8563-8_46
20. Arul, U., Prasath, A.A., Mishra, S., Shirisha, J.: IoT and Machine Learning Technology based Smart Shopping System. 2022 International Conference on Power, Energy, Control and Transmission Systems (ICPECTS), pp. 1–3. Chennai, India (2022). https://doi.org/10.1109/ICPECTS56089.2022.10047445
21. Premalatha, B., Babu, P.R., Srikanth, G.: Compact Fifth Iteration Fractal Antenna for UWB Applications. Radioelectronics and Communications Systems **64**(6), 325–329 (2021)

IoT-Based Electric Meter Application Using Convolutional Neural Networks for Electricity to Detect Cyber Attacks

G. Srikanth[1(✉)], Vootla Srisuma[1], E. N. V. Puna Chandra Rao[2], S. Muthubalaji[2], Rajesh Tiwari[3], and Radhe Shyam Panda[3]

[1] CMR Technical Campus, Hyderabad, Telangana, India
cmrtc.paper@gmail.com, gimmadisrikanth79@gmail.com
[2] CMR College of Engineering and Technology, Hyderabad, Telangana, India
muthusa15@cmrcet.ac.in
[3] CMR Engineering College, Hyderabad, Telangana, India

Abstract. Electric utility companies around the world have suffered significant financial losses as a result of the urgent problem of electricity theft. Every year, electricity worth $6 billion is stolen in the United States alone. Conventionally, physical attacks such as line tapping or meter manipulation are used to steal electricity in the consumption domain. Attacks involving electricity theft can take on new shapes thanks to the smart grid paradigm. In order to detect electricity cyber attacks, The performance of a number of deep learning algorithms is evaluated in this paper, which include convolutional neural networks (CNN), deep feedforward neural networks (DNN), and recurrent neural networks with gated recurrent units (RNN-GRU). Electricity is produced by solar panels in modern, developed countries. These customers will receive two different meters that measure production and consumption data, and they can sell their extra energy to those in need. When producing, some unscrupulous users could tamper with smart meters to increase the amount of money that can be obtained from distributed renewable energy sources and power. This attack can cost agencies a great deal of money. This initiative uses deep learning algorithms, which can identify any change and predict theft, to detect such attacks.

Keywords: Electricity theft · Cyber-attacks · IoT · deep learning · recurrent neural networks

1 Introduction

The quantity of energy used but not billed by customers is known as electricity theft. The electric utility corporations suffer significant revenue losses as a result. Electricity theft costs electric utility providers $96 billion year worldwide. Rich or poor, this problem affects all countries. For example, non-technical losses (NTLs) cost Pakistan 0.89 billion rupees a year [1], whereas India loses more than 4.8 billion rupees a year in power. Strong economies are also vulnerable to power theft; in the United States, the annual loss from

A. Bajaj et al. (Eds.): IBICA 2023, LNNS 1229, pp. 309–317, 2025.
https://doi.org/10.1007/978-3-031-78940-3_32

electricity theft is estimated to be $6 billion, and in the United Kingdom, it might reach £175 million. Furthermore, by overloading the transformers, electricity theft can disrupt power system operations and result in a voltage imbalance [2]. One of the most evident problems with both conventional and smart power systems nowadays is electric power loss. According to the data, between 1980 and 2000, distribution and transmission losses went from 11% to 16%. The amount of electricity lost varies per nation. 6%, 10%, 16%, and 18% of the world's energy production was lost by the US, Russia, Brazil, and India, correspondingly [3, 4]. Smart meters are a key component of smart grids that measure the amount of electricity lost.

2 Literature Review

In order to calculate the missing instances in the dataset based on the local values associated with the missing data point, a novel data pre-processing method was created in this study [5]. A novel CNN-RF model for detecting electricity theft was presented by Li et al. [7]. In this paradigm, the RF serves as the output classifier, while the CNN analyzes smart meter data similarly to an automatic feature extractor. During training, a fully connected layer with a dropout rate of 0.4 is constructed to mitigate the possibility of overfitting, which can be exacerbated by the adjustment of many parameters. While most other traditional classifiers primarily rely on the laborious and time-consuming process of obtaining high-quality hand-designed features, the hybrid model is capable of [8] automatically extracting features. A new model is presented by Khan et al. [9] and is based on actual electricity use data and supervised machine learning techniques. The first step in pre-processing the electricity data is to use interpolation, the three-sigma rule, and normalizing techniques. Using actual daily electricity consumption data [10] developed deep learning techniques. In order to extract meaningful results and make the information more useful, a novel approach of data reduction has been developed.

To ensure that all development facts are correct for transpositions, a chance to be tested must come before any improvements from requesting additional circles. This needs to be tested to prevent mismatches and transmission losses in telephone lines [11]. As a result, a brand-new automatic fault finder that adheres to the same criteria has been created [12]. It automatically logs the location of managed, intermittent electricity line problems [13]. This work will use radio frequency (RF) communication technology to establish a connection between the electrical board's (EB) office and industry/house energy meters [14]. The meters will measure electrical energy consumption and transmit the bill amount to the industry/house [15]. If the bills are not paid on time, the EB will disconnect the industry/house electrical power line. Once the bills are paid, the bill will reconnect the connection without entering the industry/house area [16].

3 Proposed Method

Information about the quantity of electricity utilized by each customer is included in this dataset. The dates are listed in columns, and the consumers are listed in rows. The power consumption for the year 2015 is included in this dataset.

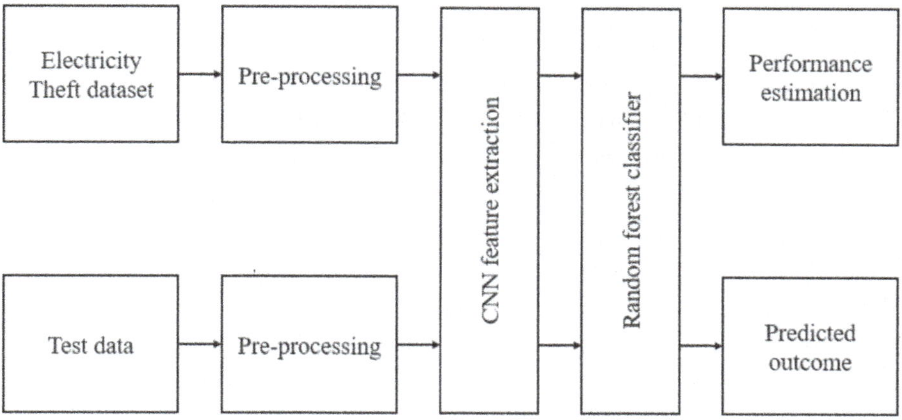

Fig. 1. System proposed model

Each source data set is fed into a series of convolution layers using a kernel or filter for ReLU, max pooling, fully connected layer, CNN training, and testing. Following that, the Soft Max layer and the classification layer are integrated to classify the objects using probabilistic values ranging from as shown in Fig. 1.

The primary layer, known as the convolution layer, uses tiny blocks of raw data to learn the properties of the image in order to extract features and maintain the link between pixels. The source image $(x,,d)$ is one of the two inputs that this mathematical function considers. The spatial coordinates, or the total number of rows and columns, are represented by the letters x and y. Figure 2 shows the correlation between an input image's having size of (d = 3) as the original picture is RGB) and the size of the output image (a filter or kernel).

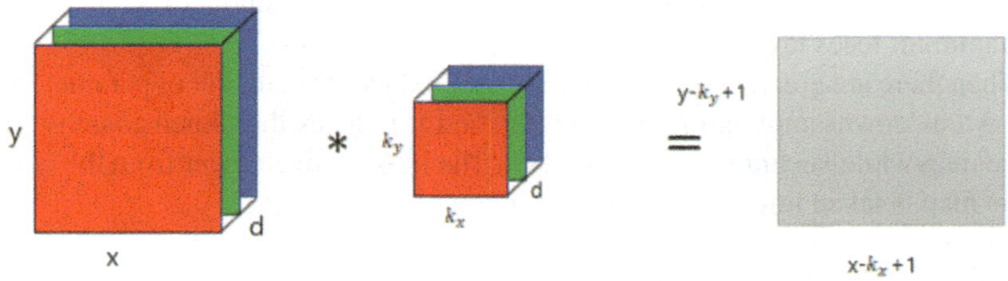

Fig. 2. Convolution layer process illustration.

Feature maps are produced when the input image and filter are convolutioned, yielding an output with a size of $((x - kx + 1),(y - ky + 1),1)$. Assume for now that the filter has a 3×3 size and the input image is 5×5. As illustrated in Fig. 3, the input image values are multiplied by the filter values to produce the feature map of the input picture.

ReLU layer

Repaired linear units (ReLUs) are networks that use the rectifier operation for the hidden

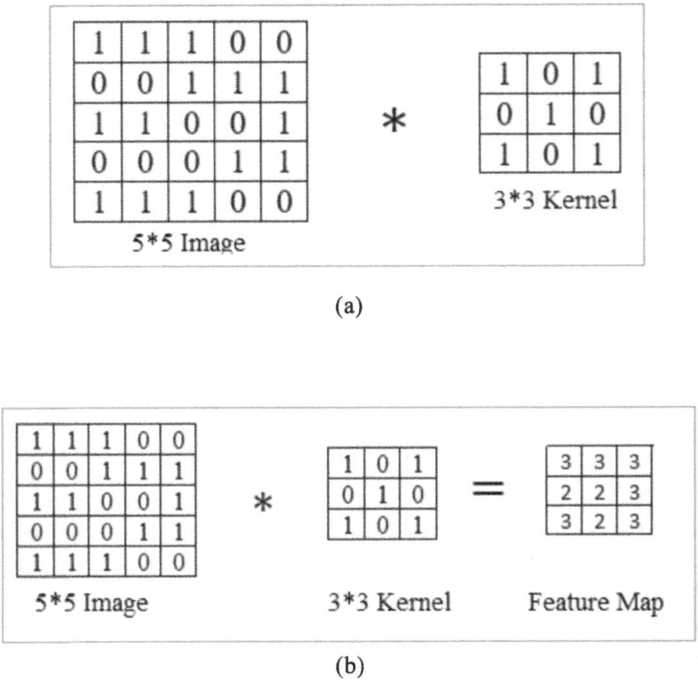

Fig. 3. An illustration of the convolution layer process is shown in (a) a 5 × 5 picture convolving with a 3 × 3 kernel and (b) a convolved feature map.

layers. This ReLU function, (•), is a straightforward calculation that, in the case where the input value is higher than zero, returns the value supplied directly; otherwise, it returns zero. The function $m(\bullet)$ over the set of 0 and the input x can be used to mathematically represent this as follows: The function $\mathcal{G}(\mathrm{x}) = \max\{0, \mu\}) = \max\{0, \}$

Maximum feces layer

When there are greater size photos, this layer reduces the amount of parameters. This is known as downsampling or subsampling, and it reduces the dimensionality of each feature map while keeping the relevant data. The maximum element from the corrected feature map is taken into account in max pooling.

Algorithm for Random Forest

One popular machine learning algorithm is Random Forest, which is used in supervised learning techniques. It can be used for machine learning problems that involve regression and classification. The foundation of this approach is in the notion of ensemble learning, a technique that involves the amalgamation of numerous classifiers to enhance the performance of the model and address complex problems. As the name suggests, "Random Forest is a classifier that contains a number of decision trees on various subsets of the given dataset and takes the average to improve the predictive accuracy of that dataset." The random forest predicts the result as shown in Fig. 4 by using the majority votes of forecasts and the predictions from each decision tree, as opposed to relying solely on one. The forest has more trees, which increases accuracy and prevents over fitting.

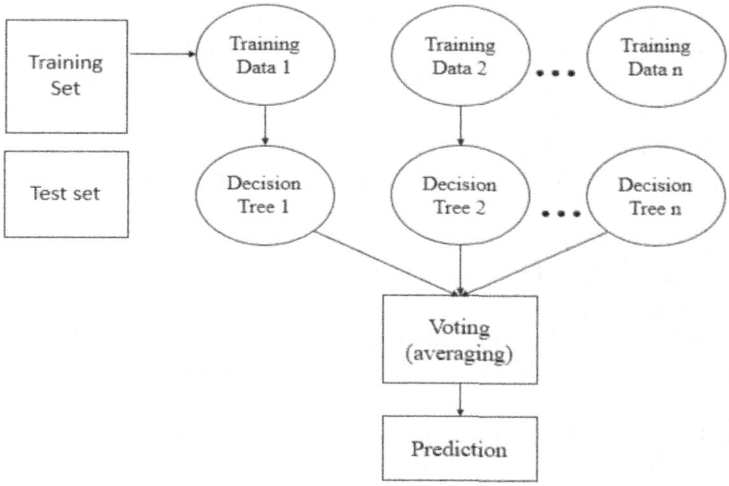

Fig. 4. Random Forest algorithm.

Random Forest (RF) algorithm

Step 1: Using Random Forest, n random records are chosen from the collection of k records.
Step 2: For every sample, a separate decision tree is built.
Step 3: An output will be produced by each decision tree.
Step 4: The final product is evaluated using either regression or classification averaging, or majority voting.

Types of Ensembles

We need to examine the ensemble technique before we can comprehend how the random forest functions. Ensemble is only a term for a grouping of models. Therefore, rather than using a single model to make predictions, a group of models is used. Ensemble employs two different approaches. Bagging: It takes sample training data and replaces it with a new training subset. The final result is determined by majority vote. Bootstrapping refers to the statistical technique of randomly selecting rows from a dataset with replacement. At now, each model undergoes separate training and generates output. When the outcomes of all the models are combined, the ultimate conclusion is determined by majority vote. Aggregation is the process of integrating all the results and producing an output according to majority voting. Boosting: It turns average learners into excellent ones by constructing increasingly precise models. XG BOOST and ADA BOOST are two such examples.

4 Results and Discussion

The details of the Smart Meter electricity recording dataset, which we used to carry out our work. In order to carry out this project, we have created the subsequent modules. The module for pre-processing the dataset is specifically intended to ingest the dataset, eliminate any instances of ignoring values, and transform all non-numeric data into numeric data. This is necessary since deep learning algorithms can only effectively analyze numeric data of the processed dataset; twenty percent will be used for testing and eighty percent for training.

Feed Forward Neural Network: The theft detection model is trained using processed train data as input to the DNN algorithm. The model is then used to test data to determine prediction accuracy.

RNN-GRU Algorithm: the trained theft detection model is trained using processed train data as input to the GRU algorithm. The prediction accuracy of this model is then determined by applying it to test data.

Profound Learning CNN Algorithm: the model for detecting theft will be trained using processed train data, and its prediction accuracy will be determined by applying the model to test data.

Predict Electricity Theft: This module allows us to upload test data, and the Extension algorithm determines if the test data is typical or includes indicators of theft.

Comparison Graph: We will create a comparison graph for each algorithm using this module.

All of the non-numeric data in the above screen has been transformed to number (Numeric) format. Click the "Feed Forward Neural Network" button to train the DNN and obtain the output shown below.

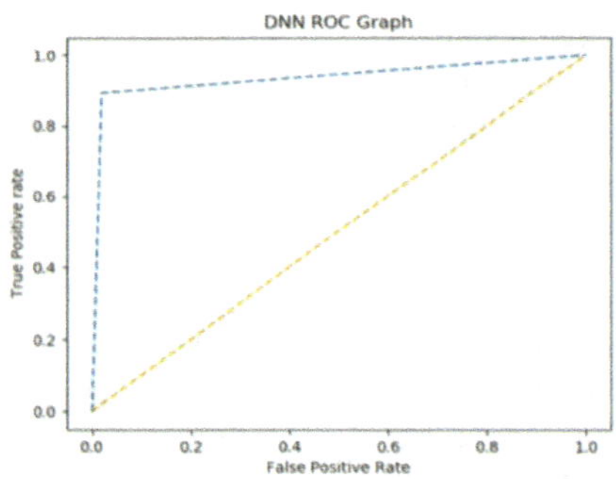

Fig. 5. DNN feed forward algorithm

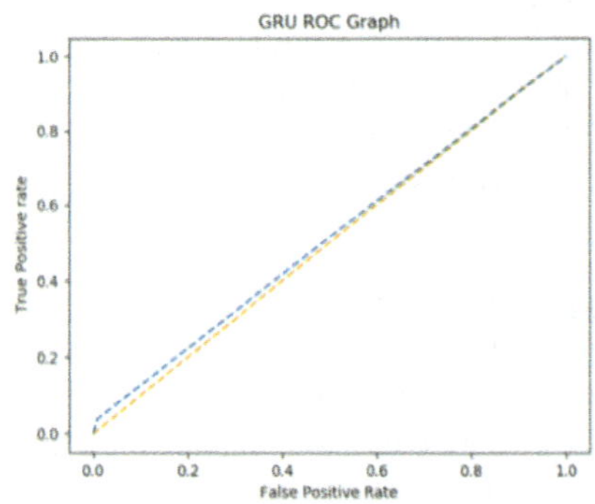

Fig. 6. GRU accuracy levels

Figure 5 illustrates the accuracy of the DNN feed forward algorithm, which was 94.62%. The ROC graph shows the True Positive Rate on the y-axis and the False Positive Rate on the x-axis. If the blue line falls below the orange line, the prediction is considered false; if it crosses the orange line, the prediction is considered correct. Close the graph above now, and then select the "RNN-GRU Algorithm" button to train GRU and obtain the output. As seen in Fig. 6, using GRU, we obtained an accuracy of 40.68%, and the blue line slightly falls below the orange line, indicating that its predictions are incorrect. Now accuracy level graphs is closed and the output shown in Fig. 8 is obtained by using the DLCNN (Depp learning CNN) button.

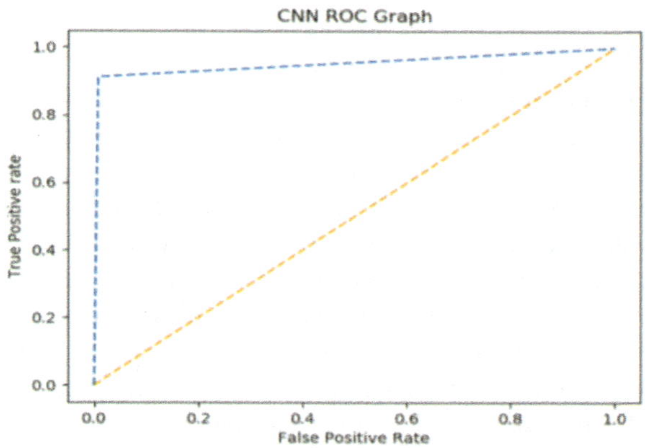

Fig. 7. CNN accuracy levels

The accuracy of 96.48% and the full alignment of the blue and orange lines indicate that the forecasts made by CNN are accurate, as depicted in Fig. 7. You may now input test data and receive prediction output by clicking the "Predict Electricity Theft" button.

TEST data, and we can observe "THEFT NOT DETECTTED" and THEFT detection following the arrow = ➔ sign. Now contrast Fig. 9 below.

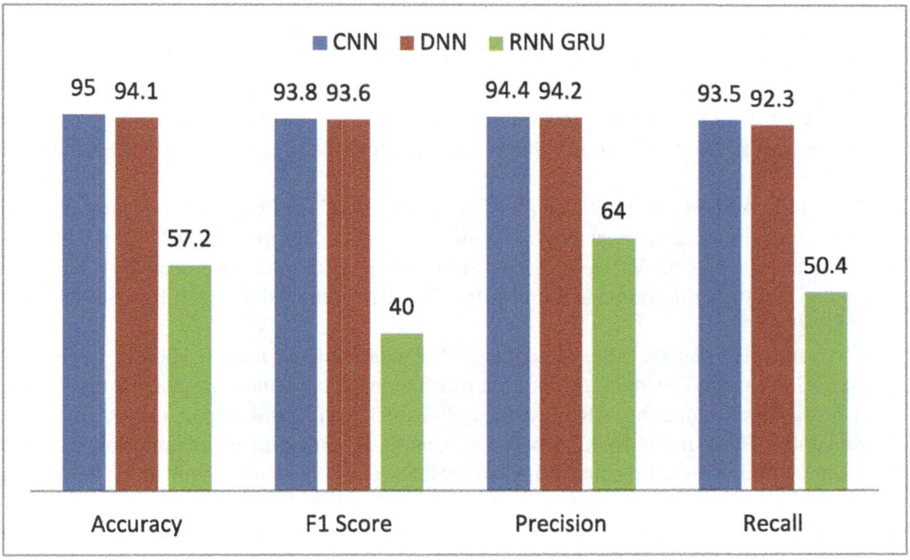

Fig. 8. Different algorithms with their accuracy, precision, recall and Fscore values

The algorithm names are shown on the x-axis and the score values are displayed on the y-axis in Fig. 8 above. As an example, the distinct color bars stand for several measures, including "accuracy, precision, recall, and F1 score." CNN did well in every algorithm.

5 Conclusion

The number of global energy issues is rising daily. Everyone is attempting to conserve energy while also focusing on producing ever-more of it. Numerous methods exist for producing electricity, which is subsequently coordinated for use on a major grid. Technical or non-technical losses are caused by weather. The calculation of technical losses can be achieved with ease, as we have covered in the mathematical modeling part. On the other hand, if technical losses are known, nontechnical losses can be assessed. Electrical theft results in non-technical losses. One can save his financial resources in order to lessen or manage theft. Smart meters might be the greatest option because of their capacity to lower electricity theft. Outstanding resistance to many types of theft, maximum effectiveness, and enhanced security over electromechanical meters. Thus, difficulties connected to theft have been the main focus of this effort.

References

1. Das, A., McFarlane, A.: Non-linear dynamics of electric power losses, electricity consumption, and GDP in Jamaica. Energy Econ. **84**, 104530 (2019)
2. Bashkari, S., Sami, A., Rastegar, M.: Outage Cause Detection in Power Distribution Systems based on Data Mining. IEEE Trans. Ind. Inf. (2020)
3. Shaik, A.S., Karsh, R.K., Islam, M.: Robust image hashing using chromatic channel. In: Nath, V., Mandal, J.K. (eds.) Proceeding of Fifth International Conference on Microelectronics, Computing and Communication Systems. Lecture Notes in Electrical Engineering, vol 748. Springer, Singapore (2021). https://doi.org/10.1007/978-981-16-0275-7_5
4. Nabil, M., Ismail, M., Mahmoud, M.M.E.A., Alasmary, W., Serpedin, E.: PPETD: privacy-preserving electricity theft detection scheme with load monitoring and billing for AMI networks. IEEE Access **7**, 96334–96348 (2019). https://doi.org/10.1109/ACCESS.2019.2925322
5. Khan, Z.A., Adil, M., Javaid, N., Saqib, M.N., Shafiq, M., Choi, J.G.: Electricity theft detection using supervised learning techniques on smart meter data. Sustainability **12**(19), 8023 (2020)
6. Kocaman, B., Tümen, V.: Detection of electricity theft using data processing and LSTM method in distribution systems. Sādhanā **45**, 286 (2020). https://doi.org/10.1007/s12046-020-01512-0
7. Godavarthi, B., Majety, V.D., Mrudula, Y., Nalajala, P.: Fault identification in power lines using GSM and IoT technology. In: First International Conference on Artificial Intelligence and Cognitive Computing: AICC 2018, pp. 647–655. Springer Singapore (2019)
8. Vankdothu, R., Hameed, M.A., Ameen, A., Unnisa, R.: Brain image identification and classification on Internet of medical things in healthcare system using support value based deep neural network. Computers and Electrical Engineering **102**, 108196 (2022)
9. David, L.G., et al.: Tool wear monitoring using improved dragonfly optimization algorithm and deep belief network. Applied Sciences **12**(16), 8130 (2022). https://doi.org/10.3390/app12168130
10. Mehbodniya, A., Varaprasad Rao, M., David, L.G., Gerard Joe Nigel, K., Vennam, P.: Online product sentiment analysis using random evolutionary whale optimization algorithm and deep belief network. Pattern Recognition Letters, **159**
11. Sunitha, D., Patra, R.K., Babu, N.V., Suresh, A., Gupta, S.C.: Twitter sentiment analysis using ensemble based deep learning model towards COVID-19 in India and European countries. Pattern Recognition Letters **158**, 164–170 (2022). ISSN 0167-8655

12. Balmuri, K.R., et al.: A long short-term memory network-based radio resource management for 5G network. Future Internet **14**(6), 184 (2022). https://doi.org/10.3390/fi14060184
13. Soma, S., Sonth, M.V., Gowre, S.C.: Design of two-dimensional photonic crystal based ultra compact optical RS flip-flop. Photon Netw. Commun. **43**, 109–115 (2022). https://doi.org/10.1007/s11107-021-00955-7
14. Sudhamani, C.: Detection probability maximization through optimization of samples in cognitive radio networks. Multimed. Tools Appl. **81**, 12275–12285 (2022). https://doi.org/10.1007/s11042-021-11089-3
15. Chinna Rao, R., et al.: Real-time implementation and testing of VoIP vocoders with asterisk PBX using wireshark packet analyzer. J. Interconnect. Netw. **22**(Supp. 01), 2141030 (2022)
16. Rao, M.V., Reddy, D.A., Ampavathi, A., Shaik, M.: Data mining for cyber-physical systems. Data Mining and Machine Learning Applications, 235–280. https://doi.org/10.1002/9781119792529.ch10
17. Laxmaiah, B., et al.: Enhanced security to MANETs using digital codes. J. Info. Secu. Appl. **66**, 103147 (2022). ISSN 2214-2126, https://doi.org/10.1016/j.jisa.2022.103147
18. Katragadda, S.: Iot enabled fog based computing with deep learning models to increase the allocation of resource. In: 2023 International Conference on Disruptive Technologies (ICDT), pp. 140–143. IEEE (2023)
19. Shaik, A.S., Ahmed, M., Suresh, M.: Low-cost irrigation and laser fencing surveillance system for the paddy fields. In: Chakravarthy, V.V.S.S.S., Flores-Fuentes, W., Bhateja, V., Biswal, B. (eds.) Advances in Micro-Electronics, Embedded Systems and IoT. Lecture Notes in Electrical Engineering, vol 838. Springer, Singapore (2022). https://doi.org/10.1007/978-981-16-8550-7_49
20. Rajesh Tiwari, M., et al.: Enhanced power quality and forecasting for PV-wind micro grid using proactive shunt power filter and neural network based time series forecasting. Electric Power Components and Systems (2023). https://doi.org/10.1080/15325008.2023.2249894

An Assessment on Convolutional Neural Networks for Recognizing and Detecting Intracranial Hemorrhage from CT Images

Jonnadula Narasimharao[1(✉)], G. Parvathidevi[1], M. Nagaraju Naik[2], Abdul Subhani Shaik[2], B. Revathi[3], and Radhe Shyam Panda[3]

[1] CMR Technical Campus, Hyderabad, Telangana, India
jonnadula.narasimharao@gmail.com
[2] CMR College of Engineering and Technology, Hyderabad, Telangana, India
{nagarajunaik1976,sasubhani}@cmrcet.ac.in
[3] Department of CSE(AI&ML), CMR Engineering College, Hyderabad, Telangana, India
bigullarevathi@cmrec.ac.in

Abstract. The number of global energy issues is rising daily. Everyone is attempting to conserve energy while also focusing on producing ever-more of it. Numerous methods exist for producing electricity, which is subsequently coordinated for use on a major grid. Technical or non-technical losses are caused by weather. The calculation of technical losses can be achieved with ease, as we have covered in the mathematical modeling part. On the other hand, if technical losses are known, nontechnical losses can be assessed. Electrical theft results in non-technical losses. One can save his financial resources in order to lessen or manage theft. Because of its superior security, optimal efficiency, and outstanding resistance to numerous theft schemes found in electromechanical meters, smart meters may be the ideal choice for reducing electricity theft. Thus, the focus of this work has primarily been on theft-related issues. In order to detect electricity cyber attacks, this project assessed the performance of many deep learning techniques, including CNN, recurrent neural networks with gated recurrent units (RNN-GRU), and deep feed forward neural networks (DNN).

Keywords: Intracranial hemorrhage · computed tomography · deep learning · and evolution measures

1 Introduction

According to a study conducted by Urden et al. [1], it has been estimated that cerebral hemorrhages contribute to around 10% of stroke cases in the United States. Compared to ischemic strokes (87% versus 87%), hemorrhagic strokes are less common but result in a higher fatality rate. Based on the existing data, it can be shown that the mortality rate during a 30-day period following a hemorrhagic stroke is from 37% to 38%. An estimated 30,000 fatal cases of cerebral hemorrhage happen annually, and 795,000 strokes happen in the US alone. As a result, cerebral bleeding is regarded as one of the most serious

A. Bajaj et al. (Eds.): IBICA 2023, LNNS 1229, pp. 318–325, 2025.
https://doi.org/10.1007/978-3-031-78940-3_33

medical situations, necessitating prompt medical attention and thorough rehabilitation. Intracranial bleeding is a neurological disorder that can be fatal. Health resources are significantly burdened as a result. It can occur for a variety of reasons, including elevated blood pressure, trauma, infarct-related bleeding, tumor hemorrhage, and many more (Fig. 1). Traumatic brain injury is one of the most frequent causes of intracranial bleeding [2]. Trauma blood irritates and swells surrounding brain tissues when it comes into touch with them. We call this cerebral edema. A hematoma is the term for the blood pool seen inside the brain parenchyma. As a result, the surrounding brain tissues are under more strain, which lowers blood flow and destroys brain cells [3].

The National Institute for Health and Care Excellence (NICE) has established guidelines for determining whether patients need CT scans when the benefits outweigh the radiation risk [5]. Upon completion of the CT scan, the radiology physician will assess the situation. Normally, other modalities in the radiology department need to be. Subdural and subarachnoid hemorrhages were the most often overlooked forms of intracranial bleeding, accounting for 39% and 33% of the cases, respectively. Either the frontal or parafalcine regions were the most frequently affected by missing subdural bleeding [6].

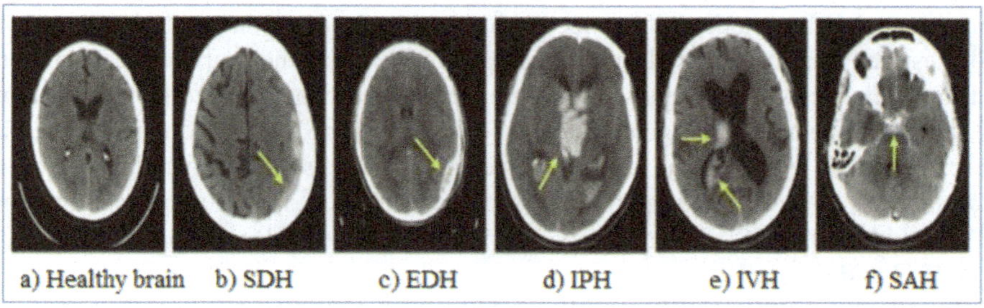

Fig. 1. Non-contrast CT sub types of intracranial hemorrhage (Color figure online)

The non-contrast CT subtypes of cerebral hemorrhage in Fig. 1 show the locations of bleeding, shown by yellow arrows. Prompt identification and patient treatment are critical since early deterioration is common in the initial hours after brain hemorrhage. In the emergency department (ED), over 20% of patients may have a decline in the Glasgow Coma Scale (GCS) of two or more points between the initial evaluation and the prehospital emergency medical services (EMS) assessment [7]. It is essential for determining whether an immediate surgical intervention is necessary and how to proceed. The objectives of the surgical procedure are to lessen the pathophysiological effect of the hematoma on the surrounding tissue, avoid herniation, and lower intracranial pressure. This is accomplished by lowering the blood product's mass affect cellular toxicity [8]. The safe administration of thrombolytic treatment in cases of acute ischemic stroke also depends on the recognition of cerebral hemorrhage in head CT scans. Since "time is brain," in these therapeutic settings, faster and less error-prone processes would be advances that could save lives.

A few medical domains have included deep learning, such as the highly accurate grading of metastasis identification in histologic sections of lymph nodes [9], skin cancer picture classification [10], and diabetic retinopathy on retinal fund photos [11]. When using convolutional neural networks (CNNs) to image processing and classification applications, it has shown to be accurate [11, 12].

The dataset is transformed by deep learning through high-dimensional spatial mapping. With its sophisticated methods, the convolutional neural network in deep learning will be of great assistance for image categorization [13]. To classify the images, CNN derives the network-fused characteristics and semantic features from the dataset [14]. The convolutional neural network provides strong support for high-resolution image categorization because of its deep architecture. This is crucial because, in order to avoid misdiagnosing illnesses that could endanger the patient and cause delays in treatment, the majority of medical data must be in high-quality resolution [15].

2 Literature Review

Chan et al. [16] employed CT scans to refine midline detection algorithms. In their study, Burduja et al. [17] introduced a systematic approach for the identification of bleeding events utilizing Convolutional Neural Networks (CNNs) in the context of intracranial hemorrhage (ICH) detection through the analysis of computed tomography (CT) images. Unfortunately, this methodology necessitates a longer duration for training and has only resulted in a reasonable level of accuracy in the assessment of picture data on the platform. Chen et al. [18] proposed strategies for the detection of bleeding using Internet of Things (IoT) smart intelligent systems. Techniques for SVM-based feature detection were introduced by Tong et al. [19]. Shuchi et al. [20] discussed the crucial variables and examined CT image-slicing approaches such as watershed, Expectation Maximization (EM), and ML-based Statistical Analysis (MLSA). DL methods were presented by Raja et al. [21] to detect ICH.

In order to identify brain hemorrhage, Agata et al. [22] suggested using consecutive slicing of the picture and intensity windows. In order to prepare the data for analysis, this method also carried out primary Region of Interest (ROI) extractions. The many image processing methods for taking into account the picture attributes, such as color, texture, size, and form of image portions, were provided by Chithra et al. [23]. An efficient computational approach to reduce the complexity of 3D image data analysis algorithms was proposed by Kamnitsas et al. [24]. Baseline margin techniques with CNN architectures for brain issue detection (CNNDS) were presented by Schmitz et al. [25].

A unique hybrid 3D/2D mask-based CNN architecture was created using the training cohort, and the trained network was used to process external validation data [26]. Using clinical reports, one board-certified radiologist's assessment of ICH positive cases (IPH, EDH/SDH, and SAH) in both cohorts was conducted [27]. Grewal et al. employed a dataset of 329 NCTCs from two hospitals for training, validation, and testing [28].

3 Proposed Method

Figure 2 shows the flow-diagram of the proposed model, which includes data collection, CNN, and transfer learning.

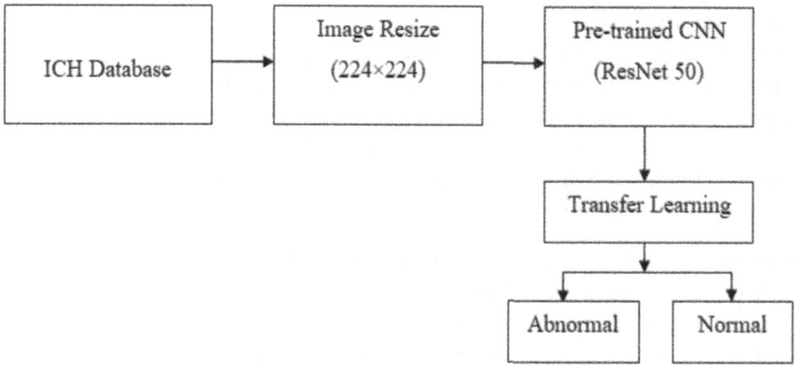

Fig. 2. Block diagram of the proposed model

Dataset

The 200 data points in this cross-sectional analysis were gathered from open sources utilizing secondary data. Abdul Kader Helwan, a faculty member et al.-Manar University in Tripoli, Lebanon, is the owner of this dataset. The owner of the dataset formally granted permission for this study's use. After being anonymized, each sample was turned into secondary data.All adult patients who had head CT scans with and without cerebral bleeding from publicly available datasets made up the study population. Both genders, CT brains of all ICH types, CT brains with normal findings, and CT of ICH with a formal report that is categorized by the dataset source are among the inclusion requirements. The CT brain with artifacts, the CT brain without anonymization into secondary data, and low-quality CT brain pictures unfit for diagnosis-making are among the exclusion criteria.The dataset provider classed the CT brain with radiological signs of intracranial hemorrhage and without intracranial hemorrhage. Two distinct files included a total of 100 CT brains—100 with hemorrhage and another 100 without. A official CT scan report served as the basis for the dataset's classification.

ResNet-50

The introduction of residual blocks is the main innovation of ResNet. A residual block is a type of network building block that enables the network to learn residual mappings rather than attempting to learn the intended underlying mapping directly. The network is able to learn residual functions by introducing shortcut connections that omit one or more levels. Training very deep networks is facilitated by these shortcut connections, which allow the network to propagate gradients effectively and mitigate the vanishing gradient problem.

ResNet-50 has been extensively utilized and shown to be extremely effective. The system has demonstrated cutting-edge results on benchmarks such as the ImageNet dataset. A collection of processes, such as convolution, batch normalization, activation (such as ReLU), and pooling, are typically carried out by each layer in ResNet-50. The network adds more channels while progressively decreasing the spatial dimensions. Global average pooling, fully connected layers, and softmax activation for classification are commonly found in the last layers of ResNet-50.

Transfer Learning (TL)

Transfer learning is based on the fundamental principle that deep neural networks learn hierarchical representations of data, with higher layers capturing more abstract and task-specific properties and lower layers capturing generic features like edges and textures. Time and computational resources may be spared by reusing and customizing these learnt representations for different applications. The following actions are commonly involved in a transfer learning workflow:

Fine-tuning: The pre-trained model serves as the basis for the new task after pre-training. The pre-trained model's last layers, which are usually task-specific, are swapped out or retrained with new layers appropriate for the intended task. The entire model is optimized on the dataset of the target job, and the additional layers are initialized at random. An alternative is to train the remaining layers of the pre-trained model while freezing parts of the layers, which would result in their weights not being updated during fine-tuning.

Managing Data Scarcity: Transfer learning allows the model to learn from the data of the source job, which can have a bigger labeled dataset, in situations where the amount of labeled data is limited. This improves the model's performance on the intended task and offers additional training examples.

4 Results and Discussions

Numerous widely-used metrics, including accuracy, F-score, positive predictive value (PPV), true positive rate (TPR), true negative rate (TNR), and area under the curve (AUC), are used to assess the effectiveness of the proposed method. TNR calculates the rate of correctly recognized non-ICH patients, whereas TPR evaluates the proportion of accurately identified ICH cases. The percentage of accurately detected CT scans that are flagged as ICH is calculated by PPV. The weighted average, or harmonic mean, of PPV and TPR is known as the F-score. AUC is a useful metric for measuring the test's overall performance. The percentage of correctly categorized CT scans—both ICH and non-ICH—over the total number of pictures is known as accuracy. The following describes the mathematical explanations of each of these parameters:

$$\text{Accuracy} = \frac{TP + TN}{TP + TN + FP + FN} \tag{1}$$

$$\text{True Positive Rate (TPR)} = \frac{TP}{TP + FN} \tag{2}$$

$$\text{True Negative Rate (TNR)} = \frac{TN}{TN + FP} \tag{3}$$

$$\text{Positive predictive Value (PPV)} = \frac{TP}{TP + FP} \tag{4}$$

$$\text{F} - \text{Score}\, 2 = \frac{PPV + TPR}{PPV * TPR} \tag{5}$$

$$\text{Area under curve (AUC)} = \frac{TPR + TNR}{2} \tag{6}$$

where, TP = True Positive; FP = False Positive, TN = True Negative, *and* FN = False Negative

This study developed a CNN model to reliably detect intracranial hypoxia in CT scans. To assess the performance of the proposed model, we split the dataset into 82% training and 18% testing.

The hyper parameters that we utilized to train the model are shown in Table 1. MATLAB 2022b is used to conduct these experiments on an Intel (R) Core (TM) i3-5005U CPU at 2 GHz.

Table 1. Hyper parameters used in the suggested work

Parameter	Name/Value
Batch Size	16
Optimizer	Adam
Learning Rate	0.0001
Epochs	30

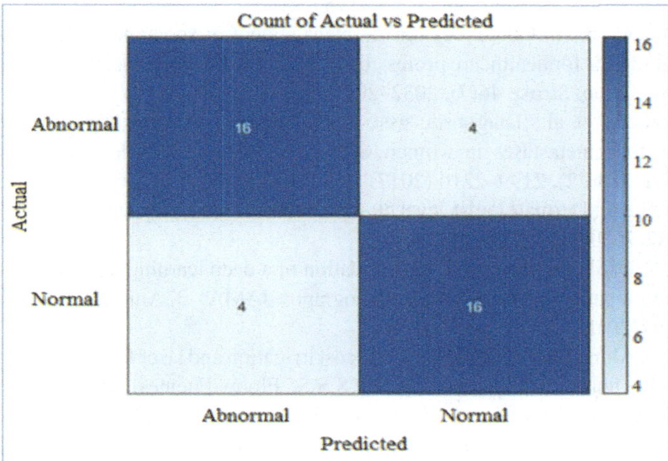

Fig. 3. Confusion matrix of the presented approach

The proposed CNN model's training progress is depicted in Fig. 2, and the associated confusion matrix is displayed in Fig. 3. We achieve 80% sensitivity, specificity, precision, F1-score, AUC, and accuracy using the proposed model.

5 Conclusion

In this work, we suggested architecture based on deep learning for brain CT image-based ICH prediction. Initially, the gathered CT images are transformed into three channels with 224 x 224 dimensions and matrices of floating-point integers normalized in the range of 0 to 1. In order to train our model with little bias for the disproportionate independent bleeding classes, binomial classification was performed using a categorical cross entropy loss function. We acquire 80% sensitivity, precision, F1-score, specificity, accuracy and AUC using the proposed model.

References

1. Urden, L.D., Stacy, K.M., Lough, M.E.: Priorities in Critical Care Nursing, 8th ed., Elsevier Health Sciences (2019)
2. Blennow, K., et al.: Traumatic brain injuries. Nat. Rev. Dis. Primers. **2**, 1–19 (2016)
3. Borsch, J.: Ovulationshemmer oder abtreibungspille? So wirken notfallkontrazeptiva. Deutsche Apotheker Zeitung **153**(9), 32–33 (2013)
4. Rath, G., Ray, B.: Head Injury: Assessment and Early Management. Practice Guidelines in Anesthesia, no. January 2014, p. 53 (2016)
5. Strub, W.M., Leach, J.L., Tomsick, T., Vagal, A.: Overnight preliminary head CT interpretations provided by residents: Locations of misidentified intracranial hemorrhage. Am. J. Neuroradiol. **28**(9), 1679–1682 (2007)
6. Moon, J.S., et al.: Prehospital neurologic deterioration in patients with intracerebral hemorrhage. Crit. Care Med. **36**(1), 172–175 (2008)
7. Hemphill, J.C., et al.: Guidelines for the management of spontaneous intracerebral hemorrhage: a guideline for healthcare professionals from the American heart association/American stroke association. Stroke **46**(7), 2032–2060 (2015)
8. Bejnordi, B.E., et al.: Diagnostic assessment of deep learning algorithms for detection of lymph node metastases in women with breast cancer. JAMA – J. American Medical Association **318**(22), 2199–2210 (2017)
9. Esteva, A., et al.: Dermatologist-level classification of skin cancer with deep neural networks. Nature **542**(7639), 115–118 (2017)
10. Gulshan, V., et al.: Development and validation of a deep learning algorithm for detection of diabetic retinopathy in retinal fundus photographs. JAMA – J. American Medi. Ass. **316**(22), 2402–2410 (2016)
11. Shaik, A.S., Ahmed, M., Suresh, M.: Low-cost irrigation and laser fencing surveillance system for the paddy fields. In: Chakravarthy, V.V.S.S.S., Flores-Fuentes, W., Bhateja, V., Biswal, B. (eds.) Advances in Micro-Electronics, Embedded Systems and IoT. Lecture Notes in Electrical Engineering, vol 838. Springer, Singapore (2022). https://doi.org/10.1007/978-981-16-8550-7_49
12. Vankdothu, R., Hameed, M.A., Ameen, A., Unnisa, R.: Brain image identification and classification on Internet of Medical Things in healthcare system using support value based deep neural network. Computers and Electrical Engineering **102**, 108196 (2022)
13. David, L.G., et al.: Tool wear monitoring using improved dragonfly optimization algorithm and deep belief network. Applied Sciences **12**(16), 8130 (2022). https://doi.org/10.3390/app12168130
14. Mehbodniya, A., Varaprasad Rao, M., David, L.G., Gerard Joe Nigel, K., Vennam, P.: Online product sentiment analysis using random evolutionary whale optimization algorithm and deep belief network. Pattern Recognition Letters, **159**

15. Sunitha, D., Patra, R.K., Babu, N.V., Suresh, A., Gupta, S.C.: Twitter sentiment analysis using ensemble based deep learning model towards COVID-19 in India and European countries. Pattern Recognition Letters **158**, 164–170 (2022). ISSN 0167-8655
16. Balmuri, K.R., et al.: A long short-term memory network-based radio resource management for 5G network. Future Internet **14**(6), 184 (2022). https://doi.org/10.3390/fi14060184
17. Ahmad, S.J., Unissa, I., Shoukath Ali, M., Kumar, A.: Enhanced security to MANETs using digital codes. J. Info. Secu. Appl. **66**, 103147 (2022). ISSN 2214-2126, https://doi.org/10.1016/j.jisa.2022.103147
18. Soma, S., Sonth, M.V., Gowre, S.C.: Design of two-dimensional photonic crystal based ultra compact optical RS flip-flop. Photon Netw. Commun. **43**, 109–115 (2022). https://doi.org/10.1007/s11107-021-00955-7
19. Sudhamani, C.: Detection probability maximization through optimization of samples in cognitive radio networks. Multimed Tools Appl **81**, 12275–12285 (2022). https://doi.org/10.1007/s11042-021-11089-3
20. Chinna Rao, R., et al.: Real-Time Implementation and Testing of VoIP Vocoders with Asterisk PBX Using Wireshark Packet Analyzer. Journal of Interconnection Networks **22**(Supp. 01), 2141030 (2022)
21. Rao, M.V., Reddy, D.A., Ampavathi, A., Shaik, M.: Data mining for cyber-physical systems. Data Mining and Machine Learning Applications, pp. 235–280. https://doi.org/10.1002/9781119792529.ch10
22. Chithra, P.L., Bhavani, P.: A study on various image processing techniques. Int. J. Emer. Technol. Innov. Eng. **5**(5), 316–322 (2019)
23. Shaik, A.S., Karsh, R.K., Islam, M.: Robust image hashing using chromatic channel. In: Nath, V., Mandal, J.K. (eds.) Proceeding of Fifth International Conference on Microelectronics, Computing and Communication Systems. Lecture Notes in Electrical Engineering, vol 748. Springer, Singapore (2021)
24. Kamnitsas, K., Ledig, C., Newcombe, V.F.J., et al.: Efficient multi-scale 3D CNN with fully connected CRF for accurate brain lesion segmentation. Med. Image Anal. **36**, 61–78 (2017)
25. Schmitz, R., Madesta, F., Nielsen, M., et al.: Multi-scale fully convolutional neural networks for histopathology image segmentation: from nuclear aberrations to the global tissue architecture. Medical Image Analysis **70**, Article ID 101996 (2021)
26. Shaik, A.S., Karsh, R.K., Islam, M.: Robust image hashing using chromatic channel. In: Nath, V., Mandal, J.K. (eds.) Proceeding of Fifth International Conference on Microelectronics, Computing and Communication Systems. Lecture Notes in Electrical Engineering, vol 748. Springer, Singapore (2021). https://doi.org/10.1007/978-981-16-0275-7_5
27. Ginbar Ensermu, M.V., Suresh, M., Shaik, A.S., Premalatha, B., Devadasu, G.: An FRLQG Controller-Based Small-Signal Stability Enhancement of Hybrid Microgrid Using the BCSSO Algorithm. Journal of Electrical and Computer Engineering **2023**, Article ID 8404457, 15 (2023). https://doi.org/10.1155/2023/8404457

PseudoRandom Registers for Random Transistor Circuit Analysis in a Sparse Neural Network Using LFSR

K. Shilpa[1](✉), M. Ravindran[1], P. Ravi Kiran[2], S. Krishnaveni[2], X. S. Asha Shiny[3], and Sayyad Rasheeduddin[3]

[1] CMR Technical Campus, Hyderabad, Telangana, India
cmrtc.paper@gmail.com, kshilpa.cse@cmrtc.ac.in
[2] CMR College of Engineering and Technology, Hyderabad, Telangana, India
{dr.pravikiran,krishnaveni}@cmrcet.org
[3] CMR Engineering College, Hyderabad, Telangana, India
{dr.ashashinyxs,srasheeduddin}@cmrec.ac.in

Abstract. This research proposes to use of LFSR to examine the random testing circuit for the sparse neural network implementation. An essential function of the LFSR is circuit testing. Data input is first directed to the LFSR block. Data storage in memory is the primary goal of LFSR. In the upcoming parallel method, data will be stored in registers in a parallel fashion using pseudo random registers. An address generator is now used to generate addresses for the stored data. The action indicated will be executed by the next command. Written data will be verified and recorded. The command generator will be used to repeat the operation if the data turns out to be flawed after verification. In contrast, testing circuits will be used to verify that the data is error-free before an output is produced. It is therefore possible to draw the conclusion that it will result in efficient outcomes in terms of area, speed, and delay based on the results.

Keywords: Testing circuit · linear feedback shift register (LFSR) · Pseudo random register · command generator

1 Introduction

By reducing its dimensions and manufacturing cost, VLSI (Very Large-Scale Integration) technology would help IC technology. The system's complexity will be significantly reduced as a result. The two main factors on which integrated circuit (IC) firms rely are the cost and performance of the systems they impact. Integration testing is greatly increased by using circuit testing. There will be limitations on the quantity of devices and available space if the traditional testing approach is used. This will make the device much better. The conventional testing needed to advance the development of integrated circuit technology will be carried out by ATE (Automatic Testing Equipment) [1].

The most popular method for testing the designed circuits is Automatic Test Pattern Generation, or Automatic Test Pattern Generation (ATPG). The circuit performance will

A. Bajaj et al. (Eds.): IBICA 2023, LNNS 1229, pp. 326–333, 2025.
https://doi.org/10.1007/978-3-031-78940-3_34

be tested using Circuit under Test (CUT), and ATPG will be used to produce input patterns. This will pinpoint the issues in a highly efficient manner. ATE circuit is used to distinguish between fault free and problematic circuits. By utilizing maximum coverage testing and minimal testing time, zero faults will be guaranteed. To test for faults, minimum test cases are used [2].

The pseudo-random test patterns that are provided to CUT (Circuit under Test) are created by TPG (Test Pattern Generation). Test patterns produced by TPG are sent into CUT's input. The CUT's output is compared to a golden signature and then using a comparator, it is saved in RAM. To achieve improved fault coverage, variables such area overhead, test storage data, test application time, and performance degradation can be adjusted [3]. Reseeding the LFSR with a finite number of gates and flip flops to generate random sequences. The recurrent synchronization of LFSRs (Linear Feedback Shift Registers), a form of shift register whose input is a linear function of its preceding bit, is also avoided by using random sequence creation.

The LFSR implementation additionally consists of the XOR gates coupled externally or in sequence with a group of D flip-flops.. Type 1 LFSR refers to an externally connected XOR, while type 2 LFSR refers to an internally connected XOR [4]. When the input bits of a single bit linear shift register are controlled by an exclusive or (XOR), the register's overall value is flip-flopped.

The shift register value in the LFSR is fully defined, mostly based on the fundamental values. The value of the shift register in the LFSR will be used to create test patterns. To achieve successful results using LFSR, test cubes are necessary. LFSR is used to produce a variety of test cases. By extending the mode of operation, LFSR will process the data. In the LFSR, periodic sequence creation makes use of non-zero state.

We take into account LFSR test patterns in our research by completing the test patterns [5]. The many surveys on LFSR designs will be displayed in the literature review. A polynomial's characteristic is the LFSR's first selection. Polynomial characteristics will be used to characterize every potential pattern.

2 Literature Review

Pattern generation is the most important module of a BIST. The ability of linear feedback shift registers (LFSR) to produce incredibly random patterns makes them one of the most often studied test pattern generators (TPG) for BIST. There are numerous enhancements available over the fundamental types of LFSR. The choice of a suitable LFSR for a certain benchmark circuit is examined in the current work. This is accomplished by taking into account a number of variables, including the To obtain high fault coverage, the reduction of wrong patterns, area overhead, and pattern generation time, characteristic polynomial and seed selection are used [11].

Reseeding LFSR to Generate Test Patterns [6] In general, Moore's Law states that as the size of integration increases, circuit testing gets harder. The standard testing methodology cannot handle the growth in density and device count. The testing process will support the investigation of errors as well as issues. Therefore, the circuit development procedure will require less time to complete the operation. ~~In the digital circuit testing system~~, the test time is the most important metric. The length of the test will have a big effect on the entire testing process.

The solution that is most affected during the process is the creation of the test pattern. This will significantly shorten the test duration. LFSR is crucial in the testing process since it generates test patterns. Therefore, this article proposes LFSR approach to generate pseudo-random test patterns. The most commonly used test pattern is the circuit test pattern. By using the test pattern generator circuits, low power is achieved.

This paper introduces reconfigurable LFSR for ASIC and FPGA VLSI IC testing.. Here, VLSI stands for very large scale integration, and LFSR stands for linear feedback shift register. In essence, ATE was first established in VLSI technology; however, the chip testing equipment has a variety of complex challenges [7]. Therefore, BIST is introduced to overcome that. Built-in self-testing is what BIST stands for. A few problems have been found in the BIST. To get around that, LFSR is presented.

A Reconfigurable Low Power LFSR [8] the need for low power design has increased in the age of nanotechnology. This work presents the implementation of a low power reconfigurable LFSR. Reconfigurable LFSR's primary goal is to increase output randomness. Low power design strategies are used in order to save energy. Security applications are therefore relevant to portable embedded systems. This paper also includes the fundamentals of LFSR. The Hspice is used to assess the LFSR's implementation as well as design. There will be a decrease in energy use as a result.

LFSR implementation on ASIC [9] with this study, the ASIC-based LFSR implementation is accomplished. Here, LFSR refers for linear feedback shift register, and ASIC is for application-specific integrated circuit.. Both of these are mostly employed in cryptography systems. Utilizing these two technologies is primarily done to speed up operations. Programmable logic flow regulators (LFSRs) with an ASIC base are used to convert data into encrypted text. Any kind of ASIC design can be created with the help of the Electronic Design Automation tools. Therefore, the cadence tool is used in this study to complete the work. The design of the LFSR is done with Verilog. This will very effectively shorten wait times and speed up operations. Design is confirmed by utilizing both timing and simulation tools.

Effective LFSR Architecture for Generating Pseudo Exhaustive Tests in BIST [10].

Pattern creation is essentially the most used module in BIST. The linear feedback shift register, or LFSR, is the most widely used test pattern generator for BIST, however several other types are also researched. Using LFSR, random patterns are produced in an extremely high way. There are simple forms available for LFSR. By choosing the proper circuit, LFSR is used to analyze the benchmark circuit. Delay, overhead time, and high fault coverage will be used to construct patterns.

In [12] using a bit-swapping technique, a modified linear feedback shift register (LFSR) is described that reduces the number of transitions at the circuit-under-test's inputs by 25%. Based on ISCAS'85 and 89 benchmark circuits, experimental results demonstrate a power reduction of up to 45% during testing. Additionally, they demonstrate how the suggested design can be used in conjunction with other methods to obtain a significant power savings of up to 63%.

Without affecting the duration of the test, the suggested architecture improves the correlation between the patterns produced by LT-LFSR." Reductions of up to 77% and 49%, respectively, in average and peak power have been confirmed by the experimental results for the ISCAS'85 and '89 standards [13].

To minimise the dynamic power consumption of the Circuit under Test (CUT), low power test vectors should be developed for the CUT.. This is accomplished by modifying a linear feedback shift register to create a low power Test Pattern Generator (TPG). Low power test patterns are produced by increasing the correlativity between succeeding vectors [14]. In order to mitigate the average and peak power of a circuit while testing, this study suggests a low transition test pattern generator (LT-LFSR) that decreases transitions both inside random test patterns and between subsequent patterns. Stated differently, transitions are only permitted between successive bits and patterns in two dimensions. LT-LFSR can be used with both scan-based and BIST architectures, and it is not dependent on the circuit being tested [15].

3 Proposed Method

Random Testing Circuits Using LFSR

Figure 1 shows the flow chart for the sparse neural network-based LFSR-based random testing circuit. The input is first feed to the LFSR block. Storing the data in memory is the main objective of LFSR [16]. The data will then be stored in registers in parallel form using pseudo random registers in the parallel process [17]. An address generator is now used to generate the address for the data that is saved.

The precise procedure will be carried out by the next command. Written information will be noted and checked. If there are any problems in the data after verification, the command generator will be used to carry out the process once more. However, if the validated data is error-free, testing circuits will be used to test it, and an output will be produced [18].

Steps by Step Process

Step 1: The LFSR block receives input data first. The primary goal of LFSR is information storage in memory.
Step 2: The data will be stored in registers in a parallel fashion using pseudo random registers to do the parallel process.
Step 3: An address generator is now used to generate the address for the data that has been saved.
Step 4: The next instruction is to carry out the particular task.
Step 5: Written data will be documented and checked.
Step 6: If there are any mistakes in the data after verification, it will execute the command generator once more.
Step 7: Testing circuits will be used to test the data and produce an output if it is deemed to be error-free.

The comparison between the LFSR testing circuit and the LFSR random testing circuit is displayed in Table 1 below. This table shows that comparing errors, delays, areas, and speeds will produce useful results. When compared to LFSR testing circuits, LFSR-based random testing circuits will effectively decrease errors, area, and delay while also increasing operating speed.

A comparison of the LFSR testing circuit and the LFSR random testing circuit is shown in Fig. 2 below. This figure shows how well the errors and delays are eliminated in the LFSR based random testing circuit.

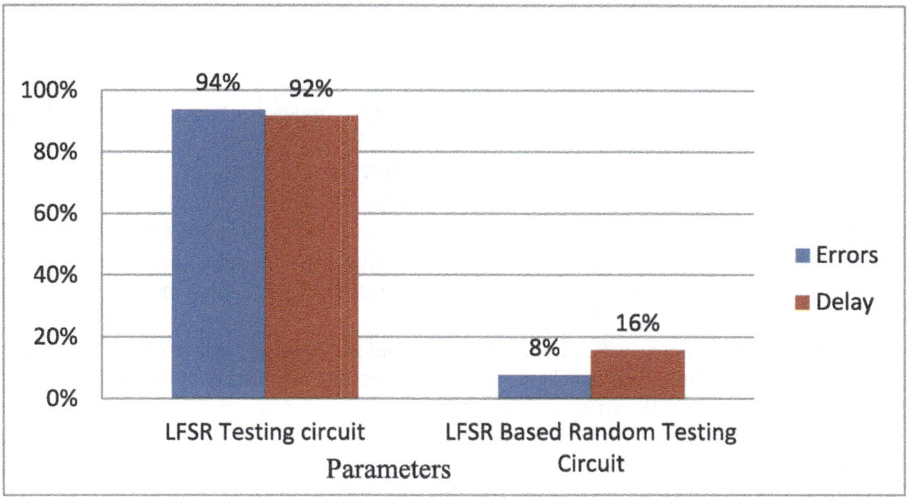

Fig. 2. Errors vs delay

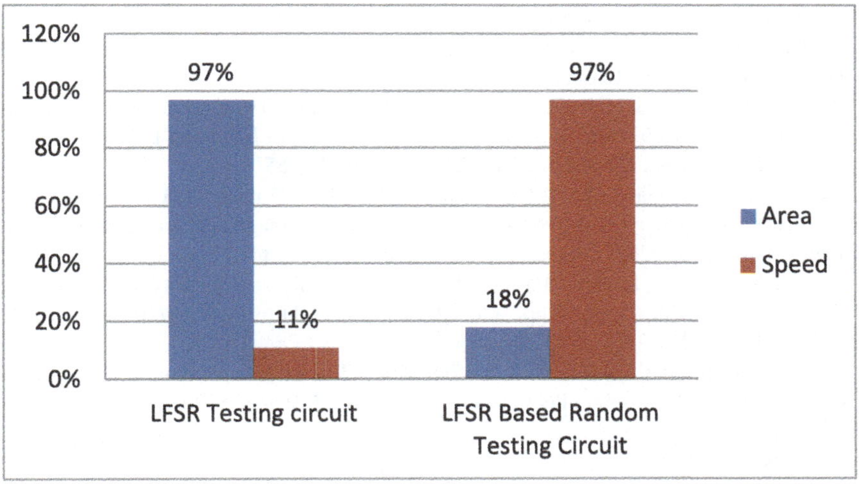

Fig. 3. Area Vs Speed

4 Conclusion

The LFSR for Sparse Neural Network Analysis of Random Testing Circuit is put into practise in this paper.Thus, self-testing of any type of circuitry with a broad range of combinations is ensured by this architecture. In pseudo random number generators, (LFSR is frequently utilized (PRNG). Unprecedented levels of generating rate acceleration can result from parallelization. In the future, sleep transistors and power gating techniques might be used to further cut down on the structure's power consumption.

References

1. Rajski, J., Tyszer, J., Mrugalski, G., Nadeau-Dostie, B.: Test generator with pre selected toggling for low power built-in self-test. In: 2019 IEEE 30th VLSITest Symposium (VTS), pp. 1–6 (2019)
2. Zhou, B., Wu, X.: A low power test-per-clock BIST scheme through selectively activating multi two-bit TRCs. In: 2019 Fourth International Conference on Instrumentation and Measurement, Computer, Communication and Control (IMCCC), pp. 505–509 (2019)
3. Abu-Issa, A.S., Quigley, S.F.: Bit swapping LFSR and scan-chain ordering: a novel technique for peak and average power reduction in scan based BIST. IEEE Trans. Comput. Aided Des. Integr. Circuits Syst. **28**(5), 755–759 (2019)
4. Abu-Issa, A.S., Quigley, S.F.: A multi-output technique for high fault coverage in test-per-scan BIST. In: 3rd International Conference on Design and Technology of Integrated Systems in Nano scale Era, DTIS 2019, pp. 1–6 (2019)
5. Vankdothu, R., Hameed, M.A., Ameen, A., Unnisa, R.: Brain image identification and classification on Internet of Medical Things in healthcare system using support value based deep neural network. Comput. Electr. Eng. **102**, 108196 (2022)
6. David, L.G., Patra, R.K., Falkowski-Gilski, P., Divakarachari, P.B., Marcilin, L.J.A.: Tool wear monitoring using improved dragonfly optimization algorithm and deep belief network. Appl. Sci. **12**(16), 8130 (2022). https://doi.org/10.3390/app12168130
7. Mehbodniya, A., Varaprasad Rao, M., David, L.G., Gerard Joe Nigel, K., Vennam, P.: Online product sentiment analysis using random evolutionary whale optimization algorithm and deep belief network. Pattern Recogn. Lett. **159**
8. Sunitha, D., Patra, R.K., Babu, N.V., Suresh, A., Gupta, S.C.: Twitter sentiment analysis using ensemble based deep learning model towards COVID-19 in India and European countries. Pattern Recogn. Lett. **158**, 164–170 (2022). ISSN 0167-8655
9. Balmuri, K.R., Konda, S., Lai, W.-C., Divakarachari, P.B., Gowda, K.M.V., Lingappa, H.K.: A long short-term memory network-based radio resource management for 5G network. Future Internet **14**(6), 184 (2022). https://doi.org/10.3390/fi14060184
10. Ahmad, S.J., Unissa, I., Shoukath Ali, M., Kumar, A.: Enhanced security to MANETs using digital codes. J. Inf. Secur. Appl. **66**, 103147 (2022). https://doi.org/10.1016/j.jisa.2022.103147. ISSN 2214-2126
11. Soma, S., Sonth, M.V., Gowre, S.C.: Design of two-dimensional photonic crystal based ultra compact optical RS flip-flop. Photon Netw. Commun. **43**, 109–115 (2022). https://doi.org/10.1007/s11107-021-00955-7
12. Sudhamani, C.: Detection probability maximization through optimization of samples in cognitive radio networks. Multimed. Tools Appl. **81**, 12275–12285 (2022). https://doi.org/10.1007/s11042-021-11089-3
13. Chinna Rao, R., et al.: Real-time implementation and testing of VoIP vocoders with asterisk PBX using wireshark packet analyzer. J. Interconnection Netw. **22**(Supp01), 2141030 (2022)
14. Rao, M.V., Reddy, D.A., Ampavathi, A., Munawar, S.: Data mining for cyber-physical systems. Data Min. Mach. Learn. Appl. 235–280. https://doi.org/10.1002/9781119792529.ch10
15. Sahu, A.K., Chandra, V.K., Sinha, G.R.: High level computation technique for characterization of sigma-delta A/D converter. In: 2017 International Conference on Microelectronic Devices, Circuits and Systems (ICMDCS), Vellore, India, pp. 1–4 (2017). https://doi.org/10.1109/ICMDCS.2017.8211601
16. Shaheen, H., et al.: An efficient classification of cirrhosis liver disease using hybrid convolutional neural network-capsule network. Biomed. Signal Process. Control **80**(Part 1), 104152 (2023). https://doi.org/10.1016/j.bspc.2022.104152. ISSN 1746-8094

17. Rajesh Tiwari, M., et al.: Enhanced power quality and forecasting for PV-wind microgrid using proactive shunt power filter and neural network-based time series forecasting. Electr. Power Compon. Syst. (2023). https://doi.org/10.1080/15325008.2023.2249894
18. Shaik, A.S., Karsh, R.K., Islam, M., et al.: A review of hashing based image authentication techniques. Multimed. Tools Appl. **81**, 2489–2516 (2022). https://doi.org/10.1007/s11042-021-11649-7

Optimizing Vision Transformer Based Generative Adversarial Networks (ViT-GAN) for Early Detection of Wheat Rust Disease Utilizing Few-Shot and Transfer Learning

K. Vishnu Vardhan(✉) and H. Ateeq Ahmed

Dr. K.V. Subba Reddy Institute of Technology, Kurnool, India
vishnukn1543@gmail.com

Abstract. Timely detection of wheat rust diseases emerges as a critical step for safeguarding wheat crops, which serve as a major source of food for the majority of population worldwide. An automated system is required for early prediction of wheat disease. This paper introduces a deep learning based model, ViT-GAN, a fusion of Vision Transformers and Generative Adversarial Networks, adopting few-shot and transfer learning techniques. ViT-GAN, optimized for wheat rust detection, utilizes the power of generative models to augment datasets with synthetic wheat leaf images, while the few-shot capabilities enable accurate disease detection. Further, ViT-GAN's optimization and integration of transfer and few-shot learning techniques facilitate the way for adaptable and efficient AI solutions in diverse domains with limited labeled data.

Keywords: Vision Transformers · Transfer Learning · Wheat-Rust disease · Few-Shot Learning · Data Augmentation

1 Introduction

Wheat most vital cereal crops globally, serving as a important food source for majority of the population. The production of this crop faces threat from several diseases, wheat rust is one such most destructive disease. This disease is caused by fungal pathogens, leads to severe yield losses if not detected and managed at early stage. These types of diseases can be detected using traditional methods such as visual inspection by experts, which is time-consuming and susceptible to errors. Recently, application of Artificial intelligence, machine learning and deep learning has shown enormous potential in agriculture field, in the early diagnosis of crop diseases. Also, deep learning model based on Generative Adversarial Networks (GANs) have gained importance for their capability to generate highly informative images. ViT-GAN an approach integrating Vision Transformers (ViT), a deep learning architecture that has exceptional performance in performing computer vision tasks, with GAN has the ability to transform the method of detecting wheat rust disease.

A. Bajaj et al. (Eds.): IBICA 2023, LNNS 1229, pp. 334–341, 2025.
https://doi.org/10.1007/978-3-031-78940-3_35

The proposed work utilizes the power of ViT-GAN, enhanced by few-shot and transfer learning techniques, to provide efficient solution for wheat rust disease detection at early stage. This approach gets the advantage of combining the strengths of both generative models and transfer learning to generate synthetic wheat leaf images that closely resemble real-world rust-infected samples. By training the ViT-GAN on areal images and incorporating few-shot learning capabilities, the proposed model aims to detect wheat rust disease with high accuracy. Considering these aspects the proposed work explores the optimization of ViT-GAN architecture, utilizing transfer and few-shot learning for wheat rust detection.

2 Background Work

This section involves reviewing existing literature related to wheat-rust disease prediction, vision transformer models, transfer learning, few-shot learning approach and deep learning architectures Wheat is an essential crop that holds important role in ensuring food security at global level. A major threat posed by wheat-rust disease cannot be under estimated. Different types of rust affecting wheat plants are caused by fungal pathogens that leads to severe yield losses if not managed properly on time. Traditional methods of disease detection can cause problems such as less accuracy, and expensive. Machine learning, deep learning, and computer vision techniques have been applied to address challenges in precision agriculture, like crop disease detection [1–7]. Generative Adversarial Networks (GANs) have emerged as a important tool for generating synthetic data that resembles real-world samples. GANs have been successfully applied to domains, such as image synthesis and data augmentation [8–11]. Vision Transformers (ViTs) is an important tool to perform computer vision based tasks. ViTs rely on self-attention mechanisms and multi-head attention to capture long-range dependencies in images. ViTs have demonstrated remarkable performance in various vision tasks, including image classification and object detection [12–15]. GANs have been used to generate synthetic images of diseased crops, used in the augmentation of training datasets. These generated images refer to be valuable in improving the performance of disease detection models. By utilizing knowledge learned from diverse datasets, transfer learning enables development of robust models for crop disease detection [24, 25], even in case of limited labeled data [16–23]. Few-shot learning techniques have gained importance for their capability to adapt models to new tasks with minimal labeled examples. In agriculture, data collection and annotation are resource-intensive process, few-shot learning can be used to enhance the generalization ability of disease detection models [26]. Considering these aspects, the proposed work addresses these challenges by optimizing the ViT-GAN architecture and integrating transfer learning and few-shot learning techniques.

3 Methodology

This section, briefs about the steps involved in the proposed architecture for predicting wheat-rust disease considering wheat leaf images utilizing hybrid VisionTransformer Generative Adversarial Network (ViT-GAN) architecture with transfer and few-shot learning. The steps for the proposed architecture for predicting wheat rust disease have been specified in Fig. 1. Initial step is to gather dataset consisting of wheat leaf images of both normal and disease affected samples. Proposed work utilizes wheat leaf dataset from kaggle for processing. Next step is to perform data annotation in order to label each image as "Healthy" or "Rust-Infected." Then perform operations like removal of duplicate, irrelevant images from the dataset and check for low-quality images and remove them. Then, resize images to a consistent resolution suitable for ViT model and a resolution of 224x224 is used for processing in the proposed work. Next is to perform normalization to reduce pixel values to a common scale to improve model training. Also, data augmentation such as random rotations, flips, brightness adjustments, and cropping were applied. Now dataset is clean, and ready for training the ViT-GAN model with both transfer learning and few-shot learning components.

Next, step in the proposed framework is to process the input wheat leaf images using Vision Transformer (ViT) component to extract high-level features from pre-processed images. ViT component represents a pre-trained Vision Transformer model trained on benchmark data such as ImageNet. ViT takes an input image and divides it as a grid consisting of fixed-size patches. Then every patch is linearly embedded into a lower-dimensional vector using input embedding component. ViT lacks the inherent spatial understanding of convolutional layers, positional encodings are combined with embedded patches to provide relative positions of the patches to the model in the original image. Multi-Head Self-Attention Mechanism is the core of ViT to capture long-range dependencies between different patches in the image. The outputs of the self-attention heads and the feedforward neural networks (MLP) are aggregated across multiple attention heads. Layer normalization (LN) specified in Fig. 1 is applied to the outputs of each layer to stabilize training and ensure that the features extracted from different layers have similar scales. The output of LN will be processed by transformer encoder block. Transformer encoder blocks are stacked several times in the model, allowing for the hierarchical extraction of features at different levels of abstraction. The output of transformer encoder is a refined representation of the input data enabling the model to process and understand complex sequences and images by capturing both local and global patterns. The final embeddings produced by ViT can be used as input to generator for image synthesis.

Generator is a neural network architecture responsible for enhancing wheat leaf images taking the feature vectors from the ViT as input and producing enhanced images. The goal of Generator is to produce high-quality images that closely resemble the input data. Generator starts with a random noise vector as its input. This vector is drawn from a simple probability distribution such as Gaussian and serves as the source of randomness for generation process. As specified in Fig. 1 random noise vector is passed through the Generator Network, which is a neural network architecture designed for the specific task with several convolution layers. The output of generator is enhanced image generated based on the patterns and features learned by the Generator Network during adversarial training. The objective of this component is to produce high-quality images that resemble the target data.

Next component discriminator evaluates the authenticity of enhanced wheat leaf images produced by Generator. It receives as input both real wheat leaf images and generated images as input. Real samples are passed through discriminator network consisting of fully connected layers. Discriminator job is to differentiate between real and fake (generated) images using probability score. In binary classification problem this score represents the likelihood that the input data is real. The goal is to assign maximum score to original data and minimum score to fake data. Discriminator is trained to become better performing this task over time, while Generator is trained to produce data difficult for the Discriminator to differentiate from real data. This type of adversarial training process helps in generation of realistic data by the Generator. The component adversarial loss, is used to train the Generator to produce images that are more indistinguishable from real images. It is used to improve the image generation of generator. The trained ViT-GAN model then produces final output enhanced leaf images. Transfer learning approach is used for training in ViT through pre-trained models. This learning mechanism helps to load the pre-trained weights and adjust the model for classification task. Also,this approach is used by the proposed ViI-GAN to get the capability of using pre-trained weights from a large ImageNet dataset used for computer vision based tasks. Optimization involves fine-tuning and training the ViT-GAN model to achieve the best performance in detecting wheat rust disease at earlier stage with limited samples. Quality of these images depends on the effectiveness of the training process and the capabilities of the Generator. Fine-tune the model using the main dataset for wheat rust detection and then using Few-shot learning dataset. Iteratively train the ViT-GAN model using the entire dataset. Few-shot learning is applied to address scarcity of labeled samples, in the context of wheat rust disease prediction then applied for final classification task. This helps the model to generalize well even with a small number of samples.

Iteratively train the proposed model using entire dataset through forward passes, loss computation and backpropagation. Forward pass mechanism helps to feed input images through the ViT-GAN model to generate enhanced images. ViT-GAN model, with few-shot learning, becomes more adaptable and capable of early detection of different rust subtypes and stages of wheat leaves. Loss computation involves calculating total loss based on the defined loss functions. Backpropagation method is used during training to compute gradients and perform parameter updation of the model using Adam optimization algorithm. Final step is to perform model evaluation using performance metrics as specified in Table 1. Also, validation of the model is performed to verify the efficiency of model to distinguish the healthy and rust-diseased wheat leaves. Optimization mechanisn in the proposed work involves a combination of transfer learning, adversarial training, few-shot learning, and fine-tuning to train ViT-GAN model utilized for early detection of wheat rust disease. Thus, ViT-GAN with transfer and few-shot learning enhances the accuracy and efficiency of wheat leaf disease detection.

4 Results and Discussions

The dataset used in proposed approach is wheat leaf dataset from Kaggle utilized for various purposes in agriculture, plant pathology, and machine learning model building process. This dataset is designed to assist in the identification and diagnosis of diseases affecting wheat plants. Wheat leaf dataset consists of digital images of wheat leaves captured using cameras various stages of growth and may exhibit different characteristics, including healthy and disease affected leaves. The image in the dataset is typically

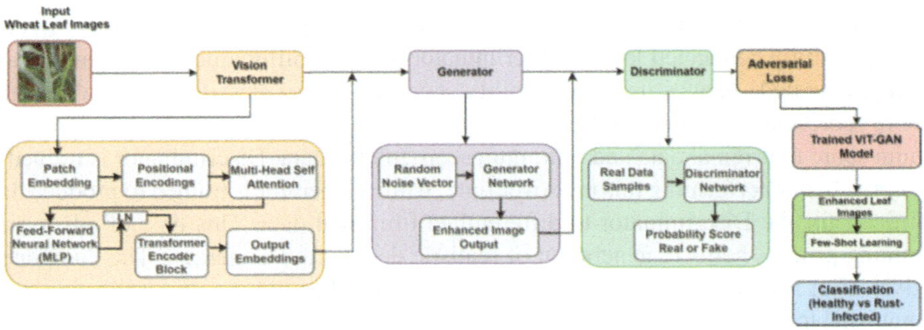

Fig. 1. Wheat-Rust Disease Prediction Using ViT-GAN Architecture

labeled or annotated to indicate its associated class such as "Healthy," "Rust-Infected," and more. The wheat leaf dataset used by the proposed ViT-GAN model from Kaggle consists of healthy, stripe rust, and detected wheat leaf samples. The healthy and rusted leaf samples are specified in Fig. 2. The dataset can be utilized for training and evaluating deep learning architectures enabling wheat-rust disease prediction and crop protection. Deep learning architectures trained using wheat leaf dataset helps with early detection of wheat leaf disease.

Fig. 2. Images of Healthy and Rust Affected Wheat Leaf Samples

Performance of ViT-GAN based approach is evaluated utilizing metrics like accuracy and more and the values are specified in Table 1. Accuracy of the ViT-GAN model is promising when compared to the methods that are in existence. The proposed ViT-GAN approach for wheat rust prediction shows better performance, compared to existing methods based on evaluation metrics. The accuracy value of 0.92 and ROC-AUC of 0.96 is remarkable. The proposed model performs well in both accurately identifying wheat rust diseased leaves and differentiating between healthy and diseased wheat leaves.

The Accuracy of ViT-GAN method during the process of training and testing is represented in Fig. 3. Increase in training accuracy based on epochs reveals the capability of model to classify the data correctly. Testing accuracy will be measured based on the performance of the model for unseen data. High test accuracy scores reveals that ViT-GAN based model can generalize the learning from training process. Figure 3 specifies the loss

Table 1. Evaluation of ViT-GAN Model Using Performance Metrics

Method	Accuracy	Precision	Recall	F1-Score	AUC-ROC
Proposed (ViT-GAN)	0.92	0.94	0.91	0.92	0.96
CNN (ResNet-50)	0.88	0.87	0.90	0.88	0.92
LSTM+CNN	0.84	0.80	0.86	0.83	0.87
GAN Only	0.86	0.85	0.89	0.87	0.90

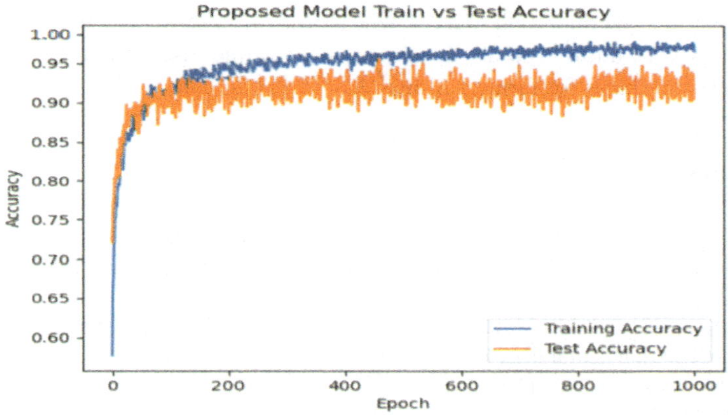

Fig. 3. Accuracy of ViT-GAN Model during Training and Testing

of ViT-GAN model during training and testing. The model adjusts the hyperparameters such as weights, bias during learning process to decrease the loss. Also, decrease in test loss specifies that the model performs better with new data. Training and test accuracy and loss are required parameters to measure the model's performance and provide information about the learning efficiency of the model on new data. Hyperparameter tuning of proposed architecture is specified in Table 2.

Table 2. Hyperparameter Tuning Values for the Proposed ViT-GAN Approach

Hyperparameter	Values	Hyperparameter	Values
Patch Size (HXW)	16x16	Learning Rate	0.0001
Number of Layers	12	Batch Size	32
Number of Attention Heads	8	Few-Shot Batch Size	16
Hidden Dimension	768	Dropout Rate	0.1

5 Conclusion

The proposed ViT-GAN based approach with transfer and few-shot learning, represents a significant advancement in early prediction of wheat-rust disease identifying the healthy and diseased leaf images properly. The proposed model performs better than the existing approaches with the accuracy of 0.94. The proposed ViT-GAN approach not only holds potential to revolutionize wheat disease diagnosis but also embodies a promising step toward more accurate, and reliable agricultural practices. Thus the proposed ViT-GAN based approach can be used by farmers and agricultural organizations to monitor and protect the wheat crops effectively. ViT-GAN contributes to this goal by helping protect one of the world's essential food sources, wheat, from disease-related yield losses. Future work is to explore multi-crop applications and real-time disease detection promoting user-friendly interfaces for farmers usage.

References

1. Goyal, L., Sharma, C.M., Singh, A., Singh, P.K.: Leaf and spike wheat disease detection & classification using an improved deep convolutional architecture. Inform. Med. Unlocked **25**, 100642 (2021)
2. Hussain, A., Ahmad, M., Mughal, I.A., Ali, H.: Automatic disease detection in wheat crop using convolution neural network. In: The 4th International Conference on Next Generation Computing, vol. 34, p. 22 (2018)
3. Sood, S., Singh, H.: An implementation and analysis of deep learning models for the detection of wheat rust disease. In: 2020 3rd International Conference on Intelligent Sustainable Systems (ICISS), pp. 341–347. IEEE (2020)
4. Paidipati, S.R., Pothuneedi, S., Gandham, V.N., Jain, L., Kumar, S., Jain, A.: A review: disease detection in wheat plant using conventional and machine learning algorithms. In: 2022 5th International Conference on Contemporary Computing and Informatics (IC3I), pp. 1436–1441. IEEE (2022)
5. Nigam, S., Jain, R., Marwaha, S., Arora, A.: Wheat rust disease identification using deep learning. De Gruyter (2021)
6. Saleem, M.H., Potgieter, J., Arif, K.M.: Plant disease detection and classification by deep learning. Plants **8**(11), 468 (2019)
7. Bao, W., Yang, X., Liang, D., Hu, G., Yang, X.: Lightweight convolutional neural network model for field wheat ear disease identification. Comput. Electron. Agric. **189**, 106367 (2021)
8. Mohapatra, S.K., Prasad, S., Nayak, S.C.: Wheat rust disease detection using deep learning. Data Science and Data Analytics: Opportunities and Challenges, p. 191 (2021)
9. Borhani, Y., Khoramdel, J., Najafi, E.: A deep learning based approach for automated plant disease classification using vision transformer. Sci. Rep. **12**(1), 11554 (2022)
10. Wang, F., Wang, R., Huang, H., Wu, S.: A shifted windows detecting method for similar plant diseases based on transform. In: 7th International Conference on Information Science, Computer Technology and Transportation, ISCTT 2022, pp. 1–4. VDE (2022)
11. Devika, K.D., Saxena, A.: Dynamic authentication using visual cryptography. In: Satapathy, S.C., Lin, J.C.W., Wee, L.K., Bhateja, V., Rajesh, T.M. (eds.) Computer Communication, Networking and IoT. LNNS, vol. 459. Springer, Singapore (2023). https://doi.org/10.1007/978-981-19-1976-3_20

12. Maithri, Y., Saxena, A.: Browser extension for digital signature. In: Satapathy, S.C., Lin, J.C.W., Wee, L.K., Bhateja, V., Rajesh, T.M. (eds.) Computer Communication, Networking and IoT. LNNS, vol. 459. Springer, Singapore (2023). https://doi.org/10.1007/978-981-19-1976-3_21
13. Singh, A., Tiwari, V., Tentu, A.N., Saxena, A.: Securing communication in IoT environment using lightweight key generation-assisted homomorphic authenticated encryption. In: Satapathy, S.C., Lin, J.C.W., Wee, L.K., Bhateja, V., Rajesh, T.M. (eds.) Computer Communication, Networking and IoT. LNNS, vol. 459. Springer, Singapore (2023). https://doi.org/10.1007/978-981-19-1976-3_26
14. Srujan Raju, K., Jagtap, V., Kulkarni, P., Varaprasad Rao, M.: Gameplay cognitive decision support using statistical and non-statistical parametric fusion. In: Satapathy, S.C., Lin, J.C.W., Wee, L.K., Bhateja, V., Rajesh, T.M. (eds.) Computer Communication, Networking and IoT. LNNS, vol. 459. Springer, Singapore (2023). https://doi.org/10.1007/978-981-19-1976-3_24
15. Kumar Apat, S., Mishra, J., Srujan Raju, K., Padhy, N.: IoT-assisted crop monitoring using machine learning algorithms for smart farming. In: Kumar, R., Pattnaik, P.K., R. S. Tavares, J.M. (eds.) Next Generation of Internet of Things. LNNS, vol. 445. Springer, Singapore (2023). https://doi.org/10.1007/978-981-19-1412-6_1
16. Abbas, A., Jain, S., Gour, M., Vankudothu, S.: Tomato plant disease detection using transfer learning with C-GAN synthetic images. Comput. Electron. Agric. **187**, 106279 (2021)
17. Zhao, Y., et al.: Plant disease detection using generated leaves based on DoubleGAN. IEEE/ACM Trans. Comput. Biol. Bioinf. **19**(3), 1817–1826 (2021)
18. Sunitha, G., Reddy Madhavi, K., Avanija, J., Tharun Kumar Reddy, S., Hitesh Sai Vittal, R.: Modeling convolutional neural network for detection of plant leaf spot diseases. In: 3rd International Conference on Electronics and Sustainable Communication Systems, pp. 1187–1192. IEEE (2022)
19. Sunitha, G., et al.: Modeling of chaotic political optimizer for crop yield prediction. Intell. Autom. Soft Comput. **34**(1), 423–437 (2022)
20. Sunitha, G., Sudeepthi, A., Sreedhar, B., Shaik, A.B., Farooq, C.: RetinaNet and vision transformer-based model for wheat head detection. In: 2023 5th International Conference on Inventive Research in Computing Applications (ICIRCA), pp. 151–156. IEEE (2023)
21. Reddy Madhavi, K., Kora, P., Venkateswara Reddy, L., Avanija, J., Soujanya, K.L.S., Prabhakar, T.: Cardiac arrhythmia detection using dual-tree wavelet transform and convolutional neural network. Soft Comput. **26**, 3561–3571 (2021)
22. Reddy Madhavi, K., Vijaya Sambhavi, Y., Sudhakara, M., Raju, S.K.: COVID-19 Isolation Monitoring System. LNDECT, vol. 63, pp. 601–609 (2021)
23. Reddy Madhavi, K., Vijaya Sambhavi, Y., Sudhakara, M., Srujan Raju, K.: COVID-19 isolation monitoring system. In: Reddy, K.A., Devi, B.R., George, B., Raju, K.S. (eds.) Data Engineering and Communication Technology. LNDECT, vol. 63, pp. 601–609. Springer, Singapore (2021). https://doi.org/10.1007/978-981-16-0081-4_60
24. Reddy Madhavi, K., Madhavi, G., Rupa Devi, B., Kora, P.: Detection of pneumonia using deep transfer learning architectures. Int. J. Adv. Trends Comput. Sci. Eng. **9**(5), 8934–8937 (2020). https://doi.org/10.30534/ijatcse/2020/292952020. ISSN 2278-3091
25. Shereesha, M., Hemavathy, C., Teja, H., Madhusudhan Reddy, G., Kumar, B.V., Sunitha, G.: Precision mango farming: using compact convolutional transformer for disease detection. In: 13th International Conference on Innovations in Bio-inspired Computing and Applications, vol. 649, pp. 458–465. Springer, Cham (2023)
26. Pan, J., Xia, L., Wu, Q., Guo, Y., Chen, Y., Tian, X.: Automatic strawberry leaf scorch severity estimation via faster R-CNN and few-shot learning. Eco. Inform. **70**, 101706 (2022)

Analysis of Surveillance Videos Using Convolutional Neural Networks for Human Action Recognition

A. Veerender[1](✉), K. Shilpa[1], Rajesh Tiwari[2], Lal Bahadur Pandey[2], B. Archana[3], and Anil Kumar[3]

[1] CMR Technical Campus, Hyderabad, Telangana, India
cmrtc.paper@gmail.com, kshilpa.cse@cmrtc.ac.in
[2] CMR Engineering College, Hyderabad, Telangana, India
lalbahadurpandey@cmrec.ac.in
[3] CMR College of Engineering and Technology, Hyderabad, Telangana, India
{archana.prakash,anilkumar}@cmrcet.ac.in

Abstract. The procedure of processing a picture, obtaining information, and evaluating the information to extract domain-specific knowledge is known as data analytics. The present development involves not only evaluating any photograph to determine data retrieval, but also evaluating live surveillance films to identify events occurring within its coverage area. These systems are going to be put into use immediately. Employing a training model like an artificial neural network makes automated face recognition from security footage easier. Estimating skin color aids in hand detection. This research project uses surveillance camera footage from exams to identify suspicious activity, including object exchange, a fresh entry, looking through someone else's answer sheet, and person transfer. These days, people are more concerned with the impartiality of monitoring, thus it is important to identify unusual conduct so as to maintain the integrity of criminal identification. The majority of approaches used now suggest models for specific abnormal conduct. We identify the optical flow of the video data in this framework and suggest a 3D convolution neural network model to address the issue. This calls for the use of facial detection, criminal detection, and the ability to distinguish among the faces of various people. The rate of mistake caused by human oversight will be reduced with the automation of "criminal identifying & detection".

Keywords: Criminal identification · artificial neural network-based sparsity learning · anomaly detection · video surveillance etc.

1 Introduction

In human features and behavior patterns comprise important impact on the detection of individuals? One important source for these determinations is visual information. Similar visual data is available through surveillance videos, that can be seen live or recorded for later use. Beyond the realm of video analytics, the recent move toward "automation"

A. Bajaj et al. (Eds.): IBICA 2023, LNNS 1229, pp. 342–350, 2025.
https://doi.org/10.1007/978-3-031-78940-3_36

having an effect. Numerous applications, including motion detection, person identification, aberrant activity recognition, vehicle counting, people counting in crowded areas, and human activity prediction, can be made with video analytics. The technical terms for the two elements employed in this domain to identify a person are gait recognition and facial recognition, respectively. Face recognition is the more flexible of these two methods for automatic human identification from security footage. Face recognition technology can be used to anticipate a person's head orientation, which helps predict their behavior. In numerous applications, such as identity verification, motion detection with facial recognition, and presence/absence detection at a certain location and time, motion recognition combined with facial recognition is highly helpful. A system that successfully detects and recognizes suspicious conduct among students in an examination hall is also created by utilizing human interactions, including delicate eye contact, head motion detection, hand gesture identification, and estimating.

This technology offers a facial recognition algorithm for the detection of questionable human activity. Research and security are the two primary areas in which video processing are employed. This kind of technology keeps an eye on live videos using clever algorithms. Time and computational complexity are two important considerations in real-time system design. For time-sensitive applications including bank robbery detection, tracking medical systems, identifying and reporting suspicious activity at train stations, exam holes, etc., the system that utilize an algorithm about a comparatively lower time difficulty, employing less hardware resources, and producing good consequences is going to be more beneficial.

2 Literature Review

Advancements in video technology have resulted in a rise in its application in numerous domains. Yet, because of the volume of data required, identifying objects in video footage is still a difficult task. Robotics, safety, surveillance, and health all depend on video monitoring. Important data including object interactions, human identity, and behavior are extracted. Video monitoring thus offers a wealth of study opportunities.

The paper presents a real-time human activity detection system that employs Principal Component Analysis (PCA), Radon transform (RT), and Linear Discriminant Analysis (LDA). Different human actions are identified by Artificial Neural Nets (ANN) [1].

A binary image is created from the data that was retrieved via optical flow. The feature vector is then extracted based on the binary pictures using the Histogram of the Oriented Gradient (HOG) descriptor. In order for creating a trained model, the Support Vector Machine (SVM) classifier receives these feature vectors as training features [2].

In this study [3], four different combination of two descriptors of features and two classifiers are used to accomplish video-based action detection on the KTH dataset. Three-dimensional Scale Invariant Feature Transform (3D SIFT) and Histogram of Oriented Gradient Descriptor (HOG) are the feature descriptors and classifiers that are utilized; they are Support Vector Machine (SVM) as well as K Nearest Neighbour (KNN). Employing descriptors, features are taken out of training video frames and clustered to create a Bag-of-words model.

Our suggestion is a completely unsupervised dynamic sparse coding method that uses online sparsely reconstructibility of query signals from a fundamentally learned event dictionary to create a sparse coding base for the purpose of identifying anomalous events in videos. Our algorithm uses a principled convex optimization formulation that enables the joint inference and updating of an online dictionary and a sparse reconstruction code. This approach is based on the intuition that regular events in a video are more probable to be reconstructible from an event dictionary, while unusual events are not [4].

Two things this paper contributes to: i) To speed up the anomaly prediction and enable parameter optimization, we suggest a TSC that can be mapped to a sRNN. ii) We create a massive dataset that, in terms of both data volume and scene diversity [5], is even larger than the total of all currently available datasets for anomaly detection.

This method [6] uses temporal pictures and convolutional neural networks (CNN) to anticipate human actions. A kind of deep learning model called CNN is capable of automatically picking up information from training footage. Modern techniques have demonstrated remarkable efficiency, despite they are computationally intensive. An additional issue is that numerous techniques need precise understanding of human postures.

Reducing a data set's dimensionality whereas preserving as much of the variance as feasible in a set of numerous interconnected variables is the main goal of principal component analysis (PCA). To do this, an additional set of uncorrelated variables called principle components (PCs) is transformed. The PCs are arranged so that the initial few preserve the majority of the disparity found in all of the unique variables [7].

Identification of Human movement A combined set of support vector machines is used to combine various features from various angles in order to improve the efficiency of classification. The outputs of individual classifiers have been combined using the Dempster-Shafer fusion and products rule generated by algebraic combiners [8].

Despite over 350 publications throughout this time, computer vision research on human motion recording is still quite active [9, 10]. As well as unique approaches to automated initialization, tracking, posture estimation, and movement detection, some noteworthy research advancements are highlighted. Reliable tracking and pose estimation in natural situations have been the focus of recent study. Additionally, advancements have been achieved in the area of automatically comprehending human behavior and activities.

Since then, the authors [11] have modified the code base multiple times. They released the code for both learning and applying these models under very broad licensing terms. This approach has a lot of intriguing natural variations. For instance, you may score component positions differently, apply the method to depth data, utilize various characteristics, quantities, or types of parts, expedite the calculations of the scores using different approximations, and so on. This article is a great place to start reading computer vision papers since it's easy to understand and "makes sense." Additionally, object detection now rests on this foundation.

An example-based learning strategy for identifying vertical frontal views of human faces in composite settings is presented by the authors in [12]. The method uses a few view-based "face" and "nonface" model clusters to simulate the distribution of human face patterns. The variation in the feature vector between the distribution-based approach and the local image pattern is calculated at each image point. Depending on the differential feature vector measurements, a trained classifier decides whether or not a human face is present at the present image spot. We demonstrate experimentally that the "nonface" clusters we incorporate in our distribution-based model and the separation metric we use to compute dissimilar feature vectors are equally essential to the effectiveness of our approach.

It proposes [13–15] an improved per-frame evaluation technique that enables us to undertake informative comparisons and probes, evaluating our results concerning occlusion and scale, and we assess the effectiveness of sixteen state-of-the-art detectors that have been trained across six data sets. This study gives us a structure for evaluating future efforts and enables us to assess the state of the art [16]. These tests demonstrate that an organization still has a lot of opportunity to grow even with notable advancements. Specifically, detection performs poorly at low frequencies and for pedestrians that are partially obscured.

3 Proposed Model

We identify the optical flow of the photographic information in this system and recommend a Convolution Neural Networks model to address the issue. As seen in Fig. 1, the suggested approach takes the temporal and spatial characteristics out of the image data and uses them to feed the classifier directly for model learning or inference. Our own dataset experiments show that the proposed model works better than current methods.

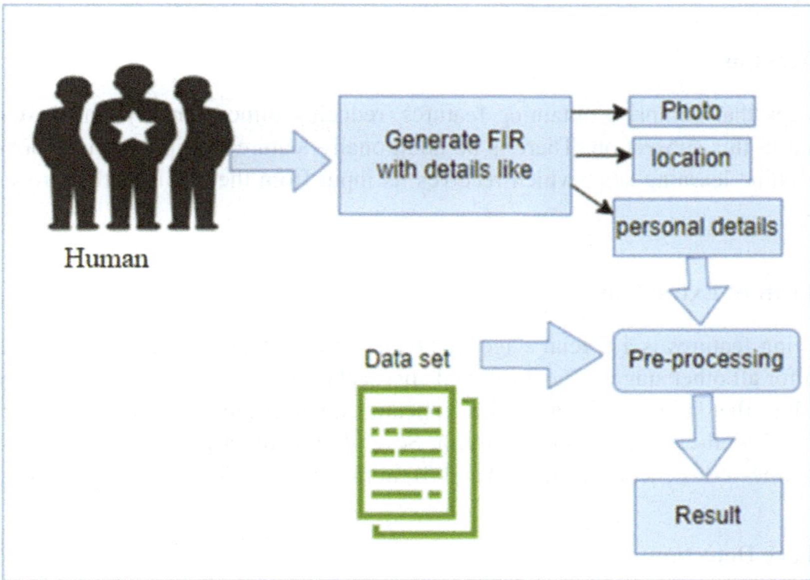

Fig. 1. Proposed System Architecture

Recently, there has been an unusual rise in both the crime rate and the number of offenders, which raises serious concerns regarding security-related issues. Although the protection of people's property and lives is the police's major priority, preventing crime and recognizing criminals are the issues that police personnel deal with first. However, the number of police personnel available to combat crime is restricted. Since the development of safety technology, numerous public and private spaces have CCTV cameras placed to enable surveillance. Suspects at the scene can be identified using the CCTV footage. This research suggested a computerized facial recognition method for criminal databases utilizing a cascade classifier based on recognized Haar features. In real time, this technology will be able to instinctively identify and recognize faces. Determining the face's precise placement remains a difficult undertaking. Researchers have frequently employed the Viola-Jones framework to locate faces and objects in an image that is given. Public communities like OpenCV share facial recognition classifiers.

3.1 Preprocessing

Preprocessing techniques like luminance normalization, registration and realignment among image sequences, and recognizing faces are usually necessary when dealing with visual inputs (images). Presumptive recordings or speech signals, for example, can also require preprocessing like segmentation. Viola and Jones suggested the face detection algorithm that is currently the most widely used. Additionally, a number of commercial facial expression analysis programs have recently been extensively utilized as preprocessing instruments, freeing up academics to concentrate on extracting high-level data.

3.2 Manipulation

The steps that comprise obtaining features, reducing dimensionality, and fusion are covered in this subsection. There is no additionally feature manipulation occurring in the machine learning step, which receives its input from the result of this processing stage.

3.3 Feature Extraction

Extracting features is a crucial stage in the analysis workflow since it forms the foundation for all other stages. The examined approaches use a variety of feature collection algorithms that fall into two categories: appearance-based and geometry-based, depending on the well-established taxonomy in. Several dynamic aspects are obtained based on the time-series encompassing (a) and (b) in the realm of depression evaluation.

3.4 Face Detection

In this instance facial features are divided into four categories: mouth/eyes, AUs, facial distinctions, and complete face features.

CNN

A basic ConvNet is a series of layers where each layer uses a distinct function to convert one volume of activations to a different one. Convolutional Layer, Pooling Layer, and Fully-Connected Layer are the three primary layer types that we use to construct ConvNet topologies (that's precisely as seen in normal neural networks). These layers will be stacked to create the whole ConvNet architecture.

INPUT[32 × 32 × 3] will save the image's raw pixel values; in this example, the image has three color channels (R, G, and B) and measures 32 by 32.

By evaluating the dot product among every one neuron's weight and the tiny input volume region to which it is attached, the CONV layer will calculate the output of neurons that are related to local regions in the input. If we choose to utilize 12 filters, this may produce a volume like [32 × 32 × 12].

An element-wise activation function, including the max(0,x) thresholding at zero, will be applied by the *RELU* layer. The volume's dimensions remain the same as a result ([32 × 32 × 12]).

Downsampling throughout the width and height of the spatial dimensions will be performed by the POOL layer, producing a volume like [16 × 16 × 12].

4 Results and Analysis

The suggested system's outcome is the use of image processing (CNN) to recognize criminals, identify and recognize crime scene elements, analyze and retrieve the optical flow of video data, and recommend a CNN model to address the issue.

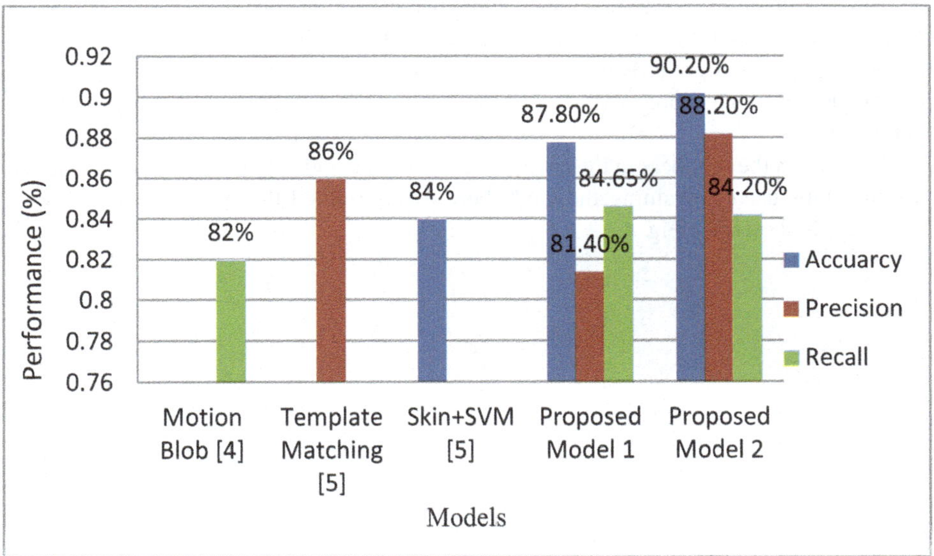

Fig. 2. Presentation of Different Methods

The input format for this system is live camera and video dataset. We are utilizing the pre-processing, feature extraction, and classification modules—all of which make use of our CNN algorithm as shown the Fig. 2 and Fig. 3. It aware that we are processing data and using a trained dataset on the system.

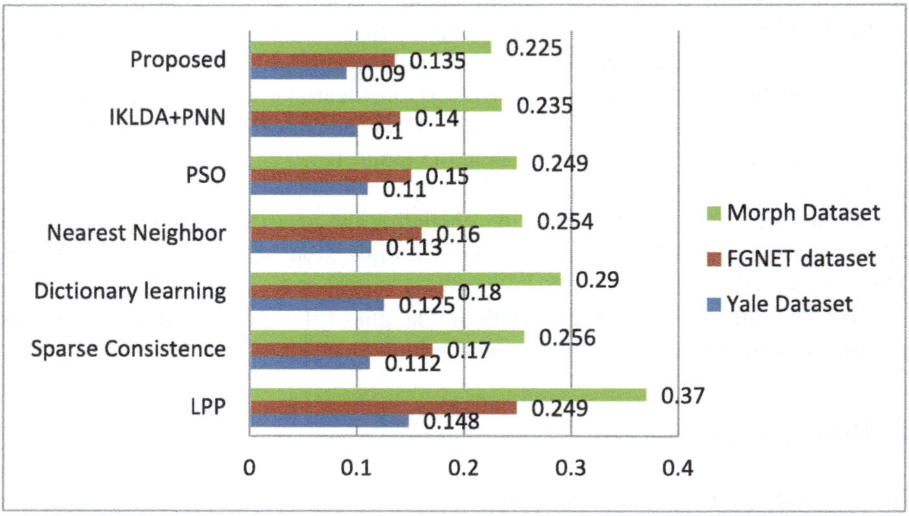

Fig. 3. A hybrid feature extraction on face for efficient face recognition

A kind of pattern recognition called "human recognition" involves storing a person's face in a delta base for use in the future. Researchers in software for human face recognition have put in a great deal of effort and diligence in developing a reliable system that focuses on both representation and recognition through the use of artificial neural networks.

Transform the numbers in the array to be between 0 and 1 by dividing by 255. These are the steps used to training the CNN here it represented the two proposed models accuracy as shown the Fig. 4.

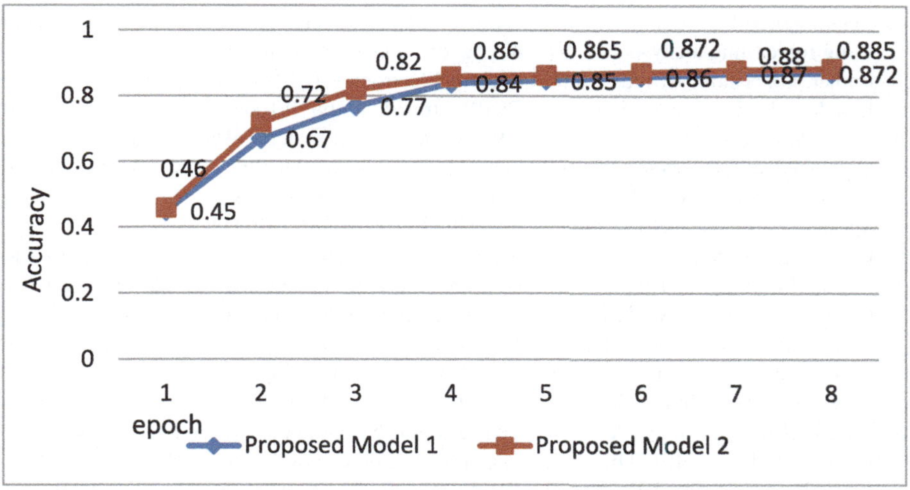

Fig. 4. The performance of two "flow image" on CNN (ROC)

5 Conclusion

We provide a unified deep learning-based architecture for CCTV surveillance-based criminal detection of events. The three pieces that make up the suggested system are intended to accomplish three goals related to criminal identification using neural networks. Crime scene incidents that occur in real life are complex and varied. It is challenging to enumerate every potential abnormal occurrence. The crime scene/dead body identity recognition method should therefore ideally not rely on any previous knowledge about the occurrences. Stated differently, anomaly detection ought to be carried out with the least amount of oversight. Techniques utilizing sparse coding are regarded as exemplary techniques that produce cutting-edge anomaly detection outcomes. These approaches generate the normal occurrence dictionary based on the assumption that only a tiny initial segment of a video comprises normal events. The fundamental tenet of crime scene identification is then that unusual occurrences cannot be reliably recreated using the regular event vocabulary. Yet, these methods result in high false alarm rates for various routine actions since the surroundings collected by security cameras could vary significantly over the years (e.g., during various periods of the day).

References

1. Khan, Z.A., Sohn, W.: Real-time human activity recognition system based on radon transform. IJCA Special Issue on Artificial Intelligence Techniques - Novel Approaches Practical Applications, AIT (2011)
2. Jagadeesh, B., Patil, C.M.: Video-based action detection and recognition human using optical flow and SVM classifier. In: IEEE International Conference on Recent Trends in Electronics Information Communication Technology, 20–21 May 2016, India (2016)
3. Raju, S., et al.: Image recognition and content retrival to build context: a survey. J. Adv. Res. Dyn. Control Syst. **11**, 1656–1666 (2019)

4. Mahmood, M.R., Patra, R.K., Raja, R., Sinha, G.R.: A novel approach for weather prediction using forecasting analysis and data mining techniques. In: Saini, H., Singh, R., Kumar, G., Rather, G., Santhi, K. (eds.) Innovations in Electronics and Communication Engineering. LNNS, vol. 65. Springer, Singapore (2019). https://doi.org/10.1007/978-981-13-3765-9_50
5. Sahu, A.K., Chandra, V.K., Sinha, G.R.: Modeling and system-level computer simulation approach for optimization of single-loop CT sigma delta ADC. In: Kolhe, M., Trivedi, M., Tiwari, S., Singh, V. (eds.) Advances in Data and Information Sciences. LNNS, vol. 39. Springer, Singapore (2019). https://doi.org/10.1007/978-981-13-0277-0_28
6. Rachapudi, V., Venkata Suryanarayana, S., Subha Mastan Rao, T.: Auto-encoder based K-means clustering algorithm. Int. J. Innov. Technol. Exploring Eng. **8**(5), 1223–1226 (2019)
7. Gothane, S., Sarode, M.V., Thakre, V.M.: Prediction for Indian road network images dataset using feature extraction method. In: Bapi, R., Rao, K., Prasad, M. (eds.) First International Conference on Artificial Intelligence and Cognitive Computing. AISC, vol. 815. Springer, Singapore (2019). https://doi.org/10.1007/978-981-13-1580-0_12
8. Luo, W., Liu, W., Gao, S.: A revisit of sparse coding based anomaly detection in stacked RNN framework. In: The IEEE International Conference on Computer Vision (ICCV) (2017)
9. Jin, C.B., Li, S., Do, T.D., Kim, H.: Real-time human action recognition using CNN over temporal images for static video surveillance cameras. In: Information and Communication Engineering, Inha University, Incheon, Korea (2015)
10. Rajesh Tiwari, M., et al.: Enhanced power quality and forecasting for PV-wind microgrid using proactive shunt power filter and neural network-based time series forecasting. Electr. Power Compon. Syst. (2023). https://doi.org/10.1080/15325008.2023.2249894
11. Moeslund, T.B., Hilton, A., Krüger, V.: A survey of advances in vision-based human motion capture and analysis. Comput. Vis. Image Underst. **104**, 90–126 (2006)
12. Felzenszwalb, P.F., Girshick, R.B., Mcallester, D., Ramanan, D.: Object detection with discriminatively trained part-based models. IEEE Trans. Pattern Anal. Mach. Intell. **32**(9), 1627 (2010)
13. Sung, K.K., Poggio, T.: Example-based learning for view-based human face detection. IEEE Trans. Pattern Anal. Mach. Intell. **20**(1), 39–51 (2002)
14. Wojek, C., Dollar, P., Schiele, B., Perona, P.: Pedestrian detection: an evaluation of the state of the art. IEEE Trans. Pattern Anal. Mach. Intell. **34**(4), 743 (2012)
15. Prashanthi, M., Chandra, M.M.: Hybrid optimization-based neural network classifier for software defect prediction. Int. J. Image Graph. **24**, 2450045 (2023). https://doi.org/10.1142/S0219467824500451
16. Sastrt, J.S.V.R.S., Suresh, V.: A survey paper on frequent itemset mining. In: Journal of Physics: Conference Series, vol. 1228, no. 1, p. 012050. IOP Publishing (2019)

An Implementation of Convolutional Neural Network-Based Architecture to Address Facial Sentiment Analysis

M. Ravindran[1](✉), G. Madhukar[1], A. Srinivasula Reddy[2], Mandapati Raja[2], R. Venkateswara Reddy[3], and R. Suhasini[3]

[1] CMR Technical Campus, Hyderabad, Telangana, India
cmrtc.paper@gmail.com
[2] CMR Engineering College, Hyderabad, Telangana, India
[3] CMR College of Engineering and Technology, Hyderabad, Telangana, India
{venkatreddyvari,suhasini}@cmrcet.ac.in

Abstract. In the field of computer vision, automatically analyzing these unsaid feelings has proven to be an intriguing and difficult problem. Its potential uses span a variety of domains, such as psychology, process automation, and product marketing. These wasn't an easy undertaking, since people vary a great deal in how they convey their feelings using their facial expressions. Computer vision is one of the many domains of research that has seen significant development thanks in large part to machine learning, and more specifically deep learning. We address the issue of facial sentiment analysis in this research study by introducing an architecture implemented with convolutional neural networks (CNNs). FER-2013 public dataset has been used by us for training and testing. This work has been done in stages: first, the data has been preprocessed; next, features have been extracted; and last, our trained model network has been used to classify the data. Our experiment's results are quite promising and represent a breakthrough in the field of automated facial sentiment analysis.

Keywords: Face identification · Convolutional Neural Network · Deep Learning · Facial Sentiment Analysis · and Network Architecture

1 Introduction

In entirety, individuals tend to have distinct facial expressions. For instance, different people may experience various emotional consequences from a smile, a conversation, or any other form of visible emotion. The algorithm used must be capable of handling the variety in facial appearance in order to recognize facial expressions. A small number of these issues have been solved by the most algorithm to date. Human emotions are significant characteristics that are necessary for productive social interactions. Verbal and nonverbal communication are both possible in humans. Emotion is a powerful tool in nonverbal communication since it expresses how people feel about a topic, and research has shown that gestures are more persuasive than spoken words in conversation [1].

A. Bajaj et al. (Eds.): IBICA 2023, LNNS 1229, pp. 351–360, 2025.
https://doi.org/10.1007/978-3-031-78940-3_37

The two main categories of feature extraction techniques for emotion recognition are appearance-based techniques and geometric feature-based techniques. The first approach takes into account the form and placement of facial features including the mouth, nose, eyebrows, and eyes, whereas the second way takes into account specific areas or the entire face. Computers still struggle with recognizing expressions since it is a complex problem to distinguish an expression's feature space. Some issues arise from the possibility that characteristics drawn from two faces with the same expression could differ, whereas features extracted from one face with two expressions could be equal, or because certain expressions, like "sad" and "fear," are extremely similar.

The proposed system's main objectives are to extract face features from the image, classify the image into seven basic emotion categories (Happy, Sad, Angry, Surprise, Fear, Disgust, and Neutral), and train the CNN using the image's facial expressions. The sentiment analysis is obtained by the use of a classifier. Through constant observation with a monocular camera and situational prediction, this analysis helps prevent disasters.

2 Literature Review

Real-time identification of emotions using the CNN model. Three datasets—CK+, JAFFE, and Own (home-brewed)—each with five photos were employed. The VGG-S network was used to classify images [1]. Following the acquisition of this freshly trained VGG-S, a regular webcam was used to link a video feed to the network. While relevant images are arranged across the face, this network concurrently identifies a fixed amount of faces per image in real time. It was shown that CK+ and own provided more training performance than the JAFFEE dataset. A review of various CNN models through varying numbers of hidden layers and filters is carried out [2]. Deeper CNN, 6-layer CNN, and shallow CNN are contrasted. By comparing the 6-layer and Shallow networks, we can observe that efficiency is improved by including an additional convolutional and max-pooling layer. Because the deeper CNN features a layer with dropouts to minimize overfitting, its testing accuracy is higher [3]. Every algorithm's detection rate has increased with more iterations, and multistep performs most effectively among those algorithms. The findings collected from various addressing techniques in testing data show that SGD may decrease error using an increase in the amount of iterations and that SGD with impetus is more effective than NAG. It appears that the shallow network has a strong recognition impact.

Paul Ekman's seminal work on emotion analysis [4, 5] revealed the identification of six main emotions: fear, happiness, surprise, anger, sadness, and disgust. Neutral had been added to this list in later publications, thereby establishing the seven fundamental human emotions for analytical purposes. A bidirectional technique has been used in the analysis of previous publications. Feature extraction from the raw pictures is the first stage, after when a classification methodology [6], is applied. These methods performed rather well on less complex image datasets in a controlled setting, but they were less effective on more difficult datasets with greater intra-class variability. Previous publications have employed a variety of image characteristics, such as local binary patterns (LBP) [7], histogram gradient orientation [8], quantification of local phase, and scale invariant features transform (SIFT) [9], for the gathering of facial features. Deep learning models have been found to be useful for analyzing emotions in recent research. Shows a Boosted Deep Belief Network (BDBN) in three recurrent training stages, as described by the researchers responsible for this investigation. In their study [10], Zhang et al. offer a novel facial feature analysis approach based on deep neural networks (DNNs) that helps identify the relationship between high-level semantic information and SIFT features.

In this work [11], we not only demonstrate that CNNs are capable of outstanding results, but we also present a method to identify the facial features that affect the CNN's predictions. Using facial expression data, we first train a zero-bias CNN and reach the highest possible accuracy on two facial expression recognition benchmarks, as far as we are aware:

It was not possible to prove that at least some facial expressions of emotion are universal despite recent studies demonstrating that members of literate cultures associated the same emotion concepts with the same facial behaviors. This is because the cultures under comparison had all been exposed to similar mass media presentations of facial expression, which may have helped the individuals in each culture identify the distinctive facial expressions of other cultures [12].

CNNs were employed by the top three competitors in the 2014 ImageNet competition to solve the given problem. Among them, the GoogleNet architecture had a startling categorization error rate of 6.66% [13, 14]. AlexNet is a notable architecture that demonstrated remarkable outcomes. It is built using the standard CNN layered technique, which consists of convolution layers, max-pooling layers, rectified linear units, or ReLUs, and finally some layers that are completely linked. Additionally, AlexNet was one of the first to propose leveraging the network's layer of dropouts to address the over-fitting issue [15, 16].

3 Projected Model

There are many techniques for identifying emotions on the face that make use of features that are highly efficient, such as Gabor filters, Scale Invariant Feature Transform (SIFT) descriptors, Local Binary Patterns (LBP), or Histograms of Oriented Gradients (HOG) features. The suggested approach use deep CNN to identify the seven different states of emotion on the face, as depicted in Fig. 1. Because the convolutional neural network is more reliable and exact, it is used. This method is being developed for emotion expression recognition and consists of the following steps: pre-processing, CNN classifier-based facial expression recognition, face extraction, face detection, and human detection.

Our proposal is to use the CNN framework as a classification technique to improve facial sentiment evaluation systems. We have used the publicly accessible FER2013 dataset since deep network training requires a substantial amount of data. In the next part, we describe our network design, list the properties of the dataset we selected, and then list the efficacy metrics we utilized in our assessment.

Face Detection: Face detection is a technique that recognizes faces in digital photos and finds uses in many different fields. It is thought to belong to a particular type of object detection. The anterior face detection algorithm relies on this information. Bit by

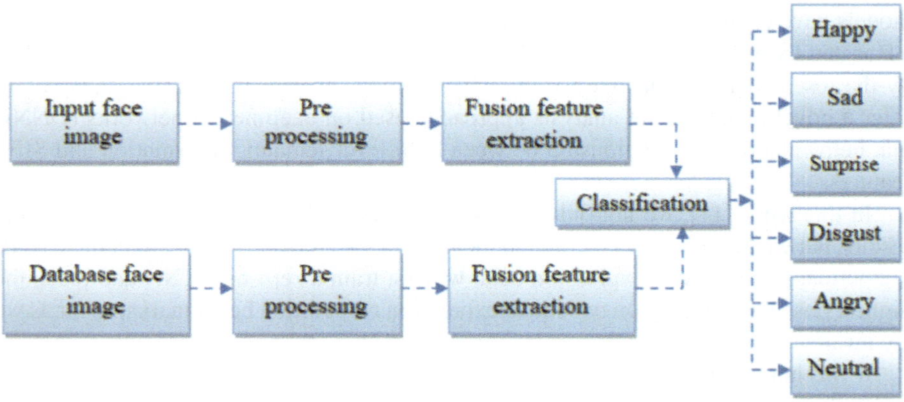

Fig. 1. Proposed System model

bit matching of a person's image is achieved, which is akin to image detection. There is a match between the photographs in the database.

Network Architecture

The structure we use is really simple, using a 48 × 48 input that goes through several convolution layers, and followed by a pooling layer, and ends as Fig. 2 illustrates.

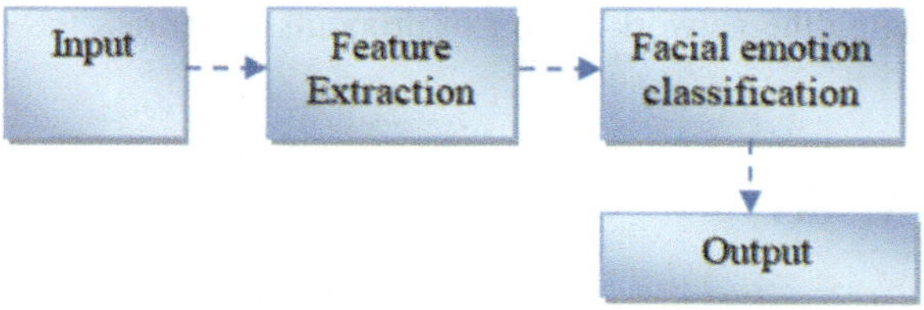

Fig. 2. Facial expression recognition system

CNN Model

CNN: As illustrated in Fig. 3, the deep convolutional network recognizes seven states of face expressions through three stages of feature learning, choosing, and categorization. It was challenging to train a model with a maximum of two layers because this one is GPU-trained and has several layers.

Layer of Convolution

The primary role of the Convolution layer is to handle the higher-level features of an image. Typically, the initial convolution layer focuses on analyzing low-level attributes such as edges, color, and gradient direction. As the architecture progresses, it starts recognizing and capturing the high-level features present in subsequent layers.

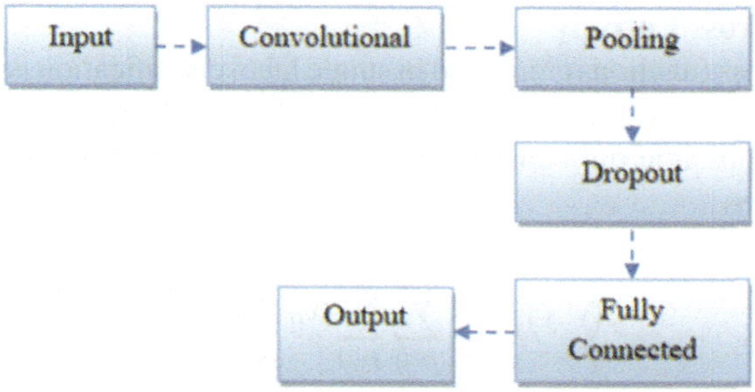

Fig. 3. CNN model

Layer of Pooling

Pooling layers reduce the spatial dimensions of the convolved features, leading to a decrease in data processing requirements through dimensionality reduction.. It can also be used to identify prominent features that exhibit positional and rotational invariance, which preserves the effectiveness of training. Different kinds of pooling techniques exist: Maximum and Mean Swimming. Whereas the AvgPool layer takes the standard deviation of the values contained within the kernel window and delivers it, MaxPool produces the maximum value from the portion of the image matrix represented by the kernel.

Thick Layer

The two-dimensional matrix undergoes a transformation into a one-dimensional form, necessitating the use of a fully connected neural layer to grasp non-linear combinations from the high-level features obtained after applying pooling and convolution processes to the input. This reduction in size and relevance regarding high-level features is achieved through weight updates and backpropagation to update these layers.The robustness of the model is maintained by removing a few random neurons to avoid overfitting. The total number of classes in the issue is equal to the number of neurons in the output layer.

Function of Activation

These functions are employed to give the model non-linear characteristics. When it comes to neural networks, they transform the input signal into the output signal that comes from a node. Numerous activation functions exist, although the following are the most popular ones:

ReLU, also known as rectified linear units, is referred to as the ramp function. Mathematically, it can be easily understood as $relu(x) = max(0, x)$.

The softmax formula takes M real numbers as input and transforms them into M probabilities that are proportional to the values of the input numbers.

Performance Metrics

Given that the task at hand involves multi-class classification, the matrix of confusion is employed to indicate efficiency and the crossed entropy is employed as a loss function.

Unconditional Cross Entropy

A number of the loss functions employed in single label classification issues is categorical crossentropy; that is, categorical crossentropy is useful in categorization issues where there is only one possible right answer. This can be expressed mathematically for cross entropy equation as

$$L(y, \hat{y}) = -\sum_{j=0}^{M}\sum_{i=0}^{N}(y_{ij} * \log(\hat{y}_{ij}))$$

Confusion Matrix

As demonstrated by the Table 1 matrix, that is employed to describe how well a classification model performs on a dataset whose true values have been determined by us.

Table 1. Confusion Matrix format

Predicted Values	Actual Values: Positive (1)	Actual Values: Negative (0)
Positive (1)	TP	FP
Negative (0)	FN	TN

Dataset

The 2013 ICML challenge on Representational Learning introduced the Facial Expression Recognition 2013 database, commonly referred to as FER2013. [11]. This collection contains an astounding 35,887 face photos, the most of which were shot in uncontrolled environments. The training set has almost 28k photos using a resolution of 48 × 48, while the validation and testing sets together have 3.5k images. The Google API for image search using faces labeled using one of the seven fundamental sentiments was used to create this database. When compared with different datasets, FER2013 is superior because it contains a greater variety of partially covered faces and photos that have low contrast, both of which improve training our model to avoid over fitting and make it more applicable.

Database Preparation

Images are saved in a certain location after being removed and reduced to a specific size. Figure 5 illustrates the way all emotion categories are represented in this image, that is then categorized into that specific emotional state. Of the data in this image, 80% is utilized to train the images and 20% is employed for testing.

The dataset's distributed graph is displayed in Fig. 4.

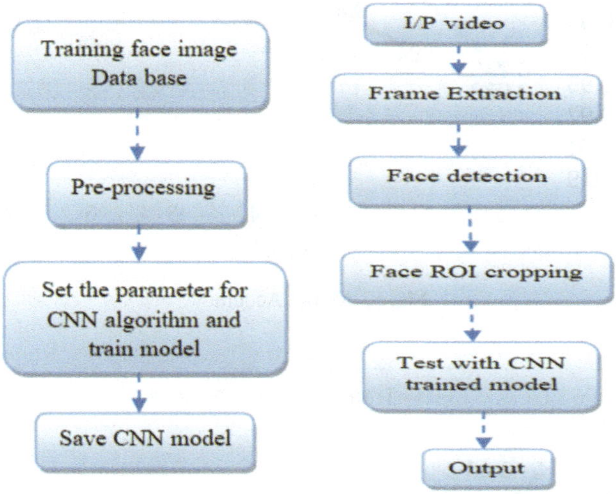

Fig. 4. a) Training data. b) testing data

4 Results and Analysis

Using its own dataset, the suggested system was developed for research on face expressions. Thirty thousand of the fifty video frames in the dataset—which have a range of emotional content—were retrieved and trained. The following expressions are represented by images in the dataset: fear, disgust, surprise, happy, angry, sad, and neutral. As seen in Figs. 5 and 6, methods for preprocessing were used in this work to enhance accuracy by extracting precisely stated emotions from the face.

The CNN model is trained using the VGG net. The model yields a precision of 97% for training and a precision of 96% for testing. It also shows extremely little loss, with less than 0.1% for education and less than 0.2% for testing. It is noted that the disgust and furious labels are correlated, and the network does an excellent job of predicting the happy, sad, and angry labels.

The subsequent assessments were made after the model had been trained and evaluated on the corresponding training and testing datasets.

metrics:

Trainingloss: 0.223031098232

Trainingaccuracy: 92.0512731201

Testingloss: 2.27945706329

Testingaccuracy: 57.4254667071

The precision in this case shouldn't convey the correct idea for a multiclass categorization task. The confusion matrix for this model is displayed here in Table 2.

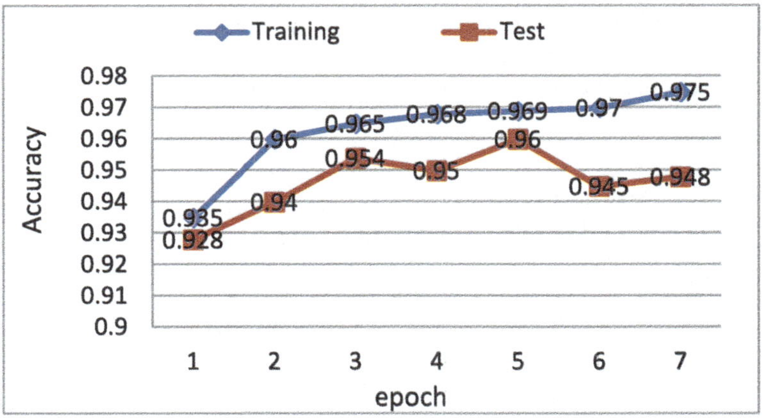

Fig. 5. Model Accuracy

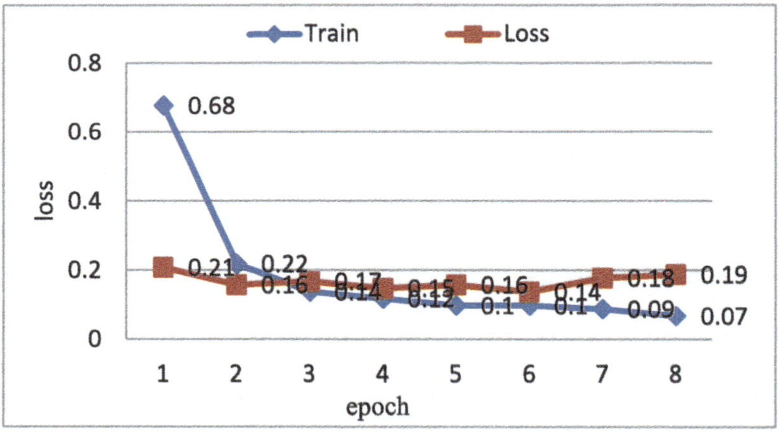

Fig. 6. Model loss

Table 2. Confusion matrix for the prediction set

	Angry	Disgust	Happy	Sad	Surprise	Neutral
Angry	215	10	30	68	9	88
Disgust	11	24	4	6	2	7
Happy	25	2	695	38	20	80
Sad	66	4	57	286	11	152
Surprise	8	2	28	10	304	28
Neutral	46	2	66	90	8	332

5 Conclusion

In this research, an emotion recognition technique based on facial feature point detection and classification into seven fundamental emotional classifications is presented. After training on the data, a convolutional neural network can identify and categorize the various emotional states on a face. Because CNN is the most accurate and exact model available, it is employed for optimal extraction of features and classification. The outcomes of the experiment demonstrate that CNN can automatically pick up on facial expression traits and improve its interpretation of facial emotions.

References

1. Wiens, N., Harper, R.G., Matarazzo, J.D.: Nonverbal Communication: The State of the Art. Wiley (1978)
2. Jones, M.J., Viola, P.: Robust real-time face detection. Int. J. Comput. Vis. **57**(2), 137–154 (2004)
3. Edwards, J., Jackson, H.J., Pattison, P.E.: Emotion recognition via facial expression and affective prosody inschizophrenia: a methodological review. Clin. Psychol. Rev. **22**(6), 789–832 (2002)
4. Chu, H.C., Tsai, W.W.J., Liao, M.J., Chen, Y.M.: Facial expression recognition with transition detection for students with high-functioning autism in adaptive e-learning. Soft Comput. 1–27 (2017)
5. Rajesh Tiwari, M., et al.: Enhanced power quality and forecasting for PV-wind microgrid using proactive shunt power filter and neural network-based time series forecasting. Electr. Power Compon. Syst. (2023). https://doi.org/10.1080/15325008.2023.2249894
6. Malyadri, M., Ravi Kumar, L.: Web innovation with IoT in social environment for sharing efficient information. J. Adv. Res. Dyn. Control Syst. **10**(13 Special Issue), 1466–1475 (2018)
7. Sulaiman, S., Hussin, S., Amir, Z.: Communication strategies among tertiary students in mlearning. Int. J. Eng. Technol. **7**(2.29), 655–659 (2018). https://doi.org/10.14419/ijet.v7i2.29.13993
8. Subha Mastan Rao, T., Malyadri, M.: A novel IOT based expert system to mitigate the problems in shrimp culture. J. Adv. Res. Dyn. Control Syst. **10**(4 Special Issue), 2031–2039 (2018)
9. Krishna, B., Amarawat, G.: Frequent item set generation using enhanced Apriori algorithm and multiple projection rule pruning algorithm. J. Adv. Res. Dyn. Control Syst. **10**(4 Special Issue), 1947–1953 (2018)
10. Krishna, B., Amarawat, G.: An enhanced scaling Apriori for association rule mining with frequent item set mining. J. Adv. Res. Dyn. Control Syst. **10**(6 Special Issue), 102–106 (2018)
11. Khorrami, P., Paine, T., Huang, T.: Do deep neural networks learn facial action units when, doing expression recognition?. In: IEEE Conference on Computer Vision (2015)
12. Paul, E., Friesen, W.V.: Constants across cultures in the face and emotion. J. Pers. Soc. Psychol. (1971)
13. LeCun, Y.: Generalization & network design strategies. Connectionism Perspect. 143–155 (1989)
14. Prashanthi, M., Chandra, M.M.: Hybrid optimization-based neural network classifier for software defect prediction. Int. J. Image Graph. **24**, 2450045 (2023). https://doi.org/10.1142/S0219467824500451

15. Ejaashwini Goud, J., Bhaskar, N., Kumar, V.N., Mubeen, S., Narasimharao, J., Unnisa, R.: A survey: classifying and predicting features based on facial analysis. In: Bhateja, V., Yang, X.S., Ferreira, M.C., Sengar, S.S., Travieso-Gonzalez, C.M. (eds.) Evolution in Computational Intelligence. FICTA 2023. Smart Innovation, Systems and Technologies, vol. 370. Springer, Singapore (2023). https://doi.org/10.1007/978-981-99-6702-5_25
16. Sapthami, I., Krishna, B.M., Bhaskar, T., Ravela, C.: Sentiment analysis using machine learning algorithms for customer product reviews. In: 2023 Second International Conference on Augmented Intelligence and Sustainable Systems (ICAISS), pp. 447–451. IEEE (2023)

An Analysis of Convolutional Neural Networks and Different Preconditioned Models for Object Detection

Bushra Tarannum[1(✉)], Ganapathi Antharam[2], B. Sree Saranya[3], Lal Bahadur Pandey[3], A. Shiva Kumar[4], and E. Gurumoorthi[4]

[1] Department of CSE (AI&ML), CMR Technical Campus, Hyderabad, Telangana, India
cmrtc.paper@gmail.com
[2] Department of CSE, CMR Technical Campus, Hyderabad, Telangana, India
[3] Department of CSE (AI&ML), CMR Engineering College, Hyderabad, Telangana, India
lalbahadurpandey@cmrec.ac.in
[4] Department of CSE, CMR College of Engineering and Technology, Hyderabad, Telangana, India
ashivakumar@cmrcet.org, dr.e.gurumoorthi@cmrcet.ac.in

Abstract. Deep learning has developed into a potent machine learning technique that combines several feature layers or data models to provide state-of-the-art outcomes. In many areas, such as object detection, the process of segmentation and image classification, deep learning has proven to function exceptionally well. In recent years, there has been a notable advancement in fine-grained picture classification using deep learning algorithms, whereby aim to distinguish between classifications at a lower level. This task has modest inter-class variance and considerable intra-class variance. In self-driving vehicle applications, identifying objects and pedestrian recognition are essential. Techniques built around convolutional neural networks have demonstrated notable improvements in precision and decision efficiency in real-time applications. The researchers of this study describe a number of cutting-edge deep learning methods, including InceptionV3, VGG-16, VGG-19, DenseNet-121, and a specially designed three-layer CNN system for object detection. The chosen model is trained and verified using a dataset of five classes of furniture that the user created. Following thorough testing, VGG-19 produced the maximum accuracy of 99.89%.

Keywords: CNN · confusion matrix · object classification · object detection · VGG

1 Introduction

The authors of the study successfully constructed a number of models that were pre-trained and five convolutional neural network techniques to handle this issue. VGG-16 outperformed the other five models evaluated in the study in terms of accuracy. Real-time photos have been employed to detect objects and for testing purposes.

A. Bajaj et al. (Eds.): IBICA 2023, LNNS 1229, pp. 361–369, 2025.
https://doi.org/10.1007/978-3-031-78940-3_38

Nowadays, object detection is used globally in many different industries, such as facial recognition, self-driving automobiles, pedestrian displays, and security cameras. An image, such as a picture, video, or webcam, is included in the Deep Learning subdiscipline of object detection [1]. These days, object recognition from videos is a crucial duty in video surveillance systems. Object recognition is a method for locating essential objects in video streams and grouping their pixels together. You can analyze the features of an image or video by using an object identification technique, which enables recognition, the localization, and detection of numerous things inside an image [2]. The act of finding and categorizing objects by utilizing geometric bounding boxes for recognizing them and putting them into groups is known as object detection. There are linkages between semantic segmentation and instance segmentation, object detection and object categorization, and more. Object detection, which is essential for both efficient industrial production and scientific research, can be used in a variety of applications, including face, text, pedestrian, logo, video, vehicle, and medical image detection [3].

From the 2014 R-CNN prediction, which was based on DNN, object detection has advanced significantly. Ultimately, R-CNN-based object identification techniques include SPP-NET, Fast-RCN, Faster RCNN, and R-FCN. These techniques had a great degree of efficiency, but a network structure is composed of many components that interact in a complex way [1]. High accuracy object recognition can be achieved using a complicated DNN model. Nevertheless, they necessitate an enormous amount of setup variables and processing, therefore they aren't truly suitable for many systems. While Convolutional Neural Networks (CNN) seem to be a promising option for systems, conventional methods can also be employed to address this difficulty. Since the SSD-MobileNet [4] paradigm cannot handle live video in an embedded device, we looked at methods that would enable us to expedite video processing times in a system while retaining the minimum amount of accuracy.

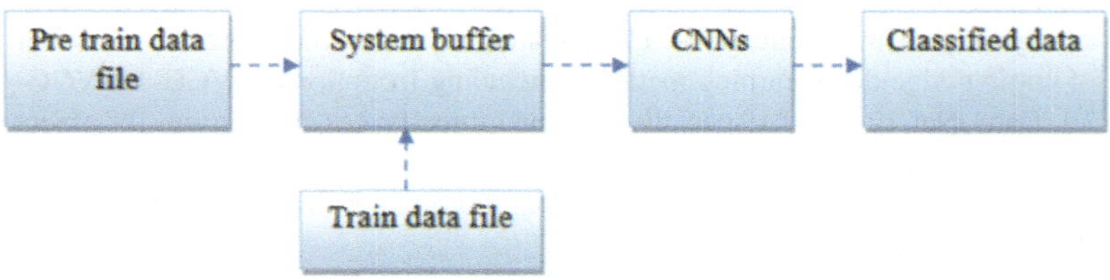

Fig. 1. Design frame work

Figure 1 shows the theoretical structure for the investigation. The pre-train as well as training datasets were the two sets of data that the framework had to work with.

The training data sample is a self-made, high-resolution dataset of five classes: bed, chair, sofa, swivel chair, and table photos with a pixel size of 1572 × 1548.

2 Literature Review

This paper [5] Single Shot Multi-Box Detection (SSD), a quicker detection technique based on a convolution neural network model, was presented in this research.With this work, the feature resampling step is removed, and all computed results are integrated into a single component.

In the present study [6], they used only single networks, single and multiple perspectives, and RGB-depth movies. Every dataset they used carries out single tasks, with one person working on a single task at a time. This article provides an overview of various state-of-the-art deep learning-based algorithms suggested for human action recognition on all three types of datasets, which are utilized in this paper [7] for face detection, mouth motions, and head position identification. This study suggests a number of methods for the online assessment and monitoring of deviant behavior using picture data. In this work, the researcher used a range of dataset models, convolutional neural networks, and deep neural networks [8]. Systems for object detection are used, including DenseNet, AlexNet, GoogleNet, ResNet, ResNeXt, and SENet. Included are various additional visual object detections, such as the identification of humans, animals, and held arms. This endeavor is not without its problems. By combining videos and pre-processing images, researchers can lower the false positive rate while also analyzing the system and using additional models. The author used three different pre-trained datasets. There are three: MobileNetv2, GoogleNet, and MobileNetv3. The findings of this study [9] indicate that MobileNetV3 [10], which offers the ideal balance between prediction speed and precision, is suitable for handgun identification. The training accuracy rate of the suggested model was 96%. The tests used real-time detection in real-time images and videos. Upon detecting the presence of a firearm, the system can notify the supervisor by text message. There are several uses for the suggested technique, one of which is the real-time identification of firearms in supermarkets. An extensive and detailed analysis of deep learning-based object detection algorithms is given in this review article [11]. It consists of three parts: loss functions, detection designs, and backbone networks. It also offers a comprehensive analysis of the challenging problems. The object detection algorithm presented in this study [12] can identify things at a maximum frame rate of 14 frames per second, meaning that even low-quality cameras can produce satisfactory results. All they have is a webcam that records at six frames per second. In our testing, the SSD approach showed both interior and outside input video frames through the camera; nonetheless, the objects' locations varied between two consecutive frames. A single frame's size is reduced to 300×300 pixels by the algorithm and the webcam's captured video [13–17].

3 Proposed Model

Identifying Objects Object detection is used to classify and localize each object within the bounding box. As simple as it sounds, object recognition in computer vision is the detection, identification, and localization of objects. Object identification is accomplished by image categorization. To lessen time-related issues in detection techniques, the most fundamental object-detection strategy is the sliding window approach. In this method, a window of appropriate size $M \times N$, also called a bounding box, is selected in order to search over the goal image. The two categories of these methods are classification-based methods and region proposal-based methods. In our model, one of the classification-based techniques is the single shot detector (SSD).

Three convolutional neural networks and a variety of pretrained models (VGG-16, VGG-19, DenseNet-121, and InceptionV3) were employed by the authors to train our image dataset; these models performed better on the real-time object detection data set previously. In order to prevent both overfitting and underfitting during model training, the author used batch normalization and set a number of hyperparameters (learning rate (lr) to 0.001, batch size to 32, beginning function ReLu and Softmax, epoch size 10). The Tensor flow and Keras frameworks were used for implementation, while the ReLu and Softmax activation functions were used for model training. A convolution layer with 64 filters makes up the first layer. Next, there is a stack of three CNN blocks, each with 32, 32, and 128 filters. Dimension reduction is done for every layer in each CNN block. DenseNet-121, VGG-19, and VGG-16 all have a single, 1024-size filter. Following the pre-processing of the data and the construction of the model, we trained our model using the Tensor Flow and Keras frameworks over ten epochs using CPU processing alone. Figure 2 shows the situation of implementation.

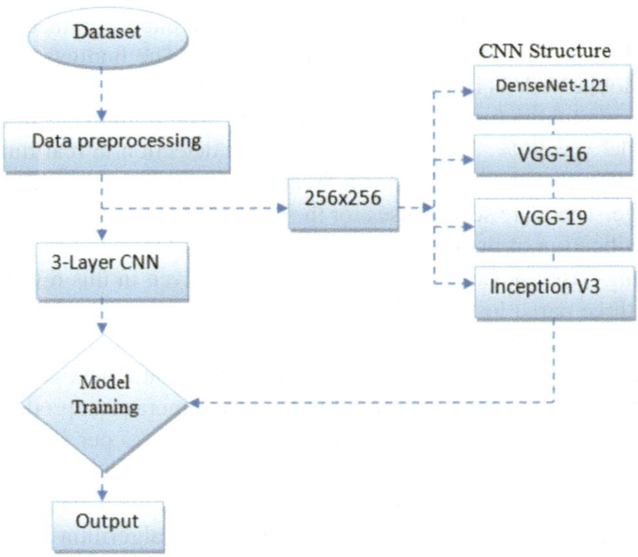

Fig. 2. Proposed model

Dataset and Models

A. Dataset

For this investigation, a self-made dataset consisting of five classes bed, chair, sofa, swivel chair, and table was utilized. In order to train our model, we used five different classes of furniture image classes: beds, chairs, sofas, swivel chairs, and tables. Each class included nine thousand training and one hundred validation images, as well as nine thousand training and one hundred validation images. The collection has 5000 image samples in total. Table 1 below illustrates our dataset's organizational structure.

Table 1. Number of classes and total images in dataset

Class	Training Images	Validation Images	Total Images
Bed	900	100	1000
Chair	900	100	1000
Sofa	900	100	1000
Swivel chair	900	100	1000
Table	900	100	1000
Total Images	4500	500	5000

Image Pre-processing

In order to reduce the background region, our dataset images are resized to 256 × 256 pixels during image pre-processing from 1572 × 1548 pixels. This size of pixel is optimal for convolutional neural networks. After that, the corrupted image is used for training and validation. DenseNet-121, InceptionV3, VGG-16, VGG-19, and CNN three-layer datasets are included. The authors specified a number of hyperparameters during the model's training, such as the training and validation split to 0.1 during picture pre-processing, the rescale to 1./255, the shear range to 0.2, and the zoom range to 0.2.

Methods

Convolutional neural networks in the modern era It is a state-of-the-art method since it can extract features from images without requiring laborious pre-processing, transfer learning, and parameter fine-tuning. These studies use InceptionV3, DenseNet-121, VGG-16, and VGG-19, which use transfer learning, which is frequently used in deep learning. Using CNN, we apply transfer learning to obtain the quality vector for arranging furniture (an object), and then separate the outcomes to determine which learning is optimal for object detection. The multiple models of convolutional neural networks over various criteria are presented in Table 2 below.

CNN layer: three layers Three layers make up CNN: convolutional, pooling, and completely connected. The convolution layer is the fundamental component of the CNN. It is primarily responsible for the network's computational load. The outputs of the network at certain places are transformed by the pooling layer using the summary statistics of nearby outputs. By doing this, the representation's spatial size is reduced, which lowers the amount of computation and load needed.

DenseNet-121: A DenseNet is a kind of convolutional neural network in which every layer is directly connected to every other layer using dense blocks.

The architecture of convolutional neural networks is called VGG-16, or Visual Geometry Group. Because 3x3 convolution layers have a lot of hyper-parameters, including

Table 2. CNN Models over different Criteria

Model Name	Size (MB)	Parameters (Millions)	Depth
VGG 16	529	138.62	24
VGG 19	550	142.82	26
DenseNet 121	32	9	122
Inception V3	92	24.2	158

filter size, they focused on using them. One straightforward and widely used convolutional neural network design is the VGG-16 architecture. VGG-16 is widely used in deep learning image categorization methods and is well-liked because of its simplicity of usage. Because of its benefits, VGG-16 is often employed in learning applications.

VGG-19: With a total of 19 layers—16 convolution layers, 5 MaxPool layers, 3 fully connected layers, and 1 softmax layer—VGG-19 is a convolutional neural network variation of the VGG model. In this convolutional neural network architecture, just 33 convolutional layers are stacked on top of one another in descending order of depth. This network has over 143 million parameters, making it extraordinarily huge.

Inception V3: Inception V3 is a highly well-liked image recognition model. Among the symmetric and asymmetric building blocks that comprise the model are convolution, average pooling, max pooling, concats, dropouts, and fully linked layers [1]. Batch normalization is applied to the activation inputs and is used throughout the model. The loss is computed using the Softmax approach.

4 Results and Analysis

The authors computed our dataset's accuracy, F1 score, and confusion matrix in order to assess our proposed method. Accuracy is the measure of how well the estimated results match the actual situation on the ground. Our suggested model's accuracy and performance were contrasted to the author's predicted 98% accuracy and performance rating.

Our suggested method, which employs a more effective neural network architecture, employs DenseNet-121, VGG-16, VGG-19, InceptionV3, and three layers to recognize the CNN pictures. The dataset's maximum accuracy, 98%, is obtained using VGG-16. The training and validation accuracy of our suggested model (VGG-19), as well as the training and validation loss, are plotted as shown in Figs. 3 and 4.

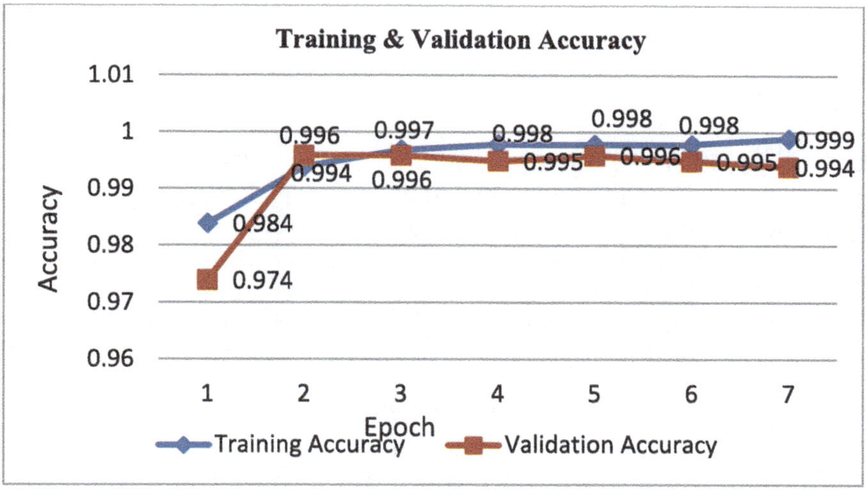

Fig. 3. Training & validation accuracy

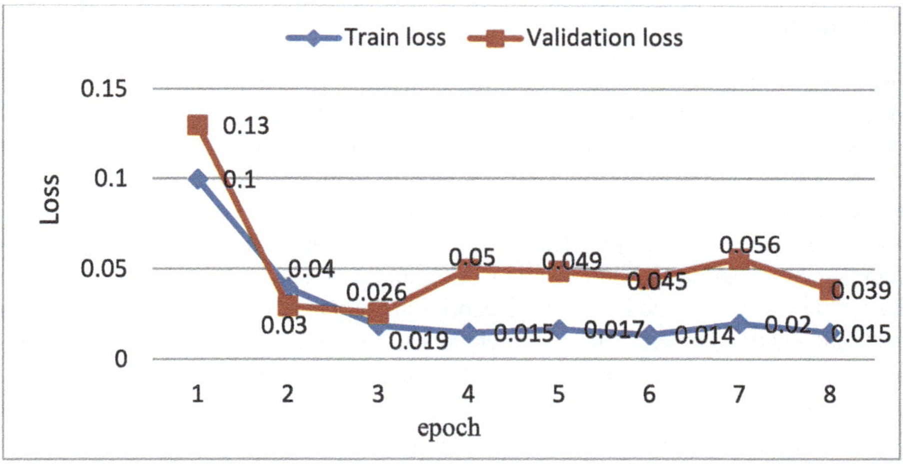

Fig. 4. Training and validation loss

The technique was able to categorize every image using the training and classification process of the CNN approach. According to the name and correctness percentage, the system makes a picture recommendation.

We calculate the validation accuracy each epoch, as seen in Fig. 5. We categorize the validation images using the optimized network, and we compute the classification accuracy, which is 98% on average.

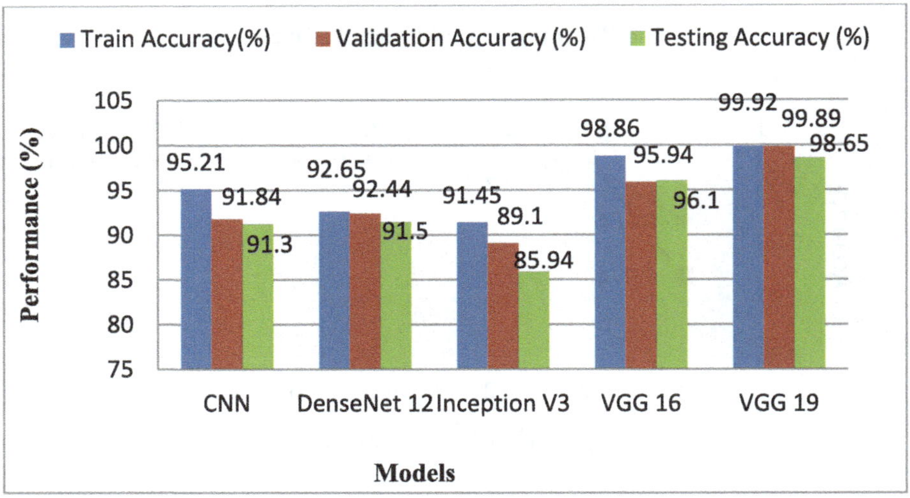

Fig. 5. Analyzing differences between different models for accuracy in training and validation

5 Conclusion

One of the most significant areas of computer science research nowadays, accurate object localization in surveillance videos has a plethora of cutting-edge applications. Due to later gadget images taken from a security video, it is now very difficult to cut modern leaves such as low resolution, light models, moving items beyond the ancient, and minor alterations in the historical past. A high-level visual development in item detecting algorithms has been presented by us. The detection method is applied to object categorization, item detection, and backdrop modeling. All item identification strategies that are now accessible are categorized into three categories in this paper: history subtraction, optical float, and spatial-temporal filter out techniques. The benefits and limitations of these techniques, which are used in many contemporary datasets, are also mentioned. Object type procedures are further divided into strategies according to three different criteria: movement, texture, and form-based thoroughness.

References

1. Younis, A., Shixin, L., Jn, S., Hai, Z.: Real-time object detection using pre-trained deep learning models MobileNet-SSD. In: Proceedings of 2020 the 6th International Conference on Computing and Data Engineering, pp. 44–48 (2020)
2. Vaishali, S.S.: Real-time object detection system using caffe model. Int. Res. J. Eng. Technol. (IRJET) **6** (2019)
3. Wu, D., Sharma, N., Blumenstein, M.: Recent advances in video-based human action recognition using deep learning: a review. In: 2017 International Joint Conference on Neural Networks (IJCNN), pp. 2865–2872 (2017). https://doi.org/10.1109/IJCNN.2017.7966210. https://ieeexplore.ieee.org/document/7966210
4. Kanimozhi, S., Gayathri, G., Mala, T.: Multiple real-time object identification using single shot multi-box detection. In: 2019 International Conference on Computational Intelligence in Data Science (ICCIDS), pp. 1–5. IEEE (2019)

5. Dhillon, A., Verma, G.K.: Convolutional neural network: a review of models, methodologies and applications to object detection. Progress Artif. Intell. **9**(2), 85–112 (2020). https://doi.org/10.1007/s13748-019-00203-0
6. Ghazal, M., Waisi, N., Abdullah, N.: The detection of handguns from live-video in real-time based on deep learning. Telkomnika **18**(6), 3026–3032 (2020)
7. Chiu, Y.C., Tsai, C.Y., Ruan, M.D., Shen, G.Y., Lee, T.T.: Mobilenet-SSDv2: an improved object detection model for embedded systems. In: 2020 International Conference on System Science and Engineering (ICSSE), pp. 1–5 (2020)
8. Younis, A., et al.: Real-time object detection using pre-trained deep learning models MobileNet-SSD. In: Proceedings of 2020 the 6th International Conference on Computing and Data Engineering (2020). https://doi.org/10.1145/3379247.3379264
9. Rafeeq, M., Sunil Kumar, C., Subhash Chandra, N.: Road traffic management system with load balancing on cloud using VM migration technique. In: Bhateja, V., Tavares, J., Rani, B., Prasad, V., Raju, K. (eds.) Proceedings of the Second International Conference on Computational Intelligence and Informatics. Advances in Intelligent Systems and Computing, vol. 712. Springer, Singapore (2018). https://doi.org/10.1007/978-981-10-8228-3_16
10. Rafeeq, Md., Sunil Kumar, C., Subhash Chandra, N.: A novel framework for road transportation systems using cloud computing. Int. J. Simul.–Syst. Sci. Technol. **19**(4) (2018)
11. Devi, B.R., Srujan Raju, K.: Energy distribution using block-based shifting in a smart grid tree network. In: Bhateja, V., Tavares, J., Rani, B., Prasad, V., Raju, K. (eds.) Proceedings of the Second International Conference on Computational Intelligence and Informatics. Advances in Intelligent Systems and Computing, vol. 712. Springer, Singapore (2018). https://doi.org/10.1007/978-981-10-8228-3_56
12. Raja, R., Sinha, T.S., Patra, R.K., Tiwari, S.: Physiological trait-based biometrical authentication of human-face using LGXP and ANN techniques. Int. J. Inf. Comput. Secur. **10**, 303 (2018). https://doi.org/10.1504/IJICS.2018.091468
13. Sinha, G.R.: Study of assessment of cognitive ability of human brain using deep learning. Int. J. Inf. Tecnol. **9**, 321–326 (2017). https://doi.org/10.1007/s41870-017-0025-8
14. Ravikanth, M., Vasumathi, D.: Improve the quality of statistical method of obtaining representative data scheme for de-duplication using fuzzy clustering and genetic algorithm. J. Theor. Appl. Inf. Technol. **95**(8) (2017)
15. Howard, A., et al.: Searching for mobilenetv3. In: Proceedings of the IEEE/CVF International Conference on Computer Vision, pp. 1314–1324 (2019)
16. Rajesh Tiwari, M., et al.: Enhanced power quality and forecasting for PV-wind microgrid using proactive shunt power filter and neural network based time series forecasting. Electr. Power Compon. Syst. (2023). https://doi.org/10.1080/15325008.2023.2249894
17. Sultanuddin, S.J., et al.: Cognitive computing and 3D facial tracking method to explore the ethical implication associated with the detection of fraudulent system in online examination. J. Intell. Fuzzy Syst. 1–15 (2023)

A Unique Method Using Deep Learning to Detect Brain Tumors and Performance Enhancement

Jonnadula Narasimharao[1(✉)], Suraya Mubeen[1], V. Santosh Kumar[2], Maisa Soujanya[2], M. Nagaraju Naik[3], and A. Veerababu[3]

[1] CMR Technical Campus, Hyderabad, Telangana, India
jonnadula.narasimharao@gmail.com
[2] CMR Engineering College, Hyderabad, Telangana, India
{santhoshkumar.v,soujanyamaisa}@cmrec.ac.in
[3] CMR College of Engineering and Technology, Hyderabad, Telangana, India
{nagarajunaik1976,a.veerababu}@cmrcet.ac.in

Abstract. The main supervisor of the anthropomorphic structure is the human brain. Brain tumors are caused by inappropriate brain development and division of cells, As and as brain tumors keep growing, brain cancer appears. Although it reduces the requirement for human judgment to obtain accurate findings, computer vision is significant in the field of human health. CT, X-ray, and MRI scans are the most dependable and secure common MRI (magnetic resonance imaging) procedures. MRI finds even the smallest items. This article focuses on the application of various methods for brain MRI-assisted brain cancer detection. In this research, we used the bilateral filter (BF) for pre-processing to eliminate noise since magnetic resonance images. For the purpose of accurately identifying the tumor location, probabilistic thresholding and Convolution Neural Network (CNN) methods for segmentation were then applied. Datasets for testing, validation, and training are used. We shall determine the possibility that the individual in question possesses a brain tumor according to our machine. Sensitivity, specificity, and accuracy are just a few of the measures of performance that may be used to evaluate the final results.

Keywords: Brain tumor · Magnetic resonance imaging · Adaptive Bilateral Filter · Convolution Neural Network

1 Introduction

The act of producing illustrations of a body's inside for clinical examination and treatment, including a visual depiction of how specific organs or tissues work, is known as medical imaging. The goals of medical imaging are to reveal internal structures hidden between both bones and skin as well as to diagnose and treat diseases. Medical imaging also builds a database of normal physiological and anatomical structures to identify anomalies.

The use of computers to handle images is referred to as medical imaging processing. Numerous methods and procedures, including picture acquisition, storage, appearance,

A. Bajaj et al. (Eds.): IBICA 2023, LNNS 1229, pp. 370–379, 2025.
https://doi.org/10.1007/978-3-031-78940-3_39

and communication, are part of this processing. The recognition and handling of disorders are pursued by this procedure. In order to make it simple to identify deviations, this procedure generates a data bank of the organs' typical form and function. This procedure combines radiological and organic imaging using temperature and isotope image processing, sonography, magnetic resonance imaging, and electromagnetic energy (gamma and X-rays). Numerous other methods are employed to capture data regarding the position and operation of the body. In contrast to systems that generate images, such methods are largely limited.

Using a computer to alter a digital image is known as an image processing technique. This method offers numerous advantages, including data storage, connectivity, flexibility, and adaptation. The development of various image scaling methods has completed it possible to preserve photos efficiently. This method requires numerous sets of rules to be applied simultaneously in the images. Multiple dimensions might be fingered for both 2D and 3D images.

One of the most prevalent and deadly brain disorders that has impacted and destroyed many lives worldwide is brain tumors. Cancer is a brain disorder where cancer cells grow and cause problems with the brain. In excess of one lakh people worldwide accept a brain tumor diagnosis respectively year, according to a recent cancer study. Statistics indicate that individuals with brain tumors do not always receive the desired outcomes, despite consistent attempts to address their consequences. To refute this, researchers are focusing on computer vision to gain a deeper comprehension of cancers in their earliest stages and how to treat them employing cutting-edge techniques.

2 Literature Review

The reliability and accurateness of procedures for prediction are dangerous in medical analysis since the consequence is critical to patient care. Forecast is made using a variety of commonly utilized clustering and classification techniques.

A K-nearest neighbor technique was applied to the MR images by K. Sudharani et al. [1] so as to locate and contain the smallest full-fledged portion private the abnormal tissues. The recommended work uses a slow procedure yet yields beautiful results. The sample retraining step is what determines the precision.

According to B.Devkota et al. [2], aberrant tissues can be identified using morphological procedures and computer-aided detection (CAD) techniques. The morphology opening and closure operations are the most effective segmentation technique available since they remove regions of tumor using the fewest errors in the shortest amount of processing time [3].

A number of clustering algorithms were specified by Jaskirat Kaur et al. (2012) [4] for the procedure of segmentation, and an evaluation of the distinctive styles of those strategies was carried out. Kaur suggested a plan to gauge the stability of particular clustering methods in extraordinary tenders.

Wavelet-based picture fusion was provided by J.T. Kwok et al. [5] to enable simple object recognition at all focal lengths. This is because bright materials in images facilitate the easier completion of many vision-related processing tasks.

In this study, Kumar and Mehta [6] proposed the texture-based technique. They accentuated the significances of segmentation in cases where the margins of tumor tissue are not sharp. These edges might cause the recommended technology's effectiveness to produce undesirable outcomes.

A method for identifying tumors in brain imaging employing artificial neural networks has been proposed by Dalia Mahmoud et al. [7]. They utilized artificial neural networks to develop aelectronic system of recognition for magnetic resonance imaging. In contrast to other ANNs systems, it was found that the Elman community's usage throughout the recognition system resulted in an extended period of both time and precision level.

Brain MRI scan computerized 3D segmentation was introduced by L. Martín et al. [8]. A distinct parameterized model will have less of an effect on a grandeur's intensities than a single multiplicative magnificent. To map the typical brain using nonrigid transformation, Brain Atlas is employed.

An operative computerized categorization technique for brain images was developed by Astina Minz et al. [9], which predicted the use of the AdaBoost gadget mastering technique. There are three primary sections to the suggested scheme. Pre-processing eliminated noise from the datasets and produced grayscale images.

Sarat Kumar Sahoo and Monica Subashini [10] have proposed a method for identifying the tumor before brain MR images are taken. Additionally, they worked on a variety of methods, such as back propagation networks for identifying brain MRI images by tumor cells and pulse-coupled neural networks and noise elimination approaches for enhancing the mind MRI images.

Image segmentation is employed to extract significant objects from an image, according to Yu et al. [11, 12]. They suggest dividing an image across three sections: white, grey, and dark. The 2D histogram is fuzzy divided using the Z- and s-functions. Next, an assortment of 12 membership parameters with a extreme value is found using QGA.

Mukambika et al. [13] suggested an approach for classifying the tumor at a later stage, assuming that it is existent. Their suggested work is a comparison of the Level set technique, discrete wavelength transformations (DWT), and K-method systems for segmentation as approaches to tumor diagnosis from MR images. Following that stage, feature extraction and SVM classification are carried out.

Magnetic resonance, according to S. Pereira et al. [14], reduces physical segment time in medical settings. Thus, the research effort had provided an automatic and trustworthy segmentation technique employing Convolutional Neural Network (CNN) for aberrant tissue identification. Because brain scans have significant three-dimensional and underneath roughness, segmenting the image is a serious problem that calls for the adoption of a reliable methodology like CNN.

The tumor region impacted was determined for proportional analysis by Roy et al. (2012) [15]. They used multiple statistical groups with different tumor sizes, intensity levels, and locations to validate the software. They demonstrated how their algorithm could identify and phase the brain tumor in the provided photo automatically.

Convolutional Neural Network (CNN) segmentation that is mostly based on the brain tumor categorization approach has been suggested by A. Sankari and S. Vigneshwari [16]. The leaky rectified linear unit (LReLU), a non-linearity activation characteristic, was utilized in the suggested work. They mainly concentrated on the required attributes, such as the image's mean and entropy, and found that the algorithm used by CNN performs better at capturing the intricate and miniature details of brain tumor tissues found in MR images.

The Dual Localization technique has been automated by brain segmentation, as demonstrated by T.U. Paul and S.K. Bandyopadhyay [17]. The first stage involves creating the skull masks using the brain MR images. The K-manner method is used to improvise the tumor regions. They assessed their suggested work's final phase using metrics like length and width.

It [18, 19] presented an approach that involves a sequential process that begins with pre-processing the images, extracts valuable objects, and then classifies the tumor region. The final step of pre-processing is enhancing the image by removing noise through the application of Gaussian filters that are authentic.

3 Recognition of Tumor in MRI Images by Means of Artificial Neural Networks

In numerous diagnostic and therapeutic applications, automatic fault identification in magnetic resonance images is crucial. In order to improve yield and efficiency while shortening diagnosis times, our work has developed one automatic approach for detecting brain tumors. The objective is to divide tissues under two groups: normal and aberrant. The MR images from both normal and aberrant brain tissues were the ones employed in this instance. Neural networks are used in this technique to accomplish the organization.

4 Contests in Tumor Organization

Tumor recognition is an extremely difficult task. Tumor location, form, and structure differ greatly between patients, making classification an extremely difficult task. The brain slices from many individuals are depicted in the picture below, showing amply illustrates how the tumors vary from one another. It is evident that each of the eight patients and photos above has a distinct tumor site. Even worse, there are differences in the intra-tumoral architecture and morphology across eight patients/images. As indicated in Fig. 1 next, there may actually be multiple tumors in one location of the tumor. This does, in fact, illustrate how intricate automated segmentation is.

This offers a structure for the system that could be created by humans. It is executed in six steps: first, a source picture representing the data set is taken; then, pre-processing, enhancement of images, binary thresholding is used for image segmentation, and convolutional neural networks are used for brain tumor categorization. Following the completion of each of the aforementioned procedures, the outcome is finally evaluated.

Figure 2 illustrates that every module is distinct in its own right. Every action has a purpose. A testing as well as training data set is additionally contained in this architecture. The algorithm is tested and trained using a data set that has been obtained through Kaggle and contains close to 2000 photos. The Sobel Filter is used to improve on the input image after it has been pre-processed employing noise filters such as the Median and Bilateral filters.

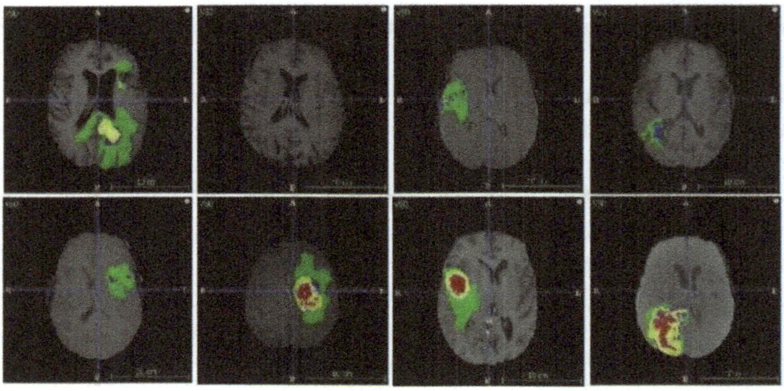

Fig. 1. Position of tumors in 8 various images.

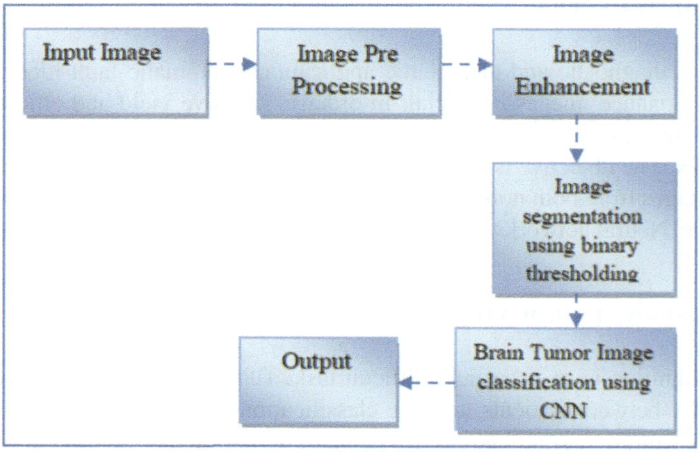

Fig. 2. Module Division

5 Experimental Analysis

We downloaded the Brain MRI image dataset through Kaggle. Approximately 1900 MRI pictures, comprising benign, aggressive, and normal ones, make up the MRI dataset. These MRI scans are obtained to provide data for the first stage. The brain MRI image is first converted hooked on its equivalent gray-scale image. The adaptable bidirectional filtering technique is utilized to eliminate undesired noise and corrupted noise from the brain image. These enhance the diagnostic process and raise the rate of categorization precision.

Here initially the sample image s taken in Fig. 3 and sent for detecting the tumor. After detecting the tumor if the tumor is detected the detected image is shown in Fig. 4.

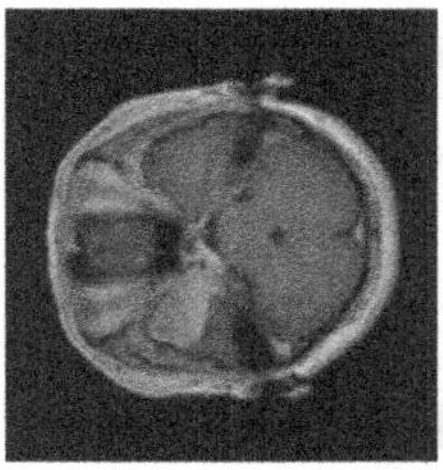

Fig. 3. Sample input

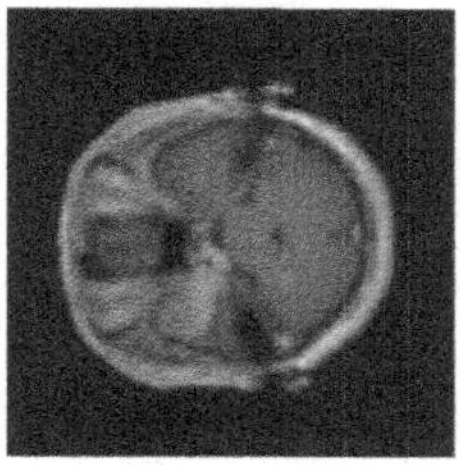

Fig. 4. Tumor detected

5.1 Testing and Performance Measures

True Positive and True Negative are among the outcomes evaluation criteria that have been used to evaluate the suggested algorithm. The first one indicates the frequency with which the suggested method can correctly identify a damaged region as such, while the second one indicates the frequency with which the suggested strategy can correctly identify a non-damaged region as such. Furthermore, the terms False Positive (FN) and False Negative (FN) indicate the frequency with which the suggested algorithm incorrectly identifies the damaged region, while the latter indicates the frequency with which the suggested approach incorrectly identifies the non-tumor region as non-tumor region. The suggested algorithm's precision, particularity, and value of sensitivity are computed created on the values of TP, TN, FP, and FN.

$$sensitivity : \frac{TP}{TP + FN}$$

$$specificity : \frac{TN}{TN + FP}$$

$$Accuracy : \frac{TN + TP}{TN + TP + FN + FP}$$

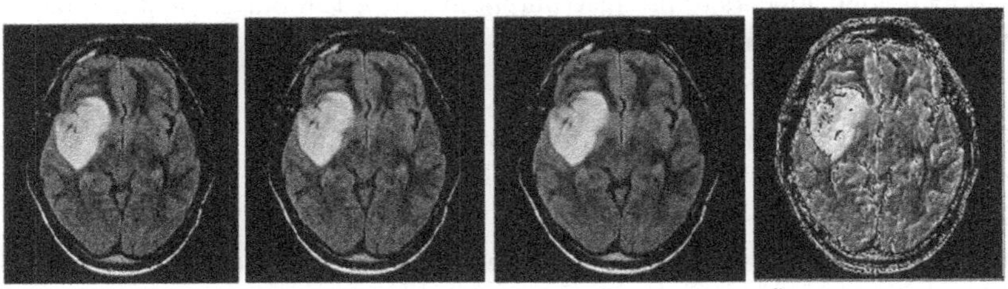

a) b) c) d)

Fig. 5. a) Input Image b) Gray-scale Image c) Filtered Image d) Image after edge detection

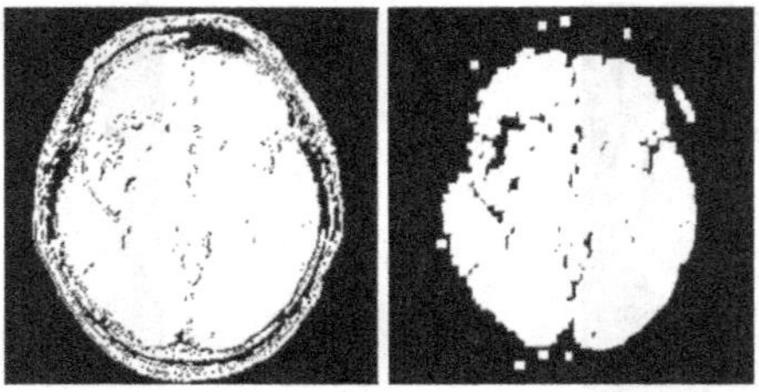

Fig. 6. a) Thresholded and Binary Image b) Morphological Image

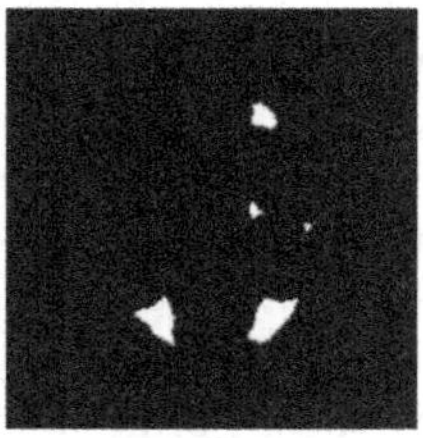

Fig. 7. Dilated Images

The total detection process is taken in the above figures. Initially the input image is taken from Fig. 5a) and then converted it into gray scale image in Fig. 5b) After applying the filtering the filtered image is shown in 5c) then after edge detection process in done as shown in Fig. 5d) now the threshold value is detected for the binary image 6a) then morphological process is done as shown in Fig. 6b) and dilated in 8th figure for detecting the effected tumor part in Fig. 7.

5.2 Performance

During testing, it was found that the suggested methodology appeared to perform better than the other sets of photos. When contrasted to its added sized images, particularly are illustrated in Table 1 and Fig. 8, the suggested Convolutional Neural Network (CNN) oriented technique appears to produce output in 128 * 128 images that is considerably superior than any other image.

Table 1. Illustrates the suggested strategy's true positive, genuine negative, false positive, and false negative outcomes for various sets of images.

Various Set of Images	True Positive (%)	True Negative (%)	False Positive (%)	False Negative (%)
128*128 Image	83.8	84.5	16.4	15.6
256*256 Image	82.5	84.2	17.5	16
512*512 Image	82.2	83.6	17.8	16.4

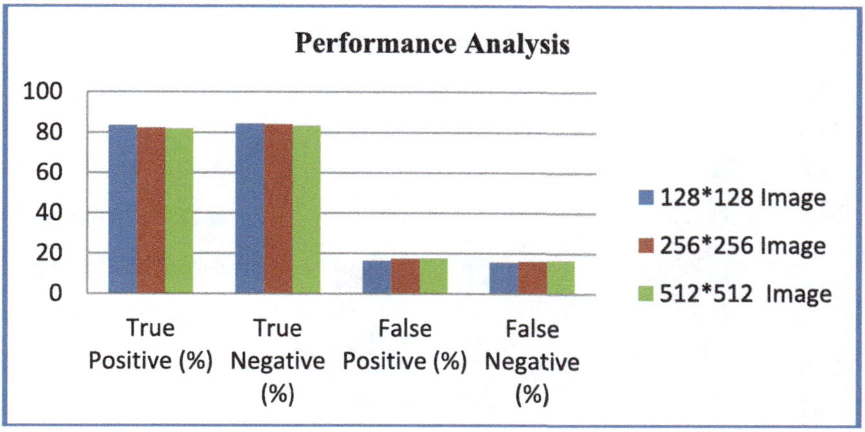

Fig. 8. The presentation investigation of CNN

Table 2 shows that, when the suggested segmented approach is applied to various sets of images, it can identify an isolated section among the MR images then is utilized to examine the dimensions and structure of the denoised image. Convolutional Neural Networks (CNNs) were employed by us for categorization, and as the figure illustrates, the results of our suggested work are satisfactory with improved sensibility, reliability, and computational efficiency as shown the Fig. 9.

Table 2. Signifies the Exactness, Sensitivity, and Specificity of the projected method for dissimilar set of images.

Various Set of Images	Accuracy (%)	Sensitivity (%)	Specificity (%)
128 * 128 Image	84.2	84.4	83.9
256 * 256 Image	83.4	83.5	82.8
512 * 512 Image	83	83.4	82.3

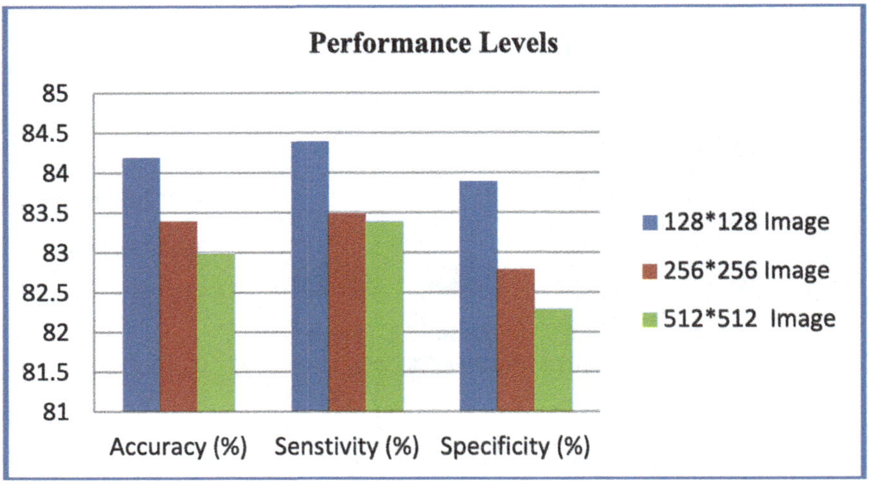

Fig. 9. The presentation of CNN

6 Conclusion

We suggested utilizing a convolution neural network to progress an automated process for the identification and detection of brain tumors. Utilizing the file location, the input MR pictures are downloaded at the current device and transformed into grayscale images. An adaptable bilateral filtering approach is used to pre-process those pictures in order to remove any noise that may have been included in the original image. So as to identify the tumor location in the MR images, binary thresholding is done to the denoised picture and Convolution Neural Network segmentation is used. Having a reliability of 84%, the suggested model produces results that are promising, error-free, and require significantly less computing time.

References

1. Sudharani, K., Sarma, T.C., Satya Rasad, K.: Intelligent brain tumor lesion classification and identification from MRI images using k-NN technique. In: 2015International Conference on Control, Instrumentation, Communication and Computational Technologies (ICCICCT), Kumaracoil, pp. 777–780 (2015)
2. Devkota, B. Alsadoon, A., Prasad, P.W.C., Singh, A.K. Elchouemi, A.: Image segmentation for early stage brain tumor detection using mathematical morphological reconstruction. Procedia Comput. Sci. **125,** 115123 (2018).https://doi.org/10.1016/j.procs.2017.12.017
3. Tiwari, R., Chaudhary, D., Diwan, T.D., Sharma, P.C.: Privacy and security solution in wireless sensor network for IoT in healthcare system. In: Next Generation Healthcare Systems Using Soft Computing Techniques, pp. 123–135. CRC Press, Boca Raton (2022). ISBN 978-1-03210-797-4
4. Kaur, J., Agrawal, S., Renu, V.: A comparative analysis of thresholding and edge detection segmentation techniques. Int. J. Comput. Appl. **39**, 29–34 (2012). https://doi.org/10.5120/4898-7432

5. Li, S., Kwok, J.T., Tsang, I.W., Wang, Y.: Fusing images with different focuses using support vector machines. In: Proceedings of the IEEE transaction on Neural Networks, China (2007)
6. Kumar, M., Mehta, K.K.: A texture based tumor detection and automatic segmentation using seeded region growing method. Int. J. Comput. Technol. Appl. **2**(4), 855–859 (2011). ISSN 2229-6093
7. Rafeeq, M., Sunil Kumar, C., Subhash Chandra, N.: Road traffic management system with load balancing on cloud using VM migration technique. In: Bhateja, V., Tavares, J., Rani, B., Prasad, V., Raju, K. (eds) Proceedings of the Second International Conference on Computational Intelligence and Informatics. AISC, vol. 712, pp. 163–173. Springer, Singapore (2018). https://doi.org/10.1007/978-981-10-8228-3_16
8. Rafeeq, M., Sunil Kumar, C., Subhash Chandra, N.: A novel framework for road transportation systems using cloud computing. Int. J. Simul.-Syst. Sci. Technol. **19**(4) (2018)
9. Devi, B.R., Srujan Raju, K.: Energy distribution using block-based shifting in a smart grid tree network. In: Bhateja, V., Tavares, J., Rani, B., Prasad, V., Raju, K. (eds.) Proceedings of the Second International Conference on Computational Intelligence and Informatics. AISC, vol. 712, pp. 609–618. Springer, Singapore (2018). https://doi.org/10.1007/978-981-10-8228-3_56
10. Raja, R., Sinha, T., Patra, R., Tiwari, S.: Physiological trait-based biometrical authentication of human-face using LGXP and ANN techniques. Int. J. Inf. Comput. Secur. **10**, 303 (2018). https://doi.org/10.1504/IJICS.2018.091468
11. Sinha, G.R.: Study of assessment of cognitive ability of human brain using deep learning. Int. J. Inf. Tecnol. **9**, 321–326 (2017). https://doi.org/10.1007/s41870-017-0025-8
12. Minz, A., Mahobiya, C.: MR image classification using adaboost for brain tumor type. In: 2017 IEEE 7th International Advance Computing Conference(IACC), pp. 701–705 (2017)
13. Monica Subashini, M, Sahoo, S.K.: Brain MR image segmentation for tumor detection using artificial neural networks. Int. J. Eng. Technol. (IJET) **5**(2) (2013)
14. Pan, Y., et al.: Brain tumor grading based on neural networks and convolutional neural networks. In: Conference Proceedings: Annual International Conference of the IEEE Engineering in Medicine and Biology Society. IEEE Engineering in Medicine and Biology Society. Conference, pp. 699–702 (2015). https://doi.org/10.1109/EMBC.2015.7318458
15. Bagga, J., Pinjarkar, L., Srivastava, S., Dewangan, O., Tiwari, R.: Latest advancement in automotive embedded system using IoT computerization. Green Comput. Appl. 131–165 (2021). ,https://doi.org/10.52305/ENYH6923.ISBN 978-1-68507-357-2
16. Mukambika, P.S., Uma Rani, K.: Segmentation and classification of MRI brain tumor. Int. Res. J. Eng. Technol. (IRJET) **4**(7), 683–688 (2017). ISSN 2395-0056
17. Pereira, S., Pinto, A., Alves, V., Silva, C.A.: Brain tumor segmentation using convolutional neural networks in MRI images. IEEE Trans. Med. Imaging **35**(5), 1240–1251 (2016)
18. Roy, S., Bandyopadhyay, S.K.: Detection and qualification of brain tumor from MRI of brain and symmetric analysis. Int. J. Inf. Commun. Technol. Res. **2**(6), 584–588 (2012)
19. Sankari, A., Vigneshwari, S.: Automatic tumor segmentation using Convolutional neural networks. In: 2017 Third International Conference on Science Technology Engineering & Management (ICONSTEM), pp. 268–272 (2017)

Analyzing Facial Recognition Using MTCNN Algorithm

Jonnadula Narasimharao[1(✉)], K. Mohana Lakshmi[1], M. Naveen Kumar[2], E. Parvathi[2], Abdul Subhani Shaik[3], and B. Venkateshwara Rao[3]

[1] CMR Technical Campus, Hyderabad, Telangana, India
jonnadula.narasimharao@gmail.com
[2] CMR Engineering College, Hyderabad, Telangana, India
naveenmds@cmrec.ac.in
[3] CMR College of Engineering & Technology, Hyderabad, Telangana, India
bvenkateshwarrao@cmrcet.ac.in

Abstract. Recognizing and detecting faces has been a popular study area in the last few years. Formerly low-level computer vision techniques HAAR Cascade, HOG, etc.—were employed for detection and recognition. However, the accessibility of affordable and effective technology has made it possible to conduct significant studies in this field and a variety of real-world applications from virtual realities and biometric verification to surveillance systems and emotion analysis have drawn a greater interest to this area of study. Convolutional neural networks (CNNs), in especially, are deep learning algorithms that have transformed face detection and identification outcomes. The algorithms created in the fields of artificial intelligence, deep learning, and computer vision in particular are fully utilized by modern CNNs. More advanced techniques have also been developed and readily accessible of large-scale datasets with annotations such as the CelebA dataset and the Labeled Faces in the Wild (LFW) dataset. We present a strategy that leverages the face detectors and extractors of features that currently exist and assesses the approach using measures like precision, recall, and F1-score. This is what we would refer to as a conventional multi-class classification issue. For the purpose of assessing process, we additionally employ an internal dataset, which presents additional difficulties for the detector in terms of face extraction and representing features learning.

Keywords: Face recognition · LFW dataset · ResNet-50 · MTCNN

1 Introduction

Many commonplace applications, including social media postings and filters, face unlocking capabilities, and facial expression analysis, depend on face detection and identification. In order to achieve optimal performance and efficiency, deep learning—especially CNNs—is recommended and should be the most effective over these systems. Facial recognition and identification have become major areas of study for numerous academic and commercial researchers. The most often used benchmarking face detection

A. Bajaj et al. (Eds.): IBICA 2023, LNNS 1229, pp. 380–387, 2025.
https://doi.org/10.1007/978-3-031-78940-3_40

datasets are [1–3]. Face recognition, categorization, and detection techniques currently come in a wide variety. The domain's research has mostly focused on CNN architectures, loss functions, dataset variety, and training methodologies.

Metrics for evaluation are crucial to accurately assess the model, in addition to the usage of CNN architectures, loss functions, varying datasets, and optimal training procedures. The majority of facial recognition techniques favor efficiency as their metric of choice for assessment [4]. However, it is not optimal to use precision alone for classification tasks.

The English court made the announcement of trying to match photos to determine offenders for the first time. Law enforcement is now using facial recognition software to detect crimes through images or videos. Forensic specialists run recognition of faces tests to compare the suspect's face image. Business processes are improved and worker efficiency is increased using automatic face recognition technology.

Our method uses mAP, Recall, and F1-score for assessing the face recognition architecture. These are the recommended evaluation metrics for categorization tasks; however, most of the time, the last two are not employed correctly [4, 5]. For the face detection and recognition challenge, [6] employed the same assessment measures. For the purposes of this work, we employ ResNet50 as the feature extractor, MTCNN [7] and the dlib library as face detectors. We compare the effectiveness of the two detectors using ResNet50-extracted features, and we also employ SVM as the classifier to assign the input photos to the appropriate classes. A bespoke in-house dataset with numerous views of the photographs we collected for evaluation presents some challenges in the feature-matching assignment.

Our principal findings are summed up as follows:

a. We compile an internal dataset, which presents certain difficulties for the model's face detection and feature extraction.
b. We conduct a number of tests with different face detectors and assess them according to mAP, Recall, and F1-score.
c. We use the LFW dataset to test our classifier. Additionally, we employ a single internal dataset for assessment, which presents greater challenges for the model.

2 Literature Review

Face detectors function similarly to object detectors in most cases. In real-world situations, face identification must handle a variety of variation issues, such as obstruction, expression respectively, makeup, scale, position, illumination, blur, etc.

Numerous face identification techniques have been presented by researchers from prestigious universities as well as independent scholars to address these issues, particularly the detection of small faces with wide variations in scale, information, and anchor order. MTCNN [7], RetinaFace [8], RefineFace [9], and one of the most current ASFD [10], MaskFace, are some of these [11].

We demonstrate significant speed improvements on three publicly accessible datasets using this method. Surprisingly, we demonstrate that when the HOG filter is learnt by MMOD, a single stiff HOG filter may outperform a state-of-the-art deformable component model on the Face Detection Data Set and Benchmark [12].

Numerous identification and recognition efforts employing both lower- and higher-level computer vision have been presented. We attempt to take a look at the most pertinent current work in this paper. The articles of [13–17] all use a multi-stage complicated system that employs an SVM for categorization and the results of a deep convolutional network using PCA for dimensionality reduction.

Google researchers developed FaceNet [4], which uses machine learning to enhance facial recognition. FaceNet is intended to use Euclidean space, where expresses the distances among similar faces as metrics of similarity, in order to immediately train facial models as shown the Fig. 1. This method contributes to increasing facial recognition accuracy.

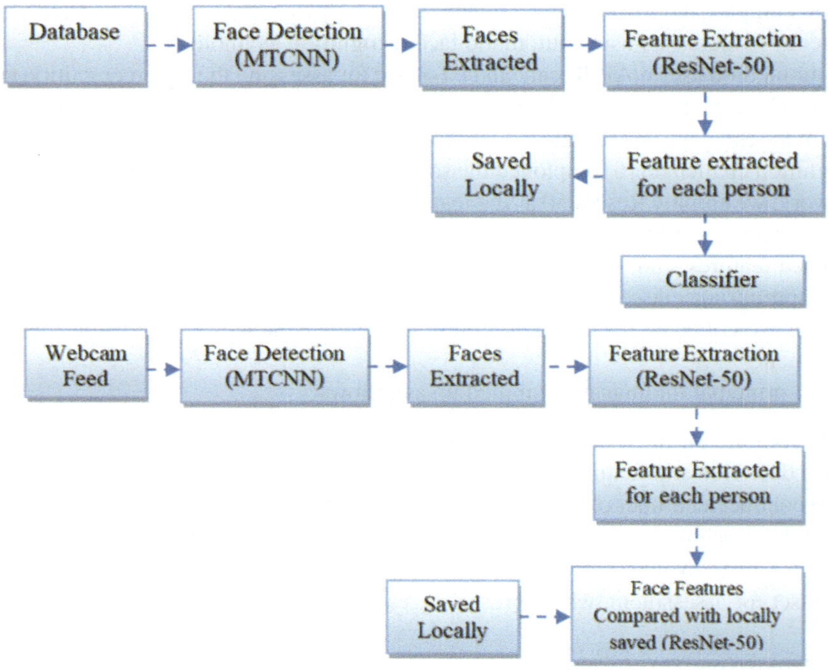

Fig. 1. Operational working model

3 Methodology

3.1 Model Experimentation

The suggested approach is shown in Fig. 1. We began our exploration employing three face detectors: Yolo5 Face, MTCNN, and DLIB. Since MTCNN and DLIB proved to be the best-performing ones, we will use them going forward in our research report. One feature extractor that is frequently used and has shown to work well in most situations is ResNet-50. The different classes and individuals have been categorized using an SVM classifier, and the model's performance has been assessed using both internal and LFW datasets.

When using a real-time execution, we save the class ids of the features that were extracted throughout the SVM classifier training process. Throughout real-time execution, these stored features are compared to the freshly obtained characteristics.

3.2 Instantaneous Execution

Regarding the sake of execution, we classify or recognize faces using an alternative method.

Initially, we capture a live stream using a webcam—in this example, a Logitech C925e. Faces in the entering video stream are identified and extracted by the MTCNN face detector.

3.3 MTCNN

Face detection is the first step in the MTCNN procedure, which then follows by face alignment and facial landmark localization.

Face Detection: A CNN known as the Proposition Network (P-Net) is used in the face detection step to provide potential face areas, or bounding boxes, in an image. This is accomplished by moving a tiny window over the picture, every single one which indicates the possibility that a face is visible in the window patch.

Localization of Facial Landmarks: The second stage, known as facial landmark localization, makes use of the Refine Network (R-Net), a deeper CNN than the P-Net, to filter and refine the candidate face areas that the P-Net produced. In addition to estimating the locations of the mouth, nose, and eyes, this aids in the prediction of more precise bounding box coordinates.

Face Alignment: The third CNN, known as the Output Network (O-Net), handles face alignment in the last stage. This network additionally refines the coordinates of the bounding box and facial landmark locations. In order to align the face, the network also determines an affine transformation, which qualifies it for tasks involving recognition.

The purpose of the obtained faces is to recover features, which should ideally be distinct for each face. However, the quality and uniqueness of features obtained from photos are greatly influenced by factors such as background noise, illumination, and other factors. ResNet50, a feature extractor that has demonstrated good generalization on unobserved data, is used in this work. The details of the face are represented by these features, which are 512-dimensional arrays.

ResNet-50

Within the ResNet (Residual Network) family is the convolutional neural network architecture known as ResNet-50. The network's total number of layers—convolutional, pooling, fully connected, and shortcut connections—is indicated by the number "50" in the network's name. The fundamental unit of ResNet-50 is called the bottleneck block, and it consists of three convolutional layers: 1x1, 3x3, and 1x1. This combination contributes to retaining excellent accuracy at a lower computing cost.

The features saved from several photos for each individual or class are compared with the features recovered with ResNet-50 from faces extracted with MTCNN. By comparison, we mean figuring out the cosine distance between the features from the webcam image and the features that have already been saved. Ideally, the distance should be larger for faces that differ and smaller for ones that are similar.

Cosine Distance

In a high-dimensional space, the Cosine Distance calculates how similar or unlike two given vectors are. Rather than calculating the vectors' absolute magnitudes, it computes their angle.

$$Cosinesimilarity = (A.B)/(//A// * //B//) \tag{1}$$

The cosine of the angle formed by the two vectors, A and B, is taken into account while calculating the cosine distance. The symbol "." denotes the dot product.

$$cosine\_distance = 1 - cosine\_similarity \tag{2}$$

According to -1 to 1, the cosine resemblance is measured. A score close to 1 indicates a high degree of resemblance, whereas a value close to -1 indicates a high degree of dissimilarity. In the event that the cosine similarity is zero, the vectors are orthogonal or wholly distinct.

Dataset

We have tested our solution for the face recognition job using two datasets: our own internal dataset and Labeled Faces in the Wild.

The LFW dataset, a common benchmark dataset, has 5425 photos over 311 classes. There are 120 photos in the internal dataset representing 12 classes. The LFW dataset is a relatively more difficult dataset that aids in our comprehension of the model's resilience and degree of generalization.

4 Experimental Results

Every picture that uses a face detector produces bounding boxes that fit the facial regions precisely. DLib and MTCNN detectors were chosen above various face detectors on the market for experimenting. Utilizing the weights that have already been trained utilizing the vgg face2 dataset on almost 3.31 million photos from roughly 9131 distinct identities, the Resnet50 model is utilized to extract features.

In the model experimentation phase, we utilize an SVM classifier to categorize the photos into the appropriate classes, and in the real-time implementation, we compute the cosine distance between the features. We have assessed the performance of the SVM classifier that uses feature embedding from Resnet50 as inputs using precision, recall, and F1-score.

We tested three face detectors in order to conduct experiments, and two of them were ultimately chosen for further consideration based on how well they performed in the trials that followed.

Assessment using DLib's facial detection feature.

DLib is a face detection framework that has shown good performance under a range of circumstances. Two face detection techniques are integrated into the Dlib framework:

HOG + Linear SVM face detector: this combination of accuracy and computational efficiency may be chosen for applications where system performance is a critical factor.

Max-Margin (MMOD) face detector: because of its durability and excellent accuracy, this detector can also recognize faces in occlusion, at varied viewing angles, and in diverse lighting situations as shown in Fig. 2.

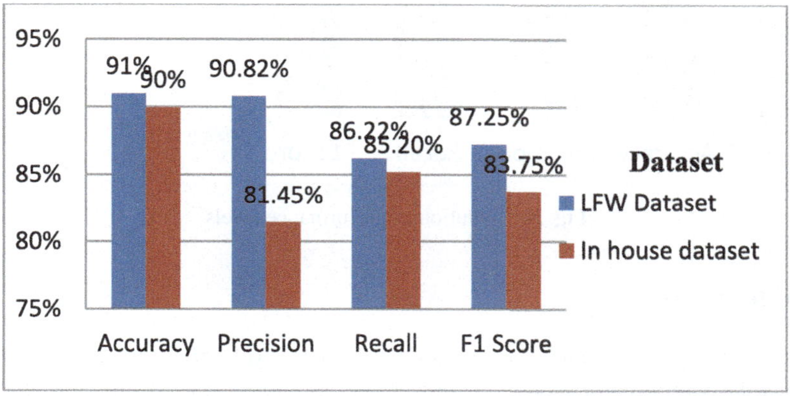

Fig. 2. Performance levels

Evaluation on MTCNN as Face Detector

The MTCNN [6] cascaded structure consists of three layers of deep convolutional networks and is intended to predict the position of landmarks and the bounding boxes of faces in images. By utilizing the benefits of multi-task learning, MTCNN has demonstrated through the author's and our own studies that it is capable of producing extremely accurate bounding box predictions as shown in Fig. 3.

Utilizing MTCNN and DLIB detectors, respectively. This illustrates the general performance of the face detectors and feature extractors even in situations when the percentage of faces in the image is smaller than that of the background and other body parts.

Better facial bounding boxes are produced by the MTCNN, which eventually aids in improved feature extraction. The fact that the feature extractor is the same in both scenarios makes this plot useful for comprehending the face detector's performance as shown the Figs. 2 and 3.

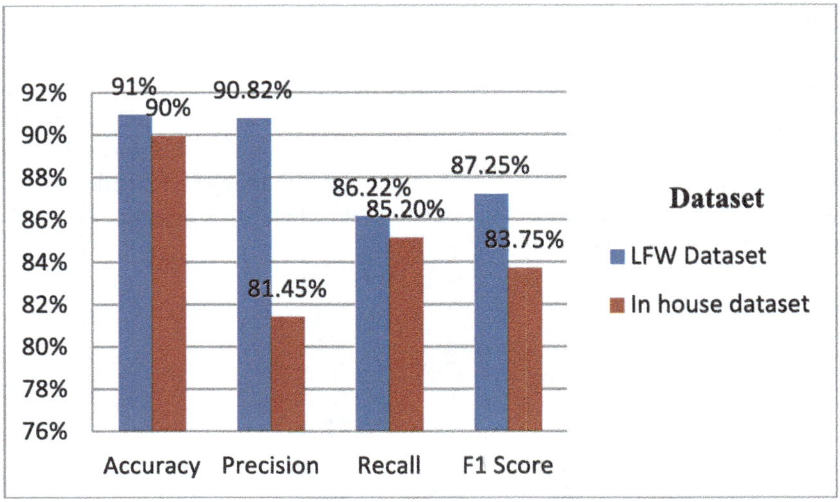

Fig. 3. Evolution of performance levels

5 Conclusion

The suggested approach functions as intended, with the MTCNN face detectors playing a major role in producing accurate feature representations. In real-time, this suggested approach also works well. This method has been evaluated on the in-house dataset, which is already a difficult dataset, and can be used to a variety of real-world applications with positive results anticipated.

References

1. Zhu, Z., Luo, P., Wang, X., Tang, X.: Recover canonical view faces in the wild with deep neural networks. CoRR, abs/1404.3543 (2014)
2. Yang, S., Luo, P., Loy, C.C., Tang, X.: Wider face: a face detection benchmark. http://shuoyang1213.me/WIDERFACE/index.html
3. Huang, G.B., Ramesh, M., Berg, T., Learned-Miller, E.: Labelled faces in the wild: a database for studying face recognition in unconstrained environments. Technical report 07-49, University of Massachusetts, Amherst (2007)
4. Schroff, F., et al.: FaceNet: a unified embedding for face recognition and clustering. In: 2015 IEEE Conference on Computer Vision and Pattern Recognition (CVPR), pp. 815–23 (2015). arXiv.org, https://doi.org/10.1109/CVPR.2015.7298682
5. Deng, J., Guo, J., Zhou, Y., Yu, J., Kotsia, I., Zafeiriou, S.: Retinaface: single-stage dense face localisation in the wild. In: CVPR (2020)
6. Qi, D., et al.: YOLO5Face: why reinventing a face detector. arXiv (2022). http://arxiv.org/abs/2105.12931
7. Zhang, K., et al.: Joint face detection and alignment using multi-task cascaded convolutional networks. IEEE Signal Process. Lett. **23**(10), 1499–1503 (2016). https://doi.org/10.1109/LSP.2016.2603342
8. Zhang, S., Chi, C., Lei, Z., Li, S.Z.: Refineface: refinement neural network for high performance face detection. ArXiv preprint ArXiv:1909.04376 (2019)

9. Gothane, S., Sarode, M.V.: Analyzing factors, construction of dataset, estimating importance of factor, and generation of association rules for Indian road accident. In: 2016 IEEE 6th International Conference on Advanced Computing (IACC), Bhimavaram, India, pp. 15–18 (2016). https://doi.org/10.1109/IACC.2016.13
10. Rafeeq, M., Sunil Kumar, C., Subhash Chandra, N.: A novel framework for road traffic data management for cloud-based storage. In: Satapathy, S., Prasad, V., Rani, B., Udgata, S., Raju, K. (eds.) Proceedings of the First International Conference on Computational Intelligence and Informatics. AISC, vol. 507, pp. 85–97. Springer, Singapore (2017). https://doi.org/10.1007/978-981-10-2471-9_9
11. Gothane, S., Sarode, M.V., Srujan Raju, K.: Design, construction and analysis of model dataset for Indian road network and performing classification to estimate accuracy of different classifier with its comparison summary evaluation. In: Panigrahi, B., Suganthan, P., Das, S., Satapathy, S. (eds.) SEMCCO 2015. LNCS, vol. 9873, pp. 50–59. Springer, Cham (2016). https://doi.org/10.1007/978-3-319-48959-9_5
12. Devi, R.N., Saikumar, T.: Application of radon transform for image segmentation on level set method using HKFCM algorithm. In: Satapathy, S., Govardhan, A., Raju, K., Mandal, J. (eds.) Emerging ICT for Bridging the Future - Proceedings of the 49th Annual Convention of the Computer Society of India (CSI) Volume 1. AISC, vol. 337, pp. 257–265. Springer, Cham (2015). https://doi.org/10.1007/978-3-319-13728-5_29
13. Kumar, T.S., Swapna, C.: Segmentation of image with DT-CWT and NLM filtering using fuzzy c-means clustering. In: 2014 IEEE International Conference on Computational Intelligence and Computing Research, Coimbatore, India, pp. 1–4 (2014). https://doi.org/10.1109/ICCIC.2014.7238367
14. Raju, K.S., Sreelatha, K., Kumari, S.: An efficient approach to improve retrieval rate in content based image retrieval using MPEG-7 features. In: Satapathy, S., Avadhani, P., Udgata, S., Lakshminarayana, S. (eds) ICT and Critical Infrastructure: Proceedings of the 48th Annual Convention of Computer Society of India- Vol II. AISC, vol. 249, pp. 337–347. Springer, Cham (2014). https://doi.org/10.1007/978-3-319-03095-1_37
15. Saikumar, T., Preetam, I.N.: Optimized kernel fuzzy c means (OKFCM) clustering algorithm on level set method for noisy images. In: 2013 IEEE International Conference on Computational Intelligence and Computing Research, Enathi, India, pp. 1–5 (2013). https://doi.org/10.1109/ICCIC.2013.6724290
16. Taigman, Y., Yang, M., Ranzato, M., Wolf, L.: Deepface: closing the gap to human-level performance in face verification. In: IEEE Conference on CVPR (2014)
17. Pranathi, K., Pingili, M., Mamatha, B.: Fundus image processing for glaucoma diagnosis using dynamic support vector machine. In: Kumar, A., Mozar, S., Haase, J. (eds) ICCCE 2023. Cognitive Science and Technology, pp. 551–558. Springer, Singapore (2023). https://doi.org/10.1007/978-981-19-8086-2_53

The Unveiling of the Hiding Accuracy: Employing Deep Convolutional Neural Networks for the Recognition and Assessment of Image Forgery

Dubbaka Navanitha[1], M. Ramasamy[1], Abdul Subhani Shaik[2], M. Vijayasanthi[2], Ramesh Chegoni[3], and Chikati Madhava Rao[3(✉)]

[1] CMR Engineering College, Hyderabad, Telangana, India
{dubbaka.navanitha,ramasamy.m}@cmrec.ac.in
[2] CMR College of Engineering and Technology, Hyderabad, Telangana, India
{sasubhani,mvijayasanthi}@cmrcet.ac.in
[3] CMR Technical Campus, Hyderabad, Telangana, India
ramesh.it@cmrtc.ac.in, cmrtc.paper@gmail.com

Abstract. The emergence of deep networks has enabled substantial advancements in the pasture of sensory technology. The proliferation of visual media, coupled with the availability of advanced editing software, has led to a considerable enlarge in the ability to modify digital content. The user's text is already academic in nature. No further rewriting is necessary. In order to detect and recognize fraudulent actions, we have put forth innovative methodologies. In this research, we discuss two fundamental aspects of employing deep convolutional neural networks in the context of image forgery detection. Firstly, an examination is conducted on different pre-processing techniques in combination with the Convolutional Neural Network (CNN) architecture. Following this, we evaluate the effectiveness of several transfers learning methods, including the utilization of pre-trained ImageNet models through fine-tuning. These techniques are applied to our dataset, CASIA V2.0. The focus of our study entails the examination of preprocessing techniques utilizing a straightforward convolutional neural network framework, while also exploring the profound impact of transfer learning models.

Keywords: Visual perception · Forgery · Deep convolutional neural networks · ImageNet

1 Introduction

Recent years, however is a notable surge of interest within the academic community about the development of methodologies aimed at validating the genuineness of documents. With the proliferation of readily available technologies for modifying multimedia content, the task of accurately verifying its reliability is growing progressively more challenging. The primary purpose of this research is to assess and evaluate the effectiveness

A. Bajaj et al. (Eds.): IBICA 2023, LNNS 1229, pp. 388–397, 2025.
https://doi.org/10.1007/978-3-031-78940-3_41

of Convolutional Neural Networks (CNNs) and their pre-processing stages. Additionally, the study aims to evaluate and contrast the efficacy of transfer learning models. A detailed study of effectiveness and achievement statistics has been presented through the utilization of multiple tables and figures.

The primary aim of the research is to highlight the potential enhancements that can be attained through either improving the pre-processing step or employing more advanced algorithms. The parameters for assessment included in this study include the Reliability and cost function, with a specific spotlight on the mean squared error (MSE). The precision can be determined by use the confusion matrix [1], where the sum of true positive and true negative is divided by the sum of true negative, true positive, false negative, and false positive.

2 Literature Review

In their study, T. J. de Carvalho et al. [2] proposed a recognition technique utilizing machine learning. The method was designed to detect splicing forgery, which involves the composition of various regions from distinct images to create a new image.

In their research, J. Ouyang [3] introduced a technique for detecting copy-move forgery utilizing deep CNNs. The approach was evaluated on three datasets: OXFORD flower [4] (as dataset1), UCID (as dataset2), and CMFD [5] (as dataset3). The UCID database was created with the primary goal of providing a dataset for the assessment of compressed domain algorithms, but it also serves as a useful benchmark set for evaluating other CBIR methods and an image set for evaluating color quantization and image compression techniques [6].J. Barad and M. M. Goswami [7] conducted an investigation and reported their result to assist other researchers in the field.. Moreover, the article provided a concise overview of image tampering detection techniques and carried out a comparative evaluation of various deep learning models, such as Deep Neural Network (DNN), Recurrent Neural Network (RNN), and Convolutional Neural Network (CNN). Sudiatmika, I.B.K. and Rahman [8] utilized a deep learning technique to differentiate authentic from manipulated images. Y. Shah et al. [9] utilized the CNN method and integrated the ELA into the pre-processing phase to address limitations in previous methods.

P. He, H. Li, and H. Wang et al. [10] utilized the GAN (Generative Adversarial Network) approach to enhance preceding performance. The present article discusses passive authentication techniques in the field of image forensics that could be achieved through two main approaches. The first approach utilizes a basic and straightforward convolutional neural network model [11] with two pre-processing techniques to generate results. The second approach involves evaluating the results obtained from various transfer learning models that have been fine-tuned. Finally, a comparison is made between all the models and a combination of techniques is selected to obtain the most accurate results using a specific dataset. The effectiveness of these approaches for image forgery detection is assessed, as well as potential avenues for further improvement.

3 Methodology

The exploration contains of two unique phases. In the early phase, pre-processing methods and a convolutional neural network (CNN) architecture, as revealed in Fig. 1 of the flowchart, are employed. The second phase utilizes a deep learning strategy, as indicated in Fig. 3 of the flowchart, to attain superior results. The dataset was trained using two alternative transfer learning approaches, and a more effective strategy was determined.

3.1 Method 1

The convolutional neural network (CNN) is widely recognized as the major neural network utilized for image identification. The CNN architecture, as depicted in Fig. 2, contains of two different components: the feature extraction process and categorization. The subject matter encompasses multiple strata [12, 13]: "The input layer receives the photos using the dataset as input". The CNN design, as depicted in Fig. 2, incorporates a division of the Feature-Extraction and categorization components, as indicated by references [14, 15]. The aforementioned layers are utilized. To improve the precision of image forgery detection, we conducted an investigation into the integration of Error Level Analysis (ELA) with a sharpening filter. The objective of the sharpening procedure is to enhance the distinction between regions of high and low brightness, hence producing more distinct features.The technique was implemented in our study using the Pillow-Python image processing library.

$$\begin{matrix} -2 & -2 & -2 \\ -2 & 32 & -2 \\ -2 & -2 & -2 \end{matrix}$$

Sharpening filter

The findings suggest that the integration of these two pre-processing steps played a significant role in enhancing the efficacy of our approach.

The process of contrasting can be observed in the tampered (false) image, as evidenced by the blurring or distortion of its borders and lines.

3.2 Method 2

To facilitate additional examination, the second approach encompasses transfer learning, a variant of active learning. Transfer learning is an approach in design that entails the application of acquired information from one activity to enhance the efficacy of another work as shown the Fig. 3. The CASIA V2.0 dataset was utilized for training purposes in this study, employing the weights that were learned of the network. In order to optimize the computation cycle of the neural network, the dataset was partitioned into distinct sets for development, validation, and testing. The allocation of data was done in accordance to the ratios of 40%, 30%, and 30% respectively. The utilization of this particular methodology offers advantages in terms of reducing the duration of the training process and exhibits a high level of compatibility using datasets that possess lower dimensions.

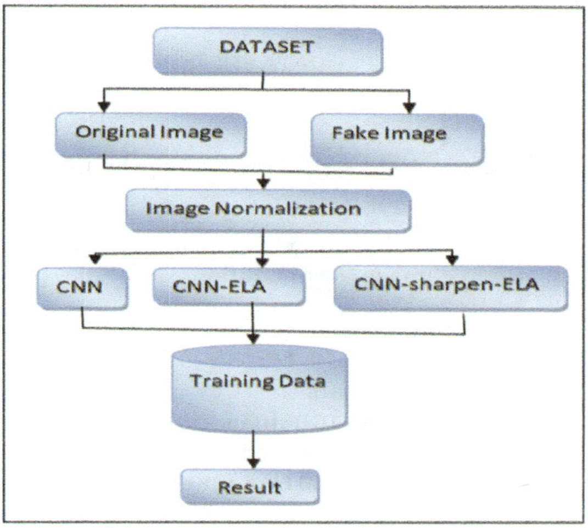

Fig. 1. Flowchart for method-1

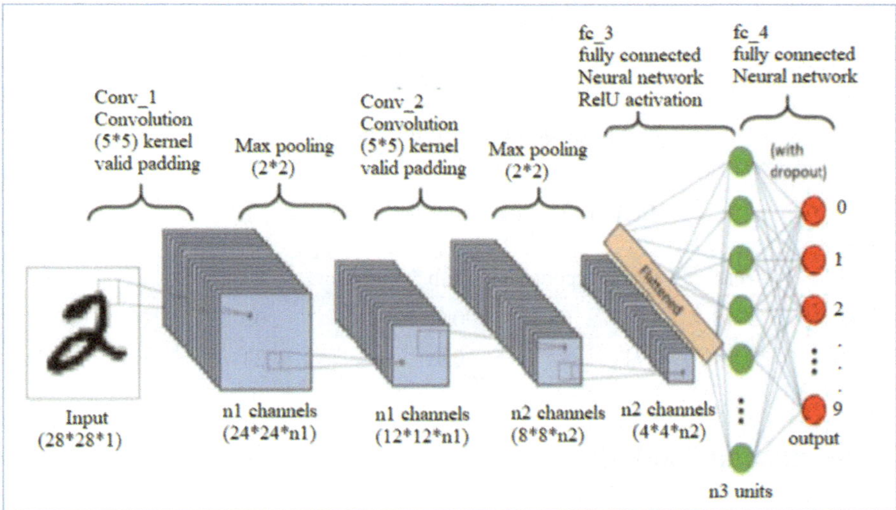

Fig. 2. Structural design of convolution neural network

The application of the filter results in the generation of a luminosity enhancement, wherein the pixels are amplified in comparison to their neighboring pixels. When implemented on the modified image, this filter will exhibit an irregular distribution.In the conducted study, two transfer learning models, namely VGG-16 and Resnet50 were employed (Table 1).

The VGG16 layout is a modified version of the VGGNet, consisting of a total of 16 layers, including 13 convolutional layers and 3 fully-connected layers. The stated parameter count of the model is 138 million, surpassing that of ResNet50.

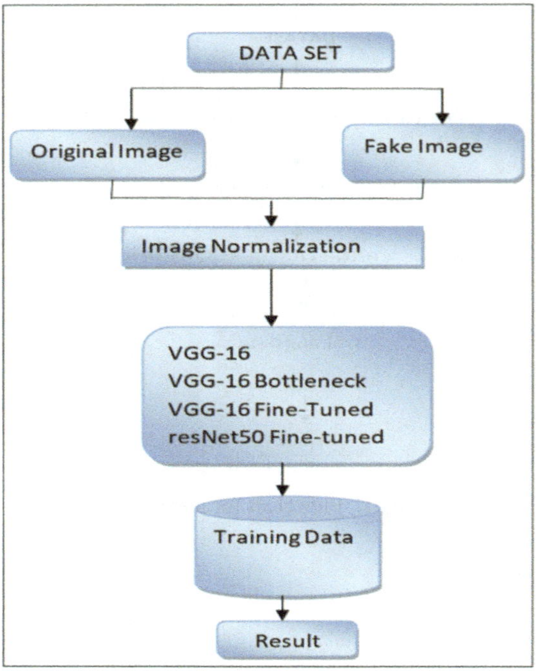

Fig. 3. Flowchart for method-2

Table 1. Details of transfer learning models

Model	Year	Top 1 Accuracy	Top 5 Accuracy	Depth	No. of Layers
ResNet50	2015	0.74	0.9	0	50
VGG16	2014	0.71	0.9	23	16

4 Results and Analysis

The following section includes a comprehensive description of the dataset employed in our investigation, as well as the outcomes and visual representations of both methodologies. The models were assessed using various metrics, including train accurateness, training loss, justification accuracy, and justification loss. Predictions were drawn through these assessments.

Dataset
The dataset utilized in this study was the CASIA V2.0 Image Tampering Detection Assessment Database.

The dataset provided is intended for the purpose of comparing and evaluating approaches used for detecting tampering, and is available to the public. The dataset used for defining the type of forgery consists of two distinct groups, namely tampered/fake

Fig. 4. Images as of the CASIA V2.0 dataset

and real/original, as illustrated in Fig. 4. The dataset contains of a total of 12,323 photos, by 7,200 images classified as authentic and 5,123 images identified as altered.

The unmodified image maintains consistent error level analysis (ELA) results, however, with each subsequent save, the image quality gradually deteriorates. On the contrary, the manipulated image demonstrates elevated Error Level Analysis (ELA) measurements, and the regions that have undergone modifications manifest a noticeable discrepancy in color.

This section presents the findings derived from the primary image illustrated in Fig. 5, along with the ELA-processed image in Fig. 5, which demonstrates a consistent value throughout. In contrast, the altered image represented in Fig. 6 exhibits incongruities that pose challenges for human visual perception. Despite the absence of color variation in the error level analysis (ELA) output, the overall image exhibits a darker appearance, as illustrated in Fig. 6.

Figure 5 and Fig. 6 depict the results obtained from the use of the sharpen filter on both the original image and the tampered image, as given in the study. This approach equally boosts the contrast of the image. Nevertheless, with the application of the ELA (Empirical Linguistic Analysis) to the aforementioned dataset, the findings indicate a noteworthy augmentation in numerical values, as seen by the tabular data and its corresponding graphical depiction.

(a) Original Image (b) Tempered Image

(c) ELA operated original Image (d) ELA operated tempered image

Fig. 5. The provided photos, both the original and altered versions, are accompanied with their corresponding Error Level Analysis (ELA) operations.

(a) Original image (b) Tampered Image

(c) Original sharpen image (d) Tampered sharpen Image

Fig. 6. Original and tampered images with respective sharpen operation

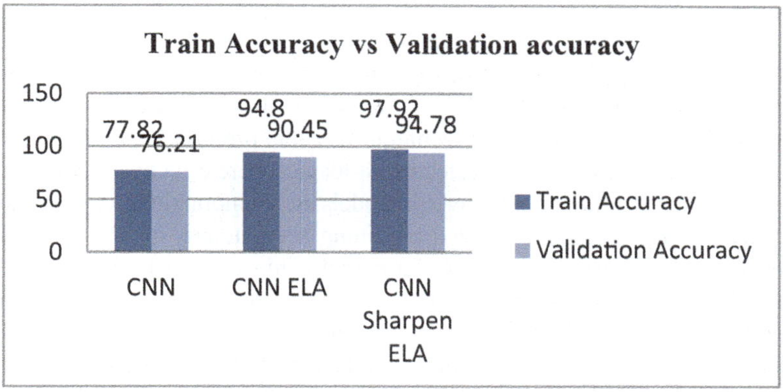

(a) compares various CNN models for train accurateness vs validation exactness

The present study undertakes a comparative analysis of different Convolutional Neural Network (CNN) models in terms of their respective training loss and validation loss. Figure 7 provides a comparative analysis of several Convolutional Neural Network (CNN) models, presenting a concise numerical comparison. The results indicate an enhancement not just in comparison to the basic CNN approach but also in comparison to the CNN_ELA model.

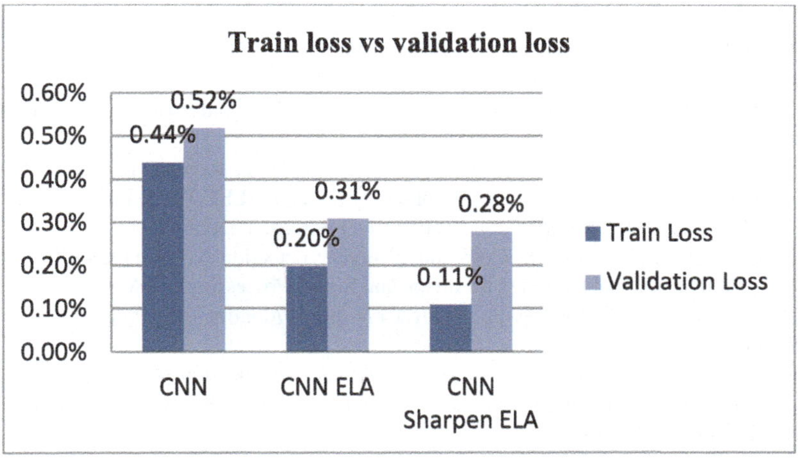

Fig. 7. Displays the outcomes obtained from several convolutional neural network models.

The Sharpen_ELA model attained a training precision of 97.92% and a loss of 0.1%, demonstrating an improvement margin of 19.87% and 0.36% correspondingly. The speed and effectiveness of an algorithm exhibited an upward trend as the total amount of layers rose, but accompanied by a growth in complication. Hence, it was determined that fine-tuning proved to be the most efficacious methodology for our research endeavor. This approach was subsequently used to ResNet50, a robust transfer learning model.

5 Conclusion

In this study, we conducted an evaluation of various deep learning techniques employed for the detection of picture fraud in the CASIA V2.0 dataset. The detection of fraud involved the utilization of two distinct techniques: (i) the pre-processing stage and (ii) several models of deep learning. The detecting strategies were effectively implemented by our team through the utilization of diverse models and combinations. As an example, the CNN_Sharpen_ELA model exhibited a training accurateness of 97% and a training loss of 0.1%. In contrast, the ResNet50 model achieved a test accuracy of 95% and a relatively low test loss of roughly 0.4%. The aforementioned methodologies are characterized by their simplicity and lack of reliance on specific datasets, rendering them applicable to a wide range of models and datasets for the purpose of investigating their impact and enhancements in diverse scenarios. Subsequent investigations will entail the undertaking of additional research endeavors aimed at discerning manipulated photographs and videos through the utilization of alternative methodologies. The numeral of convolution and pooling layers can be amplified in response to the complexity of datasets.The suggested methodologies have the potential to be used across many datasets, or alternatively, the existing model can be adapted with slight adjustments to the training methods and/or preprocessing step. In order to broaden the range of applicability, the suggested methodologies can be further extended to identify instances of tampering within video sequences comprising many frames.

References

1. Visa, S., Ramsay, B., Ralescu, A.L., Van Der Knaap, E.: Confusion matrix-based feature selection. MAICS **710**, 120–127 (2011)
2. de Carvalho, T.J., Riess, C., Angelopoulou, E., Pedrini, H., de Rezende Rocha, A.: Exposing digital image forgeries by illumination color classification. IEEE Trans. Inf. Forensics Secur. **8**(7), 1182–1194 (2013). https://doi.org/10.1109/TIFS.2013.2265677
3. Ouyang, J., Liu, Y., Liao, M.: Copy-move forgery detection based on deep learning. In: 2017 10th International Congress on Image and Signal Processing, BioMedical Engineering and Informatics (CISP-BMEI), pp. 1–5 (2017). https://doi.org/10.1109/CISP-BMEI.2017.8301940
4. Nilsback, M.-E., Zisserman, A.: Automated flower classification over a large number of classes. In: 2008 Sixth Indian Conference on Computer Vision, Graphics & Image Processing, pp. 722–729. IEEE (2008)
5. Schaefer, G., Stich, M.: UCID: an uncompressed color image database. In: 2004 Storage and Retrieval Methods and Applications for Multimedia, vol. 5307, pp. 472–480. International Society for Optics and Photonics (2003)
6. Rajesh Tiwari, M., et al.: Enhanced power quality and forecasting for PV-wind microgrid using proactive shunt power filter and neural network- based time series forecasting. Electr. Power Components Syst. (2023). https://doi.org/10.1080/15325008.2023.2249894
7. A Comprehensive guide to convolutional Neural Networks. https://towardsdatascience.com/a-comprehensive-guide-toconvolutional-neural-networks-the-eli5-way-3bd2b1164a53
8. Albawi, S., Mohammed, T.A., Al-Zawi,S.: Understanding of a convolutional neural network. In: 2017 International Conference on Engineering and Technology (ICET), pp. 1–6 (2017). https://doi.org/10.1109/ICEngTechnol.2017.8308186

9. Khan, S.H., Abbas, Z., Danish Rizvi, S.M.: Classification of diabetic retinopathy images based on customised CNN architecture. In: 2019 Amity International Conference on Artificial Intelligence (AICAI), pp. 244–248. IEEE (2019)
10. Chilakala, S., Saxena, A., Vunnava, A.: Energy-efficient reporting scheme for cognitive radio networks. IET Netw **10**, 110–117 (2021). https://doi.org/10.1049/ntw2.12019
11. Vadivel, S., Konda, S., Balmuri, K.R., Stateczny, A., Parameshachari, B.D.: Dynamic route discovery using modified grasshopper optimization algorithm in wireless ad-hoc visible light communication network. Electronics **10**(10), 1176 (2021). https://doi.org/10.3390/electronics10101176
12. Cherukuri, N., Bethapudi, N.R., Thotakura, V.S.K., Chitturi, P., Basha, C.Z., Mummidi, R.M.: Deep learning for lung cancer prediction using NSCLS patients CT Information. In: 2021 International Conference on Artificial Intelligence and Smart Systems (ICAIS), Coimbatore, India, pp. 325–330 (2021). https://doi.org/10.1109/ICAIS50930.2021.9395934
13. Samdekar, R., Ghosh, S.M., Srinivas, K.: Efficiency enhancement of intrusion detection in IoT based on machine learning through bioinspired. In: 2021 Third International Conference on Intelligent Communication Technologies and Virtual Mobile Networks (ICICV), Tirunelveli, India, pp. 383–387 (2021). https://doi.org/10.1109/ICICV50876.2021.9388392
14. Dinesh, G., Venkatakrishnan, P., Jeyanthi, K.M.A.: Modified spider monkey optimization—an enhanced optimization of spectrum sharing in cognitive radio networks. Int. J. Commun. Syst. **34**, e4658 (2021). https://doi.org/10.1002/dac.4658
15. Arunachalam, P., Venkatakrishnan, P., Janakiraman, N.: Histopathology image classification for soft tissue sarcoma in limbs using artificial neural networks. In: 2021 6th International Conference on Inventive Computation Technologies (ICICT), Coimbatore, India, pp. 778–785 (2021). https://doi.org/10.1109/ICICT50816.2021.9358635

A Machine Learning Algorithm-Based Automated Transport System for Predicting Traffic Flow

Rajesh Tiwari[1], Sheo Kumar[1], V. Narasimha[2], Golla Saidulu[2], Banothu Ramji[3], and K. Maheswari[4](✉)

[1] Department of CSE, CMR Engineering College, Hyderabad, Telangana, India
[2] Department of CSE, CMR College of Engineering and Technology, Hyderabad, Telangana, India
{narasimhacse,gsaidulu}@cmrcet.ac.in
[3] Department of CSE (DS), CMR Technical Campus, Hyderabad, Telangana, India
[4] Department of CSE, CMR Technical Campus, Hyderabad, Telangana, India
cmrtc.paper@gmail.com

Abstract. The adoption of various traffic prediction technologies has led to recent enhancements in the efficacy of the transportation system. The Intelligent Transport System (ITS) organization is dedicated to the regulation of traffic and the development of uses for future transportation systems. The forecast of traffic flow holds significant utility in the management of traffic flow, optimization of traffic signal systems, and estimation of travel durations. In this program, user input is collected and utilized to employ various machine learning algorithms for the purpose of predicting traffic flow. This study aims to conduct a comparative analysis of different traffic analysis methods, including Random Forest Regression, Support Vector Machines (SVM), K-Nearest Neighbors (KNN), and Gated Recurrent Units (GRU). The objective is to determine the most effective model by applying these methods to a comprehensive dataset of traffic. The highest performing model has the potential to be utilized for future predictions of traffic flow.

Keywords: Traffic flow · ITS · machine learning · KNN · SVM · GRU · RFR

1 Introduction

Intelligent Transport Systems (ITS) has emerged as a prominent ground of study in current years. The primary objective of ITS is to enhance transportation security, mobility, efficiency, and environmental performance for traffic planners and road users. This is achieved through the provision of advanced services pertaining to dissimilar modes of transportation and traffic management. By empowering users with comprehensive information, ITS enables them to make more informed decisions, leading to safer, more coordinated, and more efficient utilization of transport networks. The significance of vehicle identification has increased in tandem with the incessant growth of urban road networks and the extensive development of highways. The detection of vehicles is an

A. Bajaj et al. (Eds.): IBICA 2023, LNNS 1229, pp. 398–406, 2025.
https://doi.org/10.1007/978-3-031-78940-3_42

essential duty within Intelligent Transportation Systems (ITS), aiming to gather data for various purposes such as vehicle counting, vehicle speed measurement, and traffic collision recognition, among others. One notable benefit of this strategy is its reduced reliance on domain knowledge compared to parametric methods. However, it is important to note that a larger amount of data is need to accurately determine a model. This also suggests that the effectiveness of implementing data-driven models is strongly associated with the quality of the data that is accessible. On the other hand, traffic flow within urban environments exhibits a greater degree of dynamism and non-linearity, primarily attributable to the proliferation of intersections and traffic signage. In such contexts, the utilization of data-driven machine learning methodologies, including neural networks, random forests, and K-nearest neighbors (KNN), may be deemed more suitable owing to their capacity to effectively capture intricate nonlinear associations and dynamic phenomena.

2 Literature Review

In this study [1], our objective was to employ machine learning, genetic, soft computing, as well as deep learning algorithms for the analysis of huge data in the field of transportation, while substantially decreasing sophistication. Image processing algorithms play a crucial role in the recognition of traffic signs, hence contributing to the effective training of autonomous vehicles.The present study [2] proposes a method for predicting short-term traffic congestion using the random forest algorithm, with the aim of mitigating the adverse effects of traffic congestion on individuals' travel experiences.

The aforementioned research [3] presents detailed insights into the dynamics of traffic flow. The study has considered different time intervals and has implemented three specific forms of recurrent neural network architecture, including simple RNN, Long Short Term Memory (LSTM), and Gated Recurrent Unit (GRU).This study presents a suggested LSTM model that is designed to handle long as well as short time intervals. Furthermore, it can be demonstrated that the proposed technique exhibits superior performance in terms of efficiency compared to alternative schemes utilizing link state updates, whereas also producing significantly lower overhead [4].

According to the study referenced as [5, 6], the precise and real-time forecast of traffic flow is of utmost importance in the field of Intelligent Transportation Systems (ITS), specifically in the context of traffic management. To predict short-term traffic flow, Long Short Term Memory (LSTM) and Gated Recurrent Units (GRU) neural network (NN) techniques are employed. This study employs a diverse range of machine learning techniques to predict traffic patterns based on gathered data. The utilization of the collected dataset [7] has resulted in a more precise and improved traffic forecast compared to previous Machine Learning (ML) systems. The main objective of this study [8] is to address the issue of insufficient traffic speed data. To do this, the research provides a novel approach for predicting traffic speed by utilizing the multi-temporal traffic flow volume from both preceding and subsequent time periods. In this study [9], the data is preprocessed employing selenium, OSS, and Message Queue. Additionally, the traffic flow is modeled utilizing three machine learning algorithms: Linear regression, Decision Tree, and Support Vector Machine. Subsequently, an examination is conducted on the factual traffic data pertaining to Beijing as obtained from the Baidu map. The methodology centers around the utilization of a neural network model for the purpose of predicting forthcoming traffic patterns at an hourly interval [10–12]. The outcomes obtained from the simulation provided a reliable forecast of the traffic statistics. In the present study [13, 14], a novel spatial-temporal graph attention based dynamic graph convolutional network (GAGCN) model is introduced for the purpose of traffic flow prediction. The GAGCN model employs graph attention networks to automatically remove spatial relations between nodes that are concealed inside the traffic feature data. The experimental findings demonstrate that our Graph Attention Graph Convolutional Network (GAGCN) outperforms the current leading benchmark models on one out of the two datasets.

3 Proposed Model

This study aims to investigate the datasets from four junctions and develop a predictive model for traffic patterns in these areas. This has the ability to contribute to the resolution of traffic congestion by enhancing comprehension of patterns of travel, hence facilitating the development of infrastructure aimed at mitigating the issue. Various machine learning techniques, including Long Short-Term Memory (LSTM), Gated Recurrent Unit (GRU), Random Forest Regressor, and K-Nearest Neighbors (KNN), were employed in order to enhance the accurateness of the forecast outcome as shown the Fig. 1.

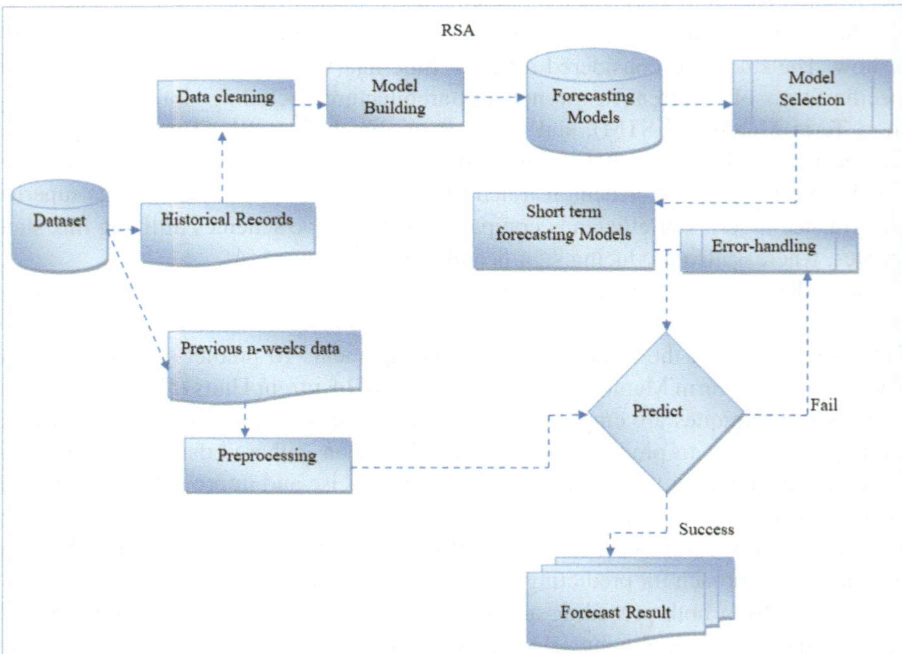

Fig. 1. Proposed System Architecture

3.1 Random Forest Regression

Random Forest is a widely recognized machine learning algorithm that has proven to be an efficient way for supervised learning. The issues of classification as well as regression in machine learning can be effectively addressed with this approach. The concept is grounded in the standard of assembly learning, involving the integration of multiple classifiers to tackle complex problems and improve the general efficacy of the model. The classification algorithm referred to as "Random Forest" derives its name from its composition of numerous decision trees, each generated using distinct subsets of the provided information as shown the Fig. 2. The Random Forest classifier seeks to improve the predictive accuracy of the dataset by taking the average of the forecasts generated by multiple decision trees as opposed to relying on a solitary decision tree, the random forest algorithm aggregates predictions from multiple trees and determines the final outcome based on the majority vote amongst these forecast.

The Random Forest Approach was utilized to predict traffic conditions in our study.

1. Select "k" characteristics randomly from the set of "m" characteristics.
2. When the value of k is much smaller than the value of m.
3. so as to identify the optimal split point, it is essential to identify the correct node "d" from a set of "k" attributes.
4. The node is divided into daughter nodes by employing the optimal split.
5. Proceed with the aforementioned stages, namely steps one through three, iteratively until the desired number of "l" nodes have been attained.
6. To generate a forest, it is necessary to iterate steps 1 to 4 "n" times, hence constructing "n" trees.

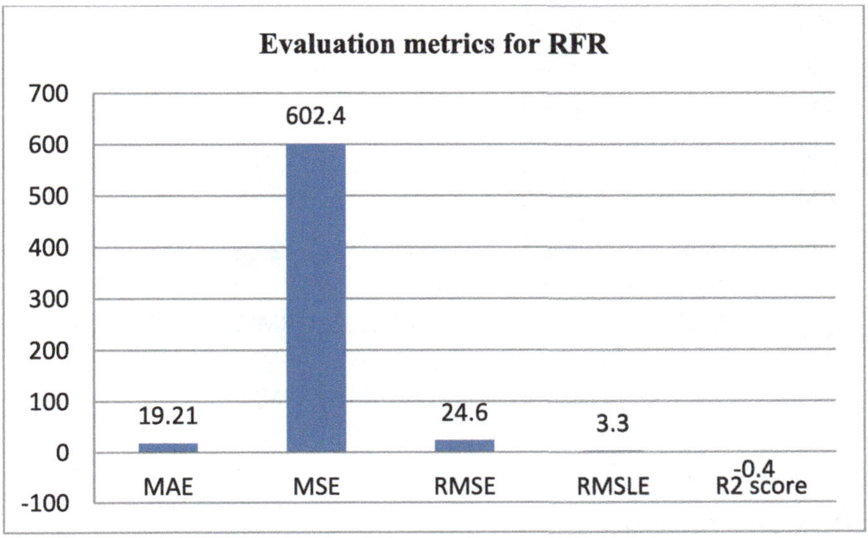

Fig. 2. Evaluation Metrics for RFR

3.2 Support Vector Machine

The Support Vector Machine (SVM) is extensively recognized as a prominent technique in the field of Supervised Learning, specifically for addressing issues with classification and regression shown in the Fig. 3. Yet, its primary application is within the field of Machine Learning, particularly in the circumstance of categorization difficulties. The objective of the SVM algorithm is to identify an optimal line or decision boundary in an n-dimensional space, with the purpose of accurately categorizing fresh data points in the future. The term "hyperplane" is used to refer to the optimal decision boundary. The Support Vector Machine (SVM) algorithm utilizes extreme areas or vectors to facilitate the construction of the hyperplane. The instances that exhibit extreme characteristics are commonly denoted as support vectors, hence giving rise to the nomenclature of the supporting vector machine algorithm. In the context of n-dimensional space, it is possible to employ many lines or decision borders for the purpose of segregating classes. However, it is imperative to identify the optimal decision boundary that facilitates the classification of data points. The optimal boundary is commonly referred to as the SVM hyperplane. The amount of features present in the dataset determines the size of the hyperplane. In the case when there are just two characteristics, the hyperplane appears as a straight line, as depicted in the accompanying image. If there exist three characteristics, the hyperplane can be described as a two-dimensional plane.

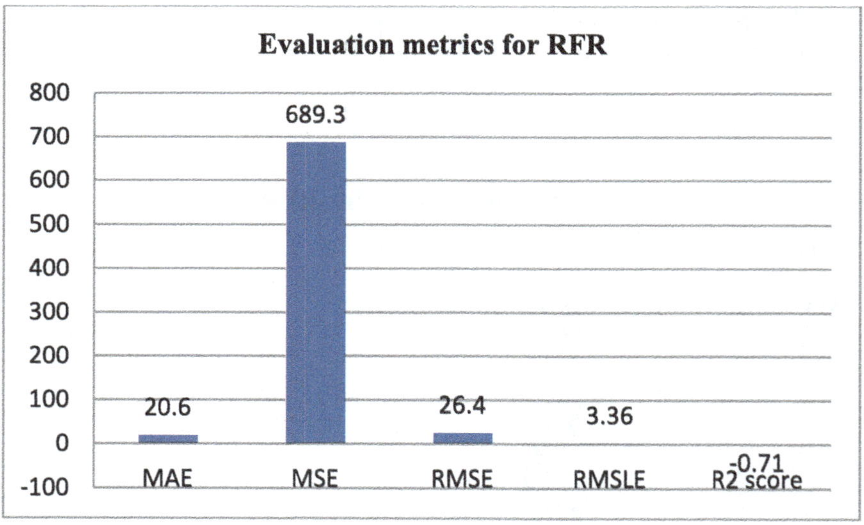

Fig. 3. Evaluation Metrics for SVM

3.3 K-Nearest Neighbor

The technique in question is a supervised learning method that falls under the non-parametric category. It leverages the concept of distance to make recommendations regarding the classification or grouping of a given data point. This approach operates on the underlying assumption that data points exhibiting similarity tend to be located in close proximity to each other. In the context of classification issues, the assignment of a class label is determined by a majority voting mechanism, whereby the label that is most commonly observed in the vicinity of a specific data point is selected. The selection of the number of neighbors (k) to be considered during prediction is a critical determinant of performance in the K-Nearest Neighbors (KNN) algorithm shown the Fig. 4. However, it should be noted that there is no universally optimal choice of k that guarantees outstanding accuracy across all datasets. Smaller values of k render the model more susceptible to noise, whereas larger values of k lead to increased generalization and longer prediction times. Hence, the selection of the number of neighbors (k) is a hyperparameter that requires optimization. To ascertain the data points that are in closest proximity to the specific query point, it is essential to compute the distance among the query area and the remaining data points. There are several ways available for performing this measurement, with Euclidean, Manhattan, and Minkowski distance being the most commonly utilized approaches. The K Nearest Neighbours algorithm enhances the efficacy of our job by acquiring knowledge of non-linear decision limits, a characteristic that is beyond the capability of numerous mathematical models to comprehend. Additionally, the system demonstrates a high level of adaptability to newly introduced data, hence minimizing downtime through the efficient storage of comprehensive training data in its memory. The training time of a lazy learning model is not reliant on a certain amount of occurrences.

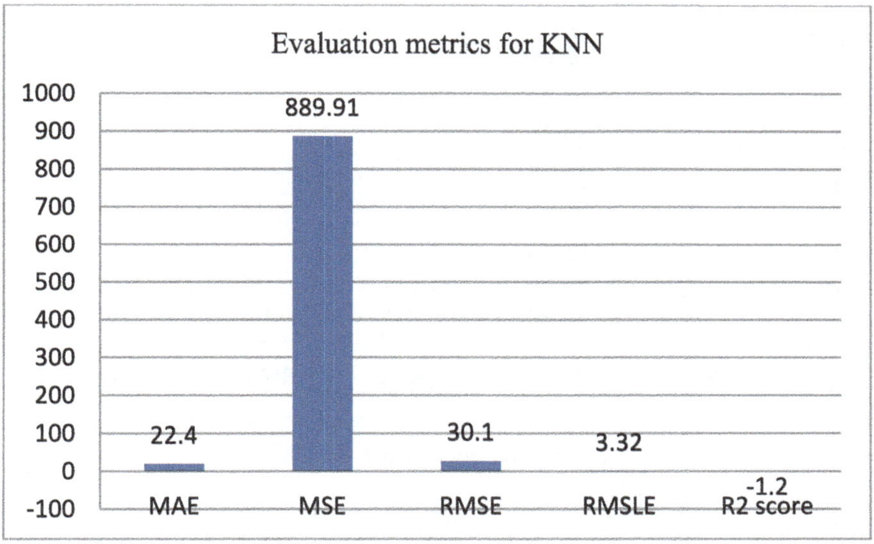

Fig. 4. Evaluation Metrics for KNN

3.4 Gated Recurrent Unit

The Gated Recurrent Unit (GRU) is a type of recurrent neural network (RNN) that can be contrasted by means of the more extensive short-term memory (LSTM). The GRU model has superior computational efficiency and memory utilization compared to the LSTM model. However, the LSTM model demonstrates enhanced accuracy when handling datasets comprising extended sequences. In addition, as shown the Fig. 5 Gated Recurrent Units (GRUs) effectively address the problem of vanishing gradients encountered by conventional recurrent neural networks, which pertains to the diminishing values utilized for updating network weights. The potential for diminished impact on learning arises when the grading process gradually diminishes in magnitude as it propagates backwards, resulting in an untrainable state of the neural network. Recurrent Neural Networks (RNNs) have the capacity to effectively disregard longer sequences of information when a layer inside a neural network fails to acquire the necessary learning. The utilization of update and reset gates in GRUs serves as a mechanism to tackle this particular challenge. The gates has the capability to undergo training processes that enable them to retain and recall knowledge from more distant temporal contexts. Additionally, they possess the capacity to discern and regulate the flow of information that is permitted to propagate to the output.

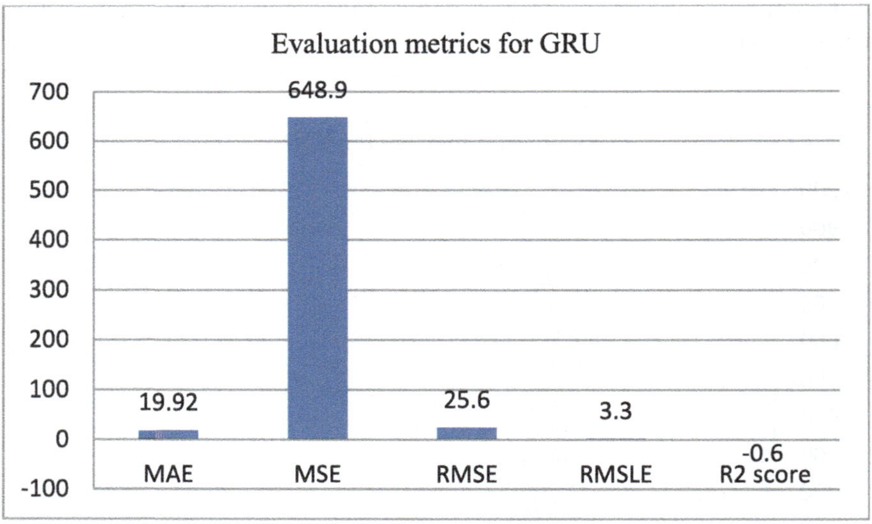

Fig. 5. Evaluation Metrics for GRU

3.5 Comparison of the Models

The results obtained from the application of multiple models on the given dataset are as following:

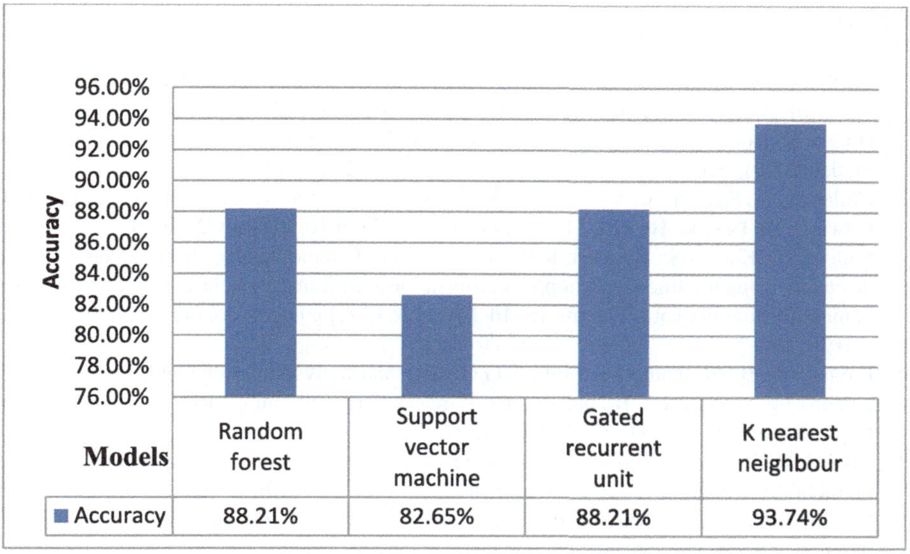

Fig. 6. Comparison of various machine leaning models

According to the data presented in Fig. 6 of the previous table, the Random Forest technique demonstrates a high level of accurateness in predicting the traffic rate, accomplish an accurateness rate of 88.21% as shown the Fig. 6.

The construction of a hyperplane in a manner that maximizes the margin, defined as the largest feasible separation among data points, is a consistent practice. The Support Vector Machine (SVM) technique has been widely employed in several applications such as face detection, picture classification, and classification of text, among others.

This capability allows for the transmission of relevant information along a series of occurrences, hence enhancing the precision of forecasts.

4 Conclusion

This study presents the development of an entire system for traffic analysis and prediction utilizing the Random Forest Regression approach, which demonstrates superior presentation compare to the individual models included in the analysis. In this study, we employed K-Nearest Neighbors (KNN) and Gated Recurrent Unit (GRU) as foundational models, while utilizing Support Vector Machine (SVM) as the meta model to aggregate the predictions generated by all of the foundational models. Across numerous iterations of stochastic cross validation, we consistently attained an accuracy rate of 88.19%. Moreover, a comprehensive examination of the traffic was conducted, leading to a thorough understanding of the effects induced by the significant factors. This study nearby an investigation into the complementary nature of distinct machine learning models and their potential to enhance system efficiency. Specifically, we examine how a random forest model might address the limitations of other models within a particular dataset, leading to an overall improvement in efficiency.

References

1. Meena, G., Sharma, D., Mahrishi, M.: Traffic prediction for intelligent transportation system using ML (2020). https://doi.org/10.35940/ijeat.D2426.0410421
2. Huang, S.: Road traffic congestion prediction based on a hybrid random forest and DBSCAN model (2020). https://doi.org/10.1109/ICSGEA51094.2020.00075
3. Chilakala, S., Saxena, A., Vunnava, A.: Energy-efficient reporting scheme for cognitive radio networks. IET Netw. **10**, 110–117 (2021). https://doi.org/10.1049/ntw2.12019
4. Vadivel, S., Konda, S., Balmuri, K.R., Stateczny, A., Parameshachari, B.D.: Dynamic route discovery using modified grasshopper optimization algorithm in wireless ad-hoc visible light communication network. Electronics **10**(10), 1176 (2021). https://doi.org/10.3390/electronics10101176
5. Premalatha, B., Srikanth, G., Abhilash, G.: Design and analysis of multi band notched MIMO antenna for portable UWB applications. Wireless Pers. Commun. **118**, 1697–1708 (2021). https://doi.org/10.1007/s11277-021-08111-0
6. Koyi, L.P., Borra, T., Prasad, G.L.V.: A research survey on state of the art heart disease prediction systems. In: 2021 International Conference on Artificial Intelligence and Smart Systems (ICAIS), Coimbatore, India, pp. 799–806 (2021). https://doi.org/10.1109/ICAIS50930.2021.9395785
7. Cherukuri, N., Bethapudi, N.R., Thotakura, V.S.K., Chitturi, P., Basha, C.Z., Mummidi, R.M.: Deep learning for lung cancer prediction using NSCLS patients CT information. In: 2021 International Conference on Artificial Intelligence and Smart Systems (ICAIS), Coimbatore, India, pp. 325–330 (2021). https://doi.org/10.1109/ICAIS50930.2021.9395934
8. Samdekar, R., Ghosh, S.M., Srinivas, K.: Efficiency enhancement of intrusion detection in IoT based on machine learning through bioinspired. In: 2021 Third International Conference on Intelligent Communication Technologies and Virtual Mobile Networks (ICICV), Tirunelveli, India, pp. 383–387 (2021). https://doi.org/10.1109/ICICV50876.2021.9388392
9. Dinesh, G., Venkatakrishnan, P., Jeyanthi, K.M.A.: Modified spider monkey optimization—an enhanced optimization of spectrum sharing in cognitive radio networks. Int. J. Commun. Syst. **34**, e4658 (2021). https://doi.org/10.1002/dac.4658
10. Arunachalam, P., Venkatakrishnan, P., Janakiraman, N.: Histopathology image classification for soft tissue sarcoma in limbs using artificial neural networks. In: 2021 6th International Conference on Inventive Computation Technologies (ICICT), Coimbatore, India, pp. 778–785 (2021). https://doi.org/10.1109/ICICT50816.2021.9358635
11. Tiwari, V., Naidu, A.S., Singh, A., Saxena, A.: Construction of 4×4 MDS matrices for lightweight block ciphers. In: 2021 IEEE International Conference on Advanced Networks and Telecommunications Systems (ANTS), Hyderabad, India, pp. 254–259 (2021). https://doi.org/10.1109/ANTS52808.2021.9936970
12. Tiwari, R., et al.: An artificial intelligence-based reactive health care system for emotion detections. Comput. Intell. Neurosci. **2022**, Article ID 8787023 (2022). https://doi.org/10.1155/2022/8787023
13. Jayakumar, T., et al.: Machine learning approach for prediction of residual energy in batteries. Energy Rep. **8**, 756–764 (2022)
14. Subramanian, M., Shankar Nayak Bhukya, R., Vijaya Prakash, K., Raju, N., Ray, S., Pandian, M.: Deploy machine learning model for effective bank telemarketing campaign. In: 2023 International Conference on Distributed Computing and Electrical Circuits and Electronics (ICDCECE), pp. 1–6. IEEE (2023)

A Comprehensive Study on Convolution Neural Networks - Architecture, Applications and Future Scope

S. Abdulkani[1], M. Shereesha[2], Nandala Narayanamma[3], A. Jyothi Babu[4(✉)], B. Sargunam[5], and P. Radhika Raju[6]

[1] Computer Science and Engineering, Indur Institute of Engineering and Technology, Siddipet, TS, India
[2] Computer Science and Engineering, B V Raju Institute of Technology, Medak, TS, India
[3] Department of CSE, Sri Venkateswara College of Engineering, Tirupati, AP, India
[4] Department of Computer Applications, School of Computing, Mohan Babu University, Tirupati, India
jyothibabuaddanki@gmail.com
[5] Avinashilingam Institute for Home Science and Higher Education for Women, School of Engineering, Coimbatore, TN, India
sargunam_ece@avinuty.ac.in
[6] Department of CSE, JNTUACEA, Anantapur, AP, India

Abstract. CNN is the most popular learning network established in the recent decades. Many variants of neural networks emerged for various applications related to image detection, recognition and classification. As images are prove input for the CNN, the CNN are more performant in machine learning and deep learning applications. CNNs are specialized in feature extraction, feature based classification. Higher the layers more intricate are the features. In this paper, enlisting various types of CNNs and the future scope is discussed.

Keywords: convolution neural networks · architectures · deep learning applications

1 Introduction

A sudden impulsive growth in the usage of machine learning is in vogue since two decades. Deep learning is one of the machine learning approaches that has become a most common name for the applications of artificial [1, 2] intelligence. With the advent of AI with Deep Learning into computational intelligence, the growth-crescent of mimicking of human brain is gradually increasing. Traditional neural network has been surpassed with the scaled-up of AI applications. Multi-layer learning models emerge as the graph transformation become demand in tandem with basic functionalities of learning models.

CNNs are the most established neural networks in deployed into learning techniques and significant subtype of artificial neural networks. Medical imaging grabbed importance with the applications of CNNs in detecting [3, 4] crucial aspects, diagnosis

A. Bajaj et al. (Eds.): IBICA 2023, LNNS 1229, pp. 407–414, 2025.
https://doi.org/10.1007/978-3-031-78940-3_43

and physiological irregularities more accurately with iterative epoch evolutions. Spatial analysis in modeling images of 2D and 3D has been of one great importance of the applications of CNNs. The CNNs greatly work on higher capacity of data, extracting features from varied resolutions of input images.

CNNs achieve many accolades of ground-breaking discoveries in the fields of pattern recognition, voice, image and natural language processing. The complexities in ANN is reduced and resolved by CNNs, which [5, 6] encourages academicians, researchers and industry developers to utilize the power of building complex models to handle difficult issues, that a conventional ANNs cannot able to redress. The main attention and objective of this review to understand the elements of CNN that make the research community usher with new comprehensions of applications.The age old adage of GIGO (Garbage-In-Garbage-Out) is no misnomer in the development of learning systems. The CNNs work efficiently how they are trained on, the training data and testing data with validation criteria sets up the effective CNN. Means as we educate the CNNs learning model with the sophisticated and positive training data, so the CNN model would be helpful to achieve the challenging problems. Quality features are extracted [7, 8] with quality training. Automatic feature extraction is a remarkable success of deep learning models in image processing and computer vision. Complexity and precision are the qualities of every CNN that develops with the sophisticated training.

2 ConvNet

The CNN is called popularly as ConvNet devised with discriminative deep learning architecture, which directly learns from the input objects. The applications of the ConvNet includes medical image analysis, image annotation, image segmentation, visual identification, classification, natural language processing and other applications that deal with 2D shapes. Automatic identification of the key elements is the important trait of the ConvNet without human intervention.

2.1 Basic Components of CNNs

The components of CNNs with respect to the applications are a critical part of comprehension and knowledge about the advancements in CNN architectures. Figure 1 shows several CNN components.

The input layer takes the input image. The image is made up of pixels, where the visual representation of pixels with all the properties in the range are 0–255 sequentially arranged as a matrix. The CNN is trained to recognize the basic patterns of the image such as lines, curves and then detail parts such as [9, 11] faces and objects, as a result making a computer can see the contents of the image using CNN framework.

The convolutional layer is an mandatory part of the CNN, contains filters called as kernels. A kernel is a rectangular matrix having width and height, which are decided based on the characteristics of the data. The kernel is assigned with weights based on the type of application where CNN is deployed. The filters or kernels represent a feature, CNN deploys a most-effective kernel to identify the features. Proper weights assigned will make the CNN learn about the features quickly and reduce the training time. The

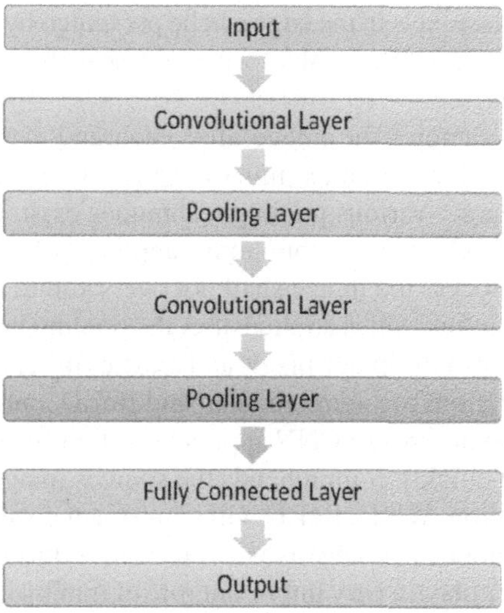

Fig. 1. CNN Components

kernel is overlapped on the image to detect where the area contains feature or not, the kernel is moved across or down which guided by the stride. As the filter moves on following the stride the pixels of the group are incident into a mathematical aggregation (Fig. 2).

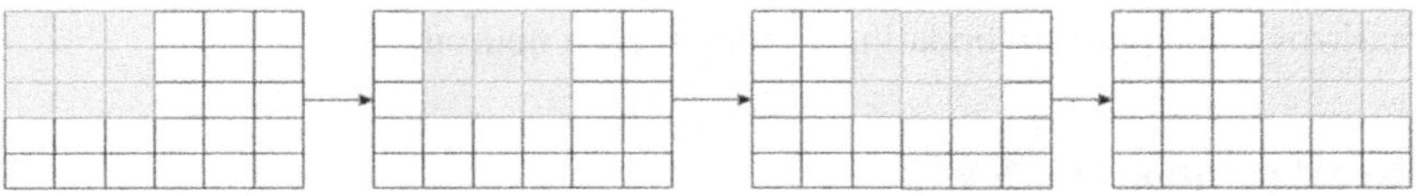

Fig. 2. Kernel or Filter moving following the stride

However there could be loss of pixels in the image edges during the stride operation, padding is taken care for all the pixels on the boundaries of the image. The following image in Fig. 3 shows the zero-padding.

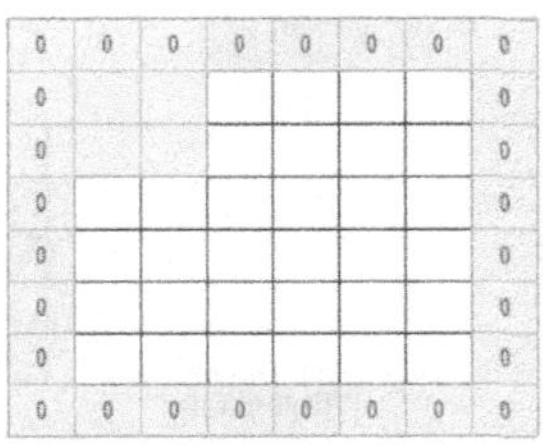

Fig. 3. Zero-padding

The effect of losing the pixels in the edge can be prevented by padding the boundaries of the image with zeros. Any other value or weight shall be considered based on the characteristics of the image.

Pooling layer down-samples the image after each and every stride operation, the mathematical aggregation forms or max, min or avg functions can be applied, to reduce the complexity of the image. Various pooling techniques exist, like GAP is the method called Global Average Pooling most commonly applied [12–14] to reduce the image complexity. While the process the image shall not oversaturate and loose the characteristics, therefore activation function is applied after the pooling layer. Activation function works on the non-linearity of the image pixels and reduces the error rates. Most popularly employed activation functions are sigmoid, tanh and ReLU, out of which ReLU is used more conveniently in vast number of CNN applications for classification. ReLU reduces the time complexity and quickly comprehends the sources image towards achieving the goal of the CNN application. While, ReLU will converge the values of the pixels during pooling that there shall not be any negative or zero values during aggregation.

Fully connected layer plays a very important role of reminiscence of neural network, where large number of parameters applied for calculation on the training samples. The fully connected layer is connected with neurons and the inputs directly as a vector from top layers. The complexity in this layer would be overcome by dropout layer which performs a dropout function, to minimize the nodes and the connections, to reduce the computational complexity.

The training data sets are therefore considered meticulously from the source data in order to reduce the complexity of learning. Training data sets consists of predefined classes and labels for a image classification problem. Back propagation is applied in the training the CNN to assess the error and based on the output values of the previous layers. Error value helps in updating weight at each neuron.

2.2 Regularization of CNN

Generalization on CNN is carefully deployed to avoid the over-fitting. The model may perform well on the training data but fails on the real data. The reason of this situation is that the characteristics of the image data [15, 16] are not properly selected at the time of training, which also leads to under-fitting. The model is said to be the best when the results are satisfactory on both training and testing data. The following figure Fig. 4 illustrates the nature of outcome for the CNN during under-fitting, balanced and over-fitting.

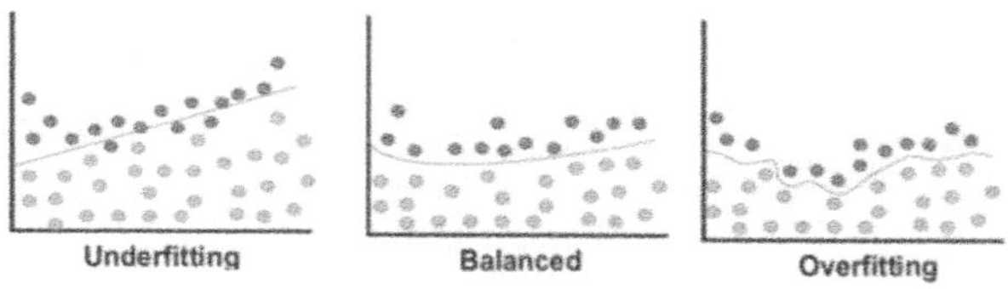

Fig. 4. Nature of CNN during Regularization

2.3 Types of CNN

Convolutional Neural Networks (CNNs) are a class of deep learning models that are particularly well-suited for tasks involving images and spatial data. There are several types of CNN architectures, each designed to address specific [17, 18] challenges and achieve better performance in various tasks. Here are some common types of CNN architectures:

Type	Layers	Application	Gap
LeNet, LeNet-5	Five to Eight layers may include three dense layers	Hand written digit recognition of MNIST databases	Low quality feature extraction, not suitable for large images
AlexNet	Twelve layers may include three dense layers	Works for large images. Image classification	Dropout layers increase the complexity of application, regularization is slow
VGGNet	Sixteen layers may include three to five dense layers	Works for large datasets of images for Image classification and localization. 4096 channels draw from two fully connected layers, able classify 1000 labels	Training time is huge (min 100 epochs)
GoogLeNet	Nineteen to Twenty two layers	Computationally efficient, it can perform even 1x1 convolutions	Regularization is slow
ResNet	Residual NN, constituted with more than sixteen layers. Constructed in terms of blocks. Also contains 48 to 50 layers	Computationally efficient, performs complex NN operations with ease	Training time is huge (min of 100 epochs)
DenseNet	Multiple convolution layers and hidden layers. Each unit possessatleast 5 dense layers, 2 fully connected layers with drop out layers	Works on highres sat images. Parallel convolutions for intense feature extraction	Training time is high and regularization is slow

MobileNet: MobileNet is designed for resource-constrained environments, such as mobile devices and embedded systems. It uses depth-wise separable convolutions to reduce the number of parameters and computations while maintaining performance.

Xception: Xception, short for "Extreme Inception," is an extension of the Inception architecture that replaces standard convolutions with depth-wise separable convolutions. This architecture aims to capture fine-grained features more efficiently.

EfficientNet: EfficientNet is a family of models that use a compound scaling method to balance model depth, width, and resolution to achieve state-of-the-art performance while being computationally efficient.

SqueezeNet: SqueezeNet is designed to have a small model size while maintaining competitive accuracy. It uses various techniques like 1x1 convolutions and aggressive compression to reduce the number of parameters.

NASNet (Neural Architecture Search Network): NASNet is not a single architecture but a methodology that uses reinforcement learning to search for optimal neural network architectures. It can discover highly efficient and effective CNN architectures.

YOLO (You Only Look Once): YOLO is an object detection architecture that can detect and classify objects in real-time. It divides the input image into a grid and predicts bounding boxes and class probabilities for each grid cell.

2.4 Types of CNN Architectures

Various types of CNN architectures exist based on the applications. CNNs are originate in wide range of application related to computer vision and image processing. The ability of automatic learning and extracting intricate features succinctly has made them indispensable in machine learning model development. Some of the popular and prominent application of CNNs are image classification, object detection, face recognition, gesture recognition, image segmentation, medical imaging, autonomous vehicle management, handwriting recognition, visual search and art generation, audio video analysis, natural language analysis, document analysis, satellite image analysis and quality control, etc,.

3 Future Directions

Amongst the application of the CNNs, classification is the most popular application which is experimented on all types of data sets. The performance of the classification in terms of accuracy, misclassification rate and the parameters like precision and recall are heavily influenced by the combination of convolutional, fully connect, dense and dropout layers. The number of pooling layers and filters, size of filters, the stride rate and the location for pooling influence the complexity of the CNN application. Therefore, the training of the CNN necessitates the utilization of the powerful hardware. Various combinations of selecting data sets into the training and testing the network need experience of selection or the selection of data set with the appropriate statistical properties.

Hyper-parameters impact the CNNs performance, sensitively to the modest shift of settings and configurations. A slight change in the hyper-parameters influences major functional objectives of the CNN. It is crucial to consider the account of the change of application mobility and the optimized performance of the CNNs. Selection of the appropriate parameter for tuning the performance of the CNN require sleight of understanding in design optimization schemes.

The number of layers in the CNN is increasingly changing from tens to hundreds, where making the CNN smaller and more effective is the challenge. ResNet, DenseNet aim at the development of less number of layers and more productivity by add the instances of the CNN into a conglomerate of symmetrical networks. These networks are tuned and the performance is ascertained with the discrete changes to each single unit of CNN. Therefore, developers and scientists opine to compose an interesting model called lightweight and compact CNN in order to reduce the latency and redundancy of outcomes.

Selecting the suitable network for feature detection application using embedded hardware strikes a great balance with respect to speed, memory usage, and accuracy. The compact models are evolve to fit into the solutions of the applications with few parameters, however the competency of detecting accuracy is decrease.

A new learning model called hint learning may be used in the future for the support of a CNN to catalyse the process of prediction and classification. The hint learning is supposed to be a knowledge distiller that improves the pre-training schemes.

Enhancements to the CNNs are multifold showing interest in multi-faceted application horizons. CNNs are able to learn with various forms with varied depths and structural modifications. Block based construction of learning models will be in vogue to overcome CNNs with maximum number of layers.

4 Conclusion

Almost a thorough overview of structure and applications of deep learning technology with CNN has been discussed, with regard to the fundamental components of artificial intelligence. The motivation starts with the history of artificial neural networks to the tandem of deep learning methods and innovations in several areas of applications. The problems of deep learning applications have multidimensional solutions. The taxonomy is accounted for the relationship among various deep learning tasks and their applications and they are annotated. The design and development of a CNN vests on the convolution layer configurations, causal component that consumes vast processing time. A network performance is affected by the number of layers and training and testing of the network with respect to the number of layers that grow according to the increasing complexities.

References

1. Zhao, Z.Q., Zheng, P., Xu, S.T., Wu, X.: Object detection with deep learning: a review. IEEE Trans. Neural Netw. Learn. Syst. **30**, 3212–3232 (2019)
2. Indrakumari, R., Poongodi, T., Singh, K.: Introduction to Deep Learning. Innov. Commun. Comput. 1–22 (2021)
3. Bezdan, T., Džakula, N.B.: Convolutional neural network layers and architectures. In: International Scientific Conference on Information Technology and Data Related Research; Singidunum University. Belgrade, pp. 445–451 (2019)
4. Zhang, J., Huang, J., Chen, X., Zhang, D.: How to fully exploit the abilities of aerial image detectors. In: Proceedings of the IEEE International Conference on Computer Vision Workshops 2019, Seoul, Republic of Korea, 27–28 October 2019
5. Rodriguez, R., Gonzalez, C.I., Martinez, G.E., Melin, P.: An improved convolutional neural network based on a parameter modification of the convolution layer. In: Fuzzy Logic Hybrid Extensions of Neural and Optimization Algorithms: Theory and Applications, pp. 125–147. Springer, Cham (2021)
6. Batmaz, Z., Yurekli, A., Bilge, A., Kaleli, C.: A review on deep learning for recommender systems: challenges and remedies. Artif. Intell. Rev. **52**, 137 (2019)

7. Ramu, G., Reddy, P.D.K., Jayanthi, A.: A survey of precision medicine strategy using cognitive computing. Int. J. Mach. Learn. Comput. **8**(6), 530–535 (2018)
8. Somasekar, J., Ramesh, G., Ramu, G., Reddy, P.D.K., Reddy, B.E., Lai, C.H.: A dataset for automatic contrast enhancement of microscopic malaria infected blood RGB images. Data Brief **27**, 104643 (2019)
9. Reddy Madhavi, K., Vijaya Sambhavi, Y., Sudhakara, M., Srujan Raju, K.: COVID-19 isolation monitoring system. In: Reddy, K.A., Devi, B.R., George, B., Raju, K.S. (eds.) Data Engineering and Communication Technology. LNDECT, vol. 63, pp. 601–609. Springer, Singapore (2021). https://doi.org/10.1007/978-981-16-0081-4_60
10. Reddy Madhavi, K., Madhavi, G., Rupa Devi, B., Kora, P.: Detection of Pneumonia Using Deep Transfer Learning architectures. Int. J. Adv. Trends Comput. Sci. Eng. **9**(5), 8934–8937 (2020). ISSN 2278–3091. https://doi.org/10.30534/ijatcse/2020/292952020
11. Rajani, A., Kora, P., Madhavi, R., Jangaraj, A.: Quality improvement of retinal optical coherence tomography, pp. 1–5 (2021). https://doi.org/10.1109/INCET51464.2021.9456151
12. Ahamed, S.F., Vijayasankar, A., Thenmozhi, M., Rajendar, S., Bindu, P., Rao, T.S.M.: Machine learning models for forecasting and estimation of business operations. J. High Technol. Manage. Res. **34**(1), 100455 (2023)
13. Alkawaz, A.N.: Training multilayer neural network based on optimal control theory for limited computational resources. Mathematics **11**(3), 778 (2023). https://doi.org/10.3390/math11030778
14. Jeyakanth, K., Venkatakrishnan, P., Chitra, C.: Optimized channel prediction and auction-based channel allocation for personal cognitive networks. Int. J. Commun. Syst. **36**(3), e5391 (2023)
15. Zaid, T., Biradar, N., Sonth, M.V., Gowre, S.C., Gadgay, B.: FDADE: Flow direction algorithm with differential evolution for measurement of intima-media thickness of the carotid artery in ultrasound images. Biomed. Sig. Process. Control, **80**, 104350 (2023). ISSN 1746-8094https://doi.org/10.1016/j.bspc.2022.104350
16. Sunitha, G., Reddy Madhavi, K., Avanija, J., Srujan Raju, K., Kirankumar, A., Raji Reddy, A.: Ensemble model for lidar data analysis and nocturnal boundary layer height estimation. In: Bhateja, V., Carroll, F., Tavares, J.M.R.S., Sengar, S.S., Peer, P. (eds.) Intelligent Data Engineering and Analytics. FICTA 2023. Smart Innovation, Systems and Technologies, vol. 371. Springer, Singapore (2023). https://doi.org/10.1007/978-981-99-6706-3_28
17. Kora, P., Rajani, A., Chinnaiah, M.C., Madhavi, K.R., Swaraja, K., Meenakshi, K.: EEG-based brain-electric activity detection during meditation using spectral estimation techniques, pp. 687–693 (2021). https://doi.org/10.1007/978-981-16-1941-0_68
18. Madhavi, K.R., Jarugula, J., Karuna, G., Kaleru, S., Raju, K.S., Sunitha, G.: Bird species recognition using deep transfer learning. In: Reddy, K.A., Devi, B.R., George, B., Raju, K.S., Sellathurai, M. (eds.) Proceedings of Fourth International Conference on Computer and Communication Technologies. Lecture Notes in Networks and Systems, vol. 606. Springer, Singapore (2023). https://doi.org/10.1007/978-981-19-8563-8_46

Hybrid Deep Learning Algorithm to Predict Leaf Diseases for Botanical Sentinals

Jonnadula Narasimharao[1], P. Megana Santhoshi[2], Pabbati Swathi[3], Ganesh Davanam[4(✉)], Naresh Tangudu[5], and P. Dileep Kumar Reddy[6]

[1] CSE, CMR Technical Campus, Medchal, Hyderabad 500020, India
[2] KLM College of Engineering for Women, Kadapa, India
[3] CSE, Annamacharya Institute of Technology and Sciences (Autonomous), Rajampet, India
[4] CSE, School of Computing, Mohan Babu University (Erstwhile Sree Vidyanikethan Engineering College (Autonomous)), Tirupati, AP, India
dgani05@gmail.com
[5] Aditya Institute of Technology and Management, Tekkali, Srikakulam, India
[6] CSE, Narsimha Reddy Engineering College (Autonomous), Secunderabad, Telangana 500100, India

Abstract. When it comes to maintaining a growing population and supporting a healthy economy, the agriculture sector is primary. Infectious plant diseases threaten biodiversity and could result in crop losses. In agriculture, early disease detection and identification from leaf photos using machine learning is a tough but crucial topic of study. Since agriculture is a significant economic factor in India (17% of GDP), there is an urgent need for studies of this nature to be conducted there. Profits for farmers and national GDP can both benefit from more efficient and better crop products. This study proposes a novel approach that combines convolutional neural networks with support vector machines to accurately diagnose 10 distinct illnesses associated with holy basil. CNNs with full connectivity, max-pooling, & convolutional layers are used in this model to automatically extract relevant features from input images. SVMs use these characteristics to make diagnoses. Evaluation measures like as precision, recall, F1-score, and total accuracy reflect the model's effectiveness. Important in practical agricultural contexts is its capacity to manage data with discrepancies, which is supported by micro-average, macro-average, and weighted-average accuracy ratings.

Keywords: plant diseases · Deep Learning (DL) · Transfer Learning (TL) · Food Security

1 Introduction

As one of the world's largest industries, agriculture provides vital support in the forms of food, income, & employment. Since there are so many farmers in India and other low and middle income nations, agriculture accounts for 18% of national income and 53% of total employment [1]. The share of agriculture's GVA in the country's total economy has risen from 17.6% up 20.2% during the past three years [2]. The majority of the

A. Bajaj et al. (Eds.): IBICA 2023, LNNS 1229, pp. 415–423, 2025.
https://doi.org/10.1007/978-3-031-78940-3_44

expansion of the economy may be attributed to this field. Therefore, the quality of food production could be negatively impacted if plant diseases and insect infections had an effect on agriculture. In the case of epidemics and endemics, prophylactic medicines are ineffective. Losses in the quality of production may be avoided through early monitoring and correct diagnosis of disease of crops utilizing an appropriate crop protection system.

It is a critical concern that plant disease types be accurately identified. The ability to detect plant diseases early on could lead to more informed choices about how to manage agricultural output. Spots, markings, or other anomalies often appear on the stems, leaves, fruit, or flowers of infected plants. In particular, the distinctive patterns left by various infections and pest conditions can be utilised for accurate diagnosis. It takes time and effort to correctly diagnose a plant disease. In addition, farmers and experts' diagnoses of plant diseases can be off-base [3] since manual examination is both subjective and time-consuming. This could result in the use of an inappropriate drug during the assessment of the plant disease, that would have a negative impact on crop quality and would ultimately pollute the environment.

Since the infected spots are first observed as spots & patterns on leaves, the development of computer vision has provided several approaches to resolving the detection challenges for plants [4]. Several methods have been developed to accurately identify and classify plant diseases by researchers. There are many who still rely on older, more conventional methods of image processing, such as those that necessitate manual feature extraction & segmentation [5]. To identify the contaminated areas of leaves, Dubey et al. [6] presented a K-means clustering technique, with multi-class support vector machine, or SVM, classification as the final step. The weather and statistical features were extracted using a probabilistic neural network by Yun et al. [7]. Cucumber plants afflicted with cucumbers downy mildew, anthrax, and blights were used in the research. In addition, numerous models employing conventional techniques, such as the one given by Liu et al. [8], who employed SVM plus K-means clustering strategies, in addition to a backward propagation neural network, have been proposed for plant disease recognition. Although disease recognition using images has made great strides forward, the technique is still laborious and time intensive. In addition, the models utilize categorization and spot segmentation that require human intervention. Following the development of AI, many studies in the field of computer vision have turned to machine learning [9] as well as deep learning [10] algorithms in an effort to improve recognition accuracy.

2 Related Work

Recently, scientists in the field of plant pathology have begun using AI tools to help spot anomalies and pest infestations in plants. The ability of these technologies to detect, analyze, and control plant diseases has the potential to revolutionize current practices. In this section, we'll discuss the numerous artificial intelligence (AI) technologies that are being proposed for recognizing plant abnormalities including infestations, as well as their benefits and drawbacks, and the effect they've had on the field of plant pathology. In plant pathology, ML is one of the most popular AI tools. Many machine learning (ML) methods have been used to classify plant diseases from digital photos, including the c4.5 classifier, tree bagger, as linear support vector machines. These algorithms are well-suited for the early-stage classification of diseases since they may be developed to identify certain patterns & symptoms of diseases. On the other hand, ML systems need a large amount of labeled data for training, which may not be available for rare diseases.

CNNs and DBNs are two examples of DL technologies that have been proposed for use in the detection of anomalies and pests in plants. The results of using these technologies to detect and identify lesions in digital pictures have been encouraging thus far [11]. Traditional image processing approaches may miss subtle indications of diseases, whereas DL models may learn these properties automatically from the photos. While Deep Learning models have the potential to revolutionize many fields, their high computational requirements and large amounts of training data with labels may prevent them from being used in certain contexts. Computer vision (CV) is another type of AI that has found use in plant pathology. Object detection or semantic segmentation are two examples of CV algorithms that can be used to locate and identify specific regions that are important in images, including plant leaves and disease signs [12]. For disease identification and classification, these techniques can be combined with ML or DL algorithms to automatically turn photos into recognized patterns or characteristics. To be sure, CV.

Algorithms may not be applicable for previously unseen diseases because to the large amount of tagged picture data required for model training. The picture is the input picture, and the second picture shows the results of the disease identification. The lesion identification findings are shown in the third image, and the segmentation of the lesion on the plant is shown in the last image. The use of AI tools for diagnosing plant problems and pest infestations has shown encouraging results. Classification and lesions segmentation of plant illnesses from digital images using ML, DL, and CV based systems has the potential to significantly alter the way plant diseases are discovered, diagnosed, and treated. However, these tools require a large amount of labeled training data and might not be applicable to previously unseen diseases. More work is required to create publically available datasets for training and testing models, and to create generalizable models that may be used across a wide range of plant species and illnesses. You may find detailed information about the equipment and methods used for identifying plant diseases. Methods for processing both small and big plant image collections are covered in detail, as are the various feature mining methods, such as those based on handmade and learning features.

3 Proposed Methodology

The four stages of the approach for the classification of holy basil leaf illness utilizing a Convolution Neural Network as well as a Support Vector Machine combination of layers of convolution, max-pooling layers, as well as fully connected layers are as follows (Fig. 1):

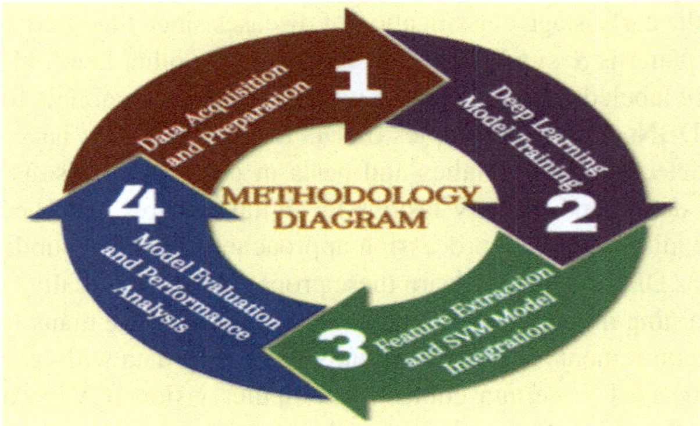

Fig. 1. Various Phases of Methodology

3.1 Data Acquisition and Preparation

In this step, the scientist begins by gathering examples for each one of the 10 illnesses they want to forecast using a large collection of religious basil leaf photos. The information gathered may originate from a number of places, including field surveys, pathology of plants labs, or internet databases. To guarantee that the pictures are in the best possible format for both the CNN as well as SVM models once they obtain the dataset, it is critical to preprocess the images. In order to improve the variety of the training dataset, this preparation includes scaling the photographs to a consistent size, normalizing pixel values to a standard scale, and using data augmentation techniques. In order to effectively assess the models' performance and prevent over fitting, they'll also need to divide the dataset into testing, validation, and training sets.

3.2 Deep Learning Model Training

The scientist designs and trains a convolutional neural network at this step. The CNN design includes layers with maximum pooling to downsample the characteristics maps and layers of convolution, each accompanied by an activation function that is non-linear. A completely interconnected layer with a reasonable number of nodes, which corresponds to the number of illness classifications, is then fed the results of the final layer that pools data. A preprocessed training dataset and an optimization technique like stochastic descent of gradients or Adam are used to train the CNN. They keep an eye on how well the model performs on the set for validation during the training phase and adjust the hyperparameters like the rate of learning and batch size to get the best accuracy and generalization.

3.3 Feature Extraction and SVM Model Integration

Following CNN training, the scientist moves on to feature extraction. In this stage, the CNN's convolution and pooling layers' learned weights and activations are used as vectors of features for the SVM model. For each picture in the training set, they extract intermediate representations of features from the CNN. The SVM, a potent linear classifier, receives these characteristics as input. Finding the best hyperplane to divide the vectors of features of various disease classes and so establish a boundary for decision-making in the area of features is the goal of SVM training. The labeled training data and their matching CNN-extracted features are used in the SVM training. The word "data" is plural, not singular.

3.4 Model Evaluation and Performance Analysis

In this last stage, the researchers examine the efficacy of the combined CNN-SVM system on the data set being tested to predict illnesses in holy basil leaves. Accuracy, precision, recall, as well as F1-score, are among the assessment metrics that are measured and reported. To comprehend the effects of adding SVM as a classifier, compare the outcomes of the CNN-SVM model combined with a single CNN model. Discuss any difficulties encountered throughout the study as well as the advantages and disadvantages of the hybrid method. In order to further increase the model's precision and usefulness in real-world circumstances, make conclusions and point out potential future improvements (Fig. 2).

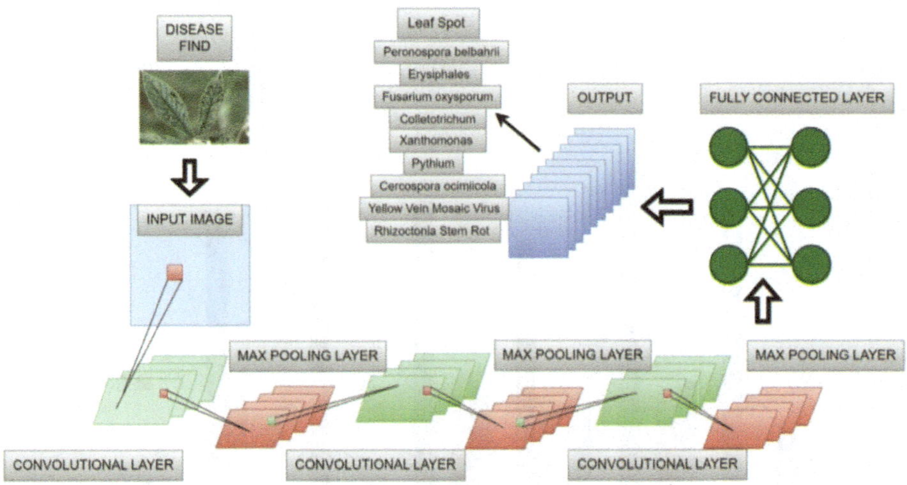

Fig. 2. Methodology Phases

4 Experimentation and Results

The Assessment metrics for a classification model used to forecast several kinds of illnesses affecting Holy Basil leaves are shown in the table. The illnesses represented by the classes include Yellow Vein Mosaic Virus, Rhizoctonia Stem Rot, Peronospora belbahrii, Erysiphales, Fusarium oxysporum, Colletotrichum, Xanthomonas, Pythium, and Leaf Spot, which all harm Holy Basil plants. Each row in the table represents a distinct disease category and includes information about the diagnostic model's precision, recall, F1-score, help, or accuracy with respect to that category. Precision measures how often a prediction was actually correct, while recall is the proportion of accurate suggestions among all correct forecasts. The F1-score provides a balanced measure of a model's efficacy by averaging recall and accuracy. The number of occurrences of each disease class in the dataset is displayed in the backing column.To get the total accuracy, recall, and F1-score, the positives, false positives, including false negatives for all classes are added together. Overall, the findings show that the categorization model performs well in predicting illnesses of Holy Basil leaves, offering insightful information for disease identification and control in agricultural practices (Fig. 3).

Table 1. Predictive Model Evaluation

Classes	Precision	Recall	F1-Score	Accuracy
Leaf Spot	94.52	93.45	92.48	99.8
Peronospora belbahrii	95.64	94.70	94.58	99.2
Erysiphales	94.92	93.45	95.32	98.9
Fusarium oxysporum	95.48	94.10	93.45	99.8
Colletotrichum	93.37	92.90	93.13	99.7
Xanthomonas	92.48	93.48	95.42	99.2

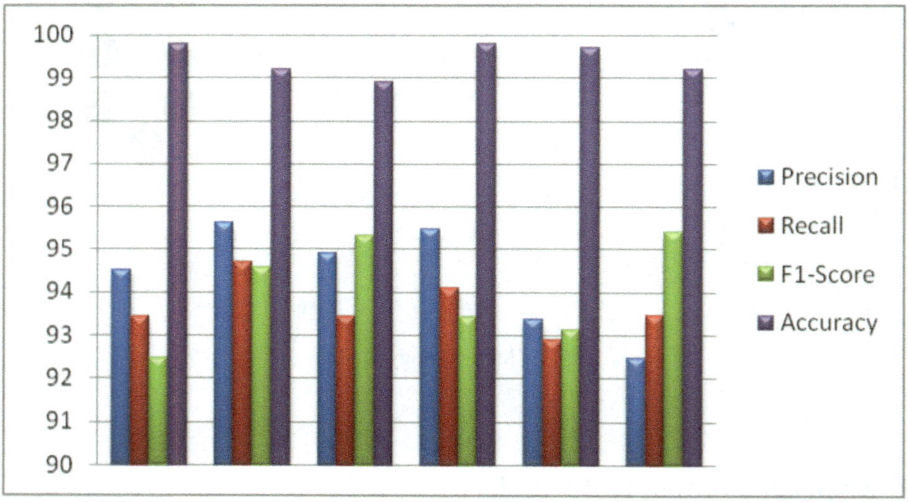

Fig. 3. Evaluated Result of the Prediction Model

The Table 1 gives a thorough explanation of the True favorable, False Positive, as well as False Negative numbers for the classification framework used to forecast several kinds of illnesses that affect Holy Basil leaves. Leaf Spot, Peronospora belbahrii,

Erysiphales, Fusarium oxysporum, Colletotrichum, Xanthomonas, Pythium, Cercospora ocimiicola, Yellow Vein Mosaic Virus, as well as Rhizoctonia Stem Rot, are among the diseases represented in each row.The number of occurrences that were correctly identified as corresponding to each illness class is represented by the True Positive value. For instance, the model successfully recognized 523 occurrences of Leaf Spot, 536 occurrences from Peronospora belbahrii, 542 occurrences of Erysiphales, & so on. To assess the effectiveness of the categorization model, it is critical to know the True favorable, False Positive, as well as False Negative values. They are used to produce several assessment metrics, including accuracy, recall, and F1-score, to determine how well the model can categorize the numerous illnesses that affect Holy Basil leaves. They can also help in determining which classes might need further clarification or attention throughout the categorization process.

The neural network's design and parameters utilized in the study article are summarised in the table given. The network is made up of three Convolutional Layers, three Max Pooling Layers that correlate to them, and a Fully Connected layer.The input size for the first layer of convolution is 2×2, while the resulting size is $127 \times 127 \times 2$. Ten parameters make up this layer. Following this convolutional layer, a $63 \times 63 \times 2$ output size is produced by the Max Pooling Layer, which has no parameters and decreases the spatial dimensions. The second layer of convolutional neural networks contains 18 parameters with input and output dimensions of 2×2 and $62 \times 62 \times 2$, respectively (Fig. 4).

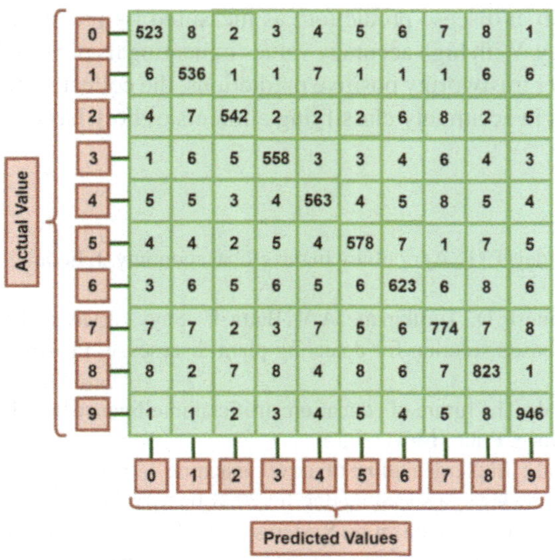

Fig. 4. Confusion Matrix

A $31 \times 31 \times 2$ output size is produced by the next Max Pooling Layer, which again does dimensionality reduction. The third layer of convolution produces a $30 \times 30 \times 2$ output with 18 parameters from a 2×2 input. The Max Pooling method shrinks the dimensions once more, resulting in a resultant size of $15 \times 15 \times 2$. In the final layer, a Fully Linked layer with a result size of 10 and a total of 4,510 attributes is fed with the information from the previous Max Pooling layer. The overall number of parameters in the neural network, which includes all of its layers, is 4,556. These variables stand in for the network's learnable weights and biases, which are modified throughout the training phase to enhance its efficacy on the specified task. The network's design and the range of parameters it uses are revealed in the table, which is useful information for comprehending the network's intricacy and possible use in resolving the current research issue.

5 Conclusion

There has been a significant advancement in the diagnosis and treatment of plant and crop diseases because to DL and ML technology. Recent developments in recognition of images have made it possible for the detection of even the most elusive of pests and diseases. The model's robustness and generalizability can be improved by collecting photos of plants at different stages of development, throughout different seasons, and in other geographical locations. Incorporating meteorological & plant health data, such as humidity and temperature, is vital for efficient identification and prediction of plant diseases and pests, as early detection is crucial in avoiding and managing their spread and growth. The study's conclusion shows how to utilize and evaluate a deep learning model to classify 10 disorders associated with holy Basil. The model, which consisted of three Convolutional Layers, three Maximum Pooling Layers, and one Fully Connected Layer, did an excellent job at disease prediction, as measured by precision, recall, F1-Score, and overall accuracy. With high accuracy values indicating its potential to eliminate false positives and deliver trustworthy positive predictions, the collected data demonstrate the model's usefulness in accurately classifying ailments of holy basil.

References

1. Alston, J.M., Pardey, P.G.: Agriculture in the global economy. J. Econ. Perspect. **28**, 121–146 (2014)
2. Kumar, M.S., Ganesh, D., Turukmane, A.V., Batta, U., Sayyadliyakat, K.K.: Deep convolution neural network based solution for detecting plant diseases. J. Pharm. Negative Results **13**, 464–471 (2022)
3. Sushama, C., et al.: The future of education: artificial intelligence based remote learning. Int. J. Early Child. Spec. Educ. **14**(3) (2022)
4. Ganesh, D., et al.: Implementation of novel machine learning methods for analysis and detection of fake reviews in social media. In: 2023 International Conference on Sustainable Computing and Data Communication Systems (ICSCDS). IEEE (2023)
5. Ganesh, D., et al.: Implementation of convolutional neural networks for detection of Alzheimer's disease. BioGecko, J. N. Z. Herpetol. **12**(01) (2023)
6. Dubey, S.R., Jalal, A.S.: Adapted approach for fruit disease identification using images. In: Image Processing: Concepts, Methodologies, Tools, and Applications; IGI Global. Hershey, pp. 1395–1409 (2013)
7. Ganesh, D., Sunil Kumar, M., Venkateswarlu Reddy, P., Kavitha, S., Sudarsana Murthy, D.: Implementation of AI pop bots and its allied applications for designing efficient curriculum in early childhood education. Int. J. Early Child. **14**(03) (2022)

8. Srinivas, K., Kavitarani, B., Kumar, M.K.: Classification and mining behavior of data. Data Min. Mach. Learn. Appl. 21–55. https://doi.org/10.1002/9781119792529.ch2\
9. Patra, R.K., Mahendar, A., Madhukar, G.: Inductive learning including decision tree and rule induction learning. Data Min. Mach. Learn. Appl. 209–234. https://doi.org/10.1002/9781119792529.ch9
10. Muni, B.R.T.V., Tata, R.K., Narasimharao, J., Murali, K., Kaur, H.: Deep learning techniques for speech emotion recognition. In: 2022 International Conference on Futuristic Technologies (INCOFT), Belgaum, India, pp. 1–5 (2022). https://doi.org/10.1109/INCOFT55651.2022.10094534
11. Hanumanthakari, S., Pullela, S.K., Bhukya, S.N., Vijayalakshmi, K., Ahmad, S.R., Kumar, N.: IoT based patients monitoring system in healthcare service. In: 2022 International Conference on Automation, Computing and Renewable Systems (ICACRS), Pudukkottai, India, pp. 1324–1329 (2022). https://doi.org/10.1109/ICACRS55517.2022.10029295
12. Gulati, K., Nayak, K.M., Priya, B.S., Venkatesh, B., Satyam, Y., Chahal, D.: An examination of how robots, artificial intelligence, and machinery learning are being applied in the medical and healthcare industries. Int. J. Recent Innov. Trends Comput. Commun. **10**(2s), 298–305 (2022). https://doi.org/10.17762/ijritcc.v10i2s.5947
13. Reddy, G., et al.: An intrusion detection using machine learning algorithm multi-layer perceptron (MlP): a classification enhancement in wireless sensor network (WSN). Int. J. Recent Innov. Trends Comput. Commun. **10**(2s), 139–145 (2022). https://doi.org/10.17762/ijritcc.v10i2s.5920
14. Mubeen, S., et al.: Linguistic based emotion detection from live social media data classification using metaheuristic deep learning techniques. Int. J. Commun. Netw. Inf. Secur. (IJCNIS), **14**(3), 176–186 (2022). https://doi.org/10.17762/ijcnis.v14i3.5604
15. Bhaskar, T., Shiney, S.A., Rani, S.B., Maheswari, K., Ray, S., Mohanavel, V.: Usage of ensemble regression technique for product price prediction. In: 2022 4th International Conference on Inventive Research in Computing Applications (ICIRCA), Coimbatore, India, pp. 1439–1445 (2022). https://doi.org/10.1109/ICIRCA54612.2022.9985521
16. Patil, S., et al.: Identification of triple negative breast cancer genes using rough set based feature selection algorithm & ensemble classifier. Hum. Centric Comput. Inf. Sci. **12** (2022)
17. Kesana, M., Rikhari, S.: Multi-modal medical image fusion using Laplacian re-decomposition (2022). https://doi.org/10.1201/9781003240310-7

Analysis on Convolutional Neural Networks for Deep Learning Intrusion Identification Using Industrial Internet of Things

K. Srujan Raju[1](✉), Mohd. Abdul Naqi[2], V. A. Narayana[3], Vivekanand Aelgani[3], Rajesh Tiwari[4], and Sheo Kumar[5]

[1] Department of CSE, CMR Technical Campus, Hyderabad, Telangana, India
ksrujanraju@gmail.com
[2] Department of ECE, CMR Technical Campus, Hyderabad, Telangana, India
[3] Department of CSE, CMR College of Engineering and Technology, Hyderabad, Telangana, India
{principal,avivekanand}@cmrcet.ac.in
[4] Department of CSE (AI&ML), CMR Engineering College, Hyderabad, Telangana, India
[5] Department of CSE, CMR Engineering College, Hyderabad, Telangana, India

Abstract. Since the Industrial Internet of Things (IIoT) is being used widely across many industries, protecting systems used in the IIoT has grown crucial. In order to protect internet of things networks from criminal activity, intrusion detection is essential. In this paper, we offer a structure for deep learning for IIoT intrusion detection that depends on Convolutional neural networks (CNNs). We make use of the reference data set NSL-KDD, which is frequently used to assess intrusion detection systems. The suggested architecture eliminates the requirement for human feature engineering by utilizing CNNs' built-in capacity to obtain attributes from network traffic data. The dataset is preprocessed, augmented with new data, and divided into training and testing sets. Next, the training set is used to train the CNN model, and the testing set is used to assess it. Our CNN-based intrusion detection system outperforms conventional machine learning techniques in regards to precision, recall, precision, and F1-score, as shown by experimental findings. The suggested architecture offers an effective way to find intrusions in IIoT systems, strengthening their resilience and safety.

Keywords: IIoT · deep learning · intrusion detection · Convolutional neural network

1 Introduction

Information and communications technology (ICT)-related systems and networks [1] are open to a variety of assaults by both internal and external threats, which increases their risk of data breaches. For instance, there was a data breach at Yahoo that cost the company about $350 million, and there was a breach at Bitcoin that cost it about $70 million. Improvements in hardware, software, and network topologies have made these

A. Bajaj et al. (Eds.): IBICA 2023, LNNS 1229, pp. 424–432, 2025.
https://doi.org/10.1007/978-3-031-78940-3_45

assaults more sophisticated and difficult to identify, particularly with the introduction of new technologies such as the Internet of Things (IoT) [2]. In order to improve network security, there has been a growing demand for a fresh and trustworthy intrusion detection system (IDS) [3, 4]. An intrusion detection system (IDS) is a proactive technology that detects and classifies intrusions, attacks, or security breaches in infrastructures at the host and network levels automatically. Intrusion detection is divided into two distinct groups: intrusion detection systems based on hosts (HIDS) [4] and network-based intrusion detection systems (NIDS) [5]. NIDS records and examines activity on network devices, including switches, routers, and network taps, in order to monitor network traffic.

This work focuses on assessing the performance of different deep neural networks (DNNs) [7] and machine learning classifiers in NIDS and HIDS in order to overcome these issues. The study combines the capabilities of host and network intrusion detection systems to model DNN as an efficient and effective technique to identify cyber-attacks beforehand [8, 9].

2 Literature Survey

The authors of [11] concentrate on applying machine learning methods to identify various kinds of assaults on host and network systems. The writers give a summary of several assaults and their distinguishing characteristics as well as talk about the security concerns that come up while utilizing machine learning for intrusion detection. They compare the efficacy of particular classifier techniques against different classifiers techniques and assess the advantages and disadvantages of various machine learning algorithms, including evolutionary tactics. The authors of [12] suggested a hybrid method for precisely identifying and categorizing dangerous software, particularly zero-day malware, that blends feature-based and data-based visualizations. They create similarity matrices by using resemblance mining for contrasting malware samples according to distance scores. The method effectively recognizes and classifies malware that was previously undetected by examining the patterns of virus behavior. The identification of botnet traffic, which presents a serious risk to people and enterprises, is covered by the writers in [13].

Adoption of analytics has been hampered by problems with information systems, people, and organizational structure. The difficulties and security of production are the subjects of this study.

The authors of [21] address the significance of incorporating machine learning methods into current security systems to enhance threat identification, categorization, and reaction. The study discusses several machine learning algorithms and how to use them to detect malware, spot attack trends, and foresee potential risks. These algorithms include decision trees, neural networks, and methods for clustering. The authors stress that in order to combat changing cyber threats, continuous surveillance and adaptive learning are essential.

3 Proposed System

Machine learning technology is being heavily utilized in the creation of an intrusion detection system (IDS) that can automatically and promptly identify and categorize assaults at the host and network levels. Both the host level and the network level are involved in this. This action is being done on the host level as well as the network level. On the contrary, a scalable system must be created as harmful attacks are happening often and in huge quantities, and they are always taking on new forms. The public domain provides access to a number of intrusion datasets that the cyber security community can use for additional research.

Figure 1 depicts the recommended method as a block diagram. The NSL-KDD dataset is initially divided into 80% and 20% sections for training and testing, respectively. To normalize the entire dataset, a dataset pre-processing step is then completed. Additionally, threat evaluation based on test samples is accomplished with the CNN classifier. The performance evaluation's goal is to show how much better the recommended strategy is. One popular algorithm with a high prediction ratio is CNN, which is used in data processing and data categorization, among other areas. As a result, the CNN model can recognize these attacks and apply dynamic attack signatures to resolve the issues they generate. There are a lot of layers in the CNN model that has been proposed. In order to continuously create the most efficient model to predict the testing class, the CNN technique uses the hidden layer as a filter during the training process. The most prevalent categories are Normal, Denial-of-Service (DOS), User to Root (U2R), Remote to user (R2L), and Probe, despite the dataset having other names. Despite the fact that the dataset includes more names, each of those names fits into one of the aforementioned categories.

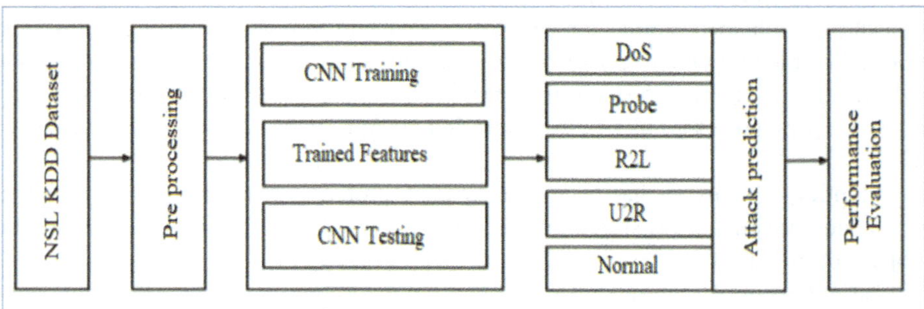

Fig. 1. Proposed model

3.1 Preprocessing

Preparing raw data for usage in a machine learning model by cleaning and structuring it is called "data preprocessing." The process of creating a model for machine learning begins with this phase, considered the first and most crucial stage in the process. Whenever we are using machine learning to construct a project, we rarely come upon data that appears clean and organized. This is because it isn't always the case. This is because it's not always accurate. Additionally, it is imperative that the data be well cleaned and organized before beginning any action using it. This needs to be completed before beginning any data-related action.

3.2 CNN Classifier

The deep CNN model utilized to detect intrusions is shown in Fig. 2. The goal of this simple graphical formalism is to depict a system in terms of the data that is put into it, all of the tasks that are done on it, and the data that the system generates. It is among the most important tools for creating models. It is used to model the different parts of the system.

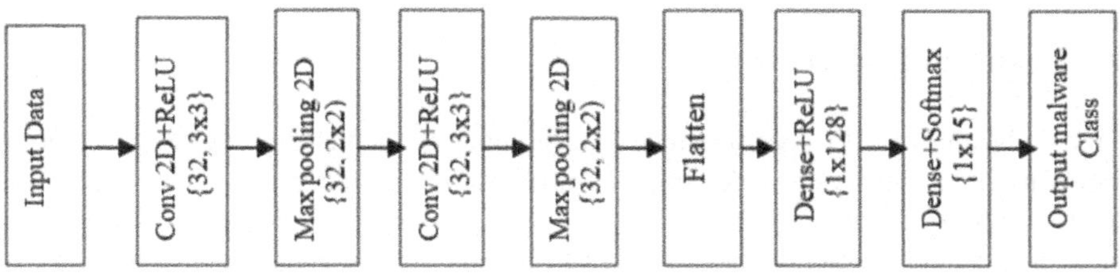

Fig. 2. Proposed deep CNN model.

Each of the source data must be passed by means of a series of convolution layers that consist of a kernel or filter, rectified linear unit (ReLU), max pooling, fully connected layer, and using SoftMax layer using classification layer to classify the objects via probabilistic values ranging from [0,1] in order to train and test a CNN [22, 23]. Figure 3 shows that the convolution layer is the primary layer utilized in the process of identifying traits from the data set in question. Additionally, through employing small blocks of source data to understand the properties of the data, this layer maintains the connection among pixels. We refer to this layer as the "convolution layer." Two inputs, which includes the source data („,), are taken into account by this mathematical function [24]. The spatial coordinates, also known as the number of rows and columns, are represented by x and y. Although the source data in this instance is three, d, the dimension of a data collection, is equivalent to three. The expression for a filter or kernel with an input data size equal to d is represented as („). And denote the input data sizes, whereas d represents the size of d [25].

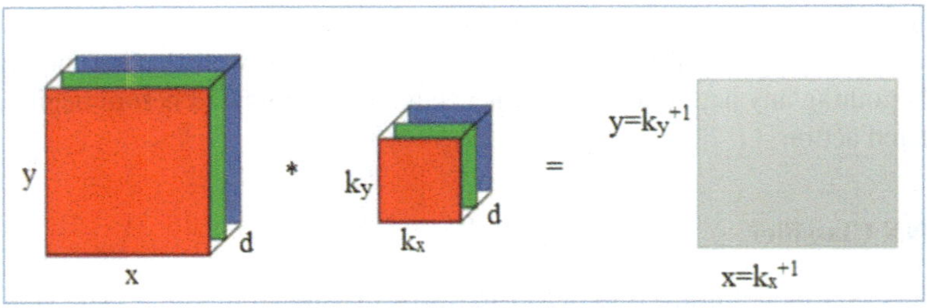

Fig. 3. Process representation for convolution layers. x = k_x+1

The output, known as the feature map, is the size of ((− + 1), (− + 1), 1) after the input data and filter are convolutional.

4 Result and Discussion

This study analyzes the performance of a number of standard algorithms, including Random Forest and SVM, in detecting harmful activity on a computer network. Nevertheless, these traditional algorithms require prior training to detect attacks of this type since they cannot predict dynamic attacks (if a perpetrator presents fresh attempts with modified attack parameters).

Furthermore, these kinds of attacks cannot be detected by these conventional techniques. In order to resolve the problem, this study evaluates the effectiveness of a CNN model that incorporates dynamic assault signatures. CNN's accuracy in detection tests has demonstrated their superiority over all other traditional procedures.

4.1 Dataset

Using the NSL-KDD dataset, we applied the SVM, Random Forest, DNN, and CNN modeling approaches in order to put this concept into practice. CNN models use secret layers to continually filter training methods and create the best effective model for testing class prediction. It is an established algorithm with a high prediction rate of accuracy across a range of fields, such as image processing and data classification.

The initials of the request signatures are displayed above in bold font and separated by commas; the initials of the attacks are marked by the column labels in the previous table.

Table 1 provides a description of the dataset. The dataset includes other names, but all of them fit into one of the following groups: Normal, R2L, DOS, U2R, DOS, or Probe. The present research examines Average, R2L, DOS, DOS, and Probe.

Table 1. Dataset description.

S.No	Category	Name
1	Denial of service (DoS)	Apache2, smurf, Neptune, back, teardrop, pod, land, mailbomb, processtable, UDPstorm
2	Remote to local (R2L)	Warez client, imap, ftp_write, named, multihop, sendmail
3	User to root (U2R) probe	Buffer_overflow, httptunnel, rootkit, loadmodule, perl, xterm, ps, sqlattack, satan, portsweep, mscan

The total amount of records for each type in the dataset is displayed in Fig. 4. In the graph above, the attack name that was found in the dataset is shown on the x-axis, and the number of instances of that attack type is shown on the y-axis. The display that was just displayed allows us to observe that there are 10137 records all together in the dataset. The program uses 2028 data to evaluate the accuracy of predictions of its algorithm and 8109 records for training.

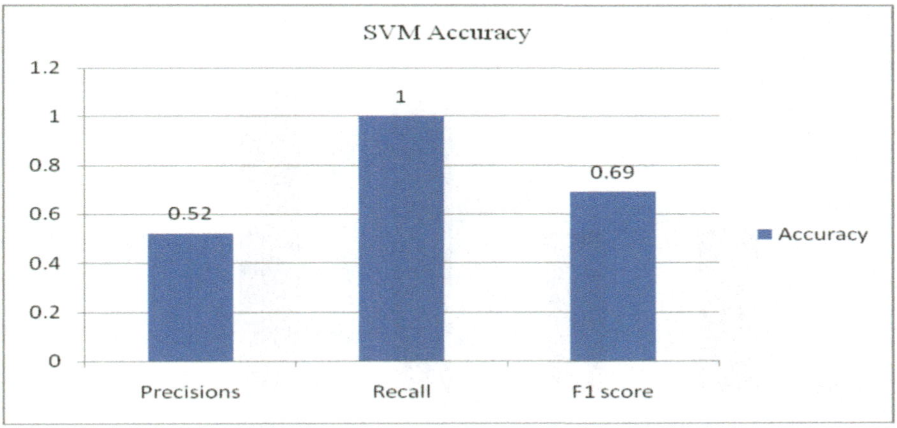

Fig. 4. Existing SVM classifier performance evaluation.

The current SVM evaluation of performance is displayed in Fig. 4. We obtained a 52.26% accuracy using the SVM. Additionally, a categorization report is displayed, with measurements for every class's F1-score, recall, and precision.

The current random forest classification system evaluation is displayed in Fig. 5. The precision of the random forest was 52.31%. Additionally, a categorization report is displayed, with measurements for every class's F1-score, recall, and accuracy.

The current DNN classifier evaluation of performance is displayed in Fig. 6. In the previous screen grab, the precision of the DNN algorithm is displayed as 89.96%; this number is higher than the average of the other two approaches.

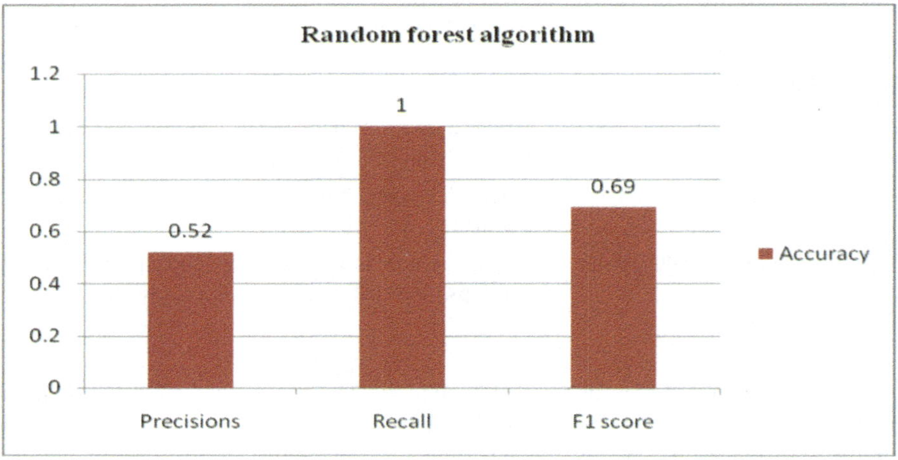

Fig. 5. Existing random forest classifier performance evaluation.

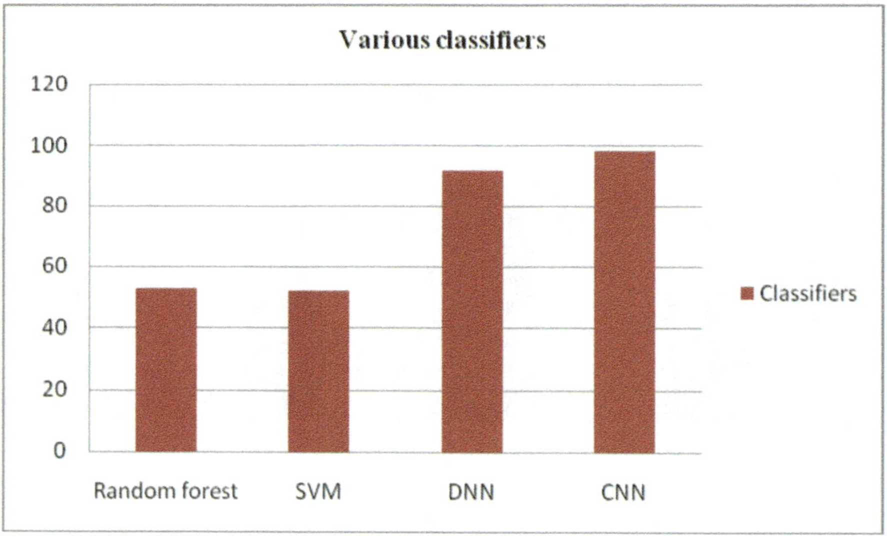

Fig. 6. Performance comparison of various classifiers

As 98%; this score is greater than that of both of the two approaches. Figure 6 compares the efficacy of several classifiers, displays the assault prediction based on test data. Neptune is the interpretation of the test findings.

5 Conclusion

To sum up, the project's goal was to create a combination of intrusion detection systems that could efficiently identify attacks across the host and network levels. The system processed and evaluated massive volumes of data in real time using a decentralized deep learning model, more precisely a CNN. Employing benchmark IDS datasets, the CNN model's performance was assessed and contrasted with that of conventional machine learning classifiers; the CNN model consistently surpassed the latter. Future research should give careful thought to how to overcome the difficulty of training intricate CNN networks using standard IDS datasets. The system's general efficiency can also be improved by investigating the combination of DNS and BGP monitoring of events and by employing more potent hardware for distributed learning. By advancing hybrid systems for intrusion detection, more precise and effective surveillance of attacks in real-time scenarios will be possible thanks to these efforts.

References

1. Chalé, M., Bastian, N.D.: Generating realistic cyber data for training and evaluating machine learning classifiers for network intrusion detection systems. Expert Syst. Appl. **207**, 117936 (2022)
2. Apruzzese, G., Andreolini, M., Ferretti, L., Marchetti, M., Colajanni, M.: Modeling realistic adversarial attacks against network intrusion detection systems. Digital Threats: Res. Pract. (DTRAP) **3**(3), 1–19 (2022)
3. Apruzzese, G., Pajola, L., Conti, M.: The cross-evaluation of machine learning-based network intrusion detection systems. IEEE Trans. Netw. Serv. Manage. (2022)
4. Zhang, C., Costa-Pérez, X., Patras, P.: Adversarial attacks against deep learning-based network intrusion detection systems and defense mechanisms. IEEE/ACM Trans. Netw. (2022)
5. Sheatsley, R., Papernot, N., Weisman, M.J., Verma, G., McDaniel, P.: Adversarial examples for network intrusion detection systems. J. Comput. Secur. (Preprint), 1–26 (2022)
6. Data, M., Aritsugi, M.: AB-HT: an ensemble incremental learning algorithm for network intrusion detection systems. In: 2022 International Conference on Data Science and Its Applications (ICoDSA), pp. 47–52. IEEE (2022)
7. Magan-Carrion, R., Urda, D., Diaz-Cano, I., Dorronsoro, B.: Improving the reliability of network intrusion detection systems through dataset aggregation. IEEE Trans. Emerg. Top. Comput. (2022)
8. Narasimharao, J., Priyanka Chowdary, P., Reddy, A.R., Swathi, G., Deepak Kumar, B.P., Batchu, S.S.: Satellite ortho image mosaic process quality verification. In: Bhateja, V., Yang, XS., Ferreira, M.C., Sengar, S.S., Travieso-Gonzalez, C.M. (eds.) Evolution in Computational Intelligence. FICTA 2023. Smart Innovation, Systems and Technologies, vol. 370. Springer, Singapore (2023). https://doi.org/10.1007/978-981-99-6702-5_26
9. Sasi Kiran, J., Avanija, J., Reddy, A.R., Naga Rama Devi, G., Charan, N.S., Fatima, T.: Early prediction of sepsis utilizing machine learning models. In: Bhateja, V., Yang, X.S., Ferreira, M.C., Sengar, S.S., Travieso-Gonzalez, C.M. (eds) Evolution in Computational Intelligence. FICTA 2023. Smart Innovation, Systems and Technologies, vol. 370. Springer, Singapore (2023). https://doi.org/10.1007/978-981-99-6702-5_27

10. Sultana, S., Reddy Madhavi, K., Lavanya, G., Swarna Latha, J., Sandhyarani, Prathyusha, B.: Detection of malicious URLs using gradient boosting classifier. In: Bhateja, V., Yang, X.S., Ferreira, M.C., Sengar, S.S., Travieso-Gonzalez, C.M. (eds.) Evolution in Computational Intelligence. FICTA 2023. Smart Innovation, Systems and Technologies, vol. 370. Springer, Singapore (2023). https://doi.org/10.1007/978-981-99-6702-5_28
11. Saravanan, P., Pandey, A., Joshi, K., Rondon, R., Narasimharao, J., Imran, A.A.: Using machine learning principles, the classification method for face spoof detection in artificial neural networks. In: 2023 3rd International Conference on Advance Computing and Innovative Technologies in Engineering (ICACITE), Greater Noida, India, pp. 2784–2788 (2023). https://doi.org/10.1109/ICACITE57410.2023.10182551
12. Bikku, T., Sree, K.S.: Deep learning approaches for classifying data: a review. J. Eng. Sci. Technol. **15**(4), 2580–2594 (2020)
13. Sunitha, G., Madhavi, K.R., Avanija, J., Reddy, S. T.K., & Vittal, R.H.S.: Modeling convolutional neural network for detection of plant leaf spot diseases. In: 3rd International Conference on Electronics and Sustainable Communication Systems, pp. 1187–1192. IEEE (2022)
14. Rajesh Tiwari, M., et al.: Enhanced power quality and forecasting for PV-Wind Microgrid using proactive shunt power filter and neural network based time series forecasting Electr. Power Compon. Syst. (2023https://doi.org/10.1080/15325008.2023.2249894
15. Muni, B.R.T.V., Tata, R.K., Narasimharao, J., Murali, K., Kaur, H.: Deep learning techniques for speech emotion recognition. In: 2022 International Conference on Futuristic Technologies (INCOFT), Belgaum, India, pp. 1–5 (2022). https://doi.org/10.1109/INCOFT55651.2022.10094534
16. Ali, M.H., et al.: A new intrusion detection system based on fast learning network and particle swarm optimization. IEEE Access **6**, 20255–20261 (2018)

A New Strategy for Categorizing Eye Diseases Analyzing, Utilizing Robust Image Processing and Deep Learning Techniques

K. Ranjith Reddy[1(✉)], Banothu Ramji[2], Borra Sivaiah[3], R. Suhasini[3], M. Kumara Swamy[4], and Nakka Venkatesh[4]

[1] Department of CSE, CMR Technical Campus, Hyderabad, Telangana, India
cmrtc.paper@gmail.com, ranjith.kssr5@gmail.com
[2] Department of CSE (DS), CMR Technical Campus, Hyderabad, Telangana, India
[3] Department of CSE, CMR College of Engineering and Technology, Hyderabad, Telangana, India
bsivaiah@cmrcet.ac.in, rsuhasini@cmrcet.org
[4] Department of CSE (AI&ML), CMR Engineering College, Hyderabad, Telangana, India
{m.kumaraswamy,venkatesh.n}@cmrec.ac.in

Abstract. Diabetic retinopathy is extensively accepted as a important etiology of visual impairment on a global scale. These particular ailment impacts individuals afflicted with diabetes, particularly those who are advanced in age. Numerous hospitals worldwide endeavor to engage in research and implement preventive measures aimed at mitigating the incidence of blindness resulting from diabetic retinopathy. Cataract, Diabetic Retinopathy, and Glaucoma are prevalent eye illnesses that are frequently encountered on a global scale. The suggested model has undergone training using a substantial dataset of fundus images in order to accomplish several tasks concurrently. These tasks encompass diabetic retinopathy (DR) categorization, Gaussian blur, and grey scaling. Through the process of combining all tasks, the algorithm is capable of generating predictions that are more informed and enhancing its overall efficacy in contrast to conventional single-task models. One of the benefits of utilizing Convolutional Neural Networks (CNN) and Transfer Learning (TL) methodologies is their ability to enhance accuracy. A dataset of 4000 photos was compiled from publicly accessible sources, namely Ocular Recognition, HRF, DRIVE, and IDRiD. This dataset was then validated using a subset of 402 images, which were selected from a larger dataset comprising 1004 images for each of the classes. The validation process resulted in a precision of 90.89%.

Keywords: CNN · Diabetic retinopathy · transfer learning · fundus image

1 Introduction

Diabetic retinopathy is a prominent etiological factor in the development of visual impairment among individuals in the working-age demographic. The timely identification of this illness is crucial for favorable prognostic outcomes. A significant proportion of adult

A. Bajaj et al. (Eds.): IBICA 2023, LNNS 1229, pp. 433–444, 2025.
https://doi.org/10.1007/978-3-031-78940-3_46

individuals afflicted by diabetes are at a heightened risk of developing diabetic retinopathy at some point in their lifespan. Diabetic retinopathy has the potential for successful treatment when identified during its initial stages. However, in cases where the patient receives a diagnosis during an advanced stage, the likelihood of achieving a cure may be significantly diminished.

According to the 2021 data provided by the World Health Organization (WHO), the estimated global population affected by eye disorders amounts to approximately 1.02 billion individuals. The primary ocular conditions that are commonly recognized include Cataract, Glaucoma, and Diabetic Retinopathy. Research has indicated that the occurrence of blindness and decreased vision loss is associated with the process of aging and can be classified into three primary groups. There are three main reason of blindness, namely Cataract, Glaucoma, and Diabetic Retinopathy. Cataract accounts for approximately 65.2 million instances, while Glaucoma and Diabetic Retinopathy have 6.9 million and 3 million cases, respectively. Collectively, these three conditions represent 2.6% of all cases of blindness worldwide [1].

The fundus image refers to the posterior section of the eye, located in its deep region. The ocular structure encompasses various components, including the retina, blood vessels, fovea, macula, and other elements [2]. Fundus cameras, also known as specialized cameras, will be employed for the reason of capturing fundus images. The process of capturing photos involves orienting the camera on the pupil. A standard fundus image consists of several components, including the Optic Disk, blood vessels, macula, discharge, and hemorrhage, as depicted in the Fig. 1.

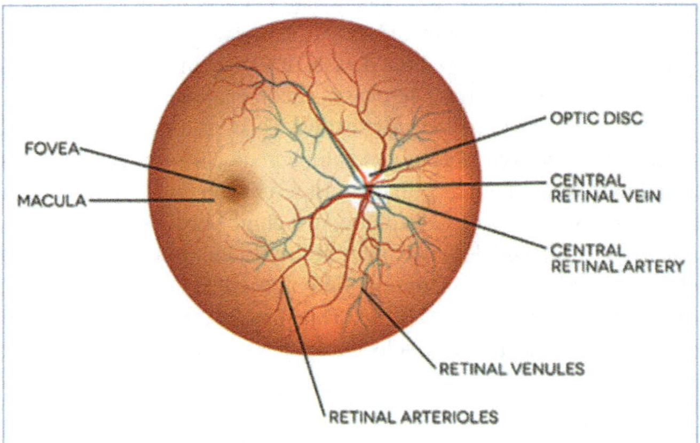

Fig. 1. Retinal fundus image.

The source of the image is cited as [3]. Fundus images used for eye diagnostics present several challenges. To commence, several prior studies on the detection and evaluation of eyes have predominantly focused on subjective feature extraction [4–8]. In recent times, deep learning (DL) has arisen as a novel field within computer vision and image processing [9]. Deep learning enables the resolution of previously intractable problems that were beyond the capabilities of traditional machine learning methods. Deep learning leverages neural networks to enhance computational operations and provide accurate outcomes.

2 Literature Review

This article explains how to build a do-it-yourself fundus camera that uses a Smartphone and can image both the pars plana and the central retina. It is a reasonably priced substitute for the fundus camera [4].

In their study [10] have presented a convolutional neural network (CNN) architecture intended for the purpose of cataract identification by the utilization of fundus imaging. The researchers put forth a Convolutional Neural Network (CNN) planning specifically designed for the detection of cataracts. The researchers utilized the dataset obtained from Eye hospitals located in the North Sumatra region. The dataset comprises 73 photos classified as normal and 326 images classified as cataract. In the initial stages, transfer learning techniques were employed to make predictions regarding cataract. The researchers employed pre-trained models such as GoogleNet, MobileNet, and ResNet.

). The researchers utilized a dataset sourced from the Kaggle platform, specifically curated by EyePacs [11]. The collection consists of a total of 35,126 photos that have been appropriately labeled.

The authors proposed a model utilizing InceptionV3 in their study [17] to detect Diabetic Retinopathy (DB) and edema. The researchers utilized the pre-trained model weights obtained from ImageNet and subsequently updated them through the application of the Stochastic Gradient Descent (SGD) optimizer. The researchers acquire a specificity of 90.3% and a sensitivity of 90.1% when evaluating the EyePACS dataset.

In their study [19] utilized five distinct convolutional neural network (CNN) architectures, including CNN, VGG19, ResNet, GoogleNet, and DENet, for the purpose of glaucoma detection in fundus images. The researchers utilized a single secret dataset obtained from the Hospital de la Esperanza in Barcelona, Spain, in addition to two publically accessible datasets (DRISHTI-GS and RIM-ONE).

The researchers conducted an investigation to examine the correlation among the amount of classes and the amount of fully connected (FC) layers. The deceleration in the learning process of a neural network can be attributed to the simultaneous increase in the amount of fully connected layers and the growing complexity of the dataset.

[21] devised a methodology that utilizes a Convolutional Neural Network (CNN) to identify and classify eight unique retinal illnesses. The aforementioned characteristic of the Convolutional Neural Network (CNN) renders it a extensively utilized model for image processing tasks. This research presents a novel methodology for the recognition and categorization of Cataract, Diabetic Retinopathy, and Glaucoma disorders based on analysis of fundus images. The planned architecture's is illustrated in the Fig. 2. The dataset utilized in this study was obtained from Kaggle's Ocular Recognition repository [3].

3 Proposed Model

Convolutional Neural Networks (CNNs) are a type of deep learning algorithms that have proven to be effective in the recognition and classification of images and objects. This is seen in Fig. 2. In contrast to conventional machine learning technique, Convolutional Neural Networks (CNNs) do not need subjective attribute removal. The utilization of convolutional neural networks (CNNs) encounters several problems in the field of retinal image analysis. These challenges comprise the limited availability of annotated data, the intricate nature of retinal characteristics and vessels, which can have an impact on adjacent vessels and are contingent upon image quality, as well as the subjective nature of manually categorized fundus images. Transfer learning is employed as a means to address and overcome the aforementioned issues.

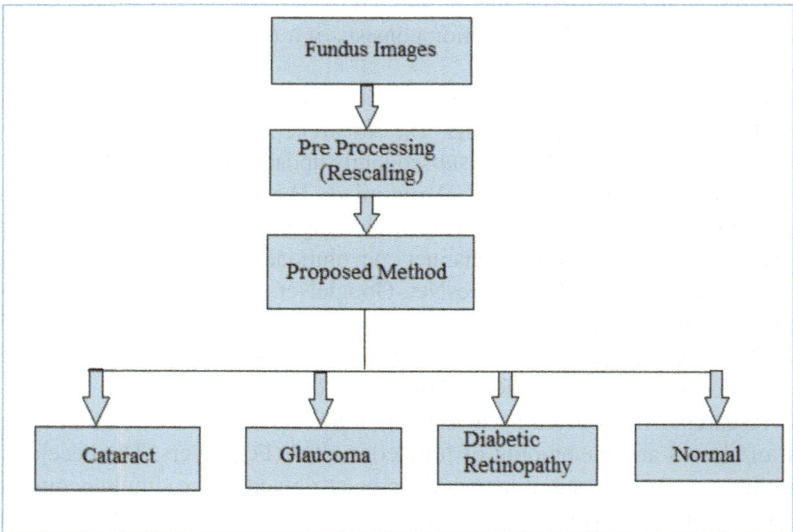

Fig. 2. Block diagram of proposed system

Sequential Model

$$Totalparams : 11{,}184{,}179$$

$$Trainableparms : 11{,}093{,}804$$

$$Nontrainableparams : 90{,}375$$

The sequential model, depicted in Table 1, accepts an input of an RGB image with dimensions of 224x224. The model utilizes a pre-trained EfficientNetB3 model with a total of 10,783,535 parameters. These variables are configured to be trainable during the training process of the model. The output of EfficientNetB3 is acquired by configuring the pooling parameter as maximum, followed by forwarding it to a Dense layer consisting of 256 neurons with ReLU activation. Prior to this, the output is subjected to a batch normalizing layer. A cancellation rate of 0.4 was implemented so as to mitigate the problem of over fitting within the model. So as to mitigate the problem of over fitting, the Dense layer incorporates the L1 and L2 regularizes as bias and kernel regularizes, respectively. Instead of utilizing conventional values for l1 and l2, the parameters in l1 and l2 are assigned the specific values of 0.006 and 0.016, respectively. Next, the output as of the dropout layer is passed into a Dense layer consisting of four neurons, each representing one of the four classes: cataract, diabetic retinopathy, glaucoma, and normal. The model is capable of generating predictions for each class by estimating their respective probabilities. The output is determined by selecting the class containing the premier probability. The Soft Max activation function is employed in the output dense layer, as depicted in Table 1.

Table 1. Sequential model

Layer Type	Output Shape	Param #
Efficientnetb3	(None, 1536)	10783536
Batch_normalization_4 (Batch normalization)	(None, 1536)	6146
Dense_3	(None, 256)	393472
Dropout_2	(None, 256)	0
Dense_4	(None, 4)	1028

The model possesses a cumulative count of 11,184,180 parameters, with 11,093,806 parameters being trainable and 90,378 parameters being non-trainable. In the last stage of the process, the model undergoes compilation using the Adamax optimizer with a learning rate of 0.002. The loss function employed is categorical cross entropy.

4 Results and Discussion

The following part presents a discussion on the efficiency of the suggested EfficientNetB3 model. The present study evaluated three different algorithms, including the model under consideration, along with two other pre-trained algorithms, using the identical dataset for the purpose of diagnosing cascade, diabetic retinopathy, and glaucoma. The investigational study The obtained consequences are evaluated to state-of-the-art methodologies for the treatment of cataracts, diabetic retinopathy, and glaucoma.

Using the identical dataset, two pre-trained models are generated, both exhibiting favorable classification outcomes. The objective is to estimate the presentation of EfficientNetB3. The models under consideration are VGG-19 and ResNet50.

This paper presents a comprehensive assessment of the existing literature on the application of deep learning method for the recognition of cataract, diabetic retinopathy, and glaucoma. The primary objective is to offer a summary of the reported investigations, focusing on the evaluation metrics of exactness, precision, recall, and f1-score. The metrics of recall, exactness, precision, and F1-score with high values have been emphasized. The performance of our model surpasses that of the earlier works and in terms of exactness, precision, recall, and f1-score. Nevertheless, in comparison to alternative models, this particular model demonstrates superior performance in terms of accuracy, recall, and f1-score. In a study conducted, Deep learning-based algorithms be employed to accurately categorize cataract by a precision rate of 92%.

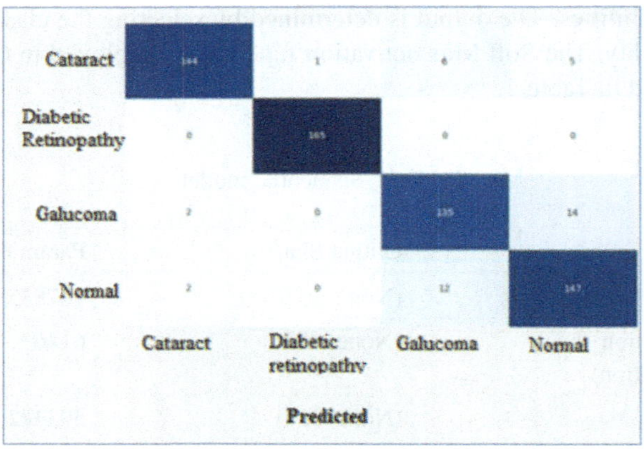

Fig. 3. The projected model Confusion Matrix

The matrix of uncertainty interpolation plot during validation of the retina pictures is depicted in Fig. 3. The data presented in the figures represents the outcomes of validation for colorful photos, grayscale images, and Gaussian filtered images, respectively [27].

Figure 3 presents the confusion matrix of EfficientNetB3. Each row and column in the table is assigned a projected label and a genuine label, respectively. Among the 633 valid photos, a total of 48 have been inaccurately classified. According to the data presented in Fig. 4, it is observed that there are misclassifications in the categorization of some photos. Specifically, 10 cataract images are incorrectly labeled as either glaucoma or normal, 2 images of diabetic retinopathy are misclassified as typical, 21 glaucoma images are inaccurately confidential as either cascade or typical, and 15 regular pictures are erroneously identified as cataract, diabetic retinopathy, or normally.

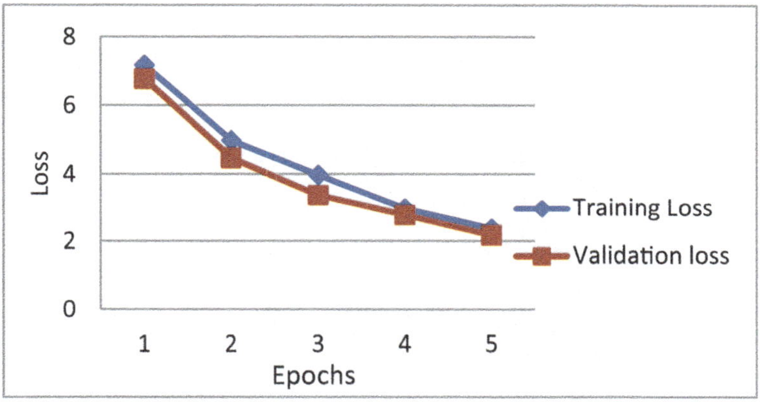

Fig. 4. Epoch and loss results for colored retina images

Depending on the information accessible in Fig. 4, it is apparent that the training loss for colored retina images is the lowest after a certain number of epochs.

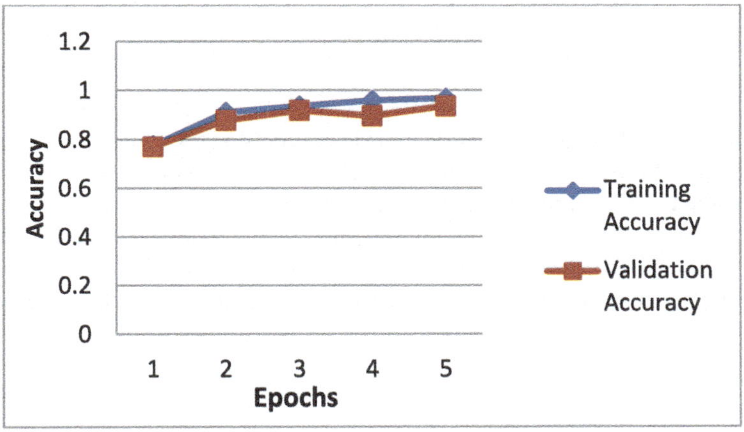

Fig. 5. Epochs and accuracy levels

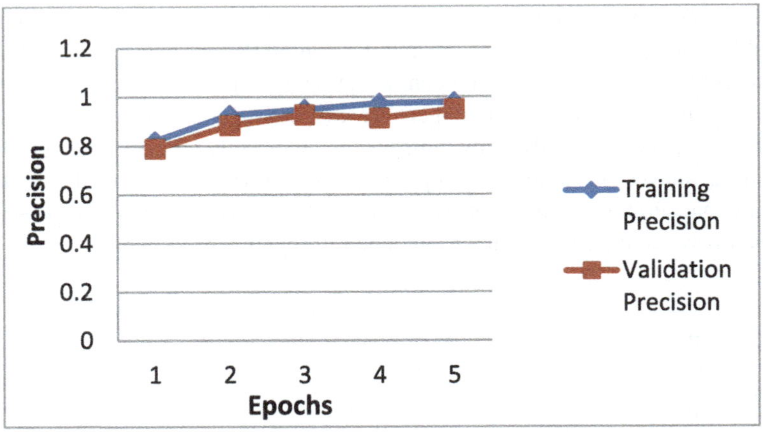

Fig. 6. Epochs and precision levels

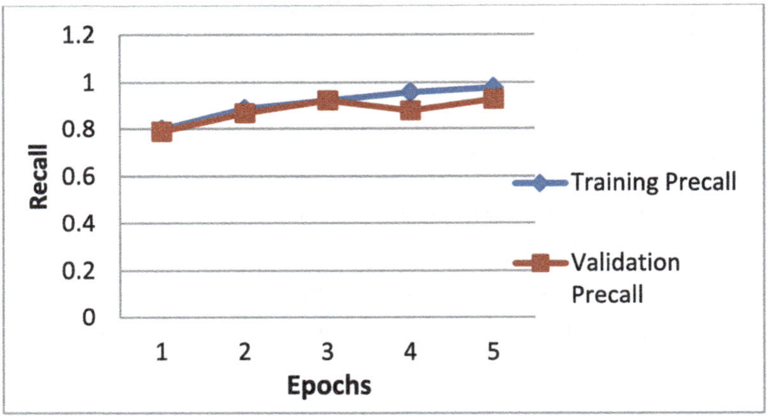

Fig. 7. Loss, Accuracy, Precision and Recall

Figures 5, 6, and 7 depict the justification loss, precision, exactness, and recall curves of the EfficientNetB3 model, accordingly. The X-axis stand for the total amount of epochs, while the Y-axis represents the values of loss, precision, exactness, and recall. The presented graphs illustrate the convergence of the assessment and training loss curves after the sixth epoch. Additionally, the exactness, precision, and recall of both the assessment and training curves exhibit parallel behavior beyond the sixth epoch.

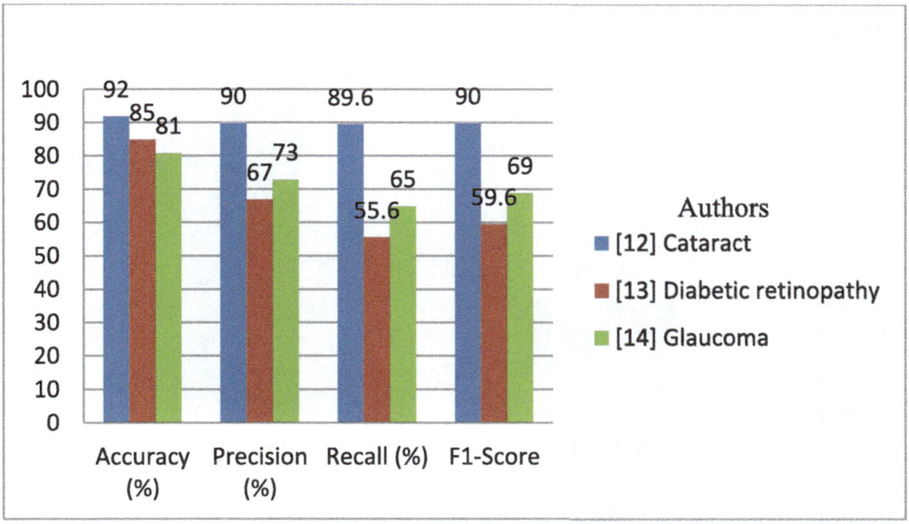

Fig. 8. Comparisons of existing models

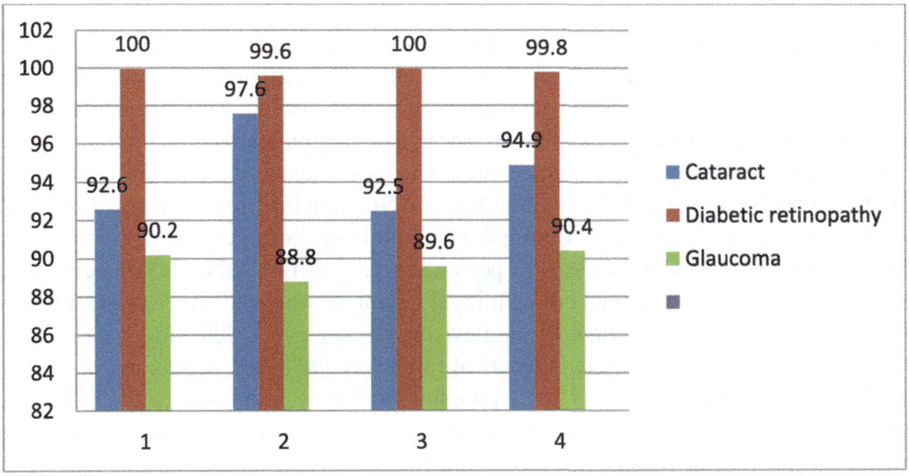

Fig. 9. Proposed model accuracy levels

Figures 8, 9, and 10 illustrate the evaluation of the presentation of our EfficientNetB3 model by respect to of exactness, precision, recall, and F1-Score, when compared to different pre-trained models.

Our proposed EfficientNetB3 model consistently achieves the top position and demonstrates superior performance across all evaluation metrics. EfficientNetB3 outperforms the other models that have been trained by a superior accuracy rate of 90.89%.

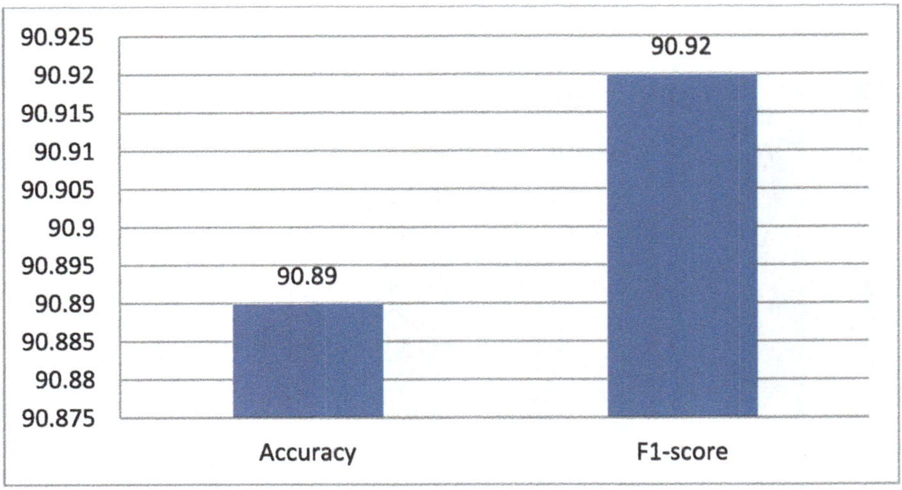

Fig. 10. Predicted proposed system

5 Conclusions

This paper presents the introduction of EfficientNet B3, a deep learning-based automated system designed for the diagnosis of glaucoma, diabetic retinopathy, and cataracts. The dataset utilized to train the deep neural network underwent initial enhancements using pre-processing techniques, including dataset reorganization and augmentation, specifically targeting a collection of fundus photos. The fundamental objective of the Efficient Net B3 algorithm is to explore dissimilar layers, activation functions, loss functions, and optimization strategies so as to diminish computational expenses whereas preserving the integrity of the model. In terms of performance, the EfficientNet B3 outperformed two out of the three pre-trained convolutional neural network models, namely VGG19 and ResNet50. Our model outperformed existing strategies for detecting cataract, diabetic retinopathy, and glaucoma in terms of reliability (90.89%), precision (92.719%), recall (91.68%), and f1-score (91.85%). Ophthalmologists can enhance the efficiency and accuracy of diagnosing cataract, diabetic retinopathy, and glaucoma by adopting EfficientNet B3 since to its exceptional exactness and cost- and time-efficiency. Our model is unable to detect diseases such as myopia and age-related macular degeneration (AMD), surrounded by others. However, it has the capability to detect cataract, glaucoma, and diabetic retinopathy. In future developments, it is possible to incorporate the degree of severity of the illness.

References

1. Bourne, R.R., et al.: Magnitude, temporal trends, and projections of the global prevalence of blindness and distance and near vision impairment: a systematic review and meta-analysis. Lancet Glob. Health **5**(9), e888–e897 (2017)
2. https://www.kaggle.com/datasets/andrewmvd/ocular-disease-recognition-odir5k

3. Raju, B., Raju, N.S.D., Akkara, J., Pathengay, A.: Do it yourself smartphone fundus camera DIYretCAM. Indian J. Ophthalmol. **64**(9), 663 (2016)
4. Dong, Y., Wang, Q., Zhang, Q., Yang, J.: Classification of cataract fundus image based on retinal vascular information. In: International Conference on Smart Health, pp. 166–173. Springer (2016)
5. Fan, W., Shen, R., Zhang, Q., Yang, J.J.: Principal component analysis-based cataract grading and classification In: International Conference on E-health Networking, Application and Services (HealthCom), pp. 459–462. IEEE (2015)
6. Manchalwar, M., Warhade, K.: Detection of cataract and conjunctivitis disease using histogram of oriented gradient. Int. J. Eng. Technol. (IJET) (2017)
7. Qiao, Z., Zhang, Q., Dong, Y., Yang, J.J.: Application of SVM based on genetic algorithm in classification of cataract fundus images In: IEEE International Conference on Imaging Systems and Techniques (IST), pp. 1–5. IEEE (2017)
8. Xiong, L., Li, H., Xu, L.: An approach to evaluate blurriness in retinal images with vitreous opacity for cataract diagnosis. J. Healthc. Eng. (2017)
9. Rajesh Tiwari, M., et al.: Enhanced power quality and forecasting for PV-wind microgrid using proactive shunt power filter and neural networkbased time series forecasting Electr. Power Compon. Syst. (2023).https://doi.org/10.1080/15325008.2023.2249894
10. Khan, Z., Khan, F.G., Khan, A., Rehman, Z.U.: Diabetic retinopathy detection using VGG-NIN a deep learning architecture. In: IEEE, vol. 9, pp.61408–61416 (2021)
11. Leite, D., Campelos, M., Fernandes, A., Batista, P., Beirão, J., Menéres, P., Cunha, A.: Machine Learning automatic assessment for glaucoma and myopia based on Corvis ST data. Procedia Comput. Sci. **196**, 454–460 (2022)
12. Rajani, A., Kora, P., Madhavi, K.R., Avanija, J.: Quality improvement of retinal optical coherence tomography, pp. 1–5 (2021).https://doi.org/10.1109/INCET51464.2021.9456151
13. Reddy Madhavi, K., et al.: COVID-19 detection using deep learning In: 20th International Conference on Hybrid Intelligent Systems-HIS 2020, at Machine Intelligence Research (MIR) Labs, USA, Springer AISC, 1375, pp 1–7 (2020)
14. Prabhakar, T., Srujan Raju, K., Reddy Madhavi, K.: Support vector machine classification of remote sensing images with the wavelet-based statistical features. In: Fifth International Conference on Smart Computing and Informatics (SCI 2021), Smart Intelligent Computing and Applications, Volume 2. Smart Innovation, Systems and Technologies, vol. 283. Springer, Singapore (2022)
15. Kora, P., Rajani, A., Chinnaiah, M.C., Madhavi, R., Swaraja, K., Kollati, M.: EEG-based brain-electric activity detection during meditation using spectral estimation techniques, pp. 687–693 (2021). https://doi.org/10.1007/978-981-16-1941-0_68
16. Gómez-Valverde, J.J., et al.: Automatic glaucoma classification using color fundus images based on convolutional neural networks and transfer learning. Biomed. Opt. Express **10**(2), 892–913 (2019)
17. Ram, A., Reyes-Aldasoro, D.C.C.: The relationship between fully connected layers and number of classes for the analysis of retinal images arXiv preprint arXiv:2004.03624 (2020)
18. Islam, M.T., Imran, S.A., Arefeen, A., Hasan, M., Shahnaz, C.: Source and camera independent ophthalmic disease recognition from fundus image using neural network. In: IEEE International Conference on Signal Processing, Information, Communication & Systems (SPICSCON), pp. 59–63. IEEE (2019)
19. He, K., Zhang, X., Ren, S., Sun, J.: Deep residual learning for image recognition. In: Proceedings of IEEE Conference on Computer Vision and Pattern Recognition (CVPR), pp. 770–778 (2016)

20. Nirmala, K., Saruladha, K., Srujan Raju, K.: Intelligent noise detection and correction with kriging on fundus images of diabetic retinopathy. In: Chowdary, P.S.R., Anguera, J., Satapathy, S.C., Bhateja, V. (eds) Evolution in Signal Processing and Telecommunication Networks. LNEE, vol. 839. Springer, Singapore (2022). https://doi.org/10.1007/978-981-16-8554-5_49
21. Patel, P., Sivaiah, B., Patel, R.: Relevance of frequent pattern (FP)-growth-based association rules on liver diseases. In: Intelligent Systems: Proceedings of ICMIB 2021, pp. 665–676. Springer, Singapore (2022)
22. Prashanthi, M., Chandra Mohan, M.: Hybrid optimization-based neural network classifier for software defect prediction. Int. J. Image Graph. **24**(04), 2450045 (2024). https://doi.org/10.1142/S0219467824500451

Enhancing Self-attention Training in Transformers for Image Captioning Using ViT and BERT

Madduri Deepika[1(✉)], N. Purushotham[2], Ch. Sarada[3], Modem Sai Kumar[4], Jyothirmai Digumarthi[5], and Gottimukkula Sravan Rao[6]

[1] Computer Science and Engineering, Malla Reddy College of Engineering for Women, Maissammaguda, Gundlapochampally, Medak, Hyderabad, India
maddurideepika9644@gmail.com

[2] CSE, School of Computing, Mohan Babu University, Tirupati, India

[3] Department of CSE, CVR College of Engineering, Hyderabad, Telangana, India

[4] MCA Department, Sri Venkateswara College of Engineering, Tirupati, India

[5] B V Raju Institute of Technology, Narsapur, Medak, Telangana, India

[6] Computer Science and Engineering, Indur Institute of Engineering and Technology, Siddipet, Telangana, India

Abstract. The convergence of Vision Transformer (ViT) and Bidirectional Encoder Representations from Transformers (BERT) has emerged as a valuable fusion mechanism for several applications. This paper introduces an automated system that utilizes ViT's in comprehending complicate visual patterns and BERT's contextual understanding of language to generate captions seamlessly integrated with the visual context. By improving the self-attention mechanisms, the proposed model learns to dynamically weigh the importance of both visual and textual information, resulting in captions that are descriptive. This paper provides a comprehensive exploration of our system's architecture, training methodology, and fusion strategies. The experimentation and evaluation, reveal that the proposed ViT + BERT based model for image captioning is efficient than the existing methods.

Keywords: Vision Transformers · Bidirectional Encoder Representations from Transformers · Image Captioning · Multi-Head Self-Attention · Layer Normalization

1 Introduction

The integration of methods from natural language processing and computer vision has led to significant enhancements in the field of image captioning.Traditional approaches utilize convolutional neural networks (CNNs) and recurrent neural networks (RNNs). Emergence of transformer-based models has altered the significance of adopting multimodal understanding. This paper introduces an effective approach that integrates two transformative architectures, Vision Transformer (ViT) and the Bidirectional Encoder Representations from Transformers (BERT). By integrating ViT's visual encoding prowess

A. Bajaj et al. (Eds.): IBICA 2023, LNNS 1229, pp. 445–453, 2025.
https://doi.org/10.1007/978-3-031-78940-3_47

with BERT's natural language understanding capabilities, the proposed model aims to redefine image captioning. It utilizes ViT's capacity to comprehend visual patterns and BERT's contextual awareness to generate captions that are not only descriptive but also deeply connected to the visual context.

The proposed model extends beyond conventional approaches by enriching self-attention mechanisms with ViT's visual insights. As the model generates captions, it dynamically assesses the significance of both visual and textual information, helping in natural language understanding. The proposed model can be applied to potential tasks ranging from assistive technologies for the visually impaired to enhanced multimedia analytics. Considering these aspects the proposed work explores the fusion of ViT + BERT architecture, utilizing a cross-modal attention mechanism for efficient image captioning.

2 Background Work

This section involves reviewing existing literature related to image captioning, vision transformer models, BERT models, fusion mechanisms and deep learning architectures. Image captioning is a fundamental task in computer vision and natural language processing, used to generate descriptive textual descriptions for visual content [1–6]. Existing approaches use convolutional neural networks (CNNs) [10] for feature extraction and recurrent neural networks (RNNs) like LSTMs[3] for sequence generation. Recent advancements in deep learning [7–12] have introduced architectures, like BERT (Bidirectional Encoder Representations from Transformers), that are successful in performing various natural language understanding tasks. Vision Transformer architecture, initially designed for sequence-to-sequence tasks, to process grid-like data, such as images. ViTs have demonstrated impressive capabilities in image classification, object detection, and segmentation by directly attending to image patches. This self-attention mechanism allows ViTs to capture dependencies and contextual information efficiently. The adoption of ViTs in image-related tasks has opened up new possibilities for image captioning, where understanding context and relationships between visual elements is critical [13–19]. BERT, which is designed for pretraining on text corpora, has found applications in multimodal tasks, where textual and visual information are combined. Fusing BERT into image captioning models enables them to leverage pretrained contextual embeddings and capture semantic relationships between words and visual elements [22–27]. Critical aspect of combining ViT and BERT in image captioning is the design of fusion mechanisms. Various fusion strategies, including concatenation, addition, multiplication, and cross-modal attention, have been explored in the literature [20–25]. Cross-modal attention mechanisms have gained popularity for their ability to capture fine-grained interactions between visual and textual information. Ensemble models, which combine multiple models or strategies, have been employed to boost image captioning performance. These ensembles often include a combination of CNNs, RNNs, and attention-based models. Considering these aspects, the proposed work addresses the challenges by enhancing the ViT + BERT architecture utilizing cross-attention mechanism to perform efficient image captioning.

3 Methodology

This section, briefs about the steps of the ViT + BERT architecture to perform image captioning utilizing Vision Transformer (ViT) and Bidirectional Encoder Representation from Transformer(BERT) architecture by enhancing the self-attention mechanism. The steps for the proposed architecture for predicting wheat rust disease have been specified in Fig. 1. Initial step is to gather dataset consisting of image samples to generate caption. Proposed work utilizes MS COCO dataset from kaggle for processing. Next step is data preparation which involves preprocessing the image performing operations like, image resizing,normalization, and applying data augmentation to increase diversity of samples used for training. This process ensures that they are suitable for input into the ViT component. Then the caption data, from datasets like COCO, is tokenized and converted into numerical form to be compatible with the BERT component. ViTcomponent is used for processing the image data and extracting visual features. Pre-trained ViT models are utilized and fine-tuned for specific image captioning task. BERT component generates descriptive captions based on visual features extracted by ViT. Pre-trained BERT models are fine-tuned on caption data to understand the textual context. Proposed model is trained using supervised learning combining image features and textual context.

Next, the ViT component takes as input the preprocessed image data. Images are typically represented as grids of image patches. ViT is a transformer-based architecture used for processing visual data. This architecture consists of a stack of transformer encoder layers to perform multi-head self-attention on the input image patches, used to extract global and local context of images. Then, the image patching process divides the images into a grid of non-overlapping patches. Next, is tp perform performed linearly embedding for each patch into a lower-dimensional feature space, resulting in patch embeddings. Then, positional encodings provide information related to spatial location of each patch. Transformer encoder layers of ViT perform multi-head self-attention and feedforward operations enabling model to extract the context within the image. Importance of different patches, considering both local and global relationships can be measures using Multi-head self-attention mechanisms. Layer normalization is applied to stabilize and normalize the outputs of multi-head self-attention and feedforward neural networks. Feedforward neural networks further process the patch embeddings, capturing complex relationships between patches. The component residual connections, also known as skip connections, allow the gradient flow and help prevent vanishing gradients during training. ViT component generates contextualized embeddings for each patch as output. This output is a fixed-length vector representation of visual content in an image also referred to as "image embedding," encapsulates the image's content in a continuous numerical format. These visual features converted into a sequence of tokens through tokenization by splitting text into individual units, such as words or subword tokens.

Next step is to process BERT component to generate image captioning.This component takes input text, which is tokenized into subword tokens or words. Token embeddings embed each token into a lower-dimensional feature space, and capture the semantic meaning of tokens. Positional encoding is performed to token embeddings in order to provide details about the ordering of tokens in input sequence. BERT model includes multiple transformer encoder layers. The encoder layer contains multi-head self-attention process to help in capturing contextual information considering the relationships between different tokens in both directions. Then, layer normalization is applied to stabilize and normalize the outputs of self-attention and feedforward neural networks within each encoder layer. Feedforward neural networks further process the patch embeddings, capturing complex relationships between tokens. The component residual connections, also known as skip connections, allow the gradient flow and help prevent vanishing gradients during training for BERT. The output of BERT is contextualized token representations capturing bidirectional context and semantic information within input text.

Next step is to apply fusion mechanism to combine information from different modalities, such as vision and text, to make predictions and generate output effectively. The fusion mechanism used in the proposed approach is cross-modal attention such as self-attention to capture complex relationships between both visual and textual based features, which is crucial for generating descriptive and contextually relevant captions. This technique is capable of dynamically measuring the importance of different modalities at each position in the sequence.This process enhances the process of image captioning using ViT and BERT architecture. ViT and BERT components are fine-tuned on a captioning dataset to the specifics of image captioning and improve their performance. It involves defining a task-specific architecture, loss function, and hyperparameter tuning while monitoring performance on validation data. Further, the models performance is enhanced by using adaptive attention mechanism to dynamically adjust the contents bases on weights and cross-modal attention integrating information from different modalities like text and image. This process is performed by the component fusion mechanism specified in Fig. 1. After iterative fine-tuning the model is deployed for generating descriptive image captions. Final step is to perform model evaluation using performance metrics as specified in Table 1. Thus, ViT and BERT based approach enhances the image captioning mechanism by generating descriptive captioning based on the image samples.

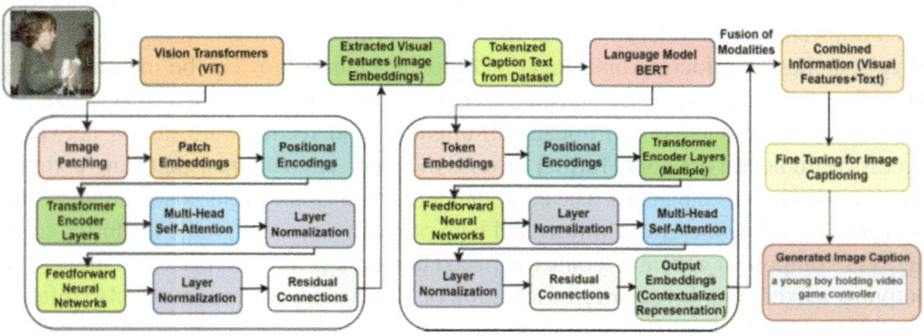

Fig. 1. Image Captioning Framework Using ViT and BERT Architecture

4 Results and Discussions

The dataset used in proposed approach is MS COCO (Microsoft Common Objects in Context) from Kaggle. This dataset is an important resource in performing computer vision tasks, over 200,000 diverse images, each annotated with object bounding boxes, segmentation masks, key points, and notably, multiple descriptive captions. MS COCO can be utilized to perform object recognition, detection, segmentation, and image captioning. The sample images in the dataset for model training is specified in Fig. 2. The important feature of the dataset is the image captioning annotations. Each image is associated with multiple descriptive captions, making it a standard benchmark for image captioning tasks. The captions provide detailed textual descriptions of the visual content in the images.

Fig. 2. Sample Images of MS COCO Dataset

Performance of the ViT-BERT approach is evaluated using metrics BLEU(Bilingual Evaluation Understudy), METEOR (Metric for Evaluation of Translation with Explicit Ordering) and CIDEr (Consensus-based Image Description Evaluation) and specified in Table 1. BLEU is used for evaluating the quality of text generated as image captions. Similarity between the generated text and reference text i.e.human-generated captions can be measured using tis metric. The specifics of BLEU are BLEU-1,2,3 and 4 to measure the precision of unigram, bigram,trigram and four-gram respectively. Higher BLEU scores specify a better degree of overlap between generated and reference text.

Table 1. Evaluation of ViT and BERT Model for Image Captioning

Model	BLEU-1	BLEU-2	BLEU-3	BLEU-4	METEOR	CIDEr
LSTM	0.65	0.48	0.35	0.28	0.62	1.72
Attention-Based	0.68	0.51	0.38	0.31	0.64	1.85
Transformer-Based	0.70	0.53	0.40	0.33	0.66	1.91
Ensemble	0.71	0.54	0.41	0.34	0.67	2.00
Proposed (ViT + BERT)	0.72	0.55	0.42	0.35	0.68	2.03

METEOR is another metric for evaluating the quality of generated text. It considers multiple aspects such as precision, recall, stemming, synonym, and word order. Higher METEOR scores indicate better quality of generated text compared to the reference text. CIDEr is specifically used for evaluating image captions by measuring the consensus between generated captions and reference captions provided by human annotators. Higher CIDEr scores suggest that the generated captions are more in line with consensus among human annotators, indicating better quality captions. As specified in Table 1 proposed approach, which combines ViT and BERT, performs well in generating image captions compared to existing methods. The results demonstrate its capability to generate high-quality, contextually relevant, and human-consensus captions.

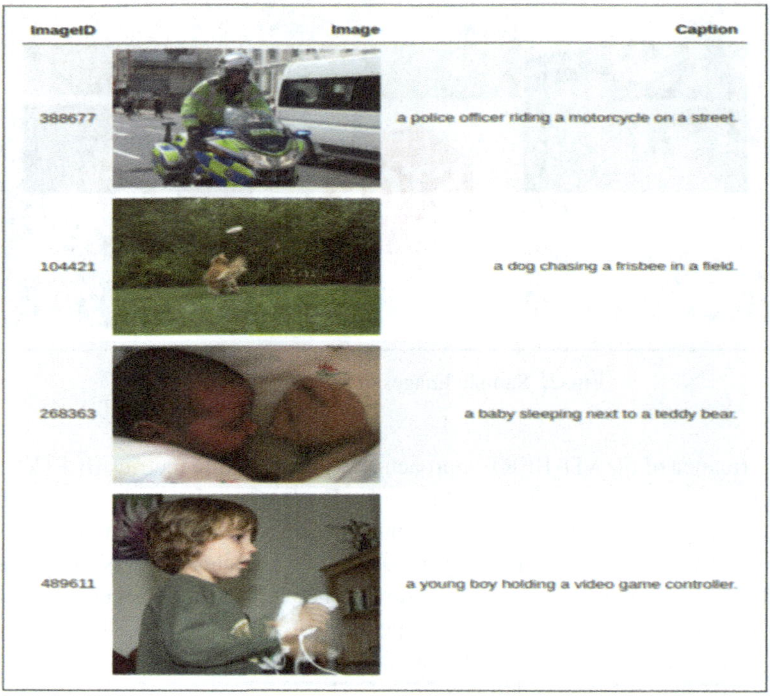

Fig. 3. Generated Captions by the Proposed ViT + BERT Model with Images

Table 2. Hyperparameter Tuning Values for the Proposed ViT + BERT Approach

Hyperparameter	Values	Hyperparameter	Values
Learning Rate	0.001	Transformer Layers (BERT)	12
Batch Size	32	Hidden Dimension (ViT)	768
Maximum Sequence Length (Captions)	128	Hidden Dimension (BERT)	768
Transformer Layers (ViT)	12	Weight Decay	1e−5

The generated captions by the proposed ViT + BERT model are represented in Fig. 3. Hyperparameter tuning is an iterative process that involves experimenting with different settings to find the configuration that yields best performance on specific image captioning task and dataset. Hyperparameter tuning of the proposed architecture is specified in Table 2.

5 Conclusion

In this paper, we propose an efficient approach to perform image captioning by integrating ViT and BERT model. The proposed model enhances self-attention mechanisms, dynamically weighting visual and textual information, resulting in contextually enriched captions. Experimental evaluation reveals that the proposed ViT + BERT based model performs well than existing state-of-art methods. The proposed work can also support diverse applications ranging from assistive technologies to multimedia analytics. Future work can explore the refinement of fusion mechanisms for more effective integration of ViT and BERT. Further, the expansion of this approach to address specific domain-related challenges and the creation of larger, specialized datasets can enhance its applicability in real-world scenarios.

References

1. Sargar, O., Kinger, S.: Image captioning methods and metrics. In: 2021 International Conference on Emerging Smart Computing and Informatics (ESCI), pp. 522–526. IEEE (2021)
2. Hossain, M.Z., Sohel, F., Shiratuddin, M.F., Laga, H.: A comprehensive survey of deep learning for image captioning. ACM Comput. Surv. (CsUR) **51**(6), 1–36 (2019)
3. Wang, C., Yang, H., Bartz, C., Meinel, C.: Image captioning with deep bidirectional LSTMs. In: Proceedings of the 24th ACM International Conference on Multimedia, pp. 988–997 (2016)
4. Sharma, H., Agrahari, M., Singh, S.K., Firoj, M., Mishra, R.K.: Image captioning: a comprehensive survey. In: 2020 International Conference on Power Electronics & IoT Applications in Renewable Energy and its Control (PARC), pp. 325–328. IEEE (2020)
5. You, Q., Jin, H., Wang, Z., Fang, C., Luo, J.: Image captioning with semantic attention. In: Proceedings of the IEEE Conference on Computer Vision and Pattern Recognition, pp. 4651–4659 (2016)
6. Yao, T., Pan, Y., Li, Y., Qiu, Z., Mei, T.: Boosting image captioning with attributes. In: Proceedings of the IEEE International Conference on Computer Vision, pp. 4894–4902 (2017)
7. Chohan, M., Khan, A., Mahar, M.S., Hassan, S., Ghafoor, A., Khan, M.: Image captioning using deep learning: a systematic. Image **11**(5) (2020)
8. Zohourianshahzadi, Z., Kalita, J.K.: Neural attention for image captioning: review of outstanding methods. Artif. Intell. Rev. **55**(5), 3833–3862 (2022)
9. Wang, Y., Xiao, B., Bouferguene, A., Al-Hussein, M., Li, H.: Vision-based method for semantic information extraction in construction by integrating deep learning object detection and image captioning. Adv. Eng. Inform. **53**, 101699 (2022)
10. Gu, J., Wang, G., Cai, J., Chen, T.: An empirical study of language cnn for image captioning. In: Proceedings of the IEEE International Conference on Computer Vision, pp. 1222–1231 (2017)

11. Periannasamy, S.M., et al.: Analysis of artificial intelligence enabled intelligent sixth generation (6G) wireless communication networks. In: 2022 IEEE International Conference on Data Science and Information System (ICDSIS), Hassan, India, pp. 1–8 (2022). https://doi.org/10.1109/ICDSIS55133.2022.9915945
12. Gothane, S.: Content summarization: journey toward word clouds. In: Kumar, A., Ghinea, G., Merugu, S., Hashimoto, T. (eds.) Proceedings of the International Conference on Cognitive and Intelligent Computing. Cognitive Science and Technology. Springer, Singapore (2022). https://doi.org/10.1007/978-981-19-2350-0_46
13. Simi Margarat, G., et al.: Early diagnosis of tuberculosis using deep learning approach for iot based healthcare applications. Comput. Intell. Neurosci. **2022** (2022)
14. Monga, C., et al.: Secure techniques for channel encryption in wireless body area network without the certificate. Wirel. Commun. Mobile Comput. **2022**(1), 2598465 (2022)
15. Patel, P., Sivaiah, B., Patel, R.: Approaches for finding optimal number of clusters using k-means and agglomerative hierarchical clustering techniques. In: 2022 International Conference on Intelligent Controller and Computing for Smart Power (ICICCSP), Hyderabad, India, pp. 1–6 (2022). https://doi.org/10.1109/ICICCSP53532.2022.9862439
16. Madala, S., Suvarna Vani, K., Jalapally, P., Srujan Raju, K., Somasekhar, G.: Framework for diabetic retinopathy classification. In: Satapathy, S.C., Bhateja, V., Favorskaya, M.N., Adilakshmi, T. (eds.) Smart Intelligent Computing and Applications, Volume 2: Proceedings of Fifth International Conference on Smart Computing and Informatics (SCI 2021), pp. 593–601. Springer Nature Singapore, Singapore (2022). https://doi.org/10.1007/978-981-16-9705-0_58
17. Kolli, C.S., Raghunath, M.P., Meenakshi, S., Maheswari, K., Britto, C.F., Kushwaha, S.: Efficient development of supervised learning algorithm for kidney stone prediction. In: 2022 International Conference on Inventive Computation Technologies (ICICT), Nepal, pp. 1373–1379 (2022). https://doi.org/10.1109/ICICT54344.2022.9850573
18. Sathishkumar, A., Majji, S., Radhika Patnala, T., Rao Karanam, S., Kumar, A., Malyadri, M.: Experimentation methodology of orthogonal frequency division multiplexing signals process using radio over fiber (RoF) system. In: 2022 International Virtual Conference on Power Engineering Computing and Control: Developments in Electric Vehicles and Energy Sector for Sustainable Future (PECCON), Chennai, India, pp. 01–05 (2022). https://doi.org/10.1109/PECCON55017.2022.9850996
19. Avanija, J., Madhavi, K.R., Sunitha, G., Sangapu, S.C., Raju, S.: Facial expression recognition using convolutional neural network. In: 2022 First International Conference on Artificial Intelligence Trends and Pattern Recognition (ICAITPR), Hyderabad, India, pp. 1–7 (2022). https://doi.org/10.1109/ICAITPR51569.2022.9844221
20. Tiwari, L., Awasthi, V., Patra, R.K., Miri, R., Raja, H., Bhaskar, N.: Lung cancer detection using deep convolutional neural networks. In: Bhateja, V., Wee, L.K., Lin, J.C.-W., Satapathy, S.C., Rajesh, T.M. (eds.) Data Engineering and Intelligent Computing: Proceedings of 5th ICICC 2021, Volume 1, pp. 373–385. Springer Nature Singapore, Singapore (2022). https://doi.org/10.1007/978-981-19-1559-8_37
21. Elbedwehy, S., Medhat, T., Hamza, T., Alrahmawy, M.F.: Efficient image captioning based on vision transformer models. Comput. Mater. Continua **73**(1) (2022)
22. Atliha, V., Šešok, D.: Text augmentation using BERT for image captioning. Appl. Sci. **10**(17), 5978 (2020)
23. Kalimuthu, M., Mogadala, A., Mosbach, M., Klakow, D.: Fusion models for improved image captioning. In: Del Bimbo, A., et al. (eds.) Pattern Recognition. ICPR International Workshops and Challenges: Virtual Event, January 10–15, 2021, Proceedings, Part VI, pp. 381–395. Springer, Cham (2021). https://doi.org/10.1007/978-3-030-68780-9_32
24. Ragab, M., Addas, A.: Low complexity encoder with multilabel classification and image captioning model. Comput. Mater. Continua **72**(3) (2022)

25. Salaberria, A., Azkune, G., de Lacalle, O.L., Soroa, A., Agirre, E.: Image captioning for effective use of language models in knowledge-based visual question answering. Expert Syst. Appl. **212**, 118669 (2023)
26. Sunitha, G., Geetha, K., Neelakandan, S., Pundir, A.K.S., Hemalatha, S., Kumar, V.: Intelligent deep learning based ethnicity recognition and classification using facial images. Image Vis. Comput. **121**, 104404 (2022)
27. Deepa, M.D.: Bidirectional encoder representations from transformers (BERT) language model for sentiment analysis task. Turk. J. Comput. Math. Educ. (TURCOMAT) **12**(7), 1708–1721 (2021)

Maize Seed Classification Using Convolutional Block Attention Networks

Rajasekhar Nennuri[1], Nakkala Geetha[2], P. Bhanu Prakash[3], M. Reddy Kedhareswar[2], Gurram Sunitha[4](✉), and Namburi Manoj Kumar[5]

[1] Department of CSE, Institute of Aeronautical Engineering, Hyderabad, Telangana, India
[2] Department of CSE, K M M Institute of Technology and Science, Tirupati, Andhra Pradesh, India
[3] Department of CSE, Annamacharya Institute of Technology and Sciences, Tirupati, Andhra Pradesh, India
[4] Department of CSE, School of Computing, Mohan Babu University, Tirupati, Andhra Pradesh, India
gurramsunitha@gmail.com
[5] CSE (Data Science), Sree Vidyanikethan Engineering College, Tirupati, Andhra Pradesh, India

Abstract. Maize seed classification is a pivotal task in modern agriculture, directly impacting crop productivity and food security. This research investigates application of learning with attention for maize seed classification, with a focus on enhancing accuracy and efficiency. A comparative analysis of four models is performed – VGG16, Inceptionv3, ResNet50, and the proposed model (ResNet50 enhanced with the Convolutional Block Attention Module). Experimentation involved training and evaluating these models on a comprehensive dataset comprising "Good" and "Bad" maize seed images. The results of experiments demonstrated the performance disparities among the models, shedding light on their strengths and weaknesses. ResNet50 with CBAM achieved an accuracy of 0.958, precision of 0.965, recall of 0.962, and F1-score of 0.968. The achieved performance can be attributed to the integration of CBAM, which enhances feature representations through attention mechanisms. CBAM allowed the model to focus on salient features within the images, leading to superior classification accuracy.

Keywords: Smart agriculture · Deep Learning · Maize crops · Seed classification · CBAM

1 Introduction

Smart agriculture has far-reaching implications for food security and crop productivity. Maize plays a pivotal role in global food security. Its cultivation and yield are intrinsically linked to the quality of maize seeds sown by farmers. Maize seed classification is the process of distinguishing between "Good" and "Bad" seeds. This task holds paramount importance in ensuring robust agricultural production. "Good" seeds yield high-quality crops, while "Bad" seeds can result in reduced yields and economic losses for farmers.

A. Bajaj et al. (Eds.): IBICA 2023, LNNS 1229, pp. 454–461, 2025.
https://doi.org/10.1007/978-3-031-78940-3_48

Accurate and efficient classification of maize seeds is therefore a critical step in agricultural practices [1]. This impacts crop productivity, food supply chains, and livelihoods worldwide.

Integration of automated learning systems into computer vision has brought forth transformative advancements in image classification tasks [2, 3]. Researchers have developed sophisticated CNN models capable of automatically learning and extracting intricate features from images [4, 5]. In this context, the Convolutional Block Attention Module (CBAM) has emerged as a valuable tool for enhancing the discriminative capacity of CNNs [6]. CBAM introduces attention mechanisms, allowing models to focus on relevant image regions and features while reducing the impact of noise. This module has demonstrated remarkable capabilities in improving image classification accuracy by augmenting the inherent strengths of CNNs.

This research paper embarks on the exploration of Maize Seed Classification Using Convolutional Block Attention Networks. This research proposes investigation of CNN architecture incorporated with CBAM to augment the feature representation and classification of maize seeds. The integration of CBAM empowers the model to attend to salient features within the maize seed images, distinguishing between "Good" and "Bad" seeds with unprecedented precision and recall. The aim is to harness the capabilities of CBAM and deep learning for maize seed classification. Thus, contributing to more reliable agricultural practices and global food security [7].

2 Literature Review

Smart agriculture driven by the integration of advanced technologies, is poised to revolutionize the global farming landscape [8]. Among the myriad technologies contributing to its transformation, deep learning for vision tasks stands out as a game-changer [9]. Deep learning has proven to be instrumental in solving complex agricultural challenges by providing unprecedented insights and automation capabilities in the domain of computer vision [10].

Maize seed classification is a crucial task in agriculture. It directly impacts crop yield and food security. Deep learning techniques are proven to be able to efficiently automate and enhance the accuracy of seed classification processes [11].

Innovative approaches are developed for maize seed classification by employing deep CNNs with extensive data augmentation techniques [12]. High classification accuracy was achieved by improving the model's generalization capabilities.

A study explored a hybrid deep learning approach for maize seed disease detection [13]. The model combines CNN with Long Short-Term Memory networks to capture both spatial and temporal patterns in seed images, facilitating disease diagnosis with high accuracy.

Accurate maize seedling detection in diverse growth stages and challenging field conditions is achieved using an enhanced Faster R-CNN model [14]. A study introduces a learning model incorporating Gabor filters for more accurate and robust maize disease classification [15].

3 Proposed Model

For maize classification, a well-established and effective deep learning CNN architecture Residual Learning Network is chosen. This architecture has proven to be effective in image classification tasks and can serve as a strong baseline for maize classification model. ResNet (Residual Networks) is known for its deep architecture while mitigating the vanishing gradient problem. ResNet-50 is a variant of ResNet that has 50 layers. It is suitable as base CNN architecture for complex image classification tasks like maize classification.

Proposed architecture uses a pre-trained ResNet-50 backbone with a CBAM incorporated to enhance feature representation (Fig. 1). CBAM consists of a Channel Attention Module (CAM) and a Spatial Attention Module (SAM). CAM employs global average pooling and convolutional layers with ReLU and sigmoid. It learns to recalibrate channel-wise feature maps to emphasize informative channels. SAM enhances spatial features by selectively attending to significant spatial regions through convolutional layers and activations. These attentively refined feature maps are combined with original feature maps from backbone. A global average pooling layer further aggregates spatial information. This layer computes the average value across both spatial dimensions, resulting in a feature vector with a size of 1x1. This vector encapsulates the essence of the entire image, effectively summarizing spatial relationships and patterns. The fully connected layer with two output units, corresponding to "Good" and "Bad" classes, produces the final predictions. This architecture integrates the strength of a pre-trained backbone with advanced attention mechanisms to facilitate accurate maize seed classification.

Pre-trained ResNet-50 Backbone (Feature Extraction). The pre-trained ResNet-50 backbone is loaded with pre-trained weights from a model trained on a large image dataset (ImageNet). It consists of convolutional layers, batch normalization, ReLU activations, and residual blocks. The pre-trained ResNet-50 backbone can be mathematically represented as follows:

Let X be the input tensor (image). The backbone can be expressed as a function $F(X)$. The output feature maps after these layers are denoted as

$$FeatureMaps = F(X) \tag{1}$$

where *FeatureMaps* is a 3D tensor.

CBAM. It enhances the feature maps produced by backbone. It contains two sub-modules:

a) Channel Attention Module (CAM)
Given *FeatureMaps*, CAM first performs global average pooling (GAP) across spatial dimensions:

$$AvgPooled = GAP(FeatureMaps) \quad (2)$$

Then, convolutional layers with ReLU activations and sigmoid activation is applied to compute channel-wise attention weights.

$$ChannelWeights = Sigmoid(Conv(ReLU(Conv(AvgPooled)))) \quad (3)$$

Finally, feature maps are recalibrated using channel-wise weights:

$$ChannelAttention = FeatureMaps \times ChannelWeights \quad (4)$$

b) Spatial Attention Module (SAM)
SAM utilizes convolutional layers and ReLU activations to produce spatial attention weights.

$$SpatialWeights = Sigmoid(Conv(ReLU(Conv(FeatureMaps)))) \quad (5)$$

These spatial attention weights are applied to the original feature maps.

$$SpatialAttention = FeatureMaps \times SpatialWeights \quad (6)$$

Global Average Pooling Layer (avgpool)
Global average pooling is applied across both spatial dimensions, reducing the feature map to a size of 1x1. It can be represented as a mathematical operation that computes the average value across both spatial dimensions. Given *FeatureMaps* from the CBAM output:

$$GlobalAvgPooled = AveragePooling(FeatureMaps) \quad (7)$$

Fully Connected Layer (fc)
Fully connected layer maps the *GlobalAvgPooled* vector to the number of classes. In this case, there are 2 classes for "Good" and "Bad" maize seeds. It computes the logits for the two classes.

$$Logits = W \times GlobalAvgPooled + b \quad (8)$$

where W represents the weight matrix and b is the bias vector specific to the fully connected layer.

The above proposed architecture combines the power of a pre-trained ResNet-50 backbone with the feature enhancement capabilities of the CBAM module, leading to improved feature representation and classification performance for maize seed classification.

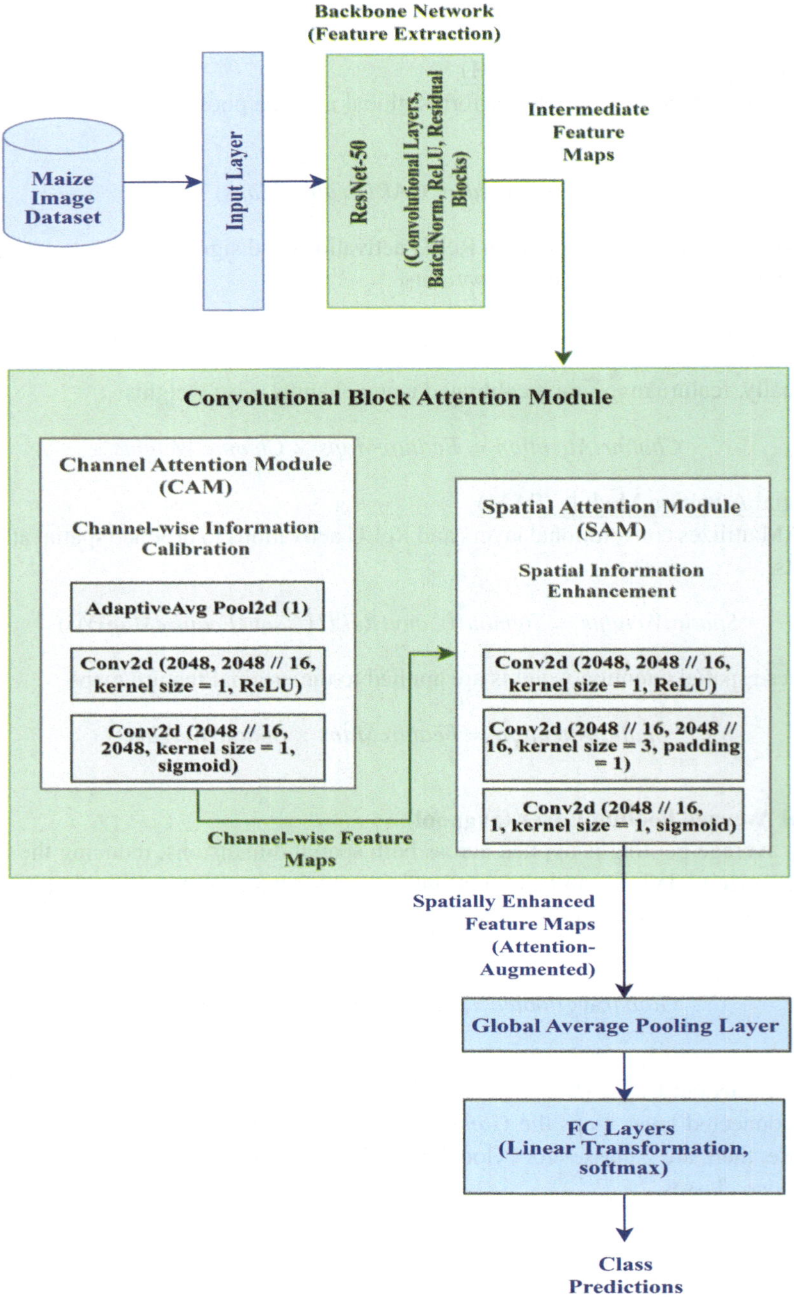

Fig. 1. Proposed CNN-CBAM Model

4 Experimental Results and Evaluation

A comprehensive dataset of maize seed images, including "Good" and "Bad" classes [16]. "Good" represents high-quality maize seeds suitable for production, and "Bad" includes damaged, infected, or low-quality seeds. A total of 4846 raw images and 28910 augmented images are available in the dataset.

The proposed model is based on a CNN architecture augmented with a Convolutional Block Attention Module (CBAM). It consists of the components – pre-trained ResNet-50 backbone for feature extraction, CBAM module to enhance feature representations, global average pooling layer for spatial aggregation, fully connected layer for classification. Proposed model is compared with the following four baseline models - VGG-16 (traditional deep CNN architecture with 16), Inception-v3 (architecture that captures multi-scale features utilizing inception modules), Simple CNN, Support Vector Machine (another traditional machine learning method, used with features extracted from the pre-trained ResNet-50 backbone). The models are trained using dam optimizer with a learning rate of 0.001. Each model is trained for 100 epochs, and early stopping based on validation loss was employed to prevent overfitting.

Table 1. Experimental Performance

Model	Accuracy	Precision	Recall	F1-Score
VGG16	0.872	0.864	0.862	0.858
Inceptionv3	0.897	0.888	0.888	0.885
ResNet50	0.923	0.918	0.913	0.923
ResNet50 with CBAM	0.958	0.965	0.962	0.968

By analyzing the experimental results for maize seed classification, it was observed that each model exhibits distinct performance characteristics (Table 1). VGG16, a classic deep CNN, achieves a reasonable accuracy of 0.872 but falls behind in precision, recall, and F1-score compared to more advanced models. Inceptionv3, with its inception modules, improves accuracy to 0.897 and maintains higher precision, indicating a reduced number of false positives. However, it still struggles with capturing all instances of "Bad" seeds, as evident from its recall score. VGG16 and Inceptionv3, while offering commendable results, lack the depth and attention mechanisms needed to discern intricate details within maize seed images. ResNet50, with its deeper architecture and residual connections, further enhances accuracy to 0.923 while excelling in both precision and recall. But it lacks attention mechanisms for finer feature discrimination. The proposed model, ResNet50 with CBAM, surpasses all others with an accuracy of 0.958. This performance can be attributed to the integration of CBAM, which enhanced feature representations through attention mechanisms, leading to improved precision and recall. The proposed model's utilization of CBAM proves instrumental in achieving superior maize seed classification performance. It allowed to focus on relevant features, underlining the importance of attention mechanisms in image classification tasks.

5 Conclusions

Smart agriculture has far-reaching implications for food security and crop productivity. This research focused on a comprehensive exploration of deep learning techniques to enhance the accuracy and efficiency of maize seed classification. Through a series of experiments and evaluations, valuable insights were gained into the performance of different models and their suitability for this critical task. The comparative analysis encompassed four models – VGG16, Inceptionv3, ResNet50, and the proposed model (ResNet50 enhanced with the CBAM). The results of experiments clearly delineate the capabilities and limitations of each model. Proposed model, ResNet50 with CBAM, demonstrated considerable performance over other models. The proposed model achieved an accuracy of 0.958, precision of 0.965, recall of 0.962, and F1-score of 0.968. Achieved performance is attributed to incorporation of CBAM into CNN. This empowered the model to focus on salient features, leading to superior classification accuracy. This research underscores the pivotal role of attention mechanisms in maize seed classification and smart agriculture at large. Integration of deep learning, particularly with attention mechanisms like CBAM, offers tremendous potential for enhancing agricultural processes. This research highlights the importance of continuous innovation in deep learning techniques to meet the ever-growing global food demands while ensuring the quality and productivity of maize crops.

References

1. Li, D., Liu, Y., Gao, L.: Research of maize seeds classification recognition based on the image processing. Int. J. Signal Process. Image Process. Pattern Recogn. **9**(11), 181–190 (2016)
2. Sunitha, G., Sudeepthi, A., Sreedhar, B., Shaik, A.B., Farooq, C.: RetinaNet and vision transformer-based model for wheat head detection. In: 2023 5th International Conference on Inventive Research in Computing Applications (ICIRCA), pp. 151–156. IEEE (2023)
3. Shereesha, M., Hemavathy, C., Teja, H., Reddy, G.M., Kumar, B.V., Sunitha, G.: Precision mango farming: using compact convolutional transformer for disease detection. In: International Conference on Innovations in Bio-Inspired Computing and Applications, pp. 458–465. Springer, Cham (2022)
4. Balakrishna, N., Sunitha, G., Karthik, A., Madhavi, K.R., Avanija, J.: Tomato leaf disease detection using deep learning: a CNN approach. In: 2022 International Conference on Data Science, Agents & Artificial Intelligence (ICDSAAI), vol. 1, pp. 1–6. IEEE (2022)
5. Sunitha, G., et al.: Modeling of chaotic political optimizer for crop yield prediction. Intell. Autom. Soft Comput. **34**(1), 423–437 (2022)
6. Rachapudi, V., Rao, K.S., Rao, T.S.M., et al.: Diabetic retinopathy detection by optimized deep learning model. Multimedia Tools Appl. **82**, 27949–27971 (2023). https://doi.org/10.1007/s11042-023-14606-8
7. Subramani, P., K, S., BD, K.R., et al.: Prediction of muscular paralysis disease based on hybrid feature extraction with machine learning technique for COVID-19 and post-COVID-19 patients. Pers. Ubiquit. Comput. **27**, 831–844 (2023). https://doi.org/10.1007/s00779-021-01531-6
8. Woo, S., Park, J., Lee, J.Y., Kweon, I.S.: CBAM: convolutional block attention module. In: Proceedings of the European Conference on Computer Vision (ECCV), pp. 3–19 (2018)

9. Nennuri, R., Kumar, R.H., Prathyusha, G., Tejaswini, K., Kanishka, G., Sunitha, G.: A multi-stage deep model for crop variety and disease prediction. In: International Conference on Soft Computing and Pattern Recognition, pp. 52–59. Springer, Cham (2022)
10. Sunitha, G., Madhavi, K.R., Avanija, J., Reddy, S.T.K., Vittal, R.H.S.: Modeling convolutional neural network for detection of plant leaf spot diseases. In: 2022 3rd International Conference on Electronics and Sustainable Communication Systems (ICESC), pp. 1187–1192. IEEE (2022)
11. Jones-Garcia, E., Krishna, V.V.: Farmer adoption of sustainable intensification technologies in the maize systems of the Global South: a review. Agron. Sustain. Dev. **41**(1), 8 (2021)
12. Hasan, M.J., Alom, M.S., Dina, U.F., Moon, M.H.: Maize diseases image identification and classification by combining CNN with bi-directional long short-term memory model. In: 2020 IEEE Region 10 Symposium (TENSYMP), pp. 1804–1807. IEEE (2020)
13. Quan, L., et al.: Maize seedling detection under different growth stages and complex field environments based on an improved Faster R-CNN. Biosys. Eng. **184**, 1–23 (2019)
14. Dhakshayani, J., Surendiran, B.: GF-CNN: an enhanced deep learning model with gabor filters for maize disease classification. SN Comput. Sci. **4**(5), 538 (2023)
15. Emmanuel A., Obed A., Peter A.: Lightweight Dataset for Maize Classification on Resource-Constrained Devices Mendeley Data. V2 (2023). https://doi.org/10.17632/r6vvm5jkh6.2
16. Allam Balaram, K., Kannan, N., Čepová, L., Kishore Kumar, M., Swaroopa Rani, B., Schindlerova, V.: Artificial intelligence for media ecological integration and knowledge management. Systems **11**(5), 222 (2023). https://doi.org/10.3390/systems11050222

Denoising Imagery: Speckle Noise Removal Using GANs

P. Bhanu Prakash[1], S. Venkata Lakshmi[1], R. Hendra Kumar[2], K. Lakshmi Kala[3], P. D. Santhi[2], and Gurram Sunitha[4](✉)

[1] Department of CSE, Annamacharya Institute of Technology and Sciences, Tirupati, Andhra Pradesh, India
[2] Department of CSE, K M M Institute of Technology and Science, Tirupati, Andhra Pradesh, India
[3] Department of Mechanical Engineering, School of Engineering, Mohan Babu University, Tirupati, Andhra Pradesh, India
[4] Department of AI & ML, School of Computing, Mohan Babu University, Tirupati, Andhra Pradesh, India
gurramsunitha@gmail.com

Abstract. In image processing, removing speckle noise from images is crucial for enhancing image quality and facilitating accurate data analysis. Speckle noise is a grainy and disruptive interference. It poses challenges especially in remote sensing applications. To address this issue, a deep learning solution is proposed, centered on a Generative Adversarial Network (GAN). This GAN-based model is purpose-built to efficiently eliminate speckle noise while retaining essential image details. Its configuration includes statistical models for each layer, optimizing noise reduction and image clarity. The model's performance is compared to other deep learning models including ResNet, VGGNet, and U-Net. The results demonstrate the GAN's superior performance in speckle noise removal. Thus, underscoring importance of advanced deep learning approaches in addressing this persistent challenge. Also, the challenge of improving image quality in aerial imagery and remote sensing applications. The proposed model offers a promising solution for more accurate data interpretation in various contexts.

Keywords: Images · Denoising · Speckle noise · Deep learning · Generative adversarial learning

1 Introduction

Speckle noise in imagery is a type of noise that manifests as granular, salt-and-pepper-like interference [1]. It most particularly occurs in images obtained through radar and ultrasound technology. Unlike other types of noise, speckle noise is characterized by its distinctive appearance. It presents unique challenges in image processing. Its multiplicative and coherent nature makes it more difficult to address using conventional denoising methods [2].

A. Bajaj et al. (Eds.): IBICA 2023, LNNS 1229, pp. 462–471, 2025.
https://doi.org/10.1007/978-3-031-78940-3_49

Researchers have developed specialized techniques to effectively remove speckle noise. Traditional noise removal techniques have several limitations when applied to the task of removing noise, such as speckle noise, from imagery [3]. Traditional filters often blur or smooth images, leading to a loss of fine image details [4]. This can be problematic in imagery, where preserving critical details is essential. Traditional techniques lack adaptability, have limited effectiveness on complex noise patterns, and are dependent on manual parameter tuning. The noise removal vs detail preservation trade-off is tough [5].

Recent advances in imagery denoising have underscored pressing need for techniques that can enhance image quality. These studies have laid groundwork for improvement of images by reducing noise. The application of deep learning in this context holds substantial potential [6]. It offers a promising alternative to traditional denoising methods. Deep learning methods denoise speckle noise while preserving critical image details in aerial and radar imagery [7].

In light of these challenges associated with speckle noise in imagery, researchers have made noteworthy progress in this area using deep learning [8]. There have been continuous efforts in applying CNNs for speckle noise reduction in remote sensing imagery [9]. Researchers have acknowledged promise of CNNs in this context. Recent research highlights their potential to address challenges posed by speckle noise in imagery [10].

This paper proposes a deep learning framework for speckle noise removal from imagery. The proposed framework is based on a type of deep model called as Generative Adversarial Networks (GAN). GANs consists of a Generator and a Discriminator. They are often used for generating realistic images, but they can also be adapted for image denoising. In this context, Generator learns to generate denoised versions of noisy SAR images, while the Discriminator distinguishes between noisy and denoised images. This adversarial training can lead to high-quality denoised images. GANs can be fine-tuned to generate noise-free imagery by training Generator to produce cleaner images and Discriminator to distinguish between real and fake patches.

2 Related Work

Speckle noise is a common artifact in images acquired by radar and other sensors. It poses a significant challenge to remote sensing applications. Researchers have been actively engaged in developing effective noise reduction techniques to enhance the quality and interpretability of imagery [11].

Traditional approaches to speckle noise removal include filters like Lee, Frost, and Kuan. These techniques exploit statistical properties of the noise [12]. While being effective up to some extent, these methods may not fully preserve image details. This situation necessitates more advanced solutions.

Rigorous research has explored the domain of deep learning for improving quality of imagery [13]. Deep learning architectures are considered as effective tools for denoising images [14]. Their capabilities extend beyond conventional methods. They offer more adaptability, effectiveness, and the potential to preserve image details while removing noise.

A comprehensive overview of deep learning techniques for image de-noising is given in [7]. It discusses the growing influence of deep learning architectures and their potential to address speckle noise. Thus, presenting a significant shift from traditional noise reduction methods.

Recent advancements in deep learning have demonstrated substantial promise for speckle noise reduction. CNNs have received notable attention for their ability to generate realistic-looking images while effectively denoising them [15]. Recent research efforts have explored application of advanced deep learning techniques for removing various types of noises in images. Many researchers proved effectiveness of deep learning models in improving the quality and interpretability of radar-derived aerial images. These methods show significant promise in providing high-quality, denoised imagery, advancing the capabilities of remote sensing applications [16].

These studies have indicated that deep learning techniques, with their ability to learn and generate cleaner imagery through adversarial training, have the potential to significantly enhance the quality of imagery by mitigating the adverse effects of speckle noise.

A comparative review of image denoising techniques has shed light on the potential of deep learning in this field. While examining various methods for noise reduction, this survey emphasizes the role of deep learning in producing high-quality, denoised images. The review underscores the need for continued re-search and development in imagery denoising, with deep models at the fore-front of this endeavor.

3 GAN-Based Deep Model for Speckle Noise Removal from Imagery

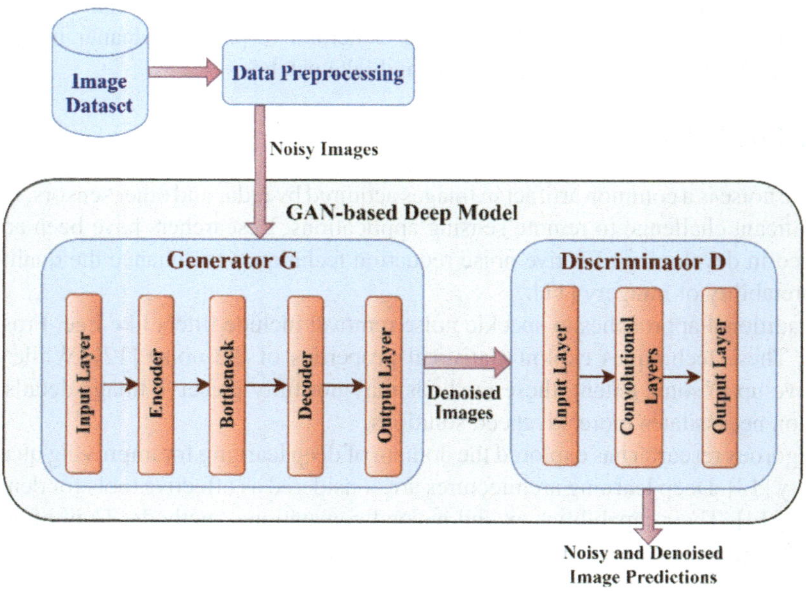

Fig. 1. Proposed GAN-based Model for Speckle Noise Removal from Imagery

The architecture of proposed GAN-based model for speckle noise removal from images is presented (Fig. 1). The GAN-based model comprises of two main modules – Generator (*G*) and Discriminator (*D*). In the GAN framework, these modules work in adversarial manner. *G* aims to generate high-quality denoised images from the noisy input. *D* learns to distinguish between real patches from clean images and fake patches produced by *G*. Adversarial training process helps *G* to improve its ability to remove speckle noise.

(1) Generator *G* – *G*'s role is to create data, in this case, denoised imagery. It takes noisy images as input and generates output images that are expected to be free of speckle noise. It learns to produce images that are as close as possible to real, noise-free images. It does this through a series of transformations and layers that map noisy inputs to cleaner outputs.

a) Input Layer: It received noisy image as input for denoising. It sets the initial size of the input image.

$$P(Input) = Gamma\,Distribution\,(shape,\ scale) \tag{1}$$

where *P(Input)* is probability distribution of input noisy image.

Gamma Distribution is often used to model noise in radar imagery.

b) Encoder: This series of layers encodes the input image's features. It learns to capture and represent relevant patterns in image using convolutional layers and leaky ReLU activation functions.

$$H(encoder) = ReLU(W * Input + b) \tag{2}$$

where *H(encoder)* is output feature map from encoder; *W* is weights, and *b* is biases of convolutional layers.

c) Bottleneck: This layer further compresses encoded features, resulting in a more compact representation. It aims to capture essential information while reducing the dimensionality.

$$H(bottleneck) = ReLU(W * H(encoder) + b) \tag{3}$$

where *H(bottleneck)* is compressed feature representation.

d) Decoder: It takes compressed features and gradually upsamples and expands them to reconstruct denoised image. It consists of linear transformations and non-linear activations, effectively transforming bottleneck's representation back into an image.

$$H(decoder) = ReLU(W * H(bottleneck) + b) \tag{4}$$

where *H(decoder)* is output feature map from decoder.

e) Output Layer: It provides the final denoised image. It uses a convolutional layer with a sigmoid activation function to ensure pixel values are in the range of [0, 1].

$$P(Output) = Sigmoid(W * H(decoder) + b) \tag{5}$$

where *P(Output)* is probability distribution of denoised image; Sigmoid ensures pixel values are in [0,1].

(2) Discriminator $D - D$'s primary function is to distinguish between real (clean) images and fake (generated) images. It takes both real, clean patches of images and generated patches as input and provides a probability score that represents how likely the input patch is to be real. It learns to classify images as either real or fake based on the features it extracts from them.

a) Input Layer: Input to this layer is a patch (small portion) of the image. This layer sets size of the image patch for classification. The input patch's statistical model corresponds to the pixel values in the patch. It follows a gamma distribution due to noise and may also be affected by the presence of real or fake patches.

$$P(Patch) = Gamma\_Distribution\,(shape,\ scale) \tag{6}$$

where *P(Patch)* is probability distribution of input patch.

Gamma_Distribution is used to model distribution of patches in real and fake images.

b) Convolutional Layers: These layers involve linear transformations followed by non-linearities (Leaky ReLU). The statistical model for these layers is a combination of linear transformations applied to the input patch.

$$H(conv) = LeakyReLU(W * Patch + b) \tag{7}$$

where *H(conv)* is output feature map from convolutional layers.

LeakyReLU is the Leaky Rectified Linear Unit activation function.

c) Output Layer: This layer uses a convolutional operation with a sigmoid activation. The statistical model here is a transformation of the features learned in the convolutional layers, mapping them to a value in the range [0, 1] to indicate the probability that the input patch is real (1) or fake (0).

$$P(Real) = Sigmoid(W * H(conv) + b) \tag{8}$$

where *P(Real)* is probability of input patch being real; *Sigmoid* ensures that output is between 0 (fake) and 1 (real).

Above statistical models represent mathematical relationships and transformations applied at each layer in the GAN-based architecture. These are specifically tailored for speckle noise removal in imagery. The specific values for the parameters *(shape, scale, W, b)* are learned during the training process.

G tries to create more convincing fake images to fool *D*, while *D* tries to become better at distinguishing real from fake images. As training progresses, *G* improves its ability to produce high-quality, denoised imagery. While *D* becomes better trained at differentiating between real and fake images. This process leads to better denoising results in *G* over time.

4 Experimental Results and Discussion

The dataset "Natural Images with Synthetic Noise" contains a collection of 3,000 images with synthetic noise artificially added to them. These images are available for access and download on Kaggle's platform [17]. The dataset includes a diverse range of natural images, encompassing various scenes, objects, and settings. Each image in the dataset has been deliberately augmented with synthetic noise. This artificial noise introduces variations in pixel values, mimicking the types of noise commonly found in real-world images, such as speckle noise, salt-and-pepper noise. The images with added noise exhibit varying levels of noise intensity. This dataset provides an opportunity to explore the effects of noise on image quality and to evaluate the performance of denoising techniques. The details of dataset are tabulated in Table 1. Figure 2 presents sample images with speckle noise.

Table 1. Specifications of Dataset

Item	Description
Data Type	Images
Provider	Flickr
Data Format	jpg
Image Size	512 × 512
Number of Images	3000
Types of Noise	Gaussian, Salt and Pepper, Poisson, Speckle
Speckle Noise Images	Training Images (#511) Validation Images (#149)
Number of Classes	15
Coverage	Natural
Availability	Public Domain (GNU)

Fig. 2. Sample Images with Speckle Noise

Figures 3 and 4 present the configuration of the proposed Generator G and Discriminator D. The input layer in both G and D are modeled using Gamma Distribution. It is a common choice for representing probabilistic distribution of noisy input images in speckle noise scenarios. Gamma distributions are suitable for modeling multiplicative noise, given the stochastic nature of speckle noise. LeakyReLU activation function is introduced to handle gradient-related challenges in the training process. Encoder, bottleneck, and decoder layers consist of standard convolutional and activation functions. Convolutional layers are used for feature extraction in image-related tasks. CNNs are

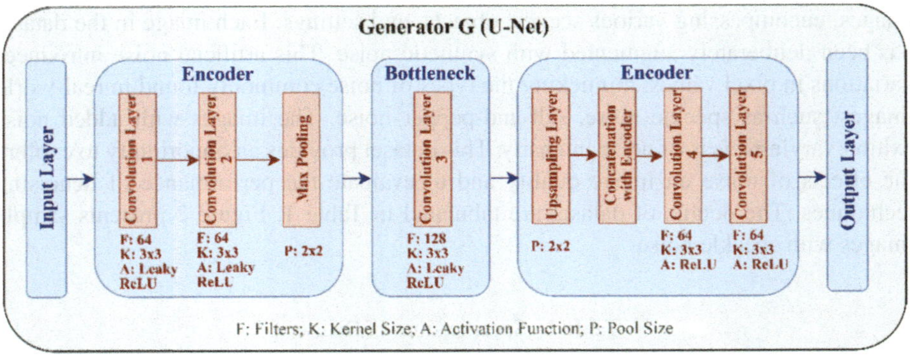

Fig. 3. Architecture of Generator G

well-suited for learning hierarchical features from images, and GANs are effective for adversarial training, producing denoised output while distinguishing between real and fake patches.

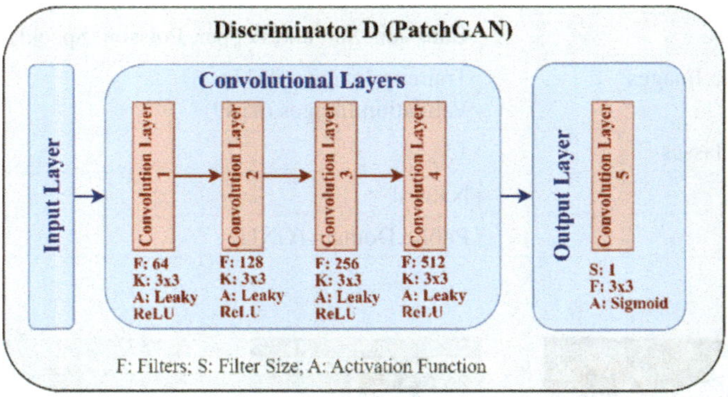

Fig. 4. Architecture of Discriminator D

Experiments were performed to prove efficiency of GAN-based deep model. Traditional CNN architectures U-Net, VGGNet, ResNet are implemented without the adversarial component to compare with the proposed GAN-based model. MSE, PSNR and ENL metrics are used for evaluating performance of models. Table 2 presents the experimental results. Figure 5 presents the MSE performance of the deep models over varying number of hidden layers.

In the context of speckle noise removal from imagery, several models were evaluated for their performance. Among these models, ResNet demonstrates moderate results, as evidenced by an MSE of 17.69, indicating a degree of reconstruction error, and a PSNR of 20.62. This suggests an improvement in image quality compared to the noisy input. The Equivalent Number of Looks (ENL) stands at 10.10, signifying enhanced image clarity. VGGNet model exhibits relatively lower performance with a higher MSE (19.45) and a reduced PSNR (17.56), indicating a decrease in image quality. The ENL

Table 2. Experimental Results of Deep Models for Speckle Noise Removal

Model	MSE	PSNR	ENL
ResNet	17.69	20.62	10.10
VGGNet	19.45	17.56	9.34
U-Net	17.32	23.27	10.26
GAN-based Model	**16.9**	**25.13**	**11.93**

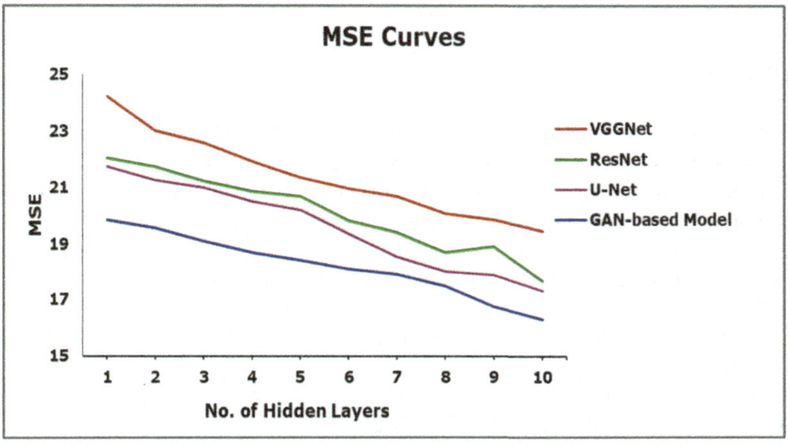

Fig. 5. Performance of Deep Models

is 9.34, reflecting a less substantial enhancement in image clarity. U-Net model excels with a low MSE (17.32) and a high PSNR (23.27), pointing to superior image quality and significant noise reduction. The ENL of 10.26 reinforces the model's efficiency in enhancing image clarity.

However, GAN-based model achieves the lowest MSE (16.9) and the highest PSNR (25.13), indicating good noise reduction capabilities and an impressive improvement in image quality. The ENL score is notably high at 11.93, emphasizing GAN-based model's efficiency in enhancing the clarity of imagery. GAN-based approach, with its use of both a generator and a discriminator in an adversarial training setup, excels in capturing complex noise patterns effectively. This, in turn, results in outstanding denoising outcomes, making it the preferred choice for speckle noise removal and the most efficient model among the options considered.

5 Conclusion

Removing speckle noise from imagery is essential for ensuring the clarity and reliability of images in remote sensing and analysis. The proposed GAN-based model, is configured for speckle noise removal. By leveraging the adversarial training paradigm, GANs have demonstrated the ability to remove speckle noise while preserving important features, contributing to clearer, more interpretable imagery. With the lowest MSE and highest PSNR, this model excels in noise reduction, significantly improving image quality. Its superior ENL score further underscores its ability to enhance image clarity. Comparative analysis against other models, including ResNet, VGGNet, and U-Net, highlights GAN's performance in speckle noise removal. This research emphasizes the importance of advanced deep learning techniques in addressing speckle noise, which is a persistent challenge in the field of aerial imagery and remote sensing. The proposed model offers a promising avenue for enhanced image quality and more precise data analysis in various applications.

References

1. Lee, J.S., Hoppel, K., Mango, S.A.: Unsupervised estimation of speckle noise in radar images. Int. J. Imaging Syst. Technol. **4**(4), 298–305 (1992)
2. Garcia-Sucerquia, J., Ramírez, J.A.H., Prieto, D.V.: Reduction of speckle noise in digital holography by using digital image processing. Optik **116**(1), 44–48 (2005)
3. Zaki, F., Wang, Y., Su, H., Yuan, X., Liu, X.: Noise adaptive wavelet thresholding for speckle noise removal in optical coherence tomography. Biomed. Opt. Express **8**(5), 2720–2731 (2017)
4. Saadia, A., Rashdi, A.: A speckle noise removal method. Circ. Syst. Signal Process. **37**, 2639–2650 (2018)
5. Tian, C., Fei, L., Zheng, W., Xu, Y., Zuo, W., Lin, C.W.: Deep learning on image denoising: an overview. Neural Netw. **131**, 251–275 (2020)
6. de Oliveira, V.A., et al.: Satellite image compression and denoising with neural networks. IEEE Geosci. Remote Sens. Lett. **19**, 1–5 (2022)
7. Ilesanmi, A.E., Ilesanmi, T.O.: Methods for image denoising using convolutional neural network: a review. Complex Intell. Syst. **7**(5), 2179–2198 (2021)
8. Rachapudi, V., Rao, K.S., Rao, T.S.M., et al.: Diabetic retinopathy detection by optimized deep learning model. Multimedia Tools Appl. **82**, 27949–27971 (2023). https://doi.org/10.1007/s11042-023-14606-8
9. Subramani, P., KS., B., K.R. et al.: Prediction of muscular paralysis disease based on hybrid feature extraction with machine learning technique for COVID-19 and post-COVID-19 patients. Pers. Ubiquit. Comput. **27**, 831–844 (2023). https://doi.org/10.1007/s00779-021-01531-6
10. Adilakshmi, J., Vinoda Reddy, G., Nidumolu, K.D., Cosme Pecho, R.D., Pasha, M.J.: A medical diagnosis system based on explainable artificial intelligence: autism spectrum disorder diagnosis. Int. J. Intell. Syst. Appl. Eng **11**(65), 385–402 (2023)
11. Balaram, A., Kannan, K.N., Čepová, L., Kumar, M.K., Rani, B.S., Schindlerova, V.: Artificial intelligence for media ecological integration and knowledge management. Systems **11**(5), 222 (2023). https://doi.org/10.3390/systems11050222
12. Alkawaz, A.N., et al.: Adaptive self-organizing map using optimal control. Mathematics **11**(9), 1995 (2023). https://doi.org/10.3390/math11091995
13. Tian, C., Xu, Y., Zuo, W., Zhang, B., Fei, L., Lin, C.W.: Coarse-to-fine CNN for image super-resolution. IEEE Trans. Multimedia **23**, 1489–1502 (2020)
14. Mehmood, A.: Image denoising using convolutional neural network. In: Pattern Recognition and Tracking XXXI, vol. 11400, pp. 22–32. SPIE (2020)

15. Ghose, S., Singh, N., Singh, P.: Image denoising using deep learning: convolutional neural network. In: 2020 10th International Conference on Cloud Computing, Data Science & Engineering (Confluence), pp. 511–517. IEEE (2020)
16. Laavanya, M., Vijayaraghavan, V.: Residual learning of transfer-learned AlexNet for image denoising. IEIE Trans. Smart Process. Comput. **9**(2), 135–141 (2020)
17. Natural Images with Synthetic Noise, Dataset of 3000 images with added synthetic noise. https://www.kaggle.com/datasets/tarunpathak/natural-images-with-synthetic-noise/data

Identification and Correction of Image Similarity using autoencoder

Karthik Kovuri[1], Nampally Vijay Kumar[2], Awatef Balobaid[3], A. V. Sriharsha[4(✉)], Bura Vijay Kumar[5], K. Reddy Madhavi[4], and A. Reni[6]

[1] School of Engineering and Technology, Kaziranga University, Rowmarikhuti, Assam, India
[2] Department of CSBS, B V Raju Institute of Technology, Narsapur, India
[3] Department of Computer Science, College of Computer Science and Information Technology, Jazan University, Jazan, Saudi Arabia
asbalobaid@jazanu.edu.sa
[4] School of Computing, Mohan Babu University, Tirupati, India
avsreeharsha@gmail.com
[5] School of Computer Science and Artificial Intelligence, SR University, Warangal, India
[6] Department of Food Processing and Preservation Technology, Avinashilingam Institute for Home Science and Higher Education for Women School of Engineering, Coimbatore, India

Abstract. In recent years, e-Health has evolved significantly due to advancements in IoT technology. This has led to the development of a system that offers improved remote healthcare services. The e-Health system based on slicing and partitions for delivering quality of service in fifth-generation. Thecurrent research study proposes an e-Health services embodied with IoT based Fog-Cloud architectures with enhancements of the quality of service with respect to the select slices, compared with the traditional medical and healthcare systems.

Keywords: IoT · e-Health · fog computing · cloud computing · 5G

1 Introduction

The Internet of Things (IoT) and e-Health services orchestrate and had transformed the traditional medical community. The Internet of Things (IoT) describes the capacity of connectivity, transmission and reception by any physical object's [1]. Cloud computing integration provides benefits and varied categories of applications, enhancing the limited capacity of computing power and storage, furthermore assist in representation of virtualization, analysis and storage about the gathered real-time patient data when used in the context of healthcare, where the e-Health services can be deployed for the purpose of remote diagnosis and treatment naming it as 'any time health service' [2, 3]. Hospitalizations, healthcare expenses, and the burden on medical staff will all decrease as a result [4].

Additionally, real-time sensor data could be transferred to enable cloud services monitor the emergency situations of the patient well in advance, which enhances the quality of service in healthcare [5]. Because many e-Health services are emergency and

A. Bajaj et al. (Eds.): IBICA 2023, LNNS 1229, pp. 472–479, 2025.
https://doi.org/10.1007/978-3-031-78940-3_50

time-critical in nature, the latency during the collection of sensory vital data through IoT devices, responding to the data centres managed in cloud (CCDC) is unsustainable. For instance, as it is fact to accept that delay due to missing times in the alarms, particularly during emergency. In the order of solution to this, the topological gap between the fog nodes as IoT devices and data centres of cloud is reduced.The quick response time shall have a positive impact on the sensitive attention of e-Health services, as the real-time processing, low latency are the considerable parameters [2, 7]. Decentralized fog/edge analytics also alleviates the payload on the cloud servers and networks, thus improving the privacy and security to health data.

Profitable network features are needed for IoT in healthcare. Low latency, high bandwidth, high reliability, and a long battery lifetime are the main performance characteristics of smart healthcare services [8]. Internet basedmedical consultations; monitoring health and unattended surgeryare the three main e-Health services trends [9]. To gain the advantages ofsystem performance, each service must meet specific requirements, where sensory data of a patient is treated as most priority for the experts in the medical institutions. Internet-based health monitoring transforms amedical institution into a virtual environment, making the process of affixing medical sensors to patients essential to real-time body health streaming, analyses, control and monitoring. Healthcare professionals in a medical institutioninvestigate the patients'statusesthat are registered with them. Because of the basic performance parameters like sensitivity and high reliability, the latency surges while sending and receiving data, where remote surgical operations required tactile application requirements.However, given that it might affect human life, any data loss can be significant.

The study's e-Health system architecture includes IoT sensors of health, fog nodes and the data centres at cloud. Three subsystems evolved logically by network slicing in Fifth-generation (5G). According to QoS, each system is used for specific healthcare services. We estimate main performance metrics for each subsystem using a queuing network. The objective of the current study is to improvise quality of service and performance of aeHealth system.

2 Related Works

The IoT devices of healthcare systems are examined thoroughly in the study. The authors of [10] has appreciated the slices of 5G network with use cases: massive IoT, critical communications and eMBB, eV2X. Various types of works are introduced into slices with different requirements of QoS parameters. An eldery-geriatric care system based on domestic setup uses only two network slices, according to the authors [11], the health care slice and the caregivers slice, building into a smart home slice, improvising the geriatric security and efficiency. Privacy preserving slice is an important selection, where the IoT device proves its efficacy and security, opined in [12].

The smart diabetes system using 5G is proposed by [13] using three layers, a sensory layer to collect the vital information of the patient, diagnostic layer to analyze data with computational intelligence algorithms,finally data layer to share socio-vital information to proceed with the treatment and further providing supervisory guidelines.

Each cell of a mobile network is constituted into a basic communicational element with respect to a radio base station. Tower companies own the sites and telecoms own the equipment for these base stations. Fog nodes such as mobiles, tablets, IoT devices and other devices of each cell communicate voice and data contingencies together through a nearest base stations,and furthered to network cores through using cables to an edge router.

A centralized cloud with a Radio Access Network is a specialized architecture called C-RAN. This technology will be crucial to 5G network deployment. RAN deployment traditionally uses proprietary hardware with embedded software to implement network functions. A Cloud RAN uses NFV to virtualize these functions and provides a platform is provided for deployment on the physical infrastructures, just akin to the servers in mobile edge computing.

Collaborating and offloading fog nodes reduced IoT node service delay. The fog nodes processed or offloaded requests to neighboring fog nodes with very feebly counted requests or cloud servers in the proposed policy. In [14] the researchers showed a work akin to reduce consumption of energy influencing the service delay for medical and healthcare systems. A smartphone with 5G enhancements utilized as a getaway processing the requests of the device while mobile edge computing is offloadedin the proposed architecture. [15] is another noteworthy work. A cloud–fogqueueing architecture for IoT nodes was presented in this paper. Their work focused on providing a dynamic scaling algorithm to handle increased middleware layer workload (fog and cloud nodes). At every "T" interval, the algorithm calculated and monitored key performance parameters. If the value exceeded a threshold, fog nodes were added. In [16], the authors proposed a fingerprint mechanism for healthcare applications. Customizing resources and meeting healthcare application requirements with this technology increased consistency. The studies on fog computing in healthcare services determine the aspects of high-response time, low latency and real-time responses. In [17, 18], authors addressed healthcare IoT privacy. Utilising a mutual privacy-preserving with k-means strategy, the cloud plays a role of third party component in the model to streamline communication.

3 Architecture

The fog servers are responsible to communicate the sensitive and socio-vital information to the components upholds the specialized services with important parameters like high-band width and low latency. An array of servers [19, 20]were deployed in the fog services in order to fulfill the objectives of the network fog computing. The data stored is processed to serve the emergencies, where fog nodes involve in transmitting the same to the CDC servers. The fog slice contains a 'C' server, representing the patient data. The diagnosis accuracy and real-time patient attendance is assured by the certified physicians and care takers (Fig. 1).

Here, the entire system and each subsystem are assessed. The suggested queueing model's key performance statistic and formulae are established. Throughput, CPU usage, average packet requests, average packets waiting in queue, average response time, and gross loss rate are among the performance measuring vectors in this section.The total system cost is computed considering the waiting time and the cost of the system. The proposed queueing system network divides the e-Health services into three classes, each with three subsystems.

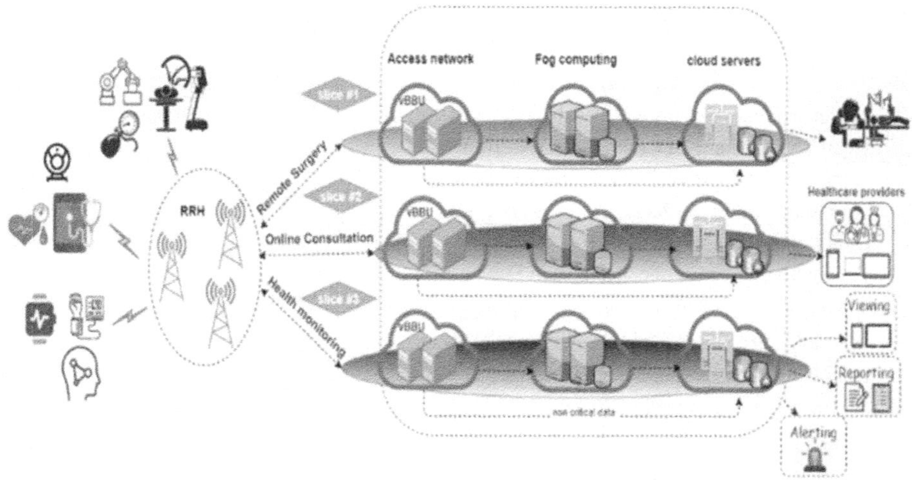

Fig. 1. A model architecture for H-IoT

4 Performance Analysis

The cloud-RAN presents the performance with respect to the computed slices. The throughput in absolute terms of the H-IoT system is proposed in a queuing model as:

$$T = T_c + T_m + T_d. \tag{1}$$

The mean of the response time in the three subsystems is computed into the mean response time of the total system. The mean of the waiting time in the three subsystems is computed into the mean waiting time of the total system. The mean number of packets of each subsystem is computed into total number of payload (packets) in the request are the waiting packets in the queue. Hence, the probability of blockage and the count of requests in the subsystem are calculatedshown below:

$$P_{block} = P_K + P_M + P_C \tag{2}$$

And the cost obtained for the total system is computed as

$$Cost_s = C_{cran}.S_{cran} + C_{fog}.S_{fog} + C_{cpu}.S_{cpu} \tag{3}$$

4.1 Simulation Parameters

A numerical example is considered for the analysis of the total system. The simulation parameters used for queuing are as follows:

The proposed queuing model was validated and improved performance measures through three scenarios. The results showed similar results for the virtual BBU cores, fog cores, and overall system cost, confirming the model's efficiency and validation (Table 1).

Table 1. Queuing parameters used in simulation

Description	Values
Packet arrival rate (λ)	[100,000 to 1000,000 packets]
Mean Service Time of Cloud-RAN (1 / μc)	0.0001 (s)
Mean Service Time of vServers in Fog (1 / μm)	0.00001 (s)
Minimum number of Packets in each core (K)	300
Total number of Fog cores in Fog system (F)	5
Maximum number of packets in each Fog core (M)	200
Probability of packets leaving from Fog core (qm)	0.4
Total number CPU cores in the subsystem (x)	20
Total number of vCPU in each core (t)	10
Maximum number of packets in each core (c)	500

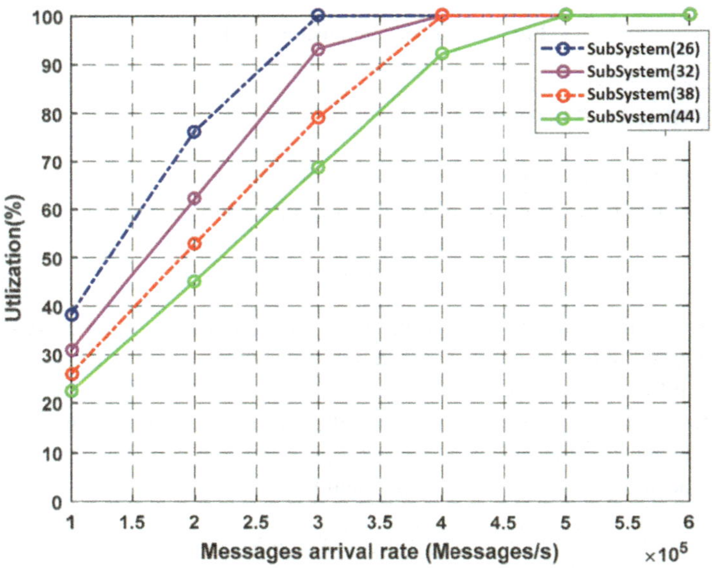

Fig. 2. Impact of Sub System cores on CPU Utilization.

The virtual subsystem cores in the experiment assess impact based on system performance indices like CPU utilization, throughput, response time, waiting time, number of requests, and system loss, with varying arrival rates.

In Fig. 2, CPU utilization is influenced by the number of virtual subsystem cores. With respect to the arrival rate increase or the load, CPU workload increases. Using virtualization and subsystem cores allows for larger requests processing, reducing CPU utilization. Using fewer cores leads to maximum CPU capacity and 100% usage, causing performance deterioration.

Figure 3illustrates the impact in the cores of virtual Subsystem with regard to the system response time. Using fewer processing units leads to faster overload, while more cores decrease response time. Using 26 cores with 300k packets/s increases response time from 0.22 to 1.28 ms. Time-sensitive services in eHealth applications require minimal response time.

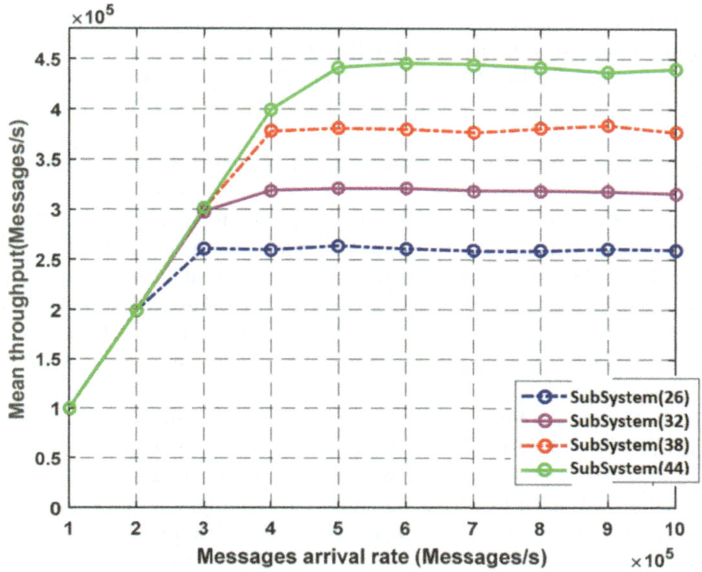

Fig. 3. Impact of Sub System cores on CPU Utilization.

The CPU's cores increased the average number of requests, resulting in a higher number of processing units operating at a higher arrival rate. The arrival rate of a system core significantly impacts the system cost, with waiting time and service costs influencing the overall system cost.

5 Conclusion

A 5G network-sliced architecture for eHealthIoT systems is shown in this study. The proposed paradigm included Cloud-RAN and fog for each eHealth slice. The queue model was used to calculate throughput, performance metrics, CPU utilisation, system blocking rate, mean number of packet requests, mean time of waiting and mean time for reponse. Aempirical example validates the queueing model and shows the nature of the system that approximates the number of computing resources required in subsystems. Simulation and analytical equation results were similar. Virtual SubSystem and fog nodes can improve eHealthQoS. AI will optimise resource requirements for system QoS in different environments in future work.

References

1. Firouzi, F., Farahani, B., Marinšek, A.: The convergence and interplay of edge, fog, and cloud in the AI-driven Internet of Things (IoT). Inf. Syst. **107**, 101840 (2022)

2. Singh, P., Singh, R.: Energy-efficient delay-aware task offloading in fog-cloud computing system for IoT sensor applications. J. Netw. Syst. Manag. **30**, 1–25 (2022)
3. Muhammad, G., Rahman, S.M.M., Alelaiwi, A., Alamri, A.: Smart health solution integrating IoT and cloud: a case study of voice pathology monitoring. IEEE Commun. Mag. **55**, 69–73 (2017)
4. Bergmo, T.S.: How to measure costs and benefits of ehealth interventions: an overview of methods and frameworks. J. Med. Internet Res. **17**, e254 (2015)
5. Qadri, Y.A., Nauman, A., Bin Zikria, Y., Vasilakos, A.V., Kim, S.W.: The future of healthcare internet of things: a survey of emerging technologies. IEEE Commun. Surv. Tutor. **22**, 1121–1167 (2020)
6. Yousefpour, A., et al.: All one needs to know about fog computing and related edge computing paradigms: a complete survey. J. Syst. Arch. **98**, 289–330 (2019)
7. Cao, H., Wachowicz, M., Renso, C., Carlini, E.: An edge-fog-cloud platform for anticipatory learning process designed for internet of mobile things. arXiv: arXiv:1711.09745 (2017)
8. Ahad, A., Tahir, M., Yau, K.-L.A.: 5G-based smart healthcare network: architecture, taxonomy, challenges and future research directions. IEEE Access **7**, 100747–100762 (2019)
9. Ullah, H., Nair, N.G., Moore, A., Nugent, C., Muschamp, P., Cuevas, M.: 5G communication: an overview of vehicle-to- everything, drones, and healthcare use-cases. IEEE Access **7**, 37251–37268 (2019)
10. Chilakala, S., Saxena, A., Vunnava, A.: Energy-efficient reporting scheme for cognitive radio networks. IET Netw **10**, 110–117 (2021). https://doi.org/10.1049/ntw2.12019
11. Vadivel, S., Konda, S., Balmuri, K.R., Stateczny, A., Parameshachari, B.D.: Dynamic route discovery using modified grasshopper optimization algorithm in wireless ad-hoc visible light communication network. Electronics **10**(10), 1176. (2021). https://doi.org/10.3390/electronics10101176
12. Premalatha, B., Srikanth, G., Abhilash, G.: Design and analysis of multi band notched MIMO antenna for portable UWB applications. Wireless Pers. Commun. **118**, 1697–1708 (2021). https://doi.org/10.1007/s11277-021-08111-0
13. Koyi, L.P., Borra, T., Prasad, G.L.V.: A research survey on state of the art heart disease prediction systems. In: 2021 International Conference on Artificial Intelligence and Smart Systems (ICAIS), Coimbatore, India, pp. 799–806 (2021). https://doi.org/10.1109/ICAIS50930.2021.9395785
14. Cherukuri, N., Bethapudi, N.R., Thotakura, V.S.K., Chitturi, P., Basha, C.Z., Mummidi, R.M.: Deep learning for lung cancer Prediction using NSCLS patients CT Information. In: 2021 International Conference on Artificial Intelligence and Smart Systems (ICAIS), Coimbatore, India, pp. 325–330 (2021). https://doi.org/10.1109/ICAIS50930.2021.9395934
15. Samdekar, R., Ghosh, S.M., Srinivas, K.: Efficiency enhancement of intrusion detection in IoT based on machine learning through bioinspired. In: 2021 Third International Conference on Intelligent Communication Technologies and Virtual Mobile Networks (ICICV), Tirunelveli, India, pp. 383–387 (2021). https://doi.org/10.1109/ICICV50876.2021.9388392
16. Dinesh, G., Venkatakrishnan, P., Jeyanthi, K.M.A.: Modified spider monkey optimization—an enhanced optimization of spectrum sharing in cognitive radio networks. Int. J. Commun. Syst. **34**, e4658 (2021). https://doi.org/10.1002/dac.4658
17. Arunachalam, P., Venkatakrishnan, P., Janakiraman, N.: Histopathology image classification for soft tissue sarcoma in limbs using artificial neural networks. In: 2021 6th International Conference on Inventive Computation Technologies (ICICT), Coimbatore, India, pp. 778–785 (2021). https://doi.org/10.1109/ICICT50816.2021.9358635
18. Arunachalam, P., Venkatakrishnan, P., Janakiraman, N.: Detection of singularity in the cell nucleus of synovial sarcoma using wavelet leaders. In: 2021 6th International Conference on Inventive Computation Technologies (ICICT), Coimbatore, India, pp. 418–425 (2021). https://doi.org/10.1109/ICICT50816.2021.9358517

19. Tiwari, V., Naidu, A.S., Singh, A., Saxena, A.: Construction of 4×4 MDS matrices for lightweight block ciphers. In: 2021 IEEE International Conference on Advanced Networks and Telecommunications Systems (ANTS), Hyderabad, India, pp. 254–259 (2021). https://doi.org/10.1109/ANTS52808.2021.9936970
20. Pala, V.C.R., Kamatagi, S., Jangiti, S., Swaraja, K., Madhavi, K.R., Kumar, G.N.: Yoga pose recognition with real time correction using deep learning. In: 2023 International Conference on Sustainable Computing and Data Communication Systems (ICSCDS), pp. 387–393. IEEE (2023)

Paddy Leaf Disease Detection Using Convolutional Neural Network

B. Naga Vardhani[1(✉)], K. Krishna Sri[2], S. D. Gousi Majahar[2], V. Siri Lakshmi[2], and A. Jyothika[2]

[1] CSE Department, Vignan's Nirula Institute of Technology and Science for Women, Guntur, India
Vardhani467@gmail.com

[2] Department of Computer Science and Engineering, Vignan's Nirula Institute of Technology and Science for Women, Guntur, India

Abstract. Rice is most basic food consumed by billions of people worldwide. it is one of the most widely grown cultivation, and leaf pathogens may possess a significant influence on yield and adequacy. Plant disease spread has increased dramatically in recent years. Plant diseases might be fungal, bacterial or viral and can cause harm to crop. The most well-known and devastating illness of paddy leaves are Black Spot Disease, Blast Disease, and Bacterial Leaf Blight, of which bacterial blight and black spot have been detected in this project. The spread of diseases to crops, as well as loss to farmers, can be significantly reduced if these diseases are identified quickly and accurately at earlier stages. The most critical element is the identification of rice leaf ailments, that possess a significant effect against financial system and food shortages. Human researchers estimate phytopathogens visually, and microscopic examination of morphological characteristics to recognize diseases, are the methods that are commonly used for the diagnosis and Crop diseases identification. This traditional procedures of manually diagnosing is time-consuming and not accurate.

Keywords: Convolutional neural networks · LeNet-5 · Tensor Flow · MATLAB · pre-processing · feature extraction · pooling · convolution layer · fully connected layer · classification · detection

1 Introduction

The finest occupation in India is farming. India's economy is heavily reliant on agriculture production. Agricultural output is one of the most significant factors that influence domestic economic conditions in our nation. About 65 percent of the populace either directly or indirectly works in agriculture or agricultural goods. Rice is One of the staple foods ingested around the globe. However, many paddy leaf Ailments are a risk to production of rice. The identification of rice leaf diseases is often quite time-consuming and challenging for the farmers due to the lack of accessibility to professionals. Even if there are experts in some disciplines, it may be difficult to detect sickness with the unaided eye, which occasionally leads to incorrect recognition. The adaptability, reliability, and

A. Bajaj et al. (Eds.): IBICA 2023, LNNS 1229, pp. 480–488, 2025.
https://doi.org/10.1007/978-3-031-78940-3_51

availability of agronomic information and equipment are still significant impediments to increasing agricultural production among nation's smaller firms. As a result of increasing population, changeable weather patterns, and ideological unpredictable nature, agro - based organizations started looking for modern digital, cost-effective innovations. Using an automated method, these problems may be minimised. This study suggests using an automated method to diagnose two common ailments (black spot and Bacterial blight). Such paddy leaf disease is detectable using Convolution Neural Network (CNN). Finding the plant's illness so that preventative measures may be made to protect it is the best way to solve the issue. This project applies the idea of using convolutional neural networks to recognize leaf aliments in rice plants. The pictures of paddy leaves were obtained out from UCI Machine Learning Database and a nearby farm. A python script was used to install the images. The obtained data source contains approximately 272 pictures divided into two categories. By default, every one of the installed and caught pictures relates to the RGB colour scheme and was saved in the unencrypted Digital format. Deep learning models that are most often employed to analyse visual data are convolutional neural networks (CNN/ConvNet). Matrix calculations generally come to our minds when we think of neural network models. It makes use of a practice known as convolution. Convolution is a mathematical method that utilizes two major functions and generates a quantifiable result that showcases how the structure of one feature is tweaked by the other. Deep learning is a type of machine language which applies self-learning strategies to massive data sets. The most recent advancements in technology and big data analytics have increased the usefulness and practicability of this approach thanks to its quick system. CNN, one of the deep learning algorithms, is used to quickly analyse and pre-process images utilizing of image processing techniques.

2 Related Work

The k-means clustering technique is employed to identify as well as segment the grey images. The Principal component analysis algorithm is employed to extract the characteristics [1]. It [2] is employing the optimized OMSVM technique. The suggested system's effectiveness is calculated using metrics such as overall accuracy, exactness, as well as false positive rates. For categorization, they employed the data sets of healthy leaves, black star, cedar rust and apple grey spot disease. For comparison and improvement, res-net models and SVM classifiers for image segmentation were utilized. The grey scale symbosis matrix is employed to obtain structural texture characteristics, contrast, correlation, energy and uniformity, mean standard deviation, entropy, etc. To identify the rice infections and nutrient deficiencies, MATLAB-based image processing [3] is used. Specifically, they have focused on two crucial vitamins: magnesium and nitrogen. The input picture is divided into three clusters with the cluster holding the Region Of interest (ROI) being selected. Raspberry Pi, DHT11 temperature and humidity sensors make up the hardware and solenoid valves, which measure humidity.

In [4] Grape leaf disease categorization using the SVM categorization strategy. The ailment ridden area is initially detected utilizing segmentation by flat clustering, and then shade and composition characteristics are reclaimed. Consequently, a categorization approach is employed in order to determine the category of plant diseases. In paper [5] they used Image segmentation approach for classifying two forms of orchid leaf disease, black leaf spot and sun burn. Selected orchid leaves are analyzed in relation to the location of interest utilizing [16] MATLAB border segmentation algorithms. In this case, filtering method, and morphological processing method will be used on the photographs. In this paper [6] Color slicing is the major part in this technique which sieve and display the color and pattern of paddy leaf. In this project [7] a novel approach for mechanized paddy leaf pathogen screening employing enhanced fuzzy logic. In this paper [8] The photograph is taken out from plants. The images are processed with a series of feature extraction mechanism using the DWT (discrete wavelet transform) and GLCM (grey level co-occurrence matrix). Paper [9] purpose of this research is to help in the identification and categorization of grape leaflet ailments using the Support Vector [17] Machine approach.

3 Proposed System

Unlike any other current paddy crop disease detection approaches, we built a model that operates in a fraction of the time and provides more accuracy for paddy crop disease diagnosis using convolutional neural networks. CNN based technique, which is a Deep neural networking methodology that uses an picture as input and prioritizes and distinguishes numerous other items in the image. A CNN, is a sort of ANN that is frequently utilized in Deep Learning for image/object detection as well as categorization. A CNN requires far less pre-processing than other classification algorithms. The benefits of utilizing CNN are that it automatically discovers essential characteristics without supervision, that it is computationally efficient, and that it is the most popular topic in the pool of deep learning, which is truly extremely large, and that this is typically due to the ConvNet0s.

Figure 1 represents the layers of Convolutional neural networks. It comprises a single input layer, three hidden layers and one output layer. A feed-forward system is the process of distributing data in a network from one layer to the next. The input layer takes the picture of paddy leaf as source and output layer gives the result identifying weather the leaf is diseased or healthy. if diseased it outputs the label.

The convolution operation is the major aspect of the CNN. It bears the majority of the network's computation burden. This is the initial layer used to draw out the different attributes from the source photograph. The above layer applies the numerical function of convolution between the input image and a MxM sieve. Dragging the sieve throughout the source images yields the dot product in between sieve and the parts of the source images. The result is known as the Feature map, and It includes data about the picture,

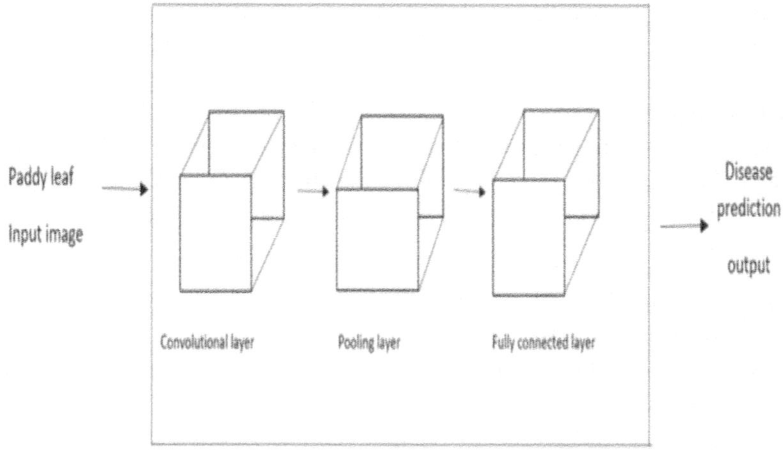

Fig. 1. Layers of CNN

for instance its edges and margins. This subset of features is then transferred on to succeeding layers, which discover various discrete characteristics deriving out of the source image. The layers of a CNN will be equivariant to translation due to property sharing. It argues that if we adjust the source, the result will likewise alter in an identical manner.

The total sieves contained in the first convoluted phase is 32, preceded by 64 and then 128 in the following block. This rise in the total number of sieves is mandatory to accommodate the reducing in dimensions of feature space. The feature map is additionally padded with zero to maintain the picture dimensions once the convolution process is applied. The scoop pooling layer is utilized to reduce the dimensions of the feature map, accelerate the technique, and make the methodology sensitive to a small extent for slight modifications in source.

Dataset Collection

The suggested system is first created utilizing the data set of sick paddy leaves. The entire dataset is then separated into two categories: Learning dataset and hold-out dataset.

Data Pre-processing

Image processing is a mechanism for making fundamental methods to enhance a picture after processing it by extracting various properties and values from it. The photos also contain noise as a result of poor lighting, weather occlusion, and other factors. Images are resized to a standard height and width to decrease computing complexity.

Feature Extraction

Image features give a wealth of information about the image's content. These qualities indicate various traits that may be utilized to distinguish across input pattern groups. Texture and color were utilized in this project. Image characteristics are used for categorization. Image texture is often expressed by contrast, homogeneity, entropy, and so on. Gray-level Co-occurrence matrix is a statistical tool for evaluating the texture of a picture (GLCM).

CNN Classification
In this study, Convolutional Deep Networks is used to distinguish between diseased and healthy leaves, as well as to detect illness in affected plants. CNN is a higher-level language of DL owing of its high precision. CNN is utilized for picture Identification and classification.

Disease Prediction
The final part of the procedure is determining whether or not the plant is healthy. If any faults are discovered in the picture, the output will provide the name of the specific illness diagnosed in the leaf.

Algorithm: Paddy leaf disease detection using Convolutional Neural Networks

Input: - The input image after pre-processing.
Output: - Final classification result.
Step 1: - Collecting the database and categorizing it according to the leaf characteristics.
Step 2: - Taking an image as the test image/ input image.
Step 3: - Conversion of the image in to arrays and storing the data
Step 4: - Training the images from datasets for the CNN classification by applying CNN classification algorithm
Step 5: - Comparison of these both test and train images.
Step 6: - Disease in the leaf is detected.

4 Results and Discussion

Over 300 training epochs, a maximum validation accuracy of 100% was achieved, and a maximum training accuracy of 94.4% was also recorded. A validation accuracy of 92.3% on average has been attained. This is a useful indicator of how well the deep learning model classified things. It is clear that the model possesses stabilized after 100 epochs, and the metrics do not indicate a noticeably higher accuracy of the model between 89% and 94%. Accuracy is somewhat down and up at a given range after 150 epochs. The outcomes demonstrate that the model works well on the dataset and can be applied to classify the two paddy leaf diseases with the least amount of resources needed.

Table 1 displays the partitioning of sample group. It is composed of 154 train images and 84 validation images of three diseases Bacterial blight, Brown spot and blast.

Table 1. Division of data set

Disease	Train	Validation	Total
Bacterial blight	50	43	93
Brown spot	79	17	96
Blast	25	24	49

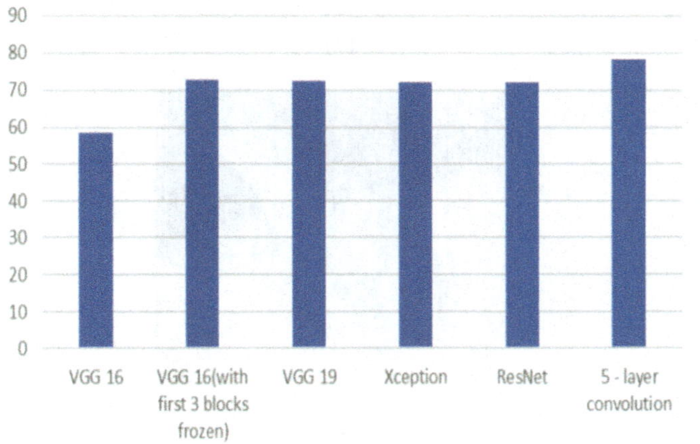

Fig. 2. Accuracy-based performance evaluation of several DL algorithms

Figure 2 shows the performance of several DL algorithms taking accuracy as the parameters. The graph consists of comparison between vgg16, vgg16 with first 3 frozen blocks, vgg16 xception, Resnet and 5 layer convolution. Among these DL algorithms 5 layer architecture has the highest accuracy and vgg16 has the lowest.

Figures 3(a), 3(b) and 3(c) displays the resultant screengrabs of rice illnesses brown spot, bacterial blight and blast diseases present in the leaflet.

(a)

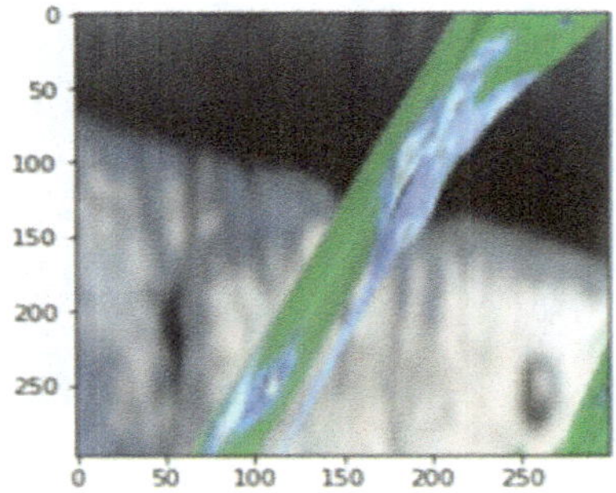

(b)

(c)

Fig. 3. (a) output screenshot of brown spot (b) output screenshot of blast (c) output screenshot of bacterial blight

5 Conclusion

The agricultural industry continues to be among the most crucial sectors on which the majority of people rely. Thus, spotting illnesses in these crops is crucial to the economy's growth. One of the major crops that is produced in enormous amounts is paddy. We suggested this technique to assist farmers in increasing production and improving agricultural systems. This technology can detect illnesses in paddy crop. As a result of this Agriculture-related technology will have a significant impact on the Indian economy. In this research, we collected healthy and sick plant leaf photos to train a CNN model to identify plant illnesses. The data utilized for model development and evaluation in this study was gathered from various online sources. CNN techniques are frequently employed to predict plant ailments. It has overcome the issues associated with standard object identification and categorization approaches. Furthermore, the recent study found that CNN is effective at detecting plant disease. The purpose of this project is to detect and identify 3 entirely distinct illnesses in the rice crop. The proposed technique classifies paddy leaf diseases using a CNN model. To categorize A straightforward CNN with a limited number of layers may be used as the design to divide paddy leaf diseases into two separate categories. These models are robust in the face of difficult situations such as images with a complex backdrop, non-uniform size, symptom variability, and inadequate datasets.

References

1. Qi, H., Liang, Y., Ding, Q., Zou, J.: Automatic identification of peanut-leaf diseases based on stack ensemble. Appl. Sci. **11**(4), 1950 (2021)
2. Nidhi, A.D., Prabhu, C.N.V., Reddy, K.C., Deepa, K.: Cluster based paddy leaf disease detection, classification and diagnosis in crop health monitoring unit. Comput. Aided Intervention Diagn. Clin. Med. Images **31**, 281–291 (2019)
3. Bikku, T.: Multi-layered deep learning perceptron approach for health risk prediction. J. Big Data **7**(1), 1–14 (2020)
4. Bikku, T., Paturi, R.: Frequency domain steganography with reversible texture combination. Traitement du Sig. **36**(1), 109–117 (2019)
5. Bikku, T., Rao, N.S., Akepogu, A.R.: Hadoop based feature selection and decision making models on big data. Indian J. Sci. Technol. **9**(10), 1–6 (2016)
6. Khin, T., Srujan Raju, K., Sinha, G.R., Khaing, K.K., Kyi, T.M.: Review of optimization methods of medical image segmentation. In: Raju, K., Govardhan, A., Rani, B., Sridevi, R., Murty, M. (eds.) Proceedings of the Third International Conference on Computational Intelligence and Informatics. Advances in Intelligent Systems and Computing, vol. 1090. Springer, Singapore (2020). https://doi.org/10.1007/978-981-15-1480-7_17
7. Mogili, R., Narsimha, G., Srinivas, K.: Early prediction of non-communicable diseases using soft computing methodology. In: Satapathy, S.C., Raju, K.S., Shyamala, K., Krishna, D.R., Favorskaya, M.N. (eds.) Advances in Decision Sciences, Image Processing, Security and Computer Vision. Learning and Analytics in Intelligent Systems, vol. 3. Springer, Cham (2020). https://doi.org/10.1007/978-3-030-24322-7_82
8. Patra, R.K., Raja, R., Sinha, T.S., Mahmood, M.R.: Image registration and rectification using background subtraction method for information security to justify cloning mechanism using high-end computing techniques. In: Raju, K., Govardhan, A., Rani, B., Sridevi, R., Murty, M. (eds.) Proceedings of the Third International Conference on Computational Intelligence and Informatics. Advances in Intelligent Systems and Computing, vol. 1090. Springer, Singapore (2020). https://doi.org/10.1007/978-981-15-1480-7_77
9. Joshi, A., Choudhury, T., Sai Sabitha, A., Srujan Raju, K.: Data mining in healthcare and predicting obesity. In: Raju, K., Govardhan, A., Rani, B., Sridevi, R., Murty, M. (eds.) Proceedings of the Third International Conference on Computational Intelligence and Informatics . Advances in Intelligent Systems and Computing, vol. 1090. Springer, Singapore (2020). https://doi.org/10.1007/978-981-15-1480-7_82

10. Khaing, K.K., Srujan Raju, K., Sinha, G.R., Swe, W.Y.: Automatic temperature control system using arduino. In: Raju, K., Govardhan, A., Rani, B., Sridevi, R., Murty, M. (eds.) Proceedings of the Third International Conference on Computational Intelligence and Informatics . Advances in Intelligent Systems and Computing, vol. 1090. Springer, Singapore (2020). https://doi.org/10.1007/978-981-15-1480-7_18
11. Gopi, A.P., Jyothi, R.N.S., Narayana, V.L., Sandeep, K.S.: Classification of tweets data based on polarity using improved RBF kernel of SVM. Int. J. Inf. Technol., 1–16 (2020)
12. Srivastava, S., et al.: The impact and potential of block chain applications based on IoT. In: 2021 International Conference on Simulation, Automation & Smart Manufacturing (SASM). IEEE (2021)
13. Chaitanya, K.: An efficient handoff scheme for uninterrupted assorted wireless networks in overlapped areas Chaitanya. J. Crit. Rev. **7**(4), 825–828 (2020)
14. Bikku, T., et al.: Machine learning identification of organic compounds using visible light. J Phys. Chem. A **127**(10), 2407–2414 (2023)
15. Bikku, T.: Fuzzy associated trust-based data security in cloud computing by mining user behaviour. Int. J. Adv. Intell. Paradigms **25**(3–4), 382–397 (2023)

IoT Based Cotton Yield Prediction Based on LSTM Model in Tropical Coastal Regions

Satya Sandeep Kanumalli(✉), G. Chandralekha, K. Niharika, J. Kavya, and B. Sruthi

Department of Computer Science and Engineering, Vignan's Nirula Institute of Technology and Science for Women, Guntur, India
satyasandeepk@gmail.com

Abstract. Soil is one of the most important resources that we must carefully manage. The fertility of the soil affects crop output by around 60%. As a result, regular soil testing is required. Many errors are committed when fertilizing and planting. To ascertain the pH level and nutritional content of the soil, soil tests are utilized. Therefore, deep learning (DL) is important since it has tools for supporting decisions related to cotton crops, as well as tools for supporting decisions related to crop yield prediction. Here, we must use sensors to test the N, P, and K values. The LSTM Model training was carried out using a deep learning algorithm., with a focus on crop yield sequence prediction issues. The findings collected demonstrate that the classification of cotton crop yield using the LSTM algorithm is equally accurate. Then, for cotton that is irrigated, the fertilizer dose is 100:50:50(NPK)kg/ha; for cotton hybrids that are fed by rain, it is 80:40:40(NPK)kg/ha.

Keywords: Deep learning · IoT · LSTM · Cotton crop

1 Introduction

One of India's most significant cash crops and sources of fiber is cotton. It is important to India's industrial and agricultural economies. It is grown on a variety of [1, 2] soil types, in rainfed or irrigated climates ranging from semi-arid to sub-humid. India now accounts for roughly 22% of global cotton consumption, making it one of the top cotton consumers. In Andhra Pradesh, five districts, Kurnool, Guntur, Anantapur, Krishna, and Prakasam, encompass 91.92% of the state's cotton acreage [3–5]and account for 93.18% of the state's cotton output. Guntur is a prominent cotton-growing region. Temperatures between 21 and 27 °C are optimum for vegetative development, and although it can withstand temperatures as high as 43 °C, the crop suffers damage at temperatures below 21 °C. In the south zone, cotton is growing a ranging of soil types, including black and mixed black and red soils. When the soil moisture is 50–70% depleted, cotton has to be watered. Before planting cotton, test the soil to determine the pH level and nutrient content so you can fertilize your crop appropriately and maintain adequate soil moisture and temperature. To forecast the cotton harvest, we use temperature, humidity, and N, P, and K values. Since nitrogen (N) is one of the primary building blocks of proteins and

A. Bajaj et al. (Eds.): IBICA 2023, LNNS 1229, pp. 489–497, 2025.
https://doi.org/10.1007/978-3-031-78940-3_52

amino acids, it is crucial for plant [6, 7] growth and development. Potassium is crucial for cotton. Lack of phosphorus causes growth buds, decreased nitrogen and magnesium uptake, and slowed plant growth. It is a successful method of supplying the growing crops with vital nutrients. N during squaring and peak flowering, P, and K at sowing are the fertilizer applications. In Andhra Pradesh, the whole NPK was applied at planting, along with half the P & K and the remaining N at blooming (for irrigated cotton). Irrigated hybrid cotton received the remaining N in four splits Every year, depending on the environment, the conditions for growth alter (weather and soil). Farmers also use [8, 9] Soil NPK Sensor. Harvest time is a great opportunity to remember this. However, it is generally a good idea to speak with a few labs and choose one before you actually collect the sample. In general, Farmers test soil and analyses are performed after a soil sample is taken. To ensure that enough samples are provided to the lab for [10–12] examination, you should discuss your issue with the consultants or managers of the soil testing lab you selected to hire. Additionally, they can aid in deciding what you want the samples to be examined or tested for. But in our experiment, we are using Deep learning techniques, to avoid that testing of soil in labs. In Deep learning we are using the Long Short-Term memory was the good technique to predict the crop yield with long dependencies to test the soil very accurately. Cotton Crop Yield Prediction using Different Deep Learning Techniques, It is based on IOT, With the use of IoT technology, new farming techniques can enhance crop productivity. The IoT is a unique idea that enables any item to send data via a network. Deep learning employs the LSTM (Long Short-Term Memory) method. The predicting was more precise, and it revealed that, although cotton output would climb gradually over the next decade [13, 14], the ratio of production to consumption will keep declining, posing risks to the economy as a whole. With sequential input, this method uses a LSTM [15, 16] network to extract characteristics from cotton vegetation indices at various phases of [17, 18] development. Warm weather is recommended for the optimum cotton production. For desi and hir-sutum kinds, the recommended fertilizer dosages are 50:25:25 NPK kg/ha and 100:50:50 NPK kg/ha for cotton produced under irrigation and 80:40:40 NPK kg/ha for cotton hybrids cultivated under rainfed circumstances.

2 Related Work

Significant study has recently been conducted in the subject of cotton crop yield prediction utilizing certain Machine Learning and Deep Learning approaches. Akinwunmi A. and colleagues [1] [2021] For the prediction, the Machine Learning Approach for a Cloud-Based Smart Farming System, three different classifiers were used: KNN, decision trees, and random forests. Nitrogen, phosphorous, potassium contents in the soil were then analyzed using the obtained data. Effects of Soil Temperature on Some Soil Properties and Plant Growth.

M.M. Darmanov et.al, [2] [2021] were all born in the same year. The biggest expense in cultivating cotton is the usage of fertilizers. Since nitrogen (N) is one of the primary building blocks of proteins and amino acids, it is crucial for plant growth and development. Potassium is crucial for cotton. N-250, P-175, K-125, and control are supplied at a rate of kg/ha when using this kind of phosphorus fertilizer. For qualities including early flowering, early ripening, enhanced output, and high-quality fiber, cotton needs potassium.

A 2019 study by Rab Nawaz Bashir et.al, [3] N, P, and K Fertilizer Applications and Cotton Yield and Quality Based on IoT and Machine Learning Based Soil estima-tion. Saline soil is a term used to describe soils that contain soluble salts in concen-trations that are above the point at which most plants become poisonous. The usefulness of N and P fertilizer nutrients mostly depends on how responsive cotton cultivars are to inputs. As was the case with K, cotton cultivars have become more responsive to N over time by focusing on early maturity and more predictable growth habits.

Precision Agriculture by Machine Learning Applications, by ABHINAV SHARMA et.al, [4] [2020]. The theoretical underpinnings of ML and DL algorithms that are used in precision agriculture have been covered. In ML using RNN and DL using LSTM algorithms. These benefits include low water consumption, the use of soluble nutrients through drip irrigation, automation, reduced soil erosion, uninterrupted operations, reduced weed concerns, and ease of double-cropping.

In the same calendar year, Rana Muhammad Saleem et.al, [5] [2021] Cotton Whitefly Prediction Using IOT Employing deep learning with the use of IoT technology, new farming techniques can enhance crop productivity. "Related work," "methodology," "implementation," and "results and discussion" are the sections of the remaining paper. One of the industries that is anticipated to be significantly impacted by IoT advancements is agriculture.

3 Proposed Methodology

In this system we demonstrate a cotton crop yield prediction with NPK values, pH, soil moisture, temperature, and humidity, which are also some crucial factors while fertilizing. This system's major objective is to develop a farming tool that employs sensors to continuously track the nutrients in soil samples over time. With the use of innovative agricultural techniques and IoT technologies, crop productivity can be boosted by predicting cotton crop yield using deep learning techniques (Internet of things) In deep learning, the LSTM (Long Short-Term Memory) method is used. Over the next ten years, cotton output will rise steadily, However, as the ratio of production to consumption declines, hazards to the economy as a whole increase. This was predicted more accurately by the prediction.

The illustration below. 1, The readings for temperature, humidity, moisture, and NPK are taken from Rasberi.pi and sent to the cloud via a MQTT broker. Utilizing the cloud, we first input the training data from the removed block before continuing with the LSM approach. The deleted block allows it to enter the hidden state or allow us to get our desired output in either favorable or unfavorable circumstances (Fig. 1).

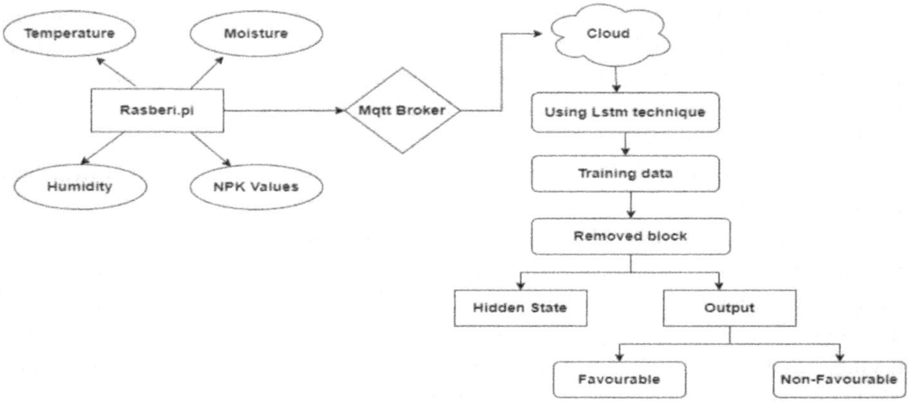

Fig. 1. Cotton crop yield prediction with LSTM

ALGORITHM FOR DATA COLLECTION

Step 1: Readings from the NPK sensors.
Step 2: Calculate the Temperature(t), humidity (h), moisture values (mv), NPK Val ues.
  a) t (5 / 9) t(°C) = (32°F - 32) *
  b) h = Sensor values for relative (h) expressed as a percentage.
  c)mv = t - ((100-h) / 5) mv
  d)N,P,K values = N:P:K – 40:20:20.
Step 3: Repeat the Step 2.
Step 4: Using a MQTT broker, upload the data that was collected during the earlier phases to the cloud.

In the initial stage, we must read the NPK sensor readings. In the subsequent phase, determine the temp, hum, mois, and N,P,K levels. In this case, we must first get the temperature using the formula (32°F-32)* (5/9) and then the humidity using The moisture is then determined using the formula T-((100-H)/5 after relative humidity in % is measured using sensors. N, P, and K are the initials of the acronym NPK. Measure temperature, humidity, and pressure once again in step 3 by repeating process 2. Values for moisture and NPK are then added, and the last step is to use a mqtt broker to export the data to the cloud.

$i_t \rightarrow$ the input gate.
$c_t \rightarrow$ the cell state.
$f_t \rightarrow$ the forget gate.
$O_t \rightarrow$ the output gate.
$h_t \rightarrow$ the hidden state.
H(t-1) $\rightarrow$ Lstm block input from the prior iteration (at timestamp t 1).
$w_i, w_f, w_O \rightarrow$ weight matrices.
$b_i, b_c, b_f, b_O \rightarrow$ the biases.
$\Sigma \rightarrow$ represents sigmoid function.

$x_t \rightarrow$ the input.

$$i_t = \sigma(w_i[h_t - 1, x_t] + b_i) \tag{1}$$

The input gate performs the subsequent actions to modify the cell state. First, the second sigmoid (t-1) is supplied the previously concealed state h and the current state X(t).

$$c_t = \tan h(w_i[h_t - 1, x_t] + b_c) \tag{2}$$

The same data from both the hidden state and the current state will then be provided to the tanh function. In order to manage the network, the tanh operator will create a vector (C(t)) holding each feasible value between -1 and 1. On a point-by-point basis, the activation functions' output values are multiplied.

$$f_t = \sigma(w_f[h_t - 1, x_t] + b_f) \tag{3}$$

The sigmoid function (t-1) receives information from the hidden state h and the current input X(t). The range of numbers produced by the Sigmoid algorithm is 0 to 1. It decides whether the output's component from before is necessary. This value of f(t) will ultimately be used by the cell for point-by-point multiplication.

$$O_t = \sigma(w_O[h_t - 1, x_t] + b_O) \tag{4}$$

The value of the following hidden state is decided by the output gate. This state includes information about previous inputs. The third sigmoid function is given the values of the previous concealed state and the current state.

$$h_t = O_t * \tan h\,(c_t) \tag{5}$$

Then, the cell state that was produced as a result of the initial cell state is put through the tanh function. The following time step inherits the just created hidden state and cell state.

Algorithm for cotton crop yield prediction using LSTM

INPUT: N, P, K values.

OUTPUT: Trained model for predicting cotton crop.

1. Take the NPK input values and the declare h(t-1) and the current x_t input and compute the function.
2. Initialize the net parameters $LSTM_{net}$;
   Convert the format of input data;
   Input the training data i_t, c_t from equation (1) and (2).
3. Calculate the detail of the removed block of the data to use Forget gate f_t from equation (3).
4. The current cell state makes it to the output to calculate the data from the sigmoid gate O_t, h_t from equation (4) and (5).

Take the input values N, P, and K in the first step, and then declare h(t-1) along with the current *xt* input and compute the function. In next step, 5nitialize the net parameters with $LSTM_{net}$ and convert it in to format of input data and input the training data i_t, c_t from equations $i_t = \sigma(w_i[h_t - 1, x_t] + b_i)$ and $c_t = \tanh(w_i\ [h_t - 1, x_t] + b_c)$.

In the next step calculate the detail of the removed block of the data to use Forget gate f_t from equation $f_t = \sigma(w_f . [h_t - 1, x_t] + b_f)$. In the final step, the current cell state makes it to the output to calculate the data from the sigmoid gate O_t, h_t from equations $O_t = \sigma\ (w_O\ [h_t{-}1, x_t]{+}b_O)$ and $h_t{=}O_t{*}\ \tanh(c_t)$.

4 Experimentation and Results

Based on the four parameters listed below, the project's results will be analyzed. Accuracy:- degree of conformity to the real value. Depending on the technique of measurement. Accuracy does not require precision.

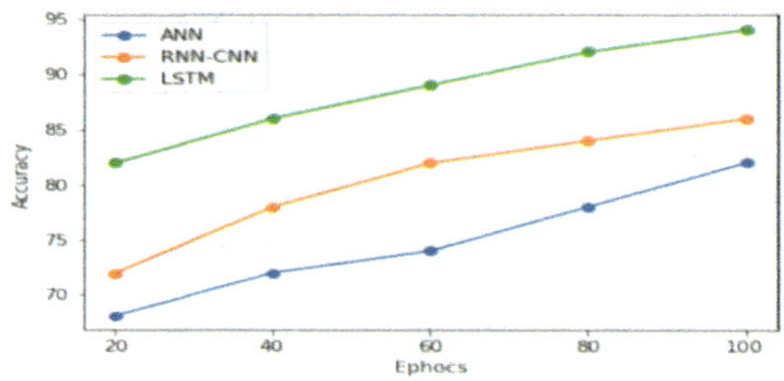

Fig. 2. Prediction accuracy of cotton crop yield at maximum, minimum, and average.

In Fig. 2 shows the LSTM Accuracy Epochs are high. The accuracy of RNN-CNN epochs is average. The ANN Accuracy Epochs are shorter. The accuracy of the LSTM epochs is greater than that of the RNN-CNN and ANN.

Precision: - The measuring tool determines how accurate your measurements are or how close they can be to one another based on the number of significant digits.

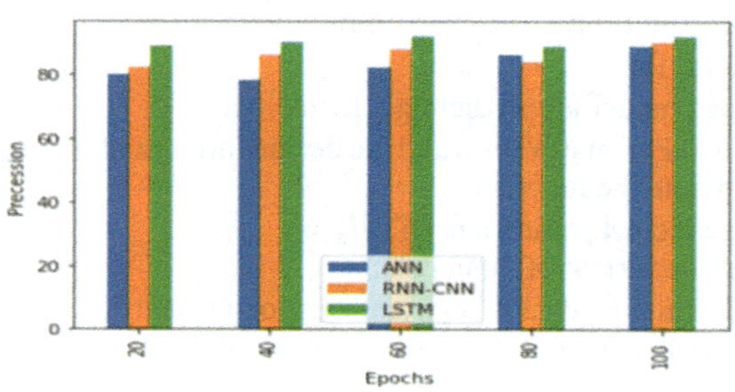

Fig. 3. Prediction precession for cotton crop yield at its highest, lowest, and average levels.

Figure 3 shows the LSTM Precision Epochs are quite high. The RNN-CNN Precision Epochs are average. The ANN Precision Epochs are shorter. Compared to RNN-CNN and ANN, the LSTM Precision has a greater epoch count.

Recall:-Recall is a measurement of how many positive examples in the data the classifier correctly predicted. In the positive class, a model makes predictions, of which 90 are correct and 10 are not.

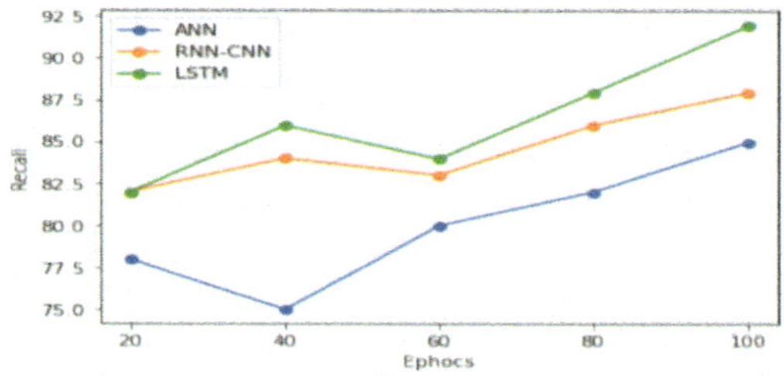

Fig. 4. The cotton crop yield prediction's maximum, minimum, and average recalls.

Figure 4 shows the LSTM Recall Epochs are quite high. The RNN-CNN Recall Epochs are average. Epochs of ANN Recall are shorter. Epochs of the LSTM Recall are greater than those of the RNN-CNN and ANN.

F-Score: -A single indicator of how effective a categorization technique is the F-score (or F-measure). The categorization process has greater predictive value when the F- score is higher.

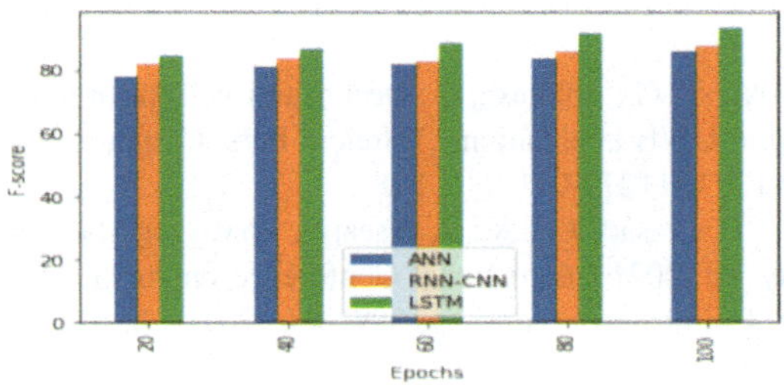

Fig. 5. Max, Min, and Ave F-scores for the predicted cotton crop production.

Figure 5 shows LSTM's F-score Epochs have a high number. The RNN-CNN F-score's average epochs. The ANN F-score epochs are shorter. The LSTM F-score has longer epochs than the RNN-CNN and ANN.

5 Conclusions

In order to predict the cotton crop and take the right action, the IoT-based agricultural field continuously analyzes the environment in terms of temperature, humidity, moisture, and NPK levels. The current field conditions update the needs in order to properly achieve the prediction's goal. The LSTM deep learning technique is trained and tested using these predictions, which is then used to optimize them for cotton crop yield. The forecast turned out to be rather correct with regard to the recommendation of fertilizers. Real-time sensor-based environmental data including temperature, humidity, moisture, and NPK levels may be accurately forecasted to increase crop production.

References

1. Akinwunmi, A.O., Aworinde, H.O., Ogunti, E.O.: Increasing agricultural productivity in Nigeria using wireless sensor network (WSN). Afr. J. Comput. ICT **8**(3), 121–128 (2015)
2. Bikku, T.: Multi-layered deep learning perceptron approach for health risk prediction. J. Big Data **7**(1), 1–14 (2020)
3. Behrens, T., Förster, H., Scholten, T., Steinrücken, U., Spies, E.D., Goldschmitt, M.: Digital soil mapping using artificial neural networks. J. Plant Nutr. Soil Sci. **168**(1), 21–33 (2005)
4. Jiang, H., Shu, H.: Optical remote-sensing data-based research on detecting soil salinity at different depth in an arid-area oasis, Xinjiang China. Earth Sci. Inf. **12**(1), 43–56 (2019)
5. Bikku, T., et al.: Machine learning identification of organic compounds using visible light. J. Phys. Chem. A **127**(10), 2407–2414 (2023)
6. Kanumalli, S.S., et al.: Automated irrigation management system using IoT. In: 2022 6th International Conference on Intelligent Computing and Control Systems (ICICCS). IEEE (2022)
7. Venkatesh, R., Chaitanya, K., Bikku, T., Paturi, R.: A review on biomedical mining. J. RNA Genomics **16**, 629–637 (2020)
8. Kumar, A., Gali, S.K., Hebsur, N.S.: Effect of different levels of NPK on growth and yield parameters of sweet corn. Karnataka J. Agric. Sci. **20**(1) (2010)
9. Chilakala, S., Saxena, A., Vunnava, A.: Energy-efficient reporting scheme for cognitive radio networks. IET Netw **10**, 110–117 (2021). https://doi.org/10.1049/ntw2.12019
10. Vadivel, S., Konda, S., Balmuri, K.R., Stateczny, A., Parameshachari, B.D.: Dynamic route discovery using modified grasshopper optimization algorithm in wireless ad-hoc visible light communication network. Electronics **10**(10), 1176 (2021). https://doi.org/10.3390/electronics10101176
11. Premalatha, B., Srikanth, G., Abhilash, G.: Design and analysis of multi band notched MIMO antenna for portable UWB applications. Wireless Pers. Commun. **118**, 1697–1708 (2021). https://doi.org/10.1007/s11277-021-08111-0
12. Koyi, L.P., Borra, T., Prasad, G.L.V.: A research survey on state of the art heart disease prediction systems. In: 2021 International Conference on Artificial Intelligence and Smart Systems (ICAIS), Coimbatore, India, pp. 799–806 (2021). https://doi.org/10.1109/ICAIS50930.2021.9395785
13. Cherukuri, N., Bethapudi, N.R., Thotakura, V.S.K., Chitturi, P., Basha, C.Z., Mummidi, R.M.: Deep learning for lung cancer prediction using NSCLS patients CT information. In: 2021 International Conference on Artificial Intelligence and Smart Systems (ICAIS), Coimbatore, India, pp. 325–330 (2021). https://doi.org/10.1109/ICAIS50930.2021.9395934
14. Samdekar, R., Ghosh, S.M., Srinivas, K.: Efficiency enhancement of intrusion detection in IoT based on machine learning through bioinspired. In: 2021 Third International Conference on Intelligent Communication Technologies and Virtual Mobile Networks (ICICV), Tirunelveli, India, pp. 383–387 (2021). https://doi.org/10.1109/ICICV50876.2021.9388392
15. Dinesh, G., Venkatakrishnan, P., Jeyanthi, K.M.A.: Modified spider monkey optimization—an enhanced optimization of spectrum sharing in cognitive radio networks. Int. J. Commun. Syst. **34**, e4658 (2021). https://doi.org/10.1002/dac.4658
16. Arunachalam, P., Venkatakrishnan, P., Janakiraman, N.: Histopathology image classification for soft tissue sarcoma in limbs using artificial neural networks. In: 2021 6th International Conference on Inventive Computation Technologies (ICICT), Coimbatore, India, pp. 778–785 (2021). https://doi.org/10.1109/ICICT50816.2021.9358635

17. Arunachalam, P., Venkatakrishnan, P., Janakiraman, N.: Detection of singularity in the cell nucleus of synovial sarcoma using wavelet leaders. In: 2021 6th International Conference on Inventive Computation Technologies (ICICT), Coimbatore, India, pp. 418–425 (2021). https://doi.org/10.1109/ICICT50816.2021.9358517
18. Tiwari, V., Naidu, A.S., Singh, A., Saxena, A.: Construction of 4×4 MDS matrices for lightweight block ciphers. In: 2021 IEEE International Conference on Advanced Networks and Telecommunications Systems (ANTS), Hyderabad, India, pp. 254–259 (2021). https://doi.org/10.1109/ANTS52808.2021.9936970

Author Index

A. Bajaj et al. (Eds.): IBICA 2023, LNNS 1229, pp. 499–501, 2025.
https://doi.org/10.1007/978-3-031-78940-3

The manufacturer's authorised representative in the EU is Springer Nature Customer Service Centre GmbH, Europaplatz 3, 69115 Heidelberg, Germany. If you have any concerns regarding our products, please contact ProductSafety@springernature.com

Printed and bound by CPI Group (UK) Ltd, Croydon, CR0 4YY
30/09/2026
02225561-0016